Chemical Biology of Glycoproteins

Chemical Biology

Editor-in-chief:
Tom Brown, *University of Oxford, UK*

Series editors:
Kira J. Weissman, *Lorraine University, France*
Sabine Flitsch, *University of Manchester, UK*
Nick J. Westwood, *University of St Andrews, UK*

Titles in the series:
1: High Throughput Screening Methods: Evolution and Refinement
2: Chemical Biology of Glycoproteins

How to obtain future titles on publication:
A standing order plan is available for this series. A standing order will bring delivery of each new volume immediately on publication.

For further information please contact:
Book Sales Department, Royal Society of Chemistry, Thomas Graham House, Science Park, Milton Road, Cambridge, CB4 0WF, UK
Telephone: +44 (0)1223 420066, Fax: +44 (0)1223 420247,
Email: booksales@rsc.org
Visit our website at www.rsc.org/books

Chemical Biology of Glycoproteins

Edited by

Zhongping Tan
University of Colorado Boulder, CO, USA
Email: zhongping.tan@colorado.edu

Lai-Xi Wang
University of Maryland, College Park, MD, USA
Email: wang518@umd.edu

ROYAL SOCIETY OF CHEMISTRY

THE QUEEN'S AWARDS FOR ENTERPRISE: INTERNATIONAL TRADE 2013

Chemical Biology No. 2

Print ISBN: 978-1-78262-333-5
PDF eISBN: 978-1-78262-382-3
EPUB eISBN: 978-1-78801-122-8
ISSN: 2055-1975

A catalogue record for this book is available from the British Library

The Royal Society of Chemistry is a charity, registered in England and Wales, Number 207890, and a company incorporated in England by Royal Charter (Registered No. RC000524), registered office: Burlington House, Piccadilly, London W1J 0BA, UK, Telephone: +44 (0) 207 4378 6556.

For further information see our web site at www.rsc.org

Printed in the United Kingdom by CPI Group (UK) Ltd, Croydon, CR0 4YY, UK

Preface

Carbohydrates have been known, for a long time now, to cover the surfaces of most cells and excreted proteins; and, from early studies with congenital disorders of glycosylation, it was shown that these complicated carbohydrate structures are extremely important for life. But despite years of constant scrutiny, the mysteries of those glycan structures still far outweigh what we know. Revealing these mysteries demands new and interdisciplinary approaches. As a result, the number of scientists and researchers using chemical biology to address the unanswered questions in the glycosciences has grown in recent years, with new discoveries emerging at an impressive rate world-wide. All of this new-found interest explains the need for the book you are now holding.

Like many of the best things in life, this book is the result of an accidental meeting and conversation. While attending a meeting for the American Chemical Society in San Francisco, Professor Virginia Cornish was discussing with Rowan Frame, a Commissioning Editor for the Royal Society of Chemistry, the potential topics for an up-coming book series on chemical biology. She suggested a book on glycoproteins and the influence chemical biology has had on the new research in that area. Naturally, we received and accepted the invitation from Rowan and began the process of asking colleagues to contribute to the tome. We have selected a diverse group of leading minds in the chemical glycobiology field to provide the reader with the best possible view of this exciting and constantly evolving field.

We begin with four chapters discussing the detailed chemistry and biology surrounding each of the major classes of protein glycosylation found in living systems. Together these chapters cover the current understanding of these important post-translational modifications from biosynthesis and regulation to tissue distribution and function. These chapters also include

Chemical Biology No. 2
Chemical Biology of Glycoproteins
Edited by Zhongping Tan and Lai-Xi Wang

Published by the Royal Society of Chemistry, www.rsc.org

state of the art chemical biology-based methods used when studying each type of protein glycosylation and an in-depth look at how interdisciplinary research can be best leveraged to design useful and informative experiments. The next five chapters cover synthetic approaches for constructing the wide variety of glycans, glycosylated peptides and proteins that exist in nature. *N*-glycoprotein synthesis is covered from both a completely synthetic and chemoenzymatic point of view in order to provide the reader with an understanding of the advantages and disadvantages of the two approaches and illustrate examples where the two complimentary synthetic methods are uniquely suited. A similar approach is also used to present the synthesis of heparan, heparan sulfate and other proteoglycan molecules, with separate chapters for totally synthetic routes and those that use the natural power of enzymes to chemoenzymatically prepare the molecules. The final chapter in this section focuses on the recent developments towards producing synthetic GPI anchors and GPI-anchored proteins. This challenging area is rapidly developing and will no doubt be of interest to aspiring chemical biologists in the glycosciences. The remaining chapters in the book cover the novel work being done at a cellular and organismal level with chemical biology. One of the prominent hurdles in the glycobiology field is getting a complete picture of exactly which glycan structures are presented on various cells across a variety of experimental and environmental conditions. Many novel approaches that rely on versatile chemical biology tools have been developed recently to address this hurdle. Glycan–glycan and glycan–protein interactions are almost without a doubt the very first occurrences during viral infection, and so glycobiology has always held a special place in the virology field. Recently, there has been a concerted effort towards developing HIV vaccines the exploit the carbohydrate structures present on the viral coat proteins. Such research is necessarily interdisciplinary and is an exciting application of much of the knowledge presented in the earlier chapters of the book. Similarly, cancer cells and their aberrant glycosylation patterns have long been of interest to cancer biologists. These special glycan structures contribute to many of the unique properties of tumor cells and have been explored for their potential use in cancer vaccines. The final chapter of the book covers a wholly different approach to exploiting the complexity of glycan structures for therapeutic advantage which is the rational engineering of such structures on therapeutic molecules. For many years, scientists have appreciated the ability of glycans to alter the physical properties and function of many biomolecules; and recently much effort has been devoted to codifying the rules that govern those effects. The grand hope for this area is that a deep understanding of those rules would allow us to rationally engineer therapeutic glycoproteins with the desired physical and functional properties by controlling the glycan structures present.

Of course this book would not have been possible without the significant efforts of the many talented researchers and editors who helped to write and edit each of the individual chapters. We would also like to extend a special thanks to Virginia Cornish for suggesting this topic for the Chemical Biology book series and Hongzhi Cao, Patrick Chaffey, Xuefei Huang, Jiaoyang Jiang, Isaac Krauss, Jian Liu, Corwin Nycholat, Richard Payne, Matthew Pratt, Daniel Varón Silva, Tadashi Suzuki, Yuhsuan Tsai and Maciej Walczak, for their efforts in reviewing the chapters of this book.

Zhongping Tan
Lai-Xi Wang

This book is dedicated to Professor Samuel Danishefsky.

Contents

Chemical Biology No. 2
Chemical Biology of Glycoproteins
Edited by Zhongping Tan and Lai-Xi Wang

Published by the Royal Society of Chemistry, www.rsc.org

CHAPTER 1

Introduction: General Aspects of the Chemical Biology of Glycoproteins

PATRICK K. CHAFFEY[a], XIAOYANG GUAN[a], LAI-XI WANG[b] AND ZHONGPING TAN*[a]

[a]Department of Chemistry and Biochemistry and BioFrontiers Institute, University of Colorado, Boulder, CO 80303, USA; [b]Department of Chemistry and Biochemistry, University of Maryland, College Park, MD 20742, USA
*E-mail: zhongping.tan@colorado.edu

1.1 Introduction

The Human Genome Project has revealed that the entire human genome comprises only about 20 000 genes, while the total number of proteins in the human proteome is estimated to be between 250 000 and one million.[1] This severe disparity has made it evident that the human proteome is far more complex than the human genome. Such vast proteomic diversity results from a variety of mechanisms that act at each stage of the protein production process. At the level of transcriptional regulation, there are genomic recombination events, differential transcription initiation and termination, and alternative transcript splicing,[2] which together can result in a single DNA-encoded gene being transcribed into more than four unique mRNA transcripts.[3] While impressive, such transcription-level events pale in comparison to

Chemical Biology No. 2
Chemical Biology of Glycoproteins
Edited by Zhongping Tan and Lai-Xi Wang

Published by the Royal Society of Chemistry, www.rsc.org

post-translational processing where most of the proteomic diversification occurs through introduction of many different post-translational modifications (PTMs).[4] The combination of these mechanisms significantly increases the number of human proteins from what would be expected based solely on the number of genes.[5]

PTMs occur after mRNA has been translated into a protein and are covalent modifications of the protein's side chains or backbone atoms. These types of modifications can be reversible or irreversible;[6] and to date, more than 200 different types of PTMs have been identified.[7] Although some modifications, such as oxidation[8] and glycation,[9] occur through loosely controlled, non-enzymatic reactions, most PTMs are introduced onto proteins through tightly controlled enzymatic processes.[4] In eukaryotes, the major types of enzyme-catalyzed PTMs are phosphorylation, glycosylation, acetylation, methylation, ubiquitylation, and SUMOylation.[10] While phosphorylation is the most common PTM regulating internal functions and activities of cellular proteins,[11] glycosylation is the most abundant modification found on extracellular membrane-associated and excreted proteins.[12] Glycosylation involves the covalent attachment of one or more glycans (also called 'saccharides' or "carbohydrates") to a protein. Over 50 years of studying protein phosphorylation has greatly clarified its roles in regulating cellular processes.[13] However, the same type of knowledge regarding the biological roles of protein glycosylation lags far behind, and the importance of this modification is still not well appreciated by the general scientific community.[14] The lack of knowledge and appreciation of protein glycosylation is a by-product of the slow progress thus far in analyzing and characterizing this type of modification, which, in turn, is due to the complex nature of this modification.[15,16]

1.1.1 Complexity of Protein Glycosylation

Similar to phosphorylation, protein glycosylation is regulated by many different factors, such as amino acid sequence, local conformational properties at potential glycosylation sites, and the availability of activated sugar substrates and enzymes involved in glycosylation. As a result, protein glycosylation is sensitive to both genetic and environmental changes. When multiple sites of glycosylation are present in a protein, the site occupancy can vary under different conditions, leading to glycosylation macroheterogeneity.[17] This level of heterogeneity is similarly observed in protein phosphorylation.[18] Glycans occupying a glycosylation site, however, are far more complex and diverse than a small phosphate group and can vary significantly in terms of size, charge state, glycosidic linkage, and branching pattern. This leads to another level of heterogeneity, namely: microheterogeneity, which is not present in protein phosphorylation.[19] The combination of two levels of heterogeneity results in glycoproteins isolated from biological sources existing as complex mixtures. Even for a single protein sequence, these mixtures can contain more than one hundred differently glycosylated isoforms (glycoforms) and the compositions can vary substantially between mixtures generated by different hosts and under different conditions.[20] Because it is difficult to exhaustively quantify individual components of the mixtures or to separate the individual glycoforms from one another, most studies of glycoproteins to date have been carried out using samples with unknown compositions. For example, one of the commonly used methods to analyze protein glycosylation is to directly compare the properties of an unglycosylated protein with its corresponding glycoform mixture.[21] This method can only provide a rough estimate of the average effects of all glycans on a given protein. Because the composition of its glycoform mixture can vary considerably, in many cases, the observed effects are sample-specific, inconsistent and sometimes even conflict with each other. The resulting inconsistency and disagreement across studies has led to significant knowledge gaps in the understanding of protein glycosylation at the levels of both fundamental science[22] and biomedical development.[23,24]

1.1.2 Strategies and Methods to Study Protein Glycosylation

To obtain a more complete understanding of protein glycosylation, it is essential to have a more detailed knowledge of the relationships between glycan structure and function, and the consequences of forming heterogeneous mixtures of glycoforms. As with studies of nucleic acids[25] and proteins,[26] such knowledge can be acquired through the use of homogeneous glycoprotein isoforms and glycopeptides with defined and well-characterized glycosylation patterns.[27,28]

The availability of homogeneous glycoforms enables a clearer and more confident view of the effects of different glycans on the same or different proteins, which then makes establishing structure–function relationships possible.[15,16,29] Through decades of research, glycosylation has been demonstrated to affect both physical and biological properties of proteins, including folding,[30] aggregation,[31] conformation and structure,[32] solubility,[33] trafficking,[34] stability,[35] binding affinity and specificity,[36] and biological activity.[37,38] These studies cover the effects of many types of glycans, differing widely in their charge, size, linkage, and branching architectures, on the properties of numerous glycoproteins.[22,35] Although it is commonly accepted that protein glycosylation is generally important, this conclusion is too vague to be practically useful for predicting the effects of protein glycosylation[24,30] or guiding the glycoengineering of proteins.[39] To reach a clearer and more definite conclusion, many questions, such as "which effects are the general impacts of protein glycosylation and which are specific to glycosylation type, glycosylation site, and glycan structure", "what are the possible magnitudes of these effects", and "what are the molecular determinants of these effects", must be answered. These answers can be achieved by systematically comparing the effects of a series of glycans on representative proteins, which, in turn, can guide more in-depth research on more glycoproteins.[29,40–44]

A better understanding of glycoprotein heterogeneity requires knowledge about both the composition of glycoform mixtures and their biological functions, both of which can be revealed by studying homogeneous glycoproteins and glycopeptides. Two broad classes of mass spectrometry (MS)-based methods, glycomics and glycoproteomics, have been widely used for investigating the compositions of glycoprotein mixtures. Glycomic methods focus on the analysis of oligosaccharides that are enzymatically or chemically cleaved from glycoproteins.[45] These methods can provide an estimate of the structure and total number of glycans that may be present in a mixture of glycoforms. However, because any information regarding their associated peptides is lost, they cannot provide quantitative measurements of the ratio of different isoforms of a glycoprotein with more than one glycan attachment site.[46,47] In theory, with the help of synthetic peptides or glycopeptides, unambiguous composition results for glycoprotein mixtures can be obtained using glycoproteomic technologies.[48] The synthetic molecules serve as standards for adjusting instrument parameters to achieve optimal sensitivity[49] and for generating calibration curves for absolute quantification.[50] After compositional analysis, the functional assays for the glycoform mixtures can be performed using either leftover samples or artificial samples, which can be prepared by mixing synthetically produced and purified glycoforms according to the measured quantitative ratios to closely mimic the truly natural mixtures. Since it is generally difficult to isolate a glycoform mixture from a relevant biological source that is not contaminated by other proteins and is of sufficient amount for both analytical and functional studies,[51] artificial mixtures made from known components are a convenient alternative that side-step such contamination and scarcity issues. After analysis, directly correlating the compositions of different mixtures to their corresponding biological activities allows researchers to elucidate structure–composition relationships of glycoform mixtures. At the same time, directly comparing biological functions with those of homogeneous glycoforms yields definitive insights regarding the advantages and disadvantages of using heterogeneous glycoform mixtures in biological systems.[52,53] Overall, these methods can provide a detailed understanding of glycoprotein heterogeneity.

Similar to the preparation of nucleic acids[54,55] and proteins,[56,57] homogeneous glycoforms can be generated using chemical methods.[27,28,58–62] However, the current methods for glycan, glycopeptide, and glycoprotein synthesis are still in their infancy and most of the essential building blocks, enzymes, and materials are either difficult or expensive to obtain.[63] Consequently, only a few laboratories in the world have the expertise, resources, and willingness to produce homogeneous glycopeptides and glycoproteins for the study of protein glycosylation, which has significantly slowed progress in this research area. In order to more quickly advance our understanding of the role of glycosylation in regulating the physical and biological properties of proteins, it is essential to inspire, educate, and attract more researchers to work towards solutions to these challenges, to develop more generally and easily applicable synthetic methods and to commercialize more research discoveries.[64–66]

1.2 Chemical Biology of Glycoproteins

This book aims to promote more research for improved understanding of protein glycosylation and its application to commercial products. In addition to providing critical educational information, including reviews of structural and functional consequences of protein glycosylation (Chapters 2–4) and reviews of current status and limitations of the methods used to prepare glycans, glycopeptides, and glycoproteins (Chapters 5–9), to analyze glycoprotein compositions (Chapters 3 and 9), and to image glycosylation (Chapter 10), this book also delivers relevant and motivational information to readers through reviews of applications of knowledge-based methods to the treatment of human diseases (Chapters 11–13).

1.2.1 Types of Protein Glycosylation

Glycans can be covalently attached to two types of biomolecules, proteins and lipids, to form glycoproteins and glycolipids. There are at least 37 different glycan–protein linkages, which together involve 13 different monosaccharides and eight different amino acids.[67] Based on how sugars are connected to amino acid side chains, protein glycosylation is classified into five major types: *N*-glycosylation, *O*-glycosylation, glypiation, *C*-glycosylation, and phosphoglycosylation.[12] These first three categories are the most common and are therefore the focus of this book.[12] In comparison, *C*-glycosylation and phosphoglycosylation, in which glycans are attached to either a carbon (C) from a Trp side chain or a phosphate group from phospho-serine or phospho-threonine, are much rarer forms of glycosylation and are not covered here. Interested readers are encouraged to read current reviews regarding these unusual types of glycosylation.[12,68–73]

In *N*-glycosylation, sugars are attached to nitrogen (N) atoms, most typically on the nitrogens of asparagine side chains within the consensus sequence Asn–Xaa–Ser/Thr/Cys, where Xaa is any amino acid except Pro (Figure 1.1). Furthermore, all known *N*-linked glycans share a common pentasaccharide core: $GlcNAc_2Man_3$.[74] This small core structure is often heavily modified and elongated after its introduction to the protein to generate at least three common structures: high-mannose, hybrid and complex *N*-glycans.[75]

In the case of *O*-glycosylation, glycans are attached to the side chain oxygens (O) of hydroxyl amino acid residues, most commonly Ser and Thr. Unlike *N*-glycosylation, there is no clear-cut consensus sequence for defining sites of *O*-glycosylation[76] and instead of using only a single monosaccharide, *O*-glycosylation is initiated by quite a few carbohydrates, with the most common ones in mammals being *O*-*N*-acetylglucosamine (*O*-GlcNAc), *O*-*N*-acetylgalactosamine (*O*-GalNAc), *O*-mannose (*O*-Man), and *O*-xylose (*O*-Xyl) (Figure 1.1).[12] β-linked *O*-GlcNAc glycosylation typically exists within the cell and the glycans remain as single monosaccharide residues that dynamically modify proteins in the nucleus and cytoplasm complimentary and somewhat analogous to protein phosphorylation.[77] On the other hand, *O*-GalNAc, *O*-Man, and

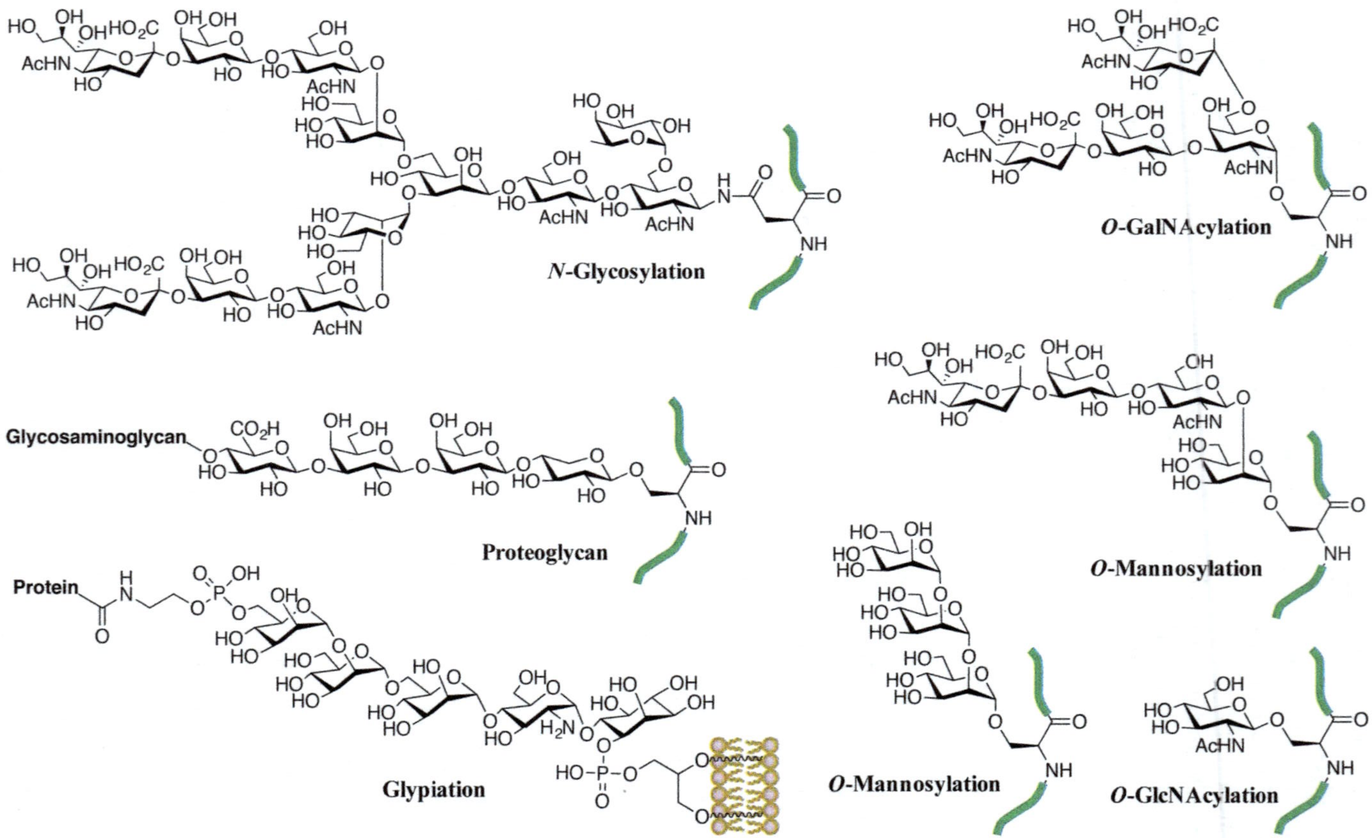

Figure 1.1 Representative structures of five major types of protein glycosylation. The bold green wavy lines represent the rest of the glycoprotein. The black wavy lines in "Glypiation" represent lipids with varied structures and are shown inserted into a light brown cartoon representation of the cell membrane.

O-Xyl glycans can be elongated and/or branched to form complex structures on par with those of *N*-glycans.[67,78–81] Protein *O*-GalNAcylation and *O*-mannosylation share many features: both glycans are α-linked to proteins and can be either densely clumped[82,83] or sparsely presented on proteins depending on the context.[84,85] Finally, β-linked *O*-Xyl residues act as the initiating monosaccharides for glycosaminoglycan (GAG) chains on proteoglycan core proteins.[86]

Glypiation is the post-translational process in which glycosylphosphatidylinositol (GPI) is attached to newly synthesized proteins (Figure 1.1). By linking proteins to lipids through glycan linkages, this modification allows proteins to be bound to membranes through so-called GPI-anchors.[87,88]

1.2.2 Biosynthesis of Glycoproteins

Glycoprotein biosynthesis is a complex process.[75,89–93] In eukaryotic species, depending on the exact glycan structure, the biosynthesis can be variously initiated in the cytosol or nucleus (*O*-GlcNAcylation),[77] endoplasmic reticulum (ER; *N*-glycosylation,[94] *O*-mannosylation,[53] *etc.*), and Golgi apparatus (*O*-GalNAcylation and formation of proteoglycans)[95,96] by specialized glycosyltransferases. Glycosyltransferases in each of these different locations utilize distinctive types of sugar donors to transfer the first glycan moieties onto proteins. In the cytosol, nucleus and Golgi, initiating glycans are transferred from nucleotide-activated monosaccharides, such as UDP–α-GalNAc and UDP–α-GlcNAc.[62] Meanwhile, the ER membrane has no nucleotide-sugar transporters and so glycosylation donors in that compartment are dolichol phosphate sugars, which are formed on the outside of the ER membrane by transferring monosaccharides from nucleotide-activated sugar substrates in the cytosol to the dolichol phosphate and subsequently "flipped" into the ER lumen (Figure 1.2).[53]

Of all types of protein glycosylation, *O*-GlcNAcylation has the simplest biosynthetic pathway. The process is completed after the attachment of the initial GlcNAc residue.[77] The biosyntheses of *O*-GalNAcylated proteins and proteoglycans are also relatively simple if only with respect to the location of the synthesis. All the steps of these two types of glycosylation occur in the Golgi apparatus.[95–97] The biosynthetic pathways of *N*-glycosylation,[92] *O*-mannosylation,[67,98] and glypiation[99] are more complex with regards to the number of locations for performing these modifications, with all three starting in the ER and finishing in the Golgi. Many features of such pathways can be illustrated using the biosynthesis of *N*-linked glycoproteins. As shown in Figure 1.2, the dolichol-activated $Glc_3Man_9GlcNAc_2$ is first synthesized on the cytosolic face of the ER membrane from nucleotide-activated monosaccharide donors. It is then flipped to the luminal side before it is used to co-translationally transfer the 14-unit oligosaccharide, $Glc_3Man_9GlcNAc_2$, to the Asn residue within the consensus sequence Asn–Xaa–Ser/Thr/Cys.[93] After protein folding, the *N*-linked glycan chain undergoes successive modifications in the late ER and Golgi, including the removal of the terminal Glc and Man and the addition of GlcNAc, Gal, Sia, and Fuc residues, to create a wealth of *N*-linked glycan structures.[34]

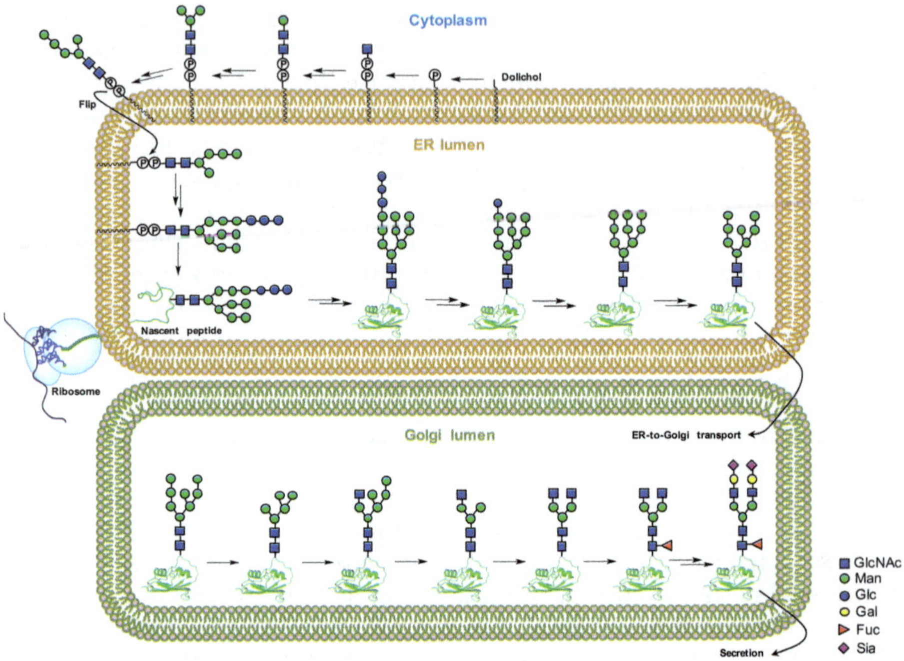

Figure 1.2 An abbreviated biosynthetic scheme for protein *N*-glycosylation in mammalian cells. First, the synthesis of $Man_5GlcNAc_2$–P–P–dolichol is completed on the cytoplasmic face of the ER membrane. It is then flipped to the ER lumen and extended to form $Glc_3Man_9GlcNAc_2$–P–P–dolichol. After $Glc_3Man_9GlcNAc_2$ is transferred to the nascent polypeptide, the *N*-linked glycans mature in a stepwise fashion in the ER and then the Golgi before the glycoprotein is secreted. P represents a phosphate group.

1.2.3 Structural and Functional Consequences of Protein Glycosylation

In eukaryotes, *N*-linked glycosylation is the most common form of protein glycosylation.[10] It is also the most frequently studied form of protein glycosylation, both from the perspective of biosynthetic pathways and its effects on the physical or functional properties of proteins.[19] It is important to keep in mind that many of the effects discussed here are not derived from the same or even closely related proteins and vary depending on size and site of glycan attachment. Nevertheless, they can still serve as guidelines for gathering more information about the structure–function relationships of glycoproteins.[35]

By systematically comparing the structure and dynamics of many glycopeptides and glycoproteins, it was found that highly conserved *N*-glycosylation may be critical for controlling protein folding and aggregation[100] and the *N*-linked glycan core can account for the entire acceleration of folding and native state stabilization.[101] In general, *N*-glycosylation does not cause major structural changes, but may lead to an increase in protein stability, which is likely to be a consequence of decreased protein dynamics after glycosylation.[32,35,102] *N*-glycosylation can also improve the solubility of many glycoproteins, possibly by increasing their solvent-accessible surfaces.[35] In several biological systems, *N*-linked glycans have been shown to have specific functional roles;[15,100,103] for example, *N*-glycosylation can assist in ligand recognition by receptors,[15] leading to both increases[99] and decreases in biological activity depending on the context and systems being considered.[38,104]

Protein *O*-GlcNAcylation has also been relatively well studied over the years since its discovery in 1983.[105] Research work on the modification has revealed a significant similarity and an exquisite interplay between protein *O*-GlcNAcylation and phosphorylation, although the former is performed by a much smaller number of enzymes.[106,107] Generally, *O*-GlcNAcylation of nuclear and cytoplasmic proteins changes in response to changes in nutritional state and cellular stress and has important implications for many cellular processes, including cellular signaling, protein stability, and transcription.[77,108]

The effects of protein *O*-GalNAcylation and mannosylation have not been so clearly established,[109,110] possibly because of the lack of consensus sequences for *O*-glycosylation and common enzymes for *O*-glycan remodeling,[111] or the greater variety of glycan structures.[98,112] Our current knowledge regarding the effects of these two types of *O*-glycosylation is mostly derived from the studies of mucin-type domains and glycoproteins.[113,114] In mucin-like regions, the extensive *O*-glycosylation leads to increased molecular rigidity and causes these domains to adopt a characteristic extended conformation.[115] This structure-stabilizing effect is mainly due to the first several saccharide units closest to the peptide backbone. Different types of carbohydrate residues can be further added to these core glycans, which can play important roles in protein protection, recognition, selection, and adhesion.[116] There are not many studies on sparse *O*-GalNAcylation and mannosylation, but the available data indicates that such modifications may have similar effects to sparsely distributed *N*-glycosylations.[29,38,104]

O-linked proteoglycans also exert diverse biological functions.[117] These functions are mainly associated with their long GAG chains.[118] For example, GAGs are one of the major components of the extracellular matrix (ECM). They can attract and bind a large amount of water, thus they have the capability to absorb compressive forces and protect tissues and cells from mechanical injury.[119] GAGs are also well known for their important role in preventing blood coagulation[120] and in interacting with many chemokines and growth factors to affect their distribution in tissues, stability, and function.[121,122]

GPI-anchored proteins are complex structures comprising three parts: a phospholipid tail, a sugar–phosphate linkage, and a protein.[123] Many types of proteins can be attached to GPI, where they perform a number of functions ranging from enzymatic catalysis to cell–cell contact and adhesion, cell wall synthesis, and cell surface protection.[124] Beyond the obvious function of providing proteins with a stable membrane anchoring device, possible biological functions for the actual GPI anchor have been difficult to elucidate.[125,126]

1.2.4 Methods to Prepare Homogeneous Glycopeptides and Glycoproteins

Although previous studies on glycoproteins have uncovered numerous links between glycan structural variation and glycoprotein functional diversity, there are still many knowledge gaps and uncertainties regarding the details of such interactions.[14,127] As discussed above, to develop a more comprehensive and quantitative awareness of the effects of protein glycosylation and the importance of glycoprotein heterogeneity, homogeneous glycopeptides and glycoproteins are necessary.[27,48,58,128]

Chemical methods have emerged as an indispensable means to prepare homogeneous glycopeptides and glycoproteins.[27,28] They can be roughly divided into two categories: chemical synthesis and enzymatic synthesis. Chemical synthesis has the notable advantage of being flexible, since chemical glycosylation is not dictated by the amino acid sequence of the substrate protein. These completely chemical methods are also very controllable and selective, which can allow for the generation of homogeneous glycoforms bearing both *O*- and *N*-linked glycans, and glycoforms containing different glycans at different positions.[129] Such advantages come with the unfortunate drawback of total chemical synthesis of large biomolecules like glycoproteins being generally difficult and tedious.[130] In comparison, enzymatic synthesis is more convenient and efficient; and once the proper enzymes are identified and optimized, they can be easily applied to the synthesis of many different glycoproteins. However, the strict substrate specificities of most glycan-synthesizing enzymes may limit their application and currently it is still difficult to use only enzymatic methods to prepare glycoproteins containing different glycan structures on the same molecule.[58,131]

Chemical synthesis of glycoproteins generally involves using a native chemical ligation strategy to assemble a set of peptide and glycopeptide fragments (Figure 1.3).[132,133] Peptides can be rapidly prepared by solid-phase peptide synthesis (SPPS),[134] while the preparation of glycopeptides is a more demanding task, mainly because of the presence of sugar moieties.[135] There are two major ways to introduce sugars into glycopeptides: one is to use glycoamino acids as building blocks in SPPS,[29,136] and the other is to couple sugars to partially protected peptides.[137] In either route, the chemical synthesis of oligosaccharides is the most challenging, time-consuming, but obviously unavoidable step. Through many years of research, synthetic routes for many *N*-linked glycans[130,138,139] as well as *O*-GalNAc[140–142] and *O*-Man glycoamino acids[143] have been developed, but a significant amount of optimization is still possible to further simplify these procedures. Although the chemical synthesis of proteoglycans[144] and GPI-anchored proteins[144] remains a daunting task, the last two decades have seen much progress made in advancing the preparation of GAG and GPI anchor glycans. Hence we are hopeful that the chemical synthesis of their conjugates with proteins will be achieved soon.[145,146]

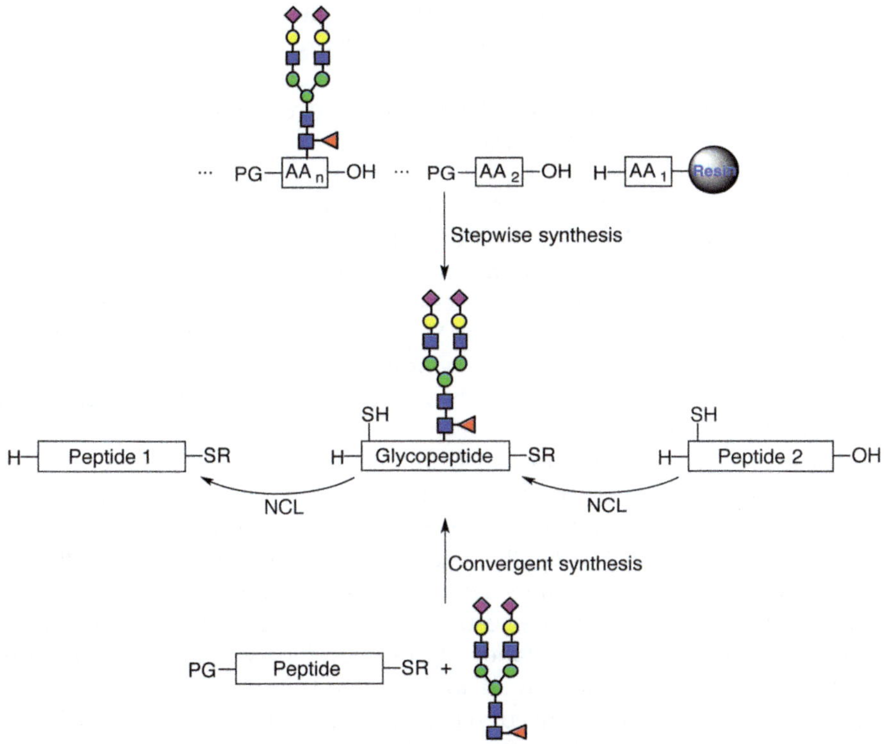

Figure 1.3 A schematic representation of the chemical synthesis of glycoproteins. As shown, glycoproteins can be generated by ligating synthetic peptide and glycopeptide fragments together. Glycopeptide fragments can be prepared either in a stepwise manner by using glycoamino acids or in a convergent manner by coupling partially protected peptides with sugar moieties. NCL = native chemical ligation, PG = protecting group, AA = amino acid, and R = alkyl or aryl groups.

In addition to chemical synthesis, enzymatic synthesis has become an important tool in the production of glycans, glycopeptides, and glycoproteins.[58,61,62] Most often, these methods are combined with chemically prepared acceptors or donors, and sometimes both simultaneously.[147,148] Akin to the development of glycoprotein chemical synthesis, the enzymatic synthesis of *N*-glycosylated,[147] *O*-GalNAcylated[149] and *O*-mannosylated proteins[150] are all relatively well investigated, while the enzymatic preparation of proteoglycans[151] and GPI-anchored proteins[146] are still in the very early stages.

1.2.5 Chemical Glycobiology and Applications

In an effort to further our understanding of protein glycosylation, many chemical biology approaches have been applied to study the roles, functions, regulation, and other aspects of glycoprotein function in living systems. Chemical biology's usefulness is exemplified by recent studies of combinatorial glycoform libraries, accessible thanks to the newly increased sophistication of glycoprotein synthesis,[129] which have given us important insights as to the effects of protein glycosylation.[29,42,44,147,152,153] The availability of collections of glyco-variants that have subtle differences in their glycan structures makes it possible to dissect detailed structure–activity relationships of glycoproteins. Besides directly studying synthetic glycoforms, chemical biologists have established numerous tools to perturb,[154,155] detect,[156] and image glycosylation[157,158] in a cellular context. Through selectively inhibiting enzymes and proteins that participate in glycoprotein biosynthesis, processing and interaction networks, or by rapidly quantifying glycosylation levels and visualizing specific glycoforms, these studies have taught us much about protein glycosylation.

All of these advances mentioned thus far indicate new opportunities to exploit glycosylation's potential in the medical field. Current work on vaccine development and therapeutic protein glycoengineering promises revolutionary new therapies.[159] Results of studies have revealed that the surfaces of pathogens, viruses, and malignant cells can be extensively modified by glycans with unique structural features, and a number of those features have already been explored by many laboratories to develop vaccines against these diseases.[160–164] Previous data have also indicated that glycoproteins with different glycan structures and glycosylation patterns can exert different *in vivo* function and effects.[41,165] These findings have attracted substantial research interest in the area of chemical glycoengineering to further improve the efficiency and specificity of antibodies,[166] cytokines,[129] and other therapeutic proteins.[24,39]

1.3 Conclusion

As the most abundant PTM of secreted proteins, glycosylation has been demonstrated to play an important role in the biosynthesis and biological actions of glycoproteins.[167] However, our current understanding of its structure–function relationships and composition–function relationships remains far from complete and yet it is exactly that knowledge which is required for any practical application of protein glycosylation such as in the development of new therapeutic products or diagnostic technologies. Even worse, the dearth of basic understanding combined with the scarcity of commercial tools for new investigations has also led to a reluctance of researchers to embrace this important field. In order to address these manifold issues, it is necessary to cultivate new easy-to-use tools, commercialize more research materials, and to achieve a more advanced understanding of protein glycosylation. Chemical biology has the potential to satisfy these needs. Indeed chemical biology has already proven its power by providing versatile tools for the detection and visualization of protein glycosylation, many different types of structurally defined glycans, glycopeptides, and glycoproteins, and a considerable amount of new information regarding protein glycosylation. It is expected that in the future these kinds of interdisciplinary approaches will further facilitate the study and application of protein glycosylation.

References

1. International Human Genome Sequencing Consortium, *Nature*, 2004, **431**, 931–945.
2. T. J. Atkinson and M. S. Halfon, *Comput. Struct. Biotechnol. J.*, 2014, **9**, e201401001.
3. L.-R. Yu, N. A. Stewart and T. D. Veenstra, in *Essentials of Genomic and Personalized Medicine*, ed. H. F. Willard, Academic Press, San Diego, 2010, pp. 89–96.
4. C. T. Walsh, S. Garneau-Tsodikova and G. J. Gatto Jr, *Angew. Chem., Int. Ed. Engl.*, 2005, **44**, 7342–7372.
5. O. N. Jensen, *Curr. Opin. Chem. Biol.*, 2004, **8**, 33–41.
6. C. Walsh, *Posttranslational Modification of Proteins : Expanding Nature's Inventory*, Roberts and Company, Scion [distributor], Eaglewood, Colorado, Bloxham, 2005.
7. G. Duan and D. Walther, *PLoS Comput. Biol.*, 2015, **11**, e1004049.
8. E. R. Stadtman, *Free Radical Res.*, 2006, **40**, 1250–1258.
9. J. Anguizola, R. Matsuda, O. S. Barnaby, K. S. Hoy, C. Wa, E. DeBolt, M. Koke and D. S. Hage, *Clin. Chim. Acta*, 2013, **425**, 64–76.
10. P. Minguez, L. Parca, F. Diella, D. R. Mende, R. Kumar, M. Helmer-Citterich, A. C. Gavin, V. van Noort and P. Bork, *Mol. Syst. Biol.*, 2012, **8**, 599.
11. B. M. Sefton, Overview of Protein Phosphorylation, in *Current Protocols in Cell Biology*, 2001, 00:14.1:14.1.1–14.1.3.
12. R. G. Spiro, *Glycobiology*, 2002, **12**, 43R–56R.
13. E. H. Fischer, *Biol. Chem.*, 2010, **391**, 131–137.
14. A. Varki, *Essentials of Glycobiology*, Cold Spring Harbor Laboratory Press, Cold Spring Harbor, N.Y., 2009.
15. R. M. Anthony, F. Wermeling and J. V. Ravetch, *Ann. N. Y. Acad. Sci.*, 2012, **1253**, 170–180.
16. J. L. Price, E. K. Culyba, W. Chen, A. N. Murray, S. R. Hanson, C. H. Wong, E. T. Powers and J. W. Kelly, *Biopolymers*, 2012, **98**, 195–211.
17. L. F. Zacchi and B. L. Schulz, *Glycoconjugate J.*, 2016, **33**, 359–376.
18. L. von Stechow, C. Francavilla and J. V. Olsen, *Expert Rev. Proteomics*, 2015, **12**, 469–487.
19. K. W. Moremen, M. Tiemeyer and A. V. Nairn, *Nat. Rev. Mol. Cell Biol.*, 2012, **13**, 448–462.
20. P. L. Storring, *Trends Biotechnol.*, 1992, **10**, 427–432.
21. B. K. Gavrilov, K. Rogers, I. J. Fernandez-Sainz, L. G. Holinka, M. V. Borca and G. R. Risatti, *Virology*, 2011, **420**, 135–145.
22. A. Varki, *Glycobiology*, 1993, **3**, 97–130.

23. M. Dicker and R. Strasser, *Expert. Opin. Biol. Ther.*, 2015, **15**, 1501–1516.
24. R. J. Sola and K. Griebenow, *BioDrugs*, 2010, **24**, 9–21.
25. S. Kosuri and G. M. Church, *Nat. Methods*, 2014, **11**, 499–507.
26. C. L. Young, Z. T. Britton and A. S. Robinson, *Biotechnol. J.*, 2012, **7**, 620–634.
27. J. R. Rich and S. G. Withers, *Nat. Chem. Biol.*, 2009, **5**, 206–215.
28. C. R. Bertozzi and L. L. Kiessling, *Science*, 2001, **291**, 2357–2364.
29. L. Chen, M. R. Drake, M. G. Resch, E. R. Greene, M. E. Himmel, P. K. Chaffey, G. T. Beckham and Z. Tan, *Proc. Natl. Acad. Sci. U. S. A.*, 2014, **111**, 7612–7617.
30. D. Shental-Bechor and Y. Levy, *Curr. Opin. Struct. Biol.*, 2009, **19**, 524–533.
31. R. Hoiberg-Nielsen, C. C. Fuglsang, L. Arleth and P. Westh, *Biochemistry*, 2006, **45**, 5057–5066.
32. H. S. Lee, Y. Qi and W. Im, *Sci. Rep.*, 2015, **5**, 8926.
33. J. N. Arnold, M. R. Wormald, R. B. Sim, P. M. Rudd and R. A. Dwek, *Annu. Rev. Immunol.*, 2007, **25**, 21–50.
34. E. Reynders, F. Foulquier, W. Annaert and G. Matthijs, *Glycobiology*, 2011, **21**, 853–863.
35. R. J. Sola and K. Griebenow, *J. Pharm. Sci.*, 2009, **98**, 1223–1245.
36. J. M. Hayes, E. F. Cosgrave, W. B. Struwe, M. Wormald, G. P. Davey, R. Jefferis and P. M. Rudd, *Curr. Top Microbiol. Immunol.*, 2014, **382**, 165–199.
37. N. Haines and K. D. Irvine, *Nat. Rev. Mol. Cell Biol.*, 2003, **4**, 786–797.
38. P. Van den Steen, P. M. Rudd, R. A. Dwek, J. Van Damme and G. Opdenakker, *Adv. Exp. Med. Biol.*, 1998, **435**, 133–143.
39. A. M. Sinclair and S. Elliott, *J. Pharm. Sci.*, 2005, **94**, 1626–1635.
40. S. E. O'Connor, J. Pohlmann, B. Imperiali, I. Saskiawan and K. Yamamoto, *J. Am. Chem. Soc.*, 2001, **123**, 6187–6188.
41. R. M. Anthony, F. Nimmerjahn, D. J. Ashline, V. N. Reinhold, J. C. Paulson and J. V. Ravetch, *Science*, 2008, **320**, 373–376.
42. J. L. Price, D. Shental-Bechor, A. Dhar, M. J. Turner, E. T. Powers, M. Gruebele, Y. Levy and J. W. Kelly, *J. Am. Chem. Soc.*, 2010, **132**, 15359–15367.
43. E. K. Culyba, J. L. Price, S. R. Hanson, A. Dhar, C. H. Wong, M. Gruebele, E. T. Powers and J. W. Kelly, *Science*, 2011, **331**, 571–575.
44. X. Guan, P. K. Chaffey, C. Zeng, E. R. Greene, L. Chen, M. R. Drake, C. Chen, A. Groobman, M. G. Resch, M. E. Himmel, G. T. Beckham and Z. Tan, *Chem. Sci.*, 2015, **6**, 7185–7189.
45. S. A. Brooks, *Mol. Biotechnol.*, 2009, **43**, 76–88.
46. Z. Shriver, S. Raguram and R. Sasisekharan, *Nat. Rev. Drug Discovery*, 2004, **3**, 863–873.
47. V. Reinhold, H. Zhang, A. Hanneman and D. Ashline, *Mol. Cell. Proteomics*, 2013, **12**, 866–873.
48. S. Pan, R. Chen, R. Aebersold and T. A. Brentnall, *Mol. Cell. Proteomics*, 2011, **10**, R110 003251.
49. H. Hinneburg, K. Stavenhagen, U. Schweiger-Hufnagel, S. Pengelley, W. Jabs, P. H. Seeberger, D. V. Silva, M. Wuhrer and D. Kolarich, *J. Am. Soc. Mass Spectrom.*, 2016, **27**, 507–519.

50. S. Pan, R. Aebersold, R. Chen, J. Rush, D. R. Goodlett, M. W. McIntosh, J. Zhang and T. A. Brentnall, *J. Proteome Res.*, 2009, **8**, 787–797.
51. K. Sparbier, A. Asperger, A. Resemann, I. Kessler, S. Koch, T. Wenzel, G. Stein, L. Vorwerg, D. Suckau and M. Kostrzewa, *J. Biomol. Tech.*, 2007, **18**, 252–258.
52. M. W. C. Hatton, L. März and E. Regoeczi, *Trends Biochem. Sci.*, 1983, 8, 287–291.
53. R. M. Bill, L. Revers and I. B. H. Wilson, *Protein Glycosylation*, Springer US, Boston, MA, 1998, pp. 85–145, DOI: 10.1007/978-1-4615-4939-0_3.
54. C. B. Reese, *Org. Biomol. Chem.*, 2005, **3**, 3851–3868.
55. B. Beckert and B. Masquida, *Methods Mol. Biol.*, 2011, **703**, 29–41.
56. B. L. Nilsson, M. B. Soellner and R. T. Raines, *Annu. Rev. Biophys. Biomol. Struct.*, 2005, **34**, 91–118.
57. I. Hunt, *Protein Expression Purif.*, 2005, **40**, 1–22.
58. L. X. Wang and M. N. Amin, *Chem. Biol.*, 2014, **21**, 51–66.
59. R. M. Schmaltz, S. R. Hanson and C. H. Wong, *Chem. Rev.*, 2011, **111**, 4259–4307.
60. A. Fernandez-Tejada, J. Brailsford, Q. Zhang, J. H. Shieh, M. A. Moore and S. J. Danishefsky, *Top. Curr. Chem.*, 2015, **362**, 1–26.
61. C. Unverzagt and Y. Kajihara, *Chem. Soc. Rev.*, 2013, **42**, 4408–4420.
62. J. B. McArthur and X. Chen, *Biochem. Soc. Trans.*, 2016, **44**, 129–142.
63. C. D. Rillahan and J. C. Paulson, *Annu. Rev. Biochem.*, 2011, **80**, 797–823.
64. M. Amblard, J. A. Fehrentz, J. Martinez and G. Subra, *Mol. Biotechnol.*, 2006, **33**, 239–254.
65. I. Zlatev, M. Manoharan, J. J. Vasseur and F. Morvan, Solid-Phase Chemical Synthesis of 5′-Triphosphate DNA, RNA, and Chemically Modified Oligonucleotides, in *Current Protocols in Nucleic Acid Chemistry*, 2012, 50:1.28:1.28.1–1.28.16.
66. P. H. Seeberger and W.-C. Haase, *Chem. Rev.*, 2000, **100**, 4349–4394.
67. M. Lommel and S. Strahl, *Glycobiology*, 2009, **19**, 816–828.
68. A. Furmanek and J. Hofsteenge, *Acta Biochim. Pol.*, 2000, **47**, 781–789.
69. Y. Ihara, Y. Inai and M. Ikezaki, *Trends Glycosci. Glycotechnol.*, 2011, **23**, 1–13.
70. D. P. Gamblin, E. M. Scanlan and B. G. Davis, *Chem. Rev.*, 2009, **109**, 131–163.
71. P. A. Haynes, *Glycobiology*, 1998, **8**, 1–5.
72. P. Lafite and R. Daniellou, *Nat. Prod. Rep.*, 2012, **29**, 729–738.
73. J. F. Vliegenthart and F. Casset, *Curr. Opin. Struct. Biol.*, 1998, **8**, 565–571.
74. K. F. Medzihradszky, *Methods Mol. Biol.*, 2008, **446**, 293–316.
75. X. Shi and D. L. Jarvis, *Curr. Drug Targets*, 2007, **8**, 1116–1125.
76. T. H. Thanka Christlet and K. Veluraja, *Biophys. J.*, 2001, **80**, 952–960.
77. M. R. Bond and J. A. Hanover, *J. Cell Biol.*, 2015, **208**, 869–880.
78. M. Loibl and S. Strahl, *Biochim. Biophys. Acta*, 2013, **1833**, 2438–2446.
79. J. Peter-Katalinic, *Methods Enzymol.*, 2005, **405**, 139–171.

80. L. Schaefer and R. M. Schaefer, *Cell Tissue Res.*, 2010, **339**, 237–246.
81. J. R. Couchman and C. A. Pataki, *J. Histochem. Cytochem.*, 2012, **60**, 885–897.
82. T. Endo, *J. Biochem.*, 2015, **157**, 1–12.
83. K. S. Bergstrom and L. Xia, *Glycobiology*, 2013, **23**, 1026–1037.
84. M. Goto, *Biosci., Biotechnol., Biochem.*, 2007, **71**, 1415–1427.
85. M. Hoglund, *Med. Oncol.*, 1998, **15**, 229–233.
86. R. V. Iozzo and L. Schaefer, *Matrix Biol.*, 2015, **42**, 11–55.
87. M. J. Nadolski and M. E. Linder, *FEBS J.*, 2007, **274**, 5202–5210.
88. S. Mayor and H. Riezman, *Nat. Rev. Mol. Cell Biol.*, 2004, **5**, 110–120.
89. A. Dell, A. Galadari, F. Sastre and P. Hitchen, *Int. J. Microbiol.*, 2010, **2010**, 148178.
90. A. Yan and W. J. Lennarz, *J. Biol. Chem.*, 2005, **280**, 3121–3124.
91. A. Varki and H. H. Freeze, *Subcell. Biochem.*, 1994, **22**, 71–100.
92. F. Schwarz and M. Aebi, *Curr. Opin. Struct. Biol.*, 2011, **21**, 576–582.
93. M. Aebi, *Biochim. Biophys. Acta*, 2013, **1833**, 2430–2437.
94. M. B. Jones, J. N. Rosenberg, M. J. Betenbaugh and S. S. Krag, *Biochim. Biophys. Acta*, 2009, **1790**, 485–494.
95. S. Rottger, J. White, H. H. Wandall, J. C. Olivo, A. Stark, E. P. Bennett, C. Whitehouse, E. G. Berger, H. Clausen and T. Nilsson, *J. Cell Sci.*, 1998, **111**(1), 45–60.
96. G. Dick, L. K. Akslen-Hoel, F. Grondahl, I. Kjos and K. Prydz, *J. Histochem. Cytochem.*, 2012, **60**, 926–935.
97. K. Prydz and K. T. Dalen, *J. Cell Sci.*, 2000, **113**(2), 193–205.
98. J. L. Praissman and L. Wells, *Biochemistry*, 2014, **53**, 3066–3078.
99. U. Loizides-Mangold, F. P. David, V. J. Nesatyy, T. Kinoshita and H. Riezman, *J. Lipid Res.*, 2012, **53**, 1522–1534.
100. S. E. O'Connor and B. Imperiali, *Chem. Biol.*, 1996, **3**, 803–812.
101. S. R. Hanson, E. K. Culyba, T. L. Hsu, C. H. Wong, J. W. Kelly and E. T. Powers, *Proc. Natl. Acad. Sci. U. S. A.*, 2009, **106**, 3131–3136.
102. B. Imperiali and S. E. O'Connor, *Curr. Opin. Chem. Biol.*, 1999, **3**, 643–649.
103. A. D. Elbein, *Trends Biotechnol.*, 1991, **9**, 346–352.
104. G. Opdenakker, P. M. Rudd, M. Wormald, R. A. Dwek and J. Van Damme, *FASEB J.*, 1995, **9**, 453–457.
105. C. R. Torres and G. W. Hart, *J. Biol. Chem.*, 1984, **259**, 3308–3317.
106. G. W. Hart, M. P. Housley and C. Slawson, *Nature*, 2007, **446**, 1017–1022.
107. K. R. Harwood and J. A. Hanover, *J. Cell Sci.*, 2014, **127**, 1857–1867.
108. M. R. Bond and J. A. Hanover, *Annu. Rev. Nutr.*, 2013, **33**, 205–229.
109. S. Wopereis, D. J. Lefeber, E. Morava and R. A. Wevers, *Clin. Chem.*, 2006, **52**, 574–600.
110. P. Van den Steen, P. M. Rudd, R. A. Dwek and G. Opdenakker, *Crit. Rev. Biochem. Mol. Biol.*, 1998, **33**, 151–208.
111. P. H. Jensen, D. Kolarich and N. H. Packer, *FEBS J.*, 2010, **277**, 81–94.
112. M. C. Rose and J. A. Voynow, *Physiol. Rev.*, 2006, **86**, 245–278.

113. H. C. Hang and C. R. Bertozzi, *Bioorg. Med. Chem.*, 2005, **13**, 5021–5034.
114. C. M. Dobson, S. J. Hempel, S. H. Stalnaker, R. Stuart and L. Wells, *Cell. Mol. Life Sci.*, 2013, **70**, 2849–2857.
115. P. Argueso, *Jpn. J. Ophthalmol.*, 2013, **57**, 150–155.
116. H. J. Gabius, *The Sugar Code: Fundamentals of Glycosciences*, Wiley, 2011.
117. R. V. Iozzo, *Proteoglycans: Structure, Biology and Molecular Interactions*, Taylor & Francis, 2000.
118. C. Noti and P. H. Seeberger, *Chem. Biol.*, 2005, **12**, 731–756.
119. M. Yanagishita, *Acta Pathol. Jpn.*, 1993, **43**, 283–293.
120. M. C. Bourin and U. Lindahl, *Biochem. J.*, 1993, **289**(2), 313–330.
121. J. Shute, *Handb. Exp. Pharmacol.*, 2012, 307–324, DOI: 10.1007/978-3-642-23056-1_13.
122. D. J. Hamel, I. Sielaff, A. E. Proudfoot and T. M. Handel, *Methods Enzymol.*, 2009, **461**, 71–102.
123. M. G. Paulick and C. R. Bertozzi, *Biochemistry*, 2008, **47**, 6991–7000.
124. F. J. Sharom and M. T. Lehto, *Biochem. Cell Biol.*, 2002, **80**, 535–549.
125. M. Fujita and T. Kinoshita, *Biochim. Biophys. Acta*, 2012, **1821**, 1050–1058.
126. J. A. Dangerfield and C. Metzner, *GPI Membrane Anchors-the Much Needed Link*, Bentham Science Publishers, 2010.
127. L. Krasnova and C. H. Wong, *Annu. Rev. Biochem.*, 2016, **85**, 599–630.
128. Y. Kajihara, N. Yamamoto, R. Okamoto, K. Hirano and T. Murase, *Chem. Rec.*, 2010, **10**, 80–100.
129. P. Wang, S. Dong, J. H. Shieh, E. Peguero, R. Hendrickson, M. A. Moore and S. J. Danishefsky, *Science*, 2013, **342**, 1357–1360.
130. M. A. Walczak, J. Hayashida and S. J. Danishefsky, *J. Am. Chem. Soc.*, 2013, **135**, 4700–4703.
131. C. Toonstra, M. N. Amin and L. X. Wang, *J. Org. Chem.*, 2016, **81**, 6176–6185.
132. P. E. Dawson, T. W. Muir, I. Clark-Lewis and S. B. Kent, *Science*, 1994, **266**, 776–779.
133. Q. Wan and S. J. Danishefsky, *Angew. Chem., Int. Ed. Engl.*, 2007, **46**, 9248–9252.
134. X. Guan, P. K. Chaffey, C. Zeng and Z. Tan, *Top. Curr. Chem.*, 2015, **363**, 155–192.
135. H. Hojo and Y. Nakahara, *Biopolymers*, 2007, **88**, 308–324.
136. L. A. Marcaurelle and C. R. Bertozzi, *Glycobiology*, 2002, **12**, 69R–77R.
137. T. Buskas, S. Ingale and G. J. Boons, *Glycobiology*, 2006, **16**, 113R–136R.
138. S. S. Shivatare, S. H. Chang, T. I. Tsai, C. T. Ren, H. Y. Chuang, L. Hsu, C. W. Lin, S. T. Li, C. Y. Wu and C. H. Wong, *J. Am. Chem. Soc.*, 2013, **135**, 15382–15391.
139. C. Unverzagt, S. Eller, S. Mezzato and R. Schuberth, *Chemistry*, 2008, **14**, 1304–1311.
140. L. Corcilius and R. J. Payne, *Org. Lett.*, 2013, **15**, 5794–5797.
141. M. R. Pratt and C. R. Bertozzi, *Chem. Soc. Rev.*, 2005, **34**, 58–68.

142. M. Pudelko, J. Bull and H. Kunz, *ChemBioChem*, 2010, **11**, 904–930.
143. J. Seifert, T. Ogawa, S. Kurono and Y. Ito, *Glycoconjugate J.*, 2000, **17**, 407–423.
144. D. B. Werz, S. Vidal and D. Crich, *Modern Synthetic Methods in Carbohydrate Chemistry: From Monosaccharides to Complex Glycoconjugates*, Wiley, 2013.
145. S. B. Dulaney and X. Huang, *Adv. Carbohydr. Chem. Biochem.*, 2012, **67**, 95–136.
146. Z. Guo, *Curr. Org. Synth.*, 2013, **10**, 366–383.
147. M. N. Amin, J. S. McLellan, W. Huang, J. Orwenyo, D. R. Burton, W. C. Koff, P. D. Kwong and L. X. Wang, *Nat. Chem. Biol.*, 2013, **9**, 521–526.
148. Z. Wang, Z. S. Chinoy, S. G. Ambre, W. Peng, R. McBride, R. P. de Vries, J. Glushka, J. C. Paulson and G. J. Boons, *Science*, 2013, **341**, 379–383.
149. V. R. Krishnamurthy, M. Y. Sardar, Y. Ying, X. Song, C. Haller, E. Dai, X. Wang, D. Hanjaya-Putra, L. Sun, V. Morikis, S. I. Simon, R. J. Woods, R. D. Cummings and E. L. Chaikof, *Nat. Commun.*, 2015, **6**, 6387.
150. R. Sardzik, A. P. Green, N. Laurent, P. Both, C. Fontana, J. Voglmeir, M. J. Weissenborn, R. Haddoub, P. Grassi, S. M. Haslam, G. Widmalm and S. L. Flitsch, *J. Am. Chem. Soc.*, 2012, **134**, 4521–4524.
151. P. L. DeAngelis, J. Liu and R. J. Linhardt, *Glycobiology*, 2013, **23**, 764–777.
152. W. Chen, S. Enck, J. L. Price, D. L. Powers, E. T. Powers, C. H. Wong, H. J. Dyson and J. W. Kelly, *J. Am. Chem. Soc.*, 2013, **135**, 9877–9884.
153. Y. S. Hsieh, L. C. Wijeyewickrema, B. L. Wilkinson, R. N. Pike and R. J. Payne, *Angew. Chem., Int. Ed. Engl.*, 2014, **53**, 3947–3951.
154. B. A. Potter, R. P. Hughey and O. A. Weisz, *Am. J. Physiol.: Cell Physiol.*, 2006, **290**, C1–C10.
155. R. T. Almaraz, M. P. Mathew, E. Tan and K. J. Yarema, *Ann. Biomed. Eng.*, 2012, **40**, 806–815.
156. S. Wang, F. Yang, D. G. Camp 2nd, K. Rodland, W. J. Qian, T. Liu and R. D. Smith, *Bioanalysis*, 2014, **6**, 2571–2580.
157. M. E. Griffin and L. C. Hsieh-Wilson, *Cell. Chem. Biol.*, 2016, **23**, 108–121.
158. S. T. Laughlin and C. R. Bertozzi, *Proc. Natl. Acad. Sci. U. S. A.*, 2009, **106**, 12–17.
159. M. Dalziel, M. Crispin, C. N. Scanlan, N. Zitzmann and R. A. Dwek, *Science*, 2014, **343**, 1235681.
160. J. Cuccui and B. Wren, *J. Pharm. Pharmacol.*, 2015, **67**, 338–350.
161. R. D. Astronomo and D. R. Burton, *Nat. Rev. Drug Discovery*, 2010, **9**, 308–324.
162. D. M. McDonald, S. N. Byrne and R. J. Payne, *Front. Chem.*, 2015, **3**, 60.

163. S. Horiya, I. S. MacPherson and I. J. Krauss, *Nat. Chem. Biol.*, 2014, **10**, 990–999.
164. M. Crispin and K. J. Doores, *Curr. Opin. Virol.*, 2015, **11**, 63–69.
165. L. C. Wasley, G. Timony, P. Murtha, J. Stoudemire, A. J. Dorner, J. Caro, M. Krieger and R. J. Kaufman, *Blood*, 1991, **77**, 2624–2632.
166. J. P. Giddens and L. X. Wang, *Methods Mol. Biol.*, 2015, **1321**, 375–387.
167. National Research Council, *Transforming Glycoscience: A Roadmap for the Future*, The National Academies Press, Washington, DC, 2012.

CHAPTER 2

Chemical Biology of Protein N-*Glycosylation*

CHAO LI[a] AND LAI-XI WANG*[a]

[a]Department of Chemistry and Biochemistry, University of Maryland, College Park, Maryland 20742, USA
*E-mail: wang518@umd.edu

2.1 Introduction

Glycosylation is a prevalent posttranslational modification (PTM) of proteins. Compelling data have demonstrated that glycosylation can profoundly affect a protein's intrinsic properties such as folding, intracellular trafficking, stability, and pharmacokinetics.[1–4] On the other hand, the oligosaccharide components of glycoproteins can participate directly in a wide variety of biological recognition processes, including cell adhesion, cell differentiation, development, host–pathogen interaction, and immune response.[5–11]

Protein glycosylation is much more complex and diverse than most other posttranslational modifications, such as protein phosphorylation, that involve only a simple functional group transfer. Protein glycosylation can be categorized into three major types: *N*-glycosylation, in which the sugar moiety is attached at the amide side chain of an Asn residue usually through a *N*-acetylglucosamine (GlcNAc)–Asn linkage; typical *O*-glycosylation, in which the sugar moiety is attached to the hydroxyl group of the Ser/Thr residues; and glypiation, where newly synthesized proteins are modified by covalent attachment of a glycosylphosphatidylinositol (GPI) anchor that helps

Chemical Biology No. 2
Chemical Biology of Glycoproteins
Edited by Zhongping Tan and Lai-Xi Wang

Published by the Royal Society of Chemistry, www.rsc.org

localize proteins to the cell membrane. Each category of glycosylation can be further divided into different subtypes in terms of sugar–amino acid linkages, the structures of the attached glycans, and the charge state. A survey of the literature has indicated that there are more than 40 different types of sugar–amino acid linkages, which involve at least 13 different proximal monosaccharides and eight different amino acid residues.[12] As a result, it is conceivable to think that the general term of "protein glycosylation" actually represents a large collection of distinct and diverse posttranslational modifications of proteins that confer a wide range of biological functions. The structural complexity and functional diversity of glycoproteins necessitate the development of novel synthetic and analytical tools to decipher the structures and functions.[13]

Protein *N*-glycosylation is a predominant posttranslational modification in eukaryotic systems. The biosynthesis of *N*-glycoproteins involves multiple and complex steps and usually ends up with mixtures of glycoforms that have the same protein backbone but differ in the pendant *N*-glycans. The diverse structures of *N*-glycans also confer glycoproteins' enormous distinct functions, ranging from signaling in protein folding/degradation and intracellular trafficking to intercellular recognition by a number of glycan-binding proteins, such as lectins and antibodies. Understanding the diverse functions of protein *N*-glycosylation requires different approaches. While genetic approaches, by knocking in or knocking out specific genes in biosynthetic pathways, have been common in glycobiology, the chemical biology approach is emerging as an exciting alternative that exploits synthetic probes, including novel chemical reporters or synthetic oligosaccharide/glycoconjugates, to decipher glycan functions in biological processes. In this chapter, we first provide a brief overview of the biosynthesis of *N*-glycoproteins and then highlight several promising chemical biology approaches to studying protein *N*-glycosylation. These include the use of natural and designer inhibitors of glycan-processing enzymes to modulate and perturb protein *N*-glycosylation; metabolic glycoengineering for cell-surface glycan labeling and imaging, and chemoenzymatic glycoprotein synthesis and glycosylation remodeling for making homogeneous glycoproteins that enables detailed structure–function relationship studies.

2.2 Biosynthesis and Intracellular Functions of *N*-Glycans of Glycoproteins

N-Linked glycosylation is a complex co- and post-translational modification involving multiple steps and several locations within the cell.[1,14] In mammalian systems, the biosynthesis of a mature *N*-glycoprotein includes the following four key steps: (a) the assembly of a lipid-linked large oligosaccharide precursor; (b) the transfer of the oligosaccharide precursor from the glycolipid to the nascent protein by an oligosaccharyltransferase (OST) at the endoplasmic reticulum (ER); (c) the initial trimming of the oligosaccharide

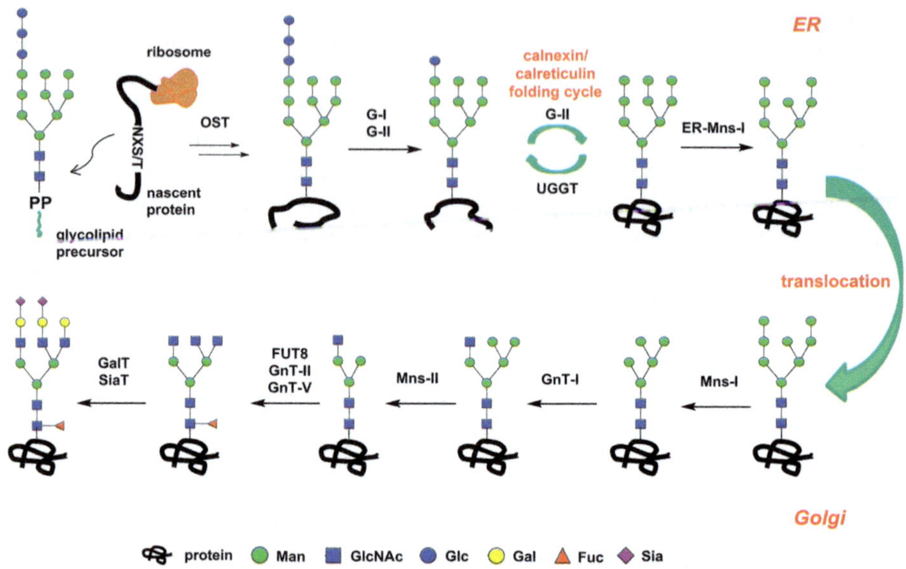

Figure 2.1 Biosynthesis of *N*-glycoproteins in eukaryotic cells. Synthesis begins with the assembly of the $Glc_3Man_9GlcNAc_2$ precursor on dolichol diphosphate through one-by-one addition of the sugar residues at ER. Then the precursor oligosaccharide is transferred by the oligosaccharyltransferase to the asparagine residue in the consensus Asn–Xaa–Ser/Thr sequence of a nascent, growing polypeptide chain. After this key co-translational step, the precursor oligosaccharide is subjected to substantial trimming, processing, and rebuild-up at the ER and the Golgi apparatus to give a broad spectrum of glycoforms. Man, mannose; GlcNAc, *N*-acetylglucosamine; Glc, glucose; Gal, galactose; Fuc, fucose; Sia, sialic acid.

precursor in the ER; and (d) the final processing of the *N*-glycan in the Golgi apparatus to yield the mature *N*-glycoprotein (Figure 2.1). The buildup of the common oligosaccharide precursor, $Glc_3Man_9GlcNAc_2$, starts on the cytosolic surface of the ER membrane by the addition of monosaccharides, one by one, on a membrane-associated dolichylphosphate moiety. When it reaches the $Man_5GlcNAc_2$-phospholipid intermediate, the oligosaccharide is flipped to the lumenal side of the membrane, and the sugar chain elongation continues to form the full-size oligosaccharide precursor. Then the $Glc_3Man_9GlcNAc_2$ is transferred from the dolichol pyrophosphate-linked glycolipid by OST to the amide nitrogen of an asparagine (Asn) side chain of a nascent polypeptide with a consensus Asn–X–Ser/Thr sequence, where X can be any amino acid except proline. After this co-translational modification, the three glucose residues are trimmed away by glucosidases I and II (G-I and G-II) to the mono-glucosylated glycoform ($Glc_1Man_9GlcNAc_2$), which is the key intermediate in the calnexin/calreticulin-mediated protein folding cycle in protein quality control.

The ER also has a UDP-glucose:glycoprotein glucosyltransferase (UGGT) that serves as a sensor of protein misfolding and is able to reglucosylate the protein to a mono-glucosylated glycoform to get into the protein folding cycle again. The correctly folded glycoprotein is subjected to further trimming by G-II and the ER α-mannosidase (ER Mns-I) to the $Man_8GlcNAc_2$ glycoform. The $Man_8GlcNAc_2$ glycoform is then translocated to the Golgi apparatus, where the glycoprotein is further trimmed by Golgi α-mannosidases (Mns-I and Mns-II) to the $Man_5GlcNAc_2$ state. The β-1,2-GlcNAc transferase I (GnT-I) catalyzed addition of a GlcNAc residue to the low α1,3Man moiety gives a key intermediate that can be elongated at the GlcNAc moiety to form hybrid glycoforms, or can be further processed to complex glycoforms through trimming of the two additional mannoses, followed by building up of diverse complex type *N*-glycans with a range of glycosyltransferases, *e.g.*, α-1,6-Fucosyltransferase (FUT8) for core-fucosylation, β-1,2-GlcNAc transferase II (GnT-II) for up-arm GlcNAc addition, GnT-III for bisecting GlcNAc addition and GnT-V for branching, and α-sialyltransferase (SiaT) for sialylation (Figure 2.1). The mature glycoforms are the outcome of a very complex glycosylation network that can be affected by many factors, such as the availability of sugar nucleotides and co-factors, the activity of the glycosidases and glycosyltransferases in the trimming and processing, and the accessibility of the glycosylation sites on the folded protein. As a result, *N*-glycoproteins are usually produced as mixtures of glycoforms that share the same polypeptide backbone but differ in the pendent *N*-glycans.

Protein *N*-glycosylation plays key roles in the secretory pathways. In the ER, the high-mannose *N*-glycans, with or without the glucose residue, are involved in promoting protein folding, quality control, trafficking, and ER-associated protein degradation signaling.[14] In addition, *N*-glycosylation can intrinsically modulate the folding kinetics and thermodynamic stability of *N*-glycoproteins *via* local glycan–peptide motif interactions.[4] After translocation to the Golgi apparatus, further processing and modification of the *N*-glycans by the Golgi enzymes confer a wide spectrum of new functions to the glycoproteins. For example, the decoration of the *N*-glycans with terminal mannose-6-phosphate (M6P) moieties provides a signal for targeting the newly synthesized hydrolases from the Golgi apparatus to lysosomes through their specific interactions with M6P receptors on lysosomes.[15,16] In addition to the intracellular functions, the *N*-glycans of secreted and membrane-associated mature *N*-glycoproteins can participate directly in a range of important intercellular molecular recognition processes such as cell adhesion, cancer metastasis, and host–pathogen interactions. The diverse functions of protein *N*-glycosylation also provide a range of notable targets for molecular design aiming at deciphering these functions.

2.3 Inhibitors of Glycan-Processing Enzymes for Controlling Protein *N*-Glycosylation

The biosynthesis of *N*-glycans and *N*-glycoproteins constitutes a very complex network involving dozens of enzymes and glycan-binding proteins, including an array of processing glycosidases, glycosyltransferases, carbohydrate-decorating enzymes, and lectins. These enzymes and lectins can serve as important targets for deciphering and perturbing the glycosylation network. In fact, specific inhibitors of *N*-glycan-processing enzymes have been discovered, either from natural sources or by design. These inhibitors can selectively block particular processing steps in the biosynthetic pathways and thus alter the glycosylation state and/or perturb the biological functions.[17,18] One particular application of the enzyme inhibitors is to control glycosylation in recombinant glycoprotein production so that well-defined and more homogeneous glycoforms can be obtained for structural and functional studies (Figure 2.2). For example, deoxynojirimycin (DNJ) and its *N*-alkyl analogs are specific inhibitors of α-glucosidase I and II (G-I and G-II) that are responsible for removing the three glucose residues in the early steps of *N*-glycan processing. It was shown that the use of *N*-butyldeoxynojirimycin (NB-DNJ) in CHO cell expression of CD58 successfully trapped the *N*-glycans at the endo H-sensitive, $Glc_3Man_{7-9}GlcNAc_2$ stage for X-ray crystallographic studies.[19] Kifunensine, an alkaloid natural product and a cyclic oxamide derivative of L-amino mannojirimycin, is a potent inhibitor of mannosidase I (ER-Mns-I) responsible for the removal of the first mannose in *N*-glycan processing.[20] The concentration giving 50% of maximum inhibition (IC_{50}) of kifunensine for Mns-I is in the range of nanomolar concentrations and

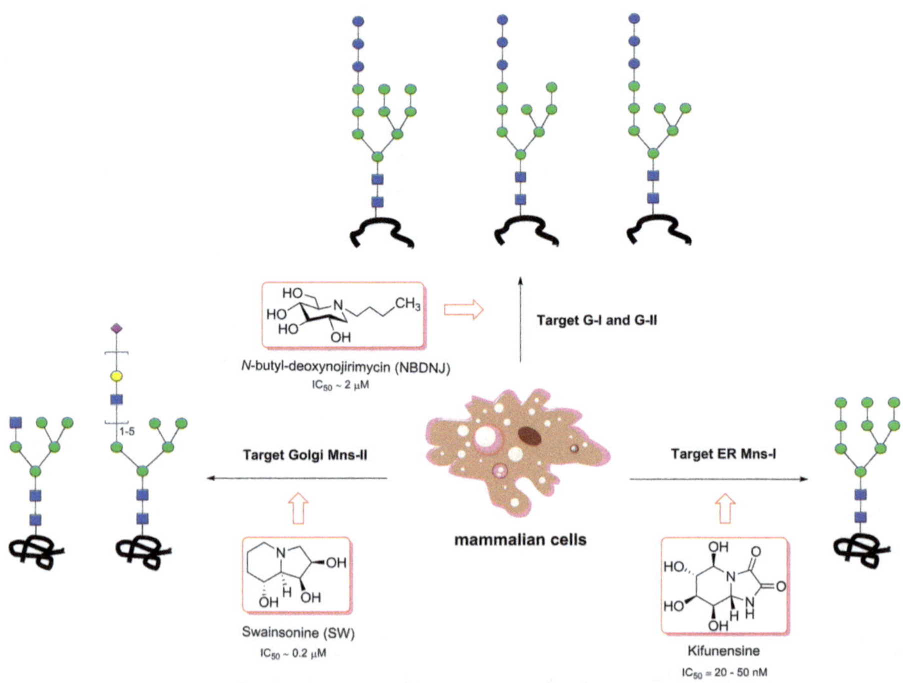

Figure 2.2 The presence of glycosidase inhibitors leads to production of well-defined *N*-glycoprotein glycoforms in mammalian cell expression systems. *N*-Butyl-deoxynojirimycin, inhibitor for ER glucosidase I and II; kifunensine: inhibitor for ER and Golgi α-mannosidase I; swainsonine, inhibitor for Golgi α-mannosidase II.

it has been demonstrated that the use of kifunensine could efficiently control *N*-glycosylation at the $Man_9GlcNAc_2$ state.[21–23] Kifunensine was successfully used to express recombinant HIV-1 gp120 and related gp120 domains in the high-mannose-type glycoforms for elucidating the antigenicity and immunogenicity of the HIV-1 envelope glycoprotein.[24–27] Swainsonine, an indolizidine alkaloid natural product, is a specific inhibitor of Golgi mannosidase II (Mns-II).[28] As a result, inhibition of this mannosidase could block the removal of the two mannose residues in the up-arm of the $Man_5GlcNAc_2$ core, thus preventing further branching of the core to complex-type *N*-glycans. This mannosidase inhibitor was successfully used to produce hybrid-type or $Man_5GlcNAc_2$ glycoforms of HIV-1 envelope glycoprotein domains for characterizing the glycan specificity of broadly HIV-neutralizing antibodies.[26,29] In addition to their application to producing defined glycoforms in recombinant glycoprotein production, these *N*-glycan-processing inhibitors have also found applications as anti-viral and anti-cancer agents. Results of early studies have demonstrated that α-glucosidase inhibitors such as NB-DNJ could inhibit HIV replication and HIV-mediated syncytium formation *in vitro*.[30,31] It was also reported that NB-DNJ was able to inhibit human hepatitis B virus (HBV)[32–34] and human hepatitis C virus (HCV) in cell-based assays.[35,36] The inhibition of the early steps in glycan processing may affect the appropriate folding and/or secretion of the viral envelope glycoproteins. On the other hand, swainsonine, a selective inhibitor of Golgi α-mannosidase II, and its derivatives showed inhibitory activities against cancer cell proliferation.[37] It should be pointed out that the success of this approach will rely on the difference in the *N*-glycan structure and function between cancer cells and normal cells, as well as the expression levels of those glycan-processing enzymes in cancer *vs.* normal cells.

2.4 Global Metabolic Enzyme Inhibitors for Perturbing Protein *N*-Glycosylation

Sialylation and core-fucosylation are two important decorations of *N*-glycans that regulate cell signaling, leukocyte trafficking, host–pathogen interactions, and other cellular recognition processes.[38–40] Recently Paulson and co-workers reported an elegant approach to global metabolic inhibition of sialylation and fucosylation in cells.[41] This approach took advantage of the relaxed substrate specificity of enzymes in the sialic acid and fucose salvage pathways to incorporate selectively fluorinated sialic acid and fucose precursors to generate the respective fluorinated sugar nucleotide substrate analogs *in situ*, which would then act as inhibitors of the sialyltransferase and fucosyltransferase, respectively (Figure 2.3). It was previously demonstrated that fluorinated sugar nucleotide substrate analogs, where the fluorine atom was placed proximal to the anomeric center, acted as competitive inhibitors of glycosyltransferases because the proximal presence of a strong electronegative fluorine would significantly retard the development of substantial

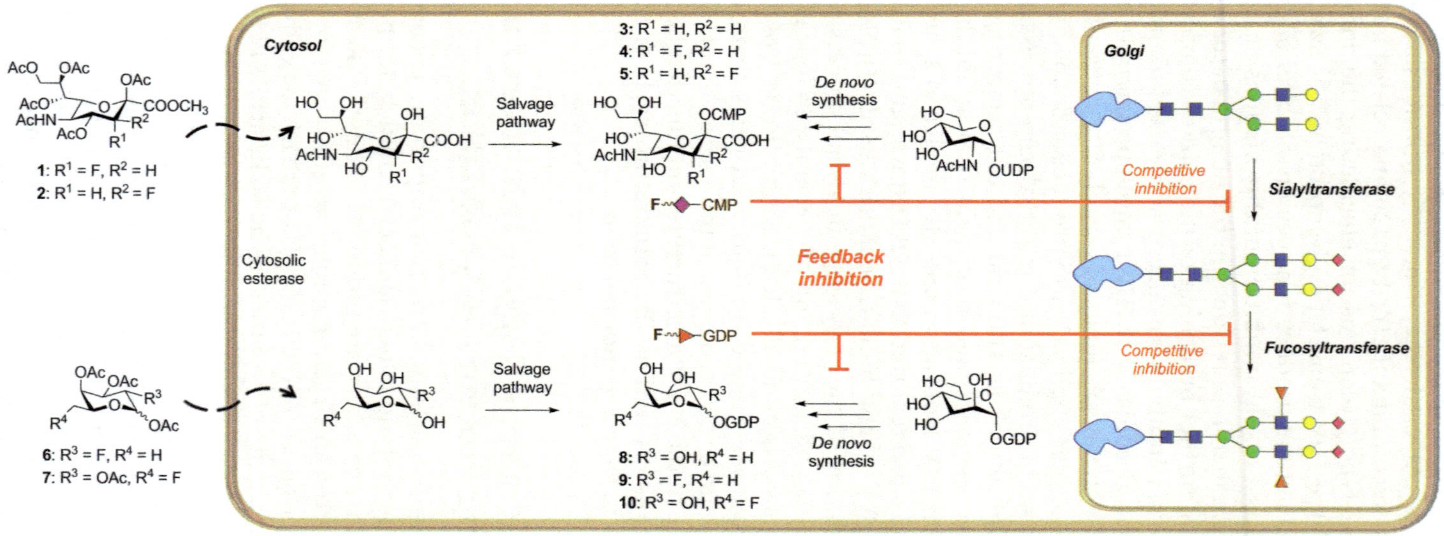

Figure 2.3 Dual inhibitory effect of fluorinated monosaccharide analogs on glycosyltransferases in mammalian cells to perturb fucosylation and sialylation of *N*-glycoproteins.

oxocarbenium ion species during the enzymatic glycosyl transfer.[42,43] However, fluorinated sugar nucleotides themselves are usually not cell-permeable and thus are less useful as probes for cellular studies. To bypass this barrier, Paulson and co-workers applied peracetylated derivatives of selectively fluorinated fucose and sialic acid as precursors, which are cell-permeable and, after *in situ* de-*O*-acetylation by the promiscuous cellular esterases, are readily converted to the corresponding sugar nucleotide substrate analogs intracellularly. It was found that the *in situ* generation of the substrate analog inhibitors could shut down effectively the synthesis of a spectrum of sialylated and fucosylated glycan epitopes, resulting in remodeling of the cell-surface glycome within several days, as shown by *N*-glycan analysis. It was further observed that feeding with the fluorinated sugar precursors alone or in combination substantially inhibited the formation of the sialyl Lewis X epitope in a human myeloid cell line (HL-60 cells), leading to abrogation of its interaction with E- and P-selectins, an event responsible for the recruitment of effector cells to the inflammatory sites. It should be mentioned that the fluorinated substrate analogs not only serve as competitive inhibitors of the respective glycosyltransferases, but that their accumulation within the cells (due to their poor cell permeability) also leads to feedback inhibition of the *de novo* synthesis of the natural sugar nucleotide substrates, due to the existence of metabolic feedback loops (Figure 2.3). More recently, it was demonstrated that administering the peracetylated 3-fluoro-*N*-acetylneuraminic acid 3F-NeuAc to mice could dramatically decrease sialylated glycans in cells of all tissues tested, including blood, spleen, liver, brain, lung, heart, kidney, and testes.[44] Furthermore, it was observed that blockade of sialylation with 3F-NeuAc had particularly significant effects on the functions of liver and kidney in mice, leading to dramatic alteration of the enzyme activities and irreversible kidney dysfunction. These experimental data confirm a critical role for sialylation in liver and kidney function and suggest an innovative chemical biological approach to *in vivo* modulation of sialyltransferase activity for glycan functions.

In line with the global metabolic inhibition of glycosyltransferases, Withers and co-workers have recently reported a new class of mechanism-based neuraminidase inhibitors that show broad-spectrum influenza antiviral activity.[45] The design of the viral neuraminidase inhibitors combines the structural features of the existing transition-state analog anti-viral drugs (Zanamivir and Oseltamivir) and the unusual properties of fluorinated sialic acid. The placement of a guanidine or amino group at the C-4 position of the 2,3-difluorosialic acid (DFSA) backbone increases its affinity and specificity to viral neuraminidases over human enzymes. For this class of mechanism-based inhibitors, the electronegative fluorine atom at C-3 destabilizes the oxocarbenium-ion-like transition states for both formation and hydrolysis of the intermediate, whereas the C-2 anomeric fluoride leaving group speeds the formation step, thus resulting in the accumulation of covalent, "dead-end" enzyme–substrate complexes (Figure 2.4). Interestingly, the 2,3-difluorosialic acid analog with an equatorial fluorine at C-3 was shown

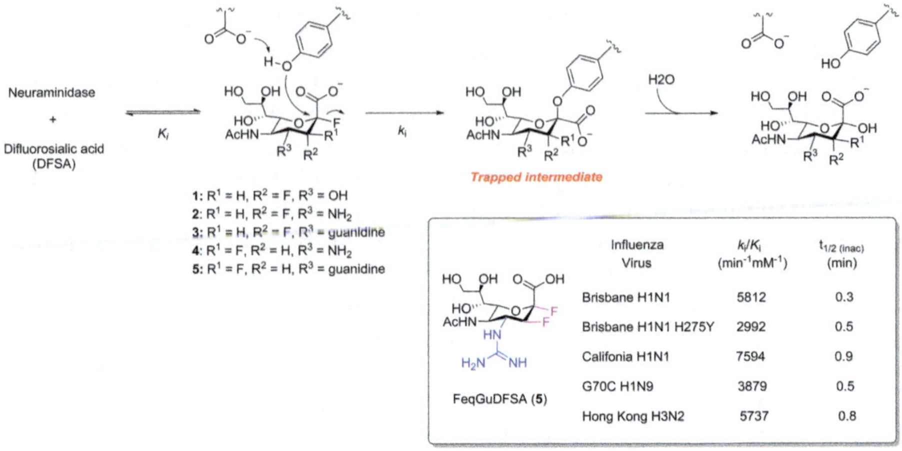

Influenza Virus	k_i/K_i ($min^{-1}mM^{-1}$)	$t_{1/2\,(inac)}$ (min)
Brisbane H1N1	5812	0.3
Brisbane H1N1 H275Y	2992	0.5
Califonia H1N1	7594	0.9
G70C H1N9	3879	0.5
Hong Kong H3N2	5737	0.8

Figure 2.4 Mechanism-based covalent neuraminidase inhibitors against influenza. The 2,3-difluorosialic acid derivatives form trapped, "dead-end" enzyme–substrate intermediates and exhibit broad-spectrum anti-influenza activity.

to be a superior inhibitor to its axial epimer, typically with an enhancement of 10-fold to 40-fold in inhibitory activity. The potency of these mechanism-based inhibitors was similar to that of the reversible and competitive inhibitor, zanamivir, but the new inhibitors exhibited broad spectrum activity against drug resistant strains with only low activities on human neuraminidases. Evaluation in a mouse model indicated that the 2,3-difluorosialic acid derivative with an equatorial fluorine at the C-3 position exhibited 100% efficacy at 10 mg kg^{-1} day^{-1} without apparent side effects. In a separate study, the difluorosialic acid analog inhibitors were shown to induce much fewer drug-resistance-associated neuraminidase mutations than does oseltamivir, indicating that influenza virus is less prone to development of resistance against the difluorosialic acid inhibitors through neuraminidase mutations.[46] The mechanism-based influenza neuraminidase inhibitors represent an exciting new class of anti-viral agents that hold promise for further development as anti-influenza drug candidates.

2.5 Metabolic Glycoengineering of Cell-Surface Glycoproteins

Metabolic cell-surface glycoengineering is emerging as a very hot area of research in chemical biology.[47–50] This approach takes advantage of glycan biosynthetic pathways to incorporate unnatural, tag-modified monosaccharides into the secreted or cell-surface glycoconjugates. Once presented in the glycoconjugates on the cell surface, the tag, *i.e.*, the chemical reporter, can serve as a handle for chemo-selective modifications *via* a bioorthogonal reaction, enabling cell-surface glycosylation remodeling and imaging.

Since the first reports on cell-surface glycoengineering through metabolic introduction of keto and azide functionality *via* the salvage sialic acid biosynthetic pathway coupled with subsequent chemoselective bio-conjugation by Bertozzi and co-workers,[51,52] new and more efficient bioorthogonal reactions have been invented, and novel imaging design and technologies have been explored.[47–50,53,54] Due to these advances, now the metabolic glycoengineering approach has been successfully extended from cell systems to living animals for glycan detection, imaging, and dynamic analysis. A number of excellent reviews have been published.[47–50,53–57] Moreover, in Chapter 10, Chen and co-workers have given a nice account of the chemical approaches to imaging glycoproteins. As a result, we only highlight here the general concepts in metabolic cell-surface glycoengineering.

The most popular approach in metabolic glycoengineering is perhaps the exploration of the sialic acid salvage biosynthetic pathway by supplying selectively modified *N*-acetylmannosamine derivatives carrying a chemical reporter (*e.g.*, an azide, an alkyne, a cyclopropene, or a 1,2,4-triazine moiety), which can still serve as substrates for the enzymes along the biosynthetic pathway and be incorporated into sialylated glycoproteins (Figure 2.5). The tagged *N*-acetylmannosamine derivative, usually provided as the peracetylated derivative for facilitating membrane permeation, is incorporated and converted to the corresponding cytidine monophosphate (CMP)–sialic acid derivative *via* the salvage sialic acid biosynthetic pathway. The CMP–sialic acid then serves as the donor substrate of sialyltransferases for the construction of sialylated glycoconjugates that would carry a specific chemical reporter. Selectively modified sialic acid derivatives can also be used directly as precursors for metabolic glycoengineering (Figure 2.5). Following the incorporation of a special chemical reporter, the success of subsequent bioconjugation for detection and imaging would rely on the efficiency and chemoselectivity of the bioorthogonal reactions. In addition to the original Cu(I)-catalyzed azide–alkyne [3+2] cycloaddition reaction (CuAAC) and the Staudinger reaction, the invention of strain-promoted, copper-free azide–alkyne cycloaddition using ring-strained cyclooctyne and its derivatives represents a major advance in the field, which addresses the problems of cytotoxicity of the copper catalyst and gives a much faster click reaction than the original CuAAC and Staudinger reactions.[48]

This area of research is still evolving. The exploration of alternative and efficient bioorthogonal click chemistry, such as broad-spectrum tetrazine ligation with a range of strained *trans*-cyclooctenes, cyclopropenes, isonitriles, and cyclooctyne derivatives, has significantly expanded the toolkits of biorthogonal click chemistry.[58–60] On the other hand, the design of a number of smart probes that become fluorescent only upon reaction permit the visualization of glycoconjugates even in the presence of excess unreacted probes.[47] Recently, Chen and coworkers reported the development of a fluorescence resonance energy transfer (FRET)-based methodology for protein-specific imaging of glycans on the surface of live cells.[61] The methodology could be very useful in the investigation of the functions and

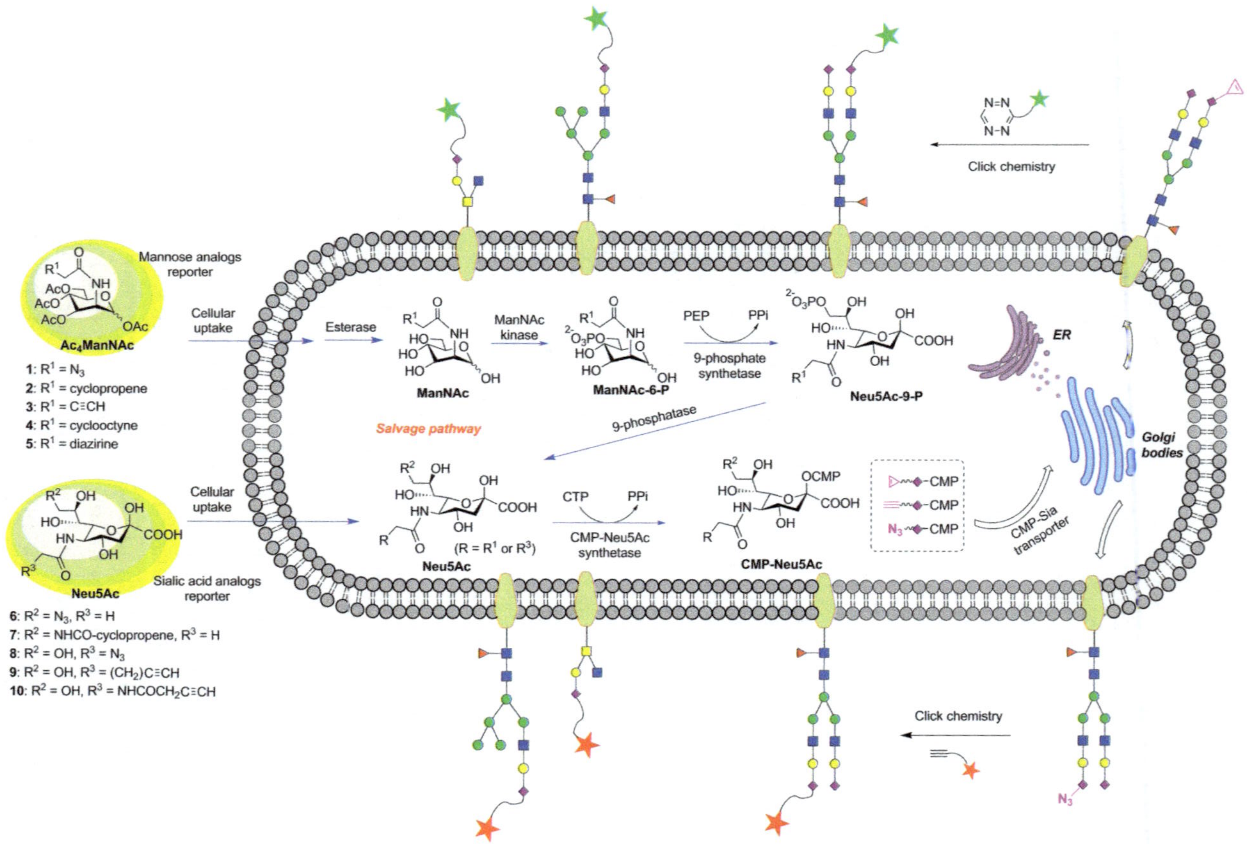

Figure 2.5 Metabolic cell-surface glycoengineering using the salvage sialic acid biosynthetic pathway. *N*-Acetylmannosamine or sialic acid derivatives carrying a chemical reporter is metabolized to form the cytidine monophosphate sialic acid (CMP–Sia) derivative, which is then incorporated into the biosynthesis of glycoconjugates including glycoproteins. The chemical reporter is visualized by a bioorthogonal click reaction for cell-surface glycan imaging.

dynamics of various cell-surface glycoproteins. In addition, a novel strategy for targeted imaging of tumor-associated glycans by using ligand-targeted liposomes encapsulating azidosugars has been demonstrated.[62] Compared with the labeling methods using azidosugars, this method offers improved labeling efficiency and high specificity. More recently, the same group has developed a liposome-assisted bioorthogonal reporter (LABOR) strategy for introducing 9-azido sialic acid into mouse brain, to metabolically label sialoglycoconjugates, including sialylated glycoproteins and glycolipids.[63] The LABOR strategy provides a promising way to efficiently visualize and monitor the biosynthesis of multicomponent sialylated biomolecules, which will help increase understanding of the roles of sialylation in the brain.

In addition to the biosynthetic pathway for sialic acid, the pathway for fucose biosynthesis has also been explored for metabolic glycoengineering (Figure 2.6). Wong and coworkers showed that either an azide or an alkyne reporting group installed at the C6 of a fucose derivative could be incorporated into cell-surface glycans through the fucose–GDP biosynthetic pathway.[64,65] Recently, Taniguchi and coworkers have demonstrated that a new fucose analog, 7-alkynyl-fucose, could be used as a highly sensitive fucose probe for labeling the core fucose moiety on cell surface *N*-glycoproteins.[66] Compared with the previously reported 6-alkynyl-fucose derivative, the newly synthesized 7-alkynyl fucose analog was found to be less cytotoxic and more efficiently incorporated into *N*-glycans.

As another biosynthetic pathway explored for glycoengineering, Bertozzi and coworkers first demonstrated that the key GlcNAc residues of *N*-glycans in the organism *Saccharomyces cerevisiae* could be metabolically targeted by an unnatural *N*-azidoacetylglucosamine (GlcNAz) derivative (Figure 2.7).[67] It was demonstrated that the GlcNAz moiety could be successfully incorporated in the core of *N*-glycans by replacing the normal GlcNAc moiety in the *N*-glycan core. More recently, Chen and coworkers reported metabolic glycan labeling in *Arabidopsis thaliana* by using GlcNAz as the chemical reporter.[68] This work demonstrates that the metabolic glycan labeling can be useful to decipher the function of *N*-glycans in plants.

2.6 Chemoenzymatic Synthesis and Glycosylation Remodeling Toward Homogeneous *N*-Glycoproteins

Natural and recombinant glycoproteins are usually produced as mixtures of glycoforms that are extremely difficult to separate into homogeneous glycoforms using current chromatographic techniques. This situation necessitates the development of efficient methods to make structurally well-defined, natural or tailor-made glycoproteins for functional studies and for biomedical applications. Many elegant synthetic strategies have been explored, including total chemical synthesis,[69–77] chemoselective ligation,[78,79] chemoenzymatic synthesis,[80–85] and glycosylation engineering of glycan biosynthetic

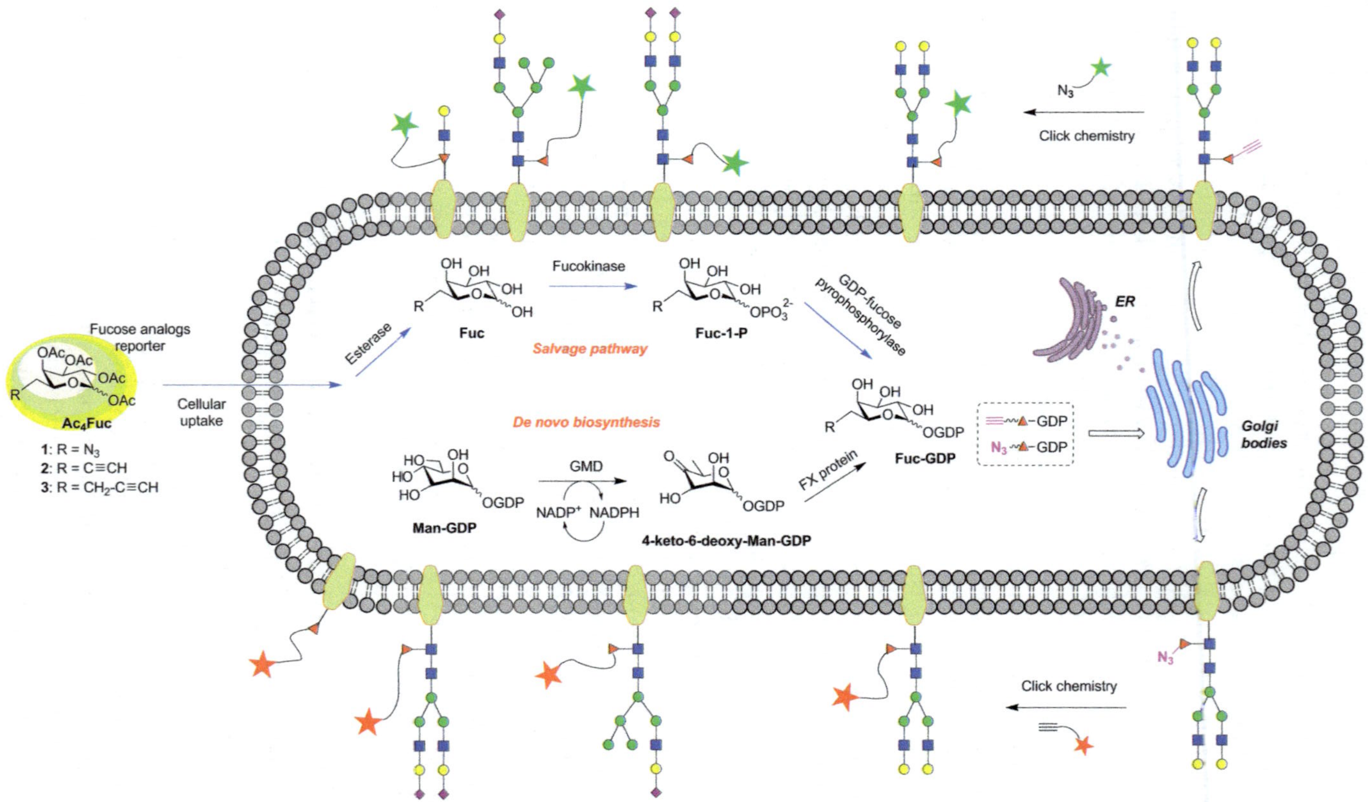

Figure 2.6 Metabolic cell-surface glycoengineering using the salvage fucose biosynthetic pathway. A tagged fucose derivative is metabolized to a guanosine diphosphate–fucose (GDP–Fuc) analog that then serves as the substrate for fucosyltransferase to incorporate it into cell-surface glycoconjugates.

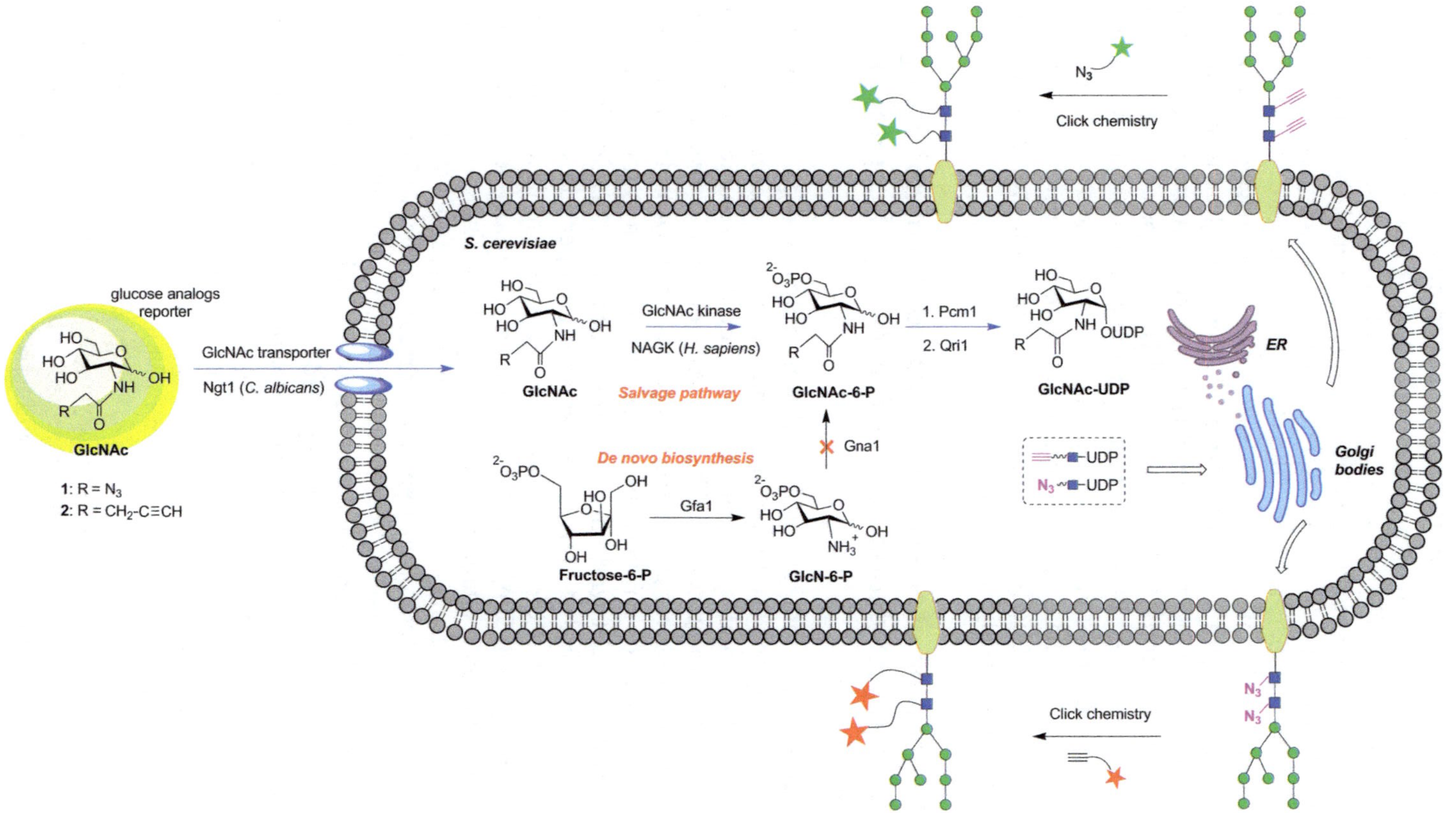

Figure 2.7 Metabolic cell-surface glycoengineering using the salvage *N*-acetylglucosamine (GlcNAc) biosynthetic pathway. A tagged GlcNAc derivative is metabolized to uridine diphosphate *N*-acetylglucosamine (UDP–GlcNAc) analogs and it then incorporated into cell-surface glycoconjugates *via* the glycosyltransferase.

pathways.[86–90] Chemical synthesis that elegantly combines glycopeptide synthesis and native chemical ligation has now made it possible to construct even some large intact *N*-glycoproteins such as human erythropoietin.[76,91] Nevertheless, pure chemical synthesis is still tedious and the efficiency of native chemical ligation is case-dependent and does not necessarily give high yields. Among other methods, the chemoenzymatic strategy that uses endo-β-*N*-acetylglucosaminidases (ENGases) catalyzed transglycosylation for convergent native ligation between pre-assembled free glycans and unprotected polypeptide or protein is emerging as a very promising method for producing homogeneous *N*-glycoproteins.[84,85] Here we gave a brief history behind the development of this chemoenzymatic method. In particular, the exploration of sugar oxazolines as activated donor substrates and the discovery of novel glycosynthases with different substrate specificities have made this chemoenzymatic method a general method of choice for synthesizing various homogeneous glycoproteins.

2.6.1 Generation of Novel ENGase-Based Glycosynthases for *N*-Glycosylation Remodeling and *N*-Glycoprotein Synthesis

ENGases are a class of endo-glycosidases that hydrolyze *N*-glycans from glycoproteins by cleaving the β-1,4-glycosidic bond in the *N*,*N*′-diacetylchitobiose core. In addition to their hydrolytic activity, some ENGases such as Endo-A from *Arthrobacter protophormiae*,[92–94] Endo-M from *Mucor hiemalis*,[95] and Endo-D from *Streptococcus pneumoniae*[96] possess transglycosylation activity and have been used for synthetic purposes. One important advance in this area is the exploration of synthetic sugar oxazolines, the transition-state substrate analogs, as activated donor substrates for glycopeptide synthesis.[80,97] It has been shown that the sugar oxazoline substrates are much more efficient for transglycosylation than the corresponding ground-state glycan or glycopeptide substrates, enabling an efficient synthesis of glycopeptides and glycoproteins carrying truncated and/or unnaturally modified *N*-glycans.[98–109] Nevertheless, product hydrolysis remains a serious problem when the target glycoproteins carry natural *N*-glycans that would be good substrates for the respective enzymes.[110–113] A rational solution to this problem would be the creation of novel glycosynthases, special glycosidase mutants that lack product hydrolysis activity but are able to use activated donor substrate for transglycosylation.[110–113] However, the conventional approach to generating glycosynthase by mutating the nucleophilic residue in a retaining glycosidase could not be applied directly to ENGases as the ENGase-catalyzed reaction proceeds *via* a substrate-assisted mechanism in which the nucleophile is the 2-acetamido group in the substrate. In 2008, Wang, Yamamoto, and co-workers reported the first case of ENGase-based glycosynthases by screening a series of Endo-M mutants generated by site-directed mutagenesis on residues in or around the putative catalytic region of wild type Endo-M, using $Man_9GlcNAc$ oxazoline as the key substrate.[114] This study led to the identification of the N175 as a key residue that promotes sugar oxazolinium ion formation during hydrolysis and transglycosylation. Thus, mutation at the N175 site resulted in diminished hydrolysis activity. Interestingly, the corresponding mutants of Endo-M, such as N175A and N175Q, could efficiently use glycan oxazoline as a donor substrate for transglycosylation, making them the first glycosynthases of the ENGase class.[114,115] Following this concept, glycosynthase mutants of Endo-A, another member of the GH family 85 endoglycosidases, were also generated that are specific for high-mannose *N*-glycans.[116]

The discovery of the glycosynthases derived from Endo-A and Endo-M has made it possible to quickly assemble a library of homogeneous glycoforms from a natural or recombinant glycoprotein, as demonstrated by a two-step glycan remodeling of bovine ribonuclease B (RNase B) (Figure 2.8). In this approach, the heterogeneous high-mannose *N*-glycans in RNase B are first removed by treatment with Endo-A or Endo-H, giving the homogeneous GlcNAc–RNase intermediate, in which the innermost GlcNAc moiety is still attached at the original glycosylation site. Then well-defined *N*-glycans are transferred from respective glycan oxazolines by the glycosynthases to give a series of homogeneous RNase glycoforms.[116–119] The Endo-M mutants can transfer both high-mannose and complex-type *N*-glycans, while the Endo-A mutant is specific for transferring high-mannose type and truncated *N*-glycan core. An interesting recent example is the efficient chemoenzymatic synthesis of phosphoglycoproteins carrying natural mannose-6-phosphate (M6P)-containing *N*-glycans.[119] The surface plasmon resonance (SPR) binding analysis of the interactions between the synthetic isomeric M6P glycoprotein glycoforms and the cation-independent mannose 6-phosphate receptor (CI-MPR) reveals that a single M6P moiety located at the low α-1,3-branch of the oligomannose context is sufficient for high-affinity binding to receptor CI-MPR, the presence of a M6P moiety at the up-branch of the oligosaccharide core is dispensable, and close proximity of two M6P-oligosaccharide ligands in the peptide/protein context is critical to achieve high affinity for the CI-MPR. The results of this study indicate that both the location and valency of the M6P moieties are critical for high-affinity binding with the major M6P receptor. Later on, this glycosynthase approach was extended to the GH18 family endoglycosidase, where a specific aspartic acid (D) residue was identified as the key residue for orientating and promoting oxazolinium ion formation during enzyme catalysis. These include the glycosynthases (D233A and D233Q mutants) derived from Endo-S of *Streptococcus pyogenes*,[120] the glycosynthases from Endo-F3 of *Elizabethkingia miricola*,[121] and the glycosynthases from Endo-S2 of *Streptococcus pyogenes*.[122] The discovery of these novel glycosynthases has also enabled efficient synthesis of large *N*-glycopeptides and glycoproteins for functional studies. These include the chemoenzymatic synthesis of different glycoforms of the CD52 glycopeptide antigens;[121,123] the synthesis of a library of homogeneous HIV-1 V1V2 glycopeptides for characterizing the glycan-reactive epitopes of anti-HIV neutralizing antibodies PG9 and PG16;[124] and the chemoenzymatic synthesis of saposin C, a hydrophobic glycoprotein using a combination of native chemical ligation and EndoM-N175Q catalyzed transglycosylation.[125] The application of glycosynthases for glycoengineering of therapeutic antibodies is discussed in more detail in the next section.

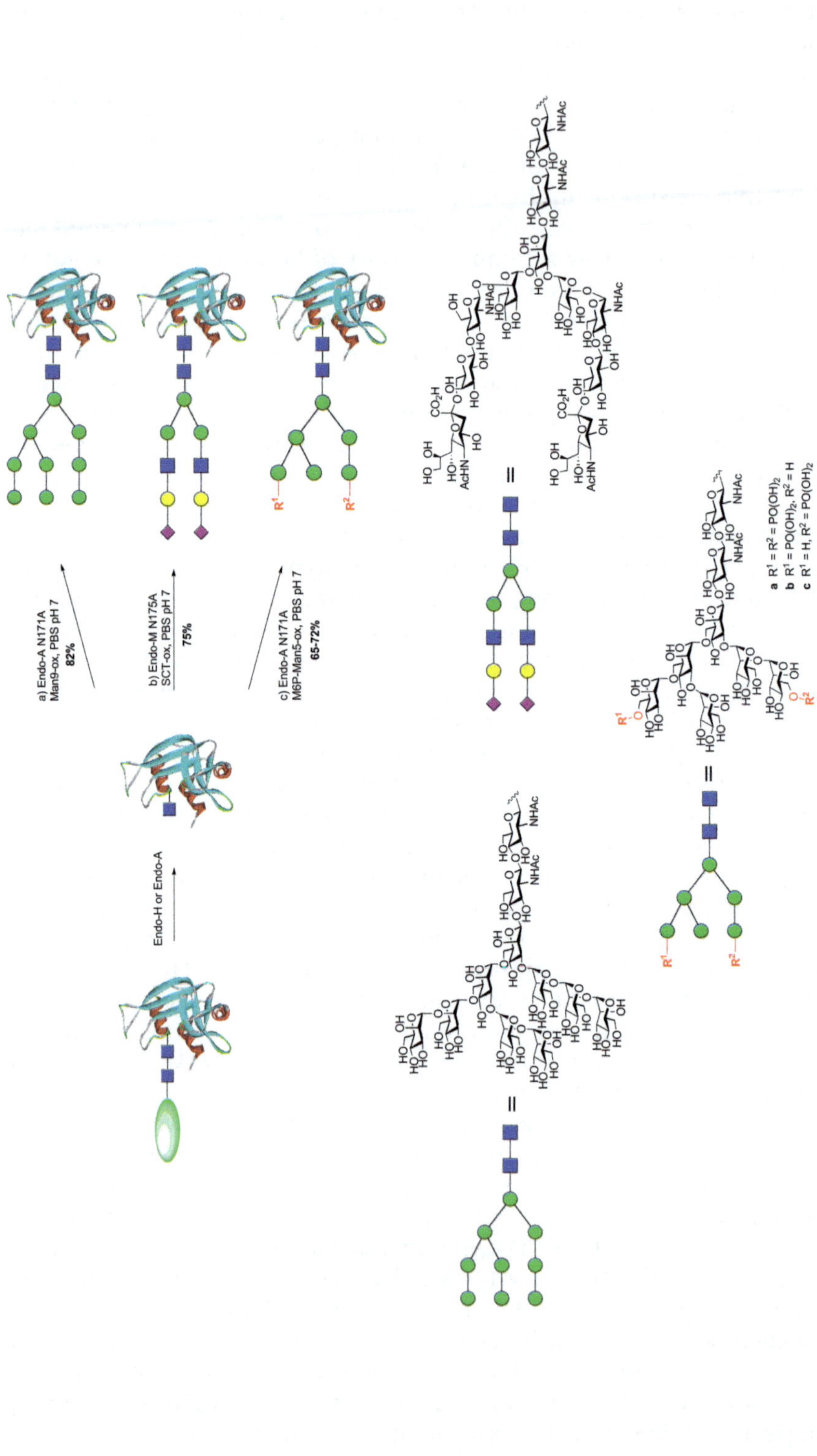

Figure 2.8 Glycosynthase-catalyzed glycosylation remodeling of ribonuclease B for producing homogeneous glycoforms. The approach includes enzymatic deglycosylation to remove the heterogeneous *N*-glycans from the starting glycoprotein and then the enzymatic transfer of pre-assembled *N*-glycans in the form of sugar oxazolines *en bloc* to provide various homogeneous glycoforms.

In parallel to the site-directed mutagenesis to generate glycosynthases, important progress was also made in the structural studies of the family GH85 and GH18 ENGases. For example, the crystal structures of Endo-A and its complex with $Man_3GlcNAc$–thiazoline, an oxazoline analog, confirmed the importance of N171 as a key residue for orientating sugar oxazolinium formation.[126,127] The crystal structure of Endo-D and its comparison with that of Endo-A indicated that the two GH family 85 enzymes have almost identical topology in folding of the catalytic domain and share a general mechanism of substrate-assisted catalysis;[128] more recently, the crystal structure of Endo-S has been solved, which reveals an unusual V-shape structure and specific Fc and Fc glycan binding domains that partially explain its specificity for immunoglobulin G Fc domain (IgG-Fc) glycosylation.[129] The structural information will be highly valuable in rational design of novel ENGase mutants with improved transglycosylation activity or altered substrate specificities.

2.6.2 Chemoenzymatic Fc Glycan Remodeling of Therapeutic Monoclonal Antibodies

One particularly attractive application of glycosynthase-catalyzed transglycosylation is the Fc glycan remodeling of therapeutic antibodies. Monoclonal antibodies (mAbs) of the IgG type are an important class of therapeutic proteins used for the treatment of cancer, autoimmune, and infectious diseases. Compelling data have shown that the fine structures of Fc *N*-glycans can modulate the effector functions of IgG antibodies.[9,10] However, controlling the glycosylation patterns during recombinant production is still a challenging task. In 2008, Wang and co-workers demonstrated the first example of glycan remodeling of a recombinant human Fc domain by endoglycosidase-catalyzed transglycosylation.[130] The method permits a quick synthesis of structurally well-defined glycoforms of IgG-Fc domains, which allows the assessment of the effect of fine glycan structures on the binding to FcγIIIa receptor.[96,131] In 2012, the same research group reported the first Fc glycosylation remodeling of an intact therapeutic monoclonal antibody, rituximab.[120] This was enabled by the discovery of two Fc-specific glycosynthase mutants (Endo-S D233A and D233Q) generated by site-directed mutagenesis of Endo-S, an endoglycosidase from *Streptococcus pyogenes*. By this approach, a variety of homogeneous glycoforms, including the sialylated complex type, non-fucosylated complex type, and azide-tagged glycoforms, could be synthesized efficiently (Figure 2.9). The synthetic non-fucosylated glycoform showed 20-fold higher affinity for FcγIIIa receptor than the commercial rituximab, indicating that the chemoenzymatic method provides an efficient approach to producing bio-better therapeutic antibody.[120] Recently this chemoenzymatic method was applied to the synthesis of specific antibody glycoforms for functional studies.[132–135] Similarly glycosynthases were also generated from Endo-F3, another GH18 family endoglycosidase.[121] It was shown that the D165A mutant of Endo-F3 was highly specific for core-fucosylated *N*-glycan and was able to transfer tri-antennary *N*-glycan to the Fc domain of intact

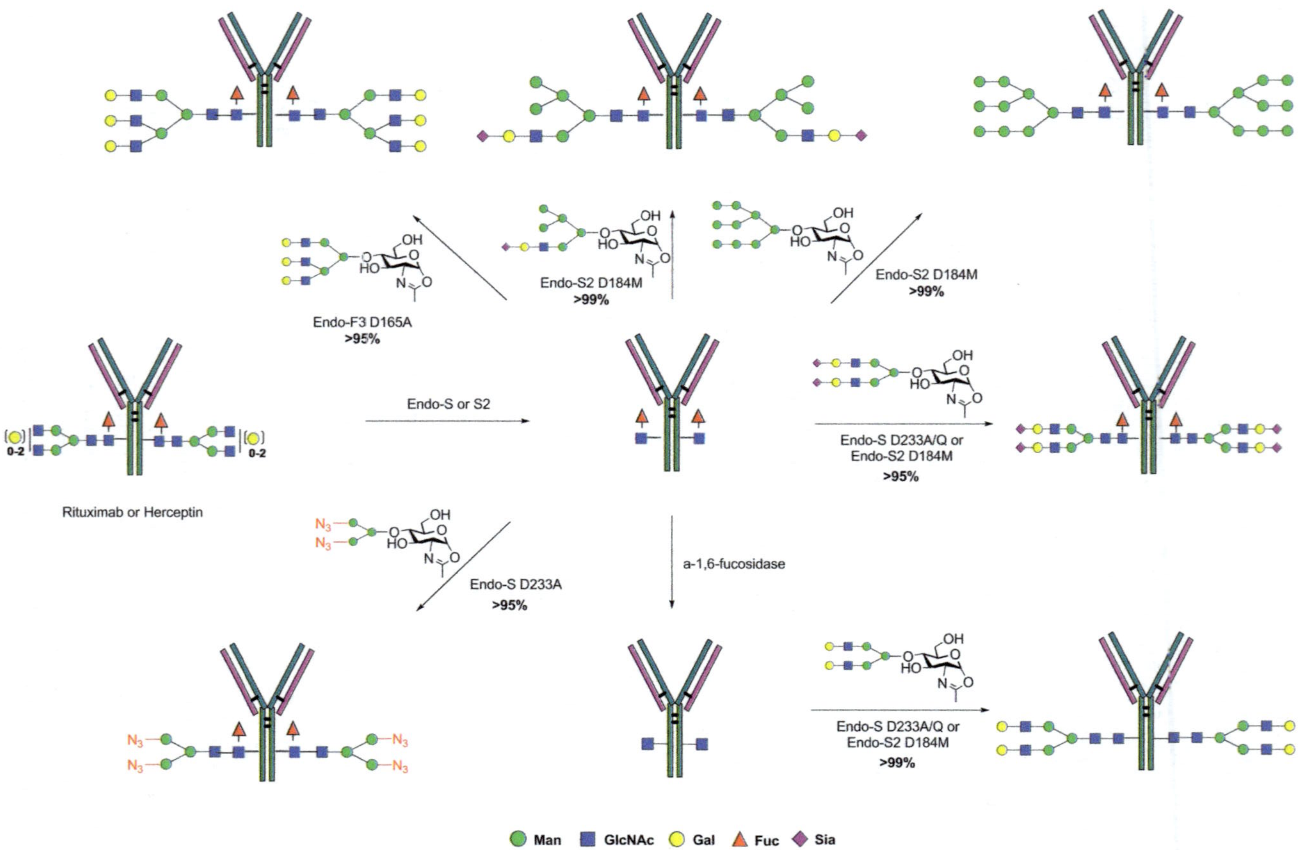

Figure 2.9 Glycosynthase-catalyzed Fc glycan remodeling of therapeutic monoclonal antibodies. Rituximab or trastuzumab is deglycosylated by wild-type Endo-S or Endo-S2. The resulting deglycosylated IgG then serves as the acceptor for the glycosynthase-catalyzed transglycosylation to provide a range of homogeneous antibody glycoforms.

antibody. Interestingly the tri-antennary glycoform of antibody, which would be difficult to generate by overexpression, showed much enhanced affinity to galectin-3. Fc glycan remodeling could gain new functions for antibodies. More recently, Wang and co-workers performed a systematic site-directed mutagenesis of Endo-S2, another GH family 18 endoglycosidase from *Streptococcus pyogenes* of serotype M49, which led to the discovery of an array of novel glycosynthases.[122] In contrast to the Endo-S-derived glycosynthase that is limited to transfer of complex type *N*-glycans, the glycosynthases derived from Endo-S2, such as the D184M and D184Q mutants, showed remarkably relaxed substrate specificity and were capable of transferring all three major types (complex, high-mannose, and hybrid type) of *N*-glycans for antibody glycosylation remodeling (Figure 2.9). In addition, the Endo-S2 glycosynthase mutants were found to be much more active in general than the Endo-S mutants for transglycosylation. It is expected that the field will continue to develop with discoveries of new and more potent glycosynthase mutants. The endoglycosidase-catalyzed glycosylation remodeling system is emerging as a general platform for producing various natural or tailor-made glycoforms of intact antibodies, which would be highly valuable for functional studies and hold promise for therapeutic applications as well.

2.6.3 Combined *E. coli* Expression and Chemoenzymatic Glycan Remodeling to Produce Humanized *N*-Glycoproteins

The endoglycosidase-catalyzed glycoprotein synthesis requires a GlcNAc–protein as the key precursor. While GlcNAc-protein can be prepared by total chemical synthesis when tandem native chemical ligation reactions are employed, it would be highly desirable if an *Escherichia coli* (system, which is biotechnology's workhorse for protein expression) could be used to produce the GlcNAc–protein precursors. However, *E. coli* lacks the protein *N*-glycosylation machinery and thus conventional *E. coli* expression is unable to produce glycoproteins. To overcome this barrier, Wang, Aebi and co-workers reported a glycoprotein synthesis method that couples *E. coli* glycoprotein expression with *in vitro* glycosylation remodeling.[136] The method involves glycosylation pathway engineering and functional transfer of the *Campylobacter jejuni* glycosylation machinery[137] into *E. coli* to express a glycoprotein, in which the Asn-linked monosaccharide bacillosamine in the Asn-linked bacterial *N*-glycan was changed to the eukaryotic GlcNAc moiety. This was achieved by deleting the genes involved in the biosynthesis of bacillosamine and by taking advantage of the *E. coli* existing biosynthetic pathway for making the GlcNAc-diphosphate–lipid precursor. It was found that the *C. jejuni* oligosaccharyl transferase PglB was able to use the resulting GlcNAc-containing glycolipid as a substrate to glycosylate the target protein. Then the bacterial *N*-glycans were trimmed down to the key GlcNAc–protein precursor. Finally, eukaryotic *N*-glycans were introduced by enzymatic transglycosylation to generate the homogeneous eukaryotic glycoproteins (Figure 2.10).

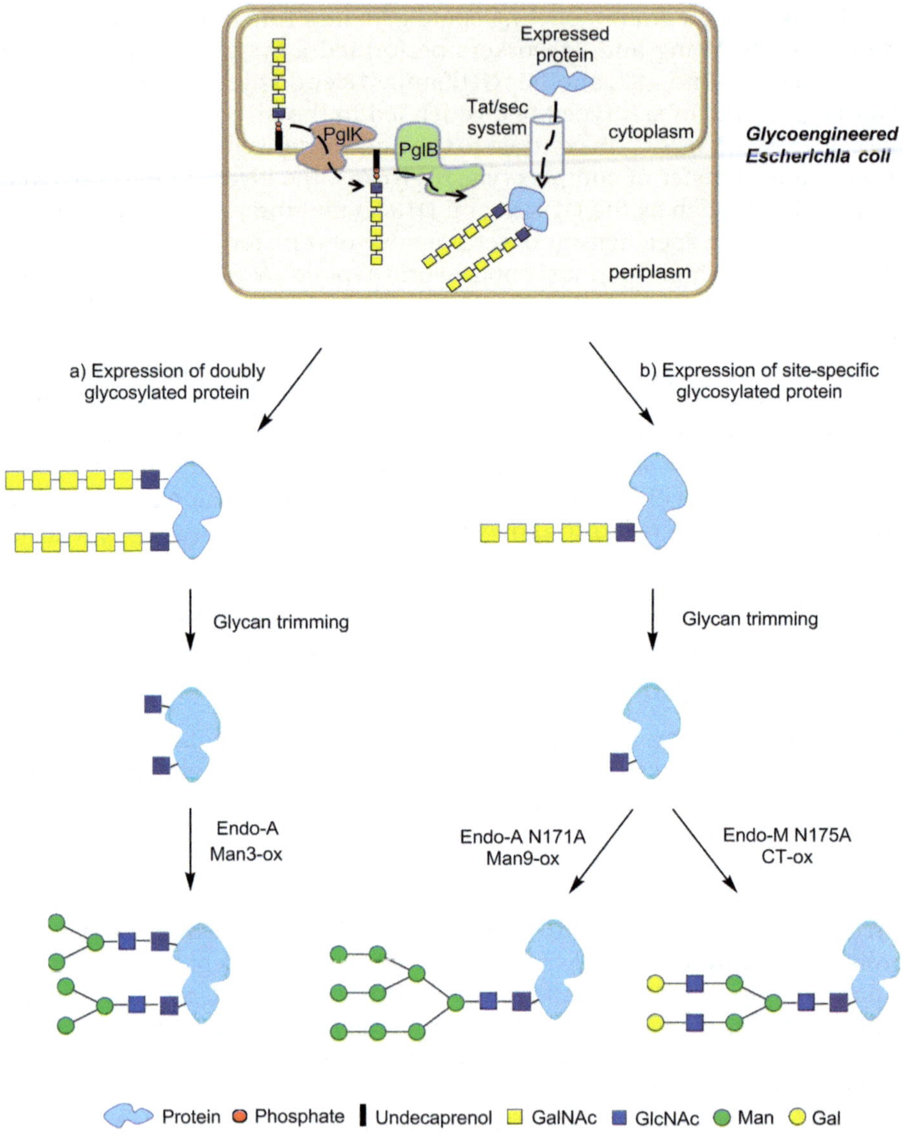

Figure 2.10 Combined *E. coli* expression and chemoenzymatic glycan remodeling to produce humanized *N*-glycoproteins. The method involves glycosylation pathway engineering and functional transfer of the *Campylobacter jejuni* glycosylation machinery into *E. coli* to express an *N*-linked glycoprotein. Then the bacterial *N*-glycans are trimmed to the GlcNAc–protein, which serves as the precursor for ENGase-catalyzed transglycosylation to produce *N*-glycoproteins carrying eukaryotic *N*-glycans.

This method was applied to making glycosylated human IgG-Fc CH2 domain and a single–chain antibody. This method combines the convenience of protein expression in *E. coli* with the flexibility of *in vitro* chemo-enzymatic glycan remodeling and therefore provides a potentially general platform for producing eukaryotic *N*-glycoproteins. Nevertheless, it should be pointed out that the efficiency of glycosylation of the exogenous eukaryotic proteins in the engineered *E. coli* system is relatively low, ranging from 5% to 45% yield. Another related study demonstrated that bottom-up engineering of a eukaryotic glycosylation pathway in *E. coli* was able to produce *N*-linked glycoprotein carrying the eukaryotic $Man_3GlcNAc_2$ *N*-glycan core.[138] This was enabled by overexpression of an array of eukaryotic glycosyltransferases involved in *N*-glycan synthesis and the bacterial oligosaccharyltransferase PglB from *C. jejuni*. However, again the yield of glycosylation was very low. Thus, despite the promise of this novel approach, future studies should be directed to increasing the efficiency of glycosylation *via* optimization of the key enzymes (such as PglB) and further engineering of the glycosylation and transportation network in the *E. coli* expression system in order to develop a truly practical method for producing humanized *N*-glycoproteins in *E. coli*.

2.7 Conclusion

Protein *N*-glycosylation presents an amazingly complex network of molecular recognition that involves dozens of enzymes and glycan-binding proteins at multiple sites for the assembly, trimming, and re-assembly of the *N*-glycans in the mature *N*-glycoproteins. The structural diversity also leads to enormously diverse functions that *N*-glycosylation can confer, ranging from protein folding and intracellular signaling to a number of important intercellular functions such as cell adhesion and host–pathogen interactions, which provides novel targets and a large chemical space for chemical biologists to explore. As described in this chapter, the design and synthesis of mechanism-based and metabolic inhibitors for perturbing protein *N*-glycosylation, the design of novel chemical reporters for cell-surface glycoengineering and imaging, and the chemoenzymatic synthesis of structurally well-defined glycoproteins including therapeutic antibodies represent only a few examples in this active area of research in chemical biology. We are only at the beginning of deciphering the glycomes of living species. It is expected that chemical biology approaches will continue to play key roles in our understanding of the diverse functions of oligosaccharides and glycoconjugates in biological processes.

Acknowledgements

We thank members of the Wang Lab for valuable discussions. This work was supported by the National Institutes of Health (NIH grants R01 GM080374 and R01 GM096973).

References

1. A. Helenius and M. Aebi, *Science*, 2001, **291**, 2364–2369.
2. A. J. Petrescu, M. R. Wormald and R. A. Dwek, *Curr. Opin. Struct. Biol.*, 2006, **16**, 600–607.
3. J. L. Price, E. K. Culyba, W. Chen, A. N. Murray, S. R. Hanson, C. H. Wong, E. T. Powers and J. W. Kelly, *Biopolymers*, 2012, **98**, 195–211.
4. L. Krasnova and C. H. Wong, *Annu. Rev. Biochem.*, 2016, **85**, 599–630.
5. A. Varki, *Glycobiology*, 1993, **3**, 97–130.
6. R. A. Dwek, *Chem. Rev.*, 1996, **96**, 683–720.
7. R. S. Haltiwanger and J. B. Lowe, *Annu. Rev. Biochem.*, 2004, **73**, 491–537.
8. D. H. Dube and C. R. Bertozzi, *Nat. Rev. Drug Discovery*, 2005, **4**, 477–488.
9. F. Nimmerjahn and J. V. Ravetch, *Nat. Rev. Immunol.*, 2008, **8**, 34–47.
10. R. Jefferis, *Nat. Rev. Drug Discovery*, 2009, **8**, 226–234.
11. G. W. Hart and R. J. Copeland, *Cell*, 2010, **143**, 672–676.
12. R. G. Spiro, *Glycobiology*, 2002, **12**, 43R–56R.
13. L. X. Wang and B. G. Davis, *Chem. Sci.*, 2013, **4**, 3381–3394.
14. M. Aebi, *Biochim. Biophys. Acta*, 2013, **1833**, 2430–2437.
15. P. Ghosh, N. M. Dahms and S. Kornfeld, *Nat. Rev. Mol. Cell Biol.*, 2003, **4**, 202–212.
16. N. M. Dahms, L. J. Olson and J. J. Kim, *Glycobiology*, 2008, **18**, 664–678.
17. N. Asano, *Glycobiology*, 2003, **13**, 93R–104R.
18. T. M. Gloster and D. J. Vocadlo, *Nat. Chem. Biol.*, 2012, **8**, 683–694.
19. T. D. Butters, L. M. Sparks, K. Harlos, S. Ikemizu, D. I. Stuart, E. Y. Jones and S. J. Davis, *Protein Sci.*, 1999, **8**, 1696–1701.
20. A. D. Elbein, J. E. Tropea, M. Mitchell and G. P. Kaushal, *J. Biol. Chem.*, 1990, **265**, 15599–15605.
21. A. D. Elbein, *FASEB J.*, 1991, **5**, 3055–3063.
22. A. D. Elbein, J. K. Kerbacher, C. J. Schwartz and E. A. Sprague, *Arch. Biochem. Biophys.*, 1991, **288**, 177–184.
23. V. T. Chang, M. Crispin, A. R. Aricescu, D. J. Harvey, J. E. Nettleship, J. A. Fennelly, C. Yu, K. S. Boles, E. J. Evans, D. I. Stuart, R. A. Dwek, E. Y. Jones, R. J. Owens and S. J. Davis, *Structure*, 2007, **15**, 267–273.
24. C. N. Scanlan, G. E. Ritchie, K. Baruah, M. Crispin, D. J. Harvey, B. B. Singer, L. Lucka, M. R. Wormald, P. Wentworth, Jr., N. Zitzmann, P. M. Rudd, D. R. Burton and R. A. Dwek, *J. Mol. Biol.*, 2007, **372**, 16–22.
25. L. Kong, N. C. Sheppard, G. B. Stewart-Jones, C. L. Robson, H. Chen, X. Xu, G. Krashias, C. Bonomelli, C. N. Scanlan, P. D. Kwong, S. A. Jeffs, I. M. Jones and Q. J. Sattentau, *J. Mol. Biol.*, 2010, **403**, 131–147.
26. K. J. Doores and D. R. Burton, *J. Virol.*, 2010, **84**, 10510–10521.
27. Q. Yang, C. Li, Y. Wei, W. Huang and L. X. Wang, *Bioconjugate Chem.*, 2010, **21**, 875–883.
28. A. D. Elbein, R. Solf, P. R. Dorling and K. Vosbeck, *Proc. Natl. Acad. Sci. U. S. A.*, 1981, **78**, 7393–7397.

29. M. Pancera, S. Shahzad-Ul-Hussan, N. A. Doria-Rose, J. S. McLellan, R. T. Bailer, K. Dai, S. Loesgen, M. K. Louder, R. P. Staupe, Y. Yang, B. Zhang, R. Parks, J. Eudailey, K. E. Lloyd, J. Blinn, S. M. Alam, B. F. Haynes, M. N. Amin, L. X. Wang, D. R. Burton, W. C. Koff, G. J. Nabel, J. R. Mascola, C. A. Bewley and P. D. Kwong, *Nat. Struct. Mol. Biol.*, 2013, **20**, 804–813.
30. P. B. Fischer, M. Collin, G. B. Karlsson, W. James, T. D. Butters, S. J. Davis, S. Gordon, R. A. Dwek and F. M. Platt, *J. Virol.*, 1995, **69**, 5791–5797.
31. P. B. Fischer, G. B. Karlsson, T. D. Butters, R. A. Dwek and F. M. Platt, *J. Virol.*, 1996, **70**, 7143–7152.
32. T. M. Block, X. Lu, F. M. Platt, G. R. Foster, W. H. Gerlich, B. S. Blumberg and R. A. Dwek, *Proc. Natl. Acad. Sci. U. S. A.*, 1994, **91**, 2235–2239.
33. T. M. Block, X. Lu, A. S. Mehta, B. S. Blumberg, B. Tennant, M. Ebling, B. Korba, D. M. Lansky, G. S. Jacob and R. A. Dwek, *Nat. Med.*, 1998, **4**, 610–614.
34. A. Mehta, T. M. Block and R. A. Dwek, *Adv. Exp. Med. Biol.*, 1998, **435**, 195–205.
35. N. Zitzmann, T. Block, A. Methta, P. Rudd, D. Burton, I. Wilson, F. Platt, T. Butters and R. A. Dwek, *Adv. Exp. Med. Biol.*, 2005, **564**, 1–2.
36. N. Zitzmann, A. S. Mehta, S. Carrouee, T. D. Butters, F. M. Platt, J. McCauley, B. S. Blumberg, R. A. Dwek and T. M. Block, *Proc. Natl. Acad. Sci. U. S. A.*, 1999, **96**, 11878–11882.
37. S. Gerber-Lemaire and L. Juillerat-Jeanneret, *Chimia (Aarau)*, 2010, **64**, 634–639.
38. D. J. Becker and J. B. Lowe, *Glycobiology*, 2003, **13**, 41R–53R.
39. P. R. Crocker, J. C. Paulson and A. Varki, *Nat. Rev. Immunol.*, 2007, **7**, 255–266.
40. A. Varki, *Nature*, 2007, **446**, 1023–1029.
41. C. D. Rillahan, A. Antonopoulos, C. T. Lefort, R. Sonon, P. Azadi, K. Ley, A. Dell, S. M. Haslam and J. C. Paulson, *Nat. Chem. Biol.*, 2012, **8**, 661–668.
42. M. D. Burkart, S. P. Vincent, A. Duffels, B. W. Murray, S. V. Ley and C. H. Wong, *Bioorg. Med. Chem.*, 2000, **8**, 1937–1946.
43. L. L. Lairson, B. Henrissat, G. J. Davies and S. G. Withers, *Annu. Rev. Biochem.*, 2008, **77**, 521–555.
44. M. S. Macauley, B. M. Arlian, C. D. Rillahan, P. C. Pang, N. Bortell, M. C. Marcondes, S. M. Haslam, A. Dell and J. C. Paulson, *J. Biol. Chem.*, 2014, **289**, 35149–35158.
45. J. H. Kim, R. Resende, T. Wennekes, H. M. Chen, N. Bance, S. Buchini, A. G. Watts, P. Pilling, V. A. Streltsov, M. Petric, R. Liggins, S. Barrett, J. L. McKimm-Breschkin, M. Niikura and S. G. Withers, *Science*, 2013, **340**, 71–75.
46. S. H. Tai, O. Agafitei, Z. Gao, R. Liggins, M. Petric, S. G. Withers and M. Niikura, *Virus Res.*, 2015, **210**, 126–132.
47. P. Shieh and C. R. Bertozzi, *Org. Biomol. Chem.*, 2014, **12**, 9307–9320.
48. E. M. Sletten and C. R. Bertozzi, *Acc. Chem. Res.*, 2011, **44**, 666–676.
49. J. C. Jewett and C. R. Bertozzi, *Chem. Soc. Rev.*, 2010, **39**, 1272–1279.

50. S. T. Laughlin and C. R. Bertozzi, *Proc. Natl. Acad. Sci. U. S. A.*, 2009, **106**, 12–17.
51. L. K. Mahal, K. J. Yarema and C. R. Bertozzi, *Science*, 1997, **276**, 1125–1128.
52. E. Saxon and C. R. Bertozzi, *Science*, 2000, **287**, 2007–2010.
53. R. Xie, S. Hong and X. Chen, *Curr. Opin. Chem. Biol.*, 2013, **17**, 747–752.
54. R. K. Lim and Q. Lin, *Acc. Chem. Res.*, 2011, **44**, 828–839.
55. W. B. Porterfield and J. A. Prescher, *Curr. Opin. Chem. Biol.*, 2015, **24**, 121–130.
56. D. M. Patterson and J. A. Prescher, *Curr. Opin. Chem. Biol.*, 2015, **28**, 141–149.
57. H. W. Shih, D. N. Kamber and J. A. Prescher, *Curr. Opin. Chem. Biol.*, 2014, **21**, 103–111.
58. M. L. Blackman, M. Royzen and J. M. Fox, *J. Am. Chem. Soc.*, 2008, **130**, 13518–13519.
59. N. K. Devaraj and R. Weissleder, *Acc. Chem. Res.*, 2011, **44**, 816–827.
60. M. Grammel and H. C. Hang, *Nat. Chem. Biol.*, 2013, **9**, 475–484.
61. W. Lin, Y. Du, Y. Zhu and X. Chen, *J. Am. Chem. Soc.*, 2014, **136**, 679–687.
62. R. Xie, L. Dong, R. Huang, S. Hong, R. Lei and X. Chen, *Angew. Chem., Int. Ed.*, 2014, **53**, 14082–14086.
63. R. Xie, L. Dong, Y. Du, Y. Zhu, R. Hua, C. Zhang and X. Chen, *Proc. Natl. Acad. Sci. U. S. A.*, 2016, **113**, 5173–5178.
64. M. Sawa, T. L. Hsu, T. Itoh, M. Sugiyama, S. R. Hanson, P. K. Vogt and C. H. Wong, *Proc. Natl. Acad. Sci. U. S. A.*, 2006, **103**, 12371–12376.
65. T. L. Hsu, S. R. Hanson, K. Kishikawa, S. K. Wang, M. Sawa and C. H. Wong, *Proc. Natl. Acad. Sci. U. S. A.*, 2007, **104**, 2614–2619.
66. Y. Kizuka, S. Funayama, H. Shogomori, M. Nakano, K. Nakajima, R. Oka, S. Kitazume, Y. Yamaguchi, M. Sano, H. Korekane, T. L. Hsu, H. Y. Lee, C. H. Wong and N. Taniguchi, *Cell Chem. Biol.*, 2016, **23**, 782–792.
67. M. A. Breidenbach, J. E. Gallagher, D. S. King, B. P. Smart, P. Wu and C. R. Bertozzi, *Proc. Natl. Acad. Sci. U. S. A.*, 2010, **107**, 3988–3993.
68. Y. Zhu, J. Wu and X. Chen, *Angew. Chem., Int. Ed.*, 2016, **55**, 9301–9305.
69. M. J. Grogan, M. R. Pratt, L. A. Marcaurelle and C. R. Bertozzi, *Annu. Rev. Biochem.*, 2002, **71**, 593–634.
70. M. R. Pratt and C. R. Bertozzi, *Chem. Soc. Rev.*, 2005, **34**, 58–68.
71. Z. Guo and N. Shao, *Med. Res. Rev.*, 2005, **25**, 655–678.
72. T. Buskas, S. Ingale and G. J. Boons, *Glycobiology*, 2006, **16**, 113R–136R.
73. D. P. Gamblin, E. M. Scanlan and B. G. Davis, *Chem. Rev.*, 2009, **109**, 131–163.
74. Y. Yuan, J. Chen, Q. Wan, R. M. Wilson and S. J. Danishefsky, *Biopolymers*, 2010, **94**, 373–384.
75. N. Gaidzik, U. Westerlind and H. Kunz, *Chem. Soc. Rev.*, 2013, **42**, 4421–4442.
76. C. Unverzagt and Y. Kajihara, *Chem. Soc. Rev.*, 2013, **42**, 4408–4420.
77. R. M. Wilson, S. Dong, P. Wang and S. J. Danishefsky, *Angew. Chem., Int. Ed.*, 2013, **52**, 7646–7665.

78. H. C. Hang and C. R. Bertozzi, *Acc. Chem. Res.*, 2001, **34**, 727–736.
79. J. M. Chalker, G. J. Bernardes and B. G. Davis, *Acc. Chem. Res.*, 2011, **44**, 730–741.
80. L. X. Wang, *Carbohydr. Res.*, 2008, **343**, 1509–1522.
81. J. R. Rich and S. G. Withers, *Nat. Chem. Biol.*, 2009, **5**, 206–215.
82. R. M. Schmaltz, S. R. Hanson and C. H. Wong, *Chem. Rev.*, 2011, **111**, 4259–4307.
83. L. X. Wang and J. V. Lomino, *ACS Chem. Biol.*, 2012, **7**, 110–122.
84. L. X. Wang and M. N. Amin, *Chem. Biol.*, 2014, **21**, 51–66.
85. P. M. Danby and S. G. Withers, *ACS Chem. Biol.*, 2016, **11**, 1784–1794.
86. P. Stanley, *Glycobiology*, 1992, **2**, 99–107.
87. S. Wildt and T. U. Gerngross, *Nat. Rev. Microbiol.*, 2005, **3**, 119–128.
88. S. R. Hamilton and T. U. Gerngross, *Curr. Opin. Biotechnol.*, 2007, **18**, 387–392.
89. P. P. Jacobs, S. Geysens, W. Vervecken, R. Contreras and N. Callewaert, *Nat. Protoc.*, 2009, **4**, 58–70.
90. A. Castilho and H. Steinkellner, *Biotechnol. J.*, 2012, **7**, 1088–1098.
91. P. Wang, S. Dong, J. H. Shieh, E. Peguero, R. Hendrickson, M. A. Moore and S. J. Danishefsky, *Science*, 2013, **342**, 1357–1360.
92. K. Takegawa, S. Yamaguchi, A. Kondo, H. Iwamoto, M. Nakoshi, I. Kato and S. Iwahara, *Biochem. Int.*, 1991, **24**, 849–855.
93. K. Takegawa, S. Yamaguchi, A. Kondo, I. Kato and S. Iwahara, *Biochem. Int.*, 1991, **25**, 829–835.
94. J. Q. Fan, K. Takegawa, S. Iwahara, A. Kondo, I. Kato, C. Abeygunawardana and Y. C. Lee, *J. Biol. Chem.*, 1995, **270**, 17723–17729.
95. K. Yamamoto, S. Kadowaki, J. Watanabe and H. Kumagai, *Biochem. Biophys. Res. Commun.*, 1994, **203**, 244–252.
96. S. Q. Fan, W. Huang and L. X. Wang, *J. Biol. Chem.*, 2012, **287**, 11272–11281.
97. A. J. Fairbanks, *Pure Appl. Chem.*, 2013, **85**, 1847–1863.
98. B. Li, Y. Zeng, S. Hauser, H. Song and L. X. Wang, *J. Am. Chem. Soc.*, 2005, **127**, 9692–9693.
99. H. Li, B. Li, H. Song, L. Breydo, I. V. Baskakov and L. X. Wang, *J. Org. Chem.*, 2005, **70**, 9990–9996.
100. B. Li, H. Song, S. Hauser and L. X. Wang, *Org. Lett.*, 2006, **8**, 3081–3084.
101. H. Ochiai, W. Huang and L. X. Wang, *J. Am. Chem. Soc.*, 2008, **130**, 13790–13803.
102. P. Priyanka and A. J. Fairbanks, *Carbohydr. Res.*, 2016, **426**, 40–45.
103. Y. Tomabechi, M. A. Squire and A. J. Fairbanks, *Org. Biomol. Chem.*, 2014, **12**, 942–955.
104. T. B. Parsons, J. W. Moir and A. J. Fairbanks, *Org. Biomol. Chem.*, 2009, **7**, 3128–3140.
105. C. D. Heidecke, T. B. Parsons and A. J. Fairbanks, *Carbohydr. Res.*, 2009, **344**, 2433–2438.
106. T. W. Rising, C. D. Heidecke, J. W. Moir, Z. Ling and A. J. Fairbanks, *Chem.–Eur. J.*, 2008, **14**, 6444–6464.

107. T. W. Rising, T. D. Claridge, J. W. Moir and A. J. Fairbanks, *ChemBioChem*, 2006, 7, 1177–1180.
108. T. W. Rising, T. D. Claridge, N. Davies, D. P. Gamblin, J. W. Moir and A. J. Fairbanks, *Carbohydr. Res.*, 2006, **341**, 1574–1596.
109. D. P. Gamblin, P. Garnier, S. J. Ward, N. J. Oldham, A. J. Fairbanks and B. G. Davis, *Org. Biomol. Chem.*, 2003, **1**, 3642–3644.
110. L. F. MacKenzie, Q. Wang, R. A. J. Warren and S. G. Withers, *J. Am. Chem. Soc.*, 1998, **120**, 5583–5584.
111. H. D. Ly and S. G. Withers, *Annu. Rev. Biochem.*, 1999, **68**, 487–522.
112. G. Perugino, A. Trincone, M. Rossi and M. Moracci, *Trends Biotechnol.*, 2004, **22**, 31–37.
113. S. M. Hancock, M. D. Vaughan and S. G. Withers, *Curr. Opin. Chem. Biol.*, 2006, **10**, 509–519.
114. M. Umekawa, W. Huang, B. Li, K. Fujita, H. Ashida, L. X. Wang and K. Yamamoto, *J. Biol. Chem.*, 2008, **283**, 4469–4479.
115. M. Umekawa, C. Li, T. Higashiyama, W. Huang, H. Ashida, K. Yamamoto and L. X. Wang, *J. Biol. Chem.*, 2010, **285**, 511–521.
116. W. Huang, C. Li, B. Li, M. Umekawa, K. Yamamoto, X. Zhang and L. X. Wang, *J. Am. Chem. Soc.*, 2009, **131**, 2214–2223.
117. M. N. Amin, W. Huang, R. M. Mizanur and L. X. Wang, *J. Am. Chem. Soc.*, 2011, **133**, 14404–14417.
118. P. Priyanka, T. B. Parsons, A. Miller, F. M. Platt and A. J. Fairbanks, *Angew. Chem., Int. Ed.*, 2016, **55**, 5058–5061.
119. T. Yamaguchi, M. N. Amin, C. Toonstra and L. X. Wang, *J. Am. Chem. Soc.*, 2016, **138**, 12472–12485.
120. W. Huang, J. Giddens, S. Q. Fan, C. Toonstra and L. X. Wang, *J. Am. Chem. Soc.*, 2012, **134**, 12308–12318.
121. J. P. Giddens, J. V. Lomino, M. N. Amin and L. X. Wang, *J. Biol. Chem.*, 2016, **291**, 9356–9370.
122. T. Li, X. Tong, Q. Yang, J. P. Giddens and L. X. Wang, *J. Biol. Chem.*, 2016, **291**, 16508–16518.
123. W. Huang, X. Zhang, T. Ju, R. D. Cummings and L. X. Wang, *Org. Biomol. Chem.*, 2010, **8**, 5224–5233.
124. M. N. Amin, J. S. McLellan, W. Huang, J. Orwenyo, D. R. Burton, W. C. Koff, P. D. Kwong and L. X. Wang, *Nat. Chem. Biol.*, 2013, **9**, 521–526.
125. H. Hojo, H. Tanaka, M. Hagiwara, Y. Asahina, A. Ueki, H. Katayama, Y. Nakahara, A. Yoneshige, J. Matsuda, Y. Ito and Y. Nakahara, *J. Org. Chem.*, 2012, **77**, 9437–9446.
126. J. Yin, L. Li, N. Shaw, Y. Li, J. K. Song, W. Zhang, C. Xia, R. Zhang, A. Joachimiak, H. C. Zhang, L. X. Wang, Z. J. Liu and P. Wang, *PLoS One*, 2009, **4**, e4658.
127. Z. Ling, M. D. Suits, R. J. Bingham, N. C. Bruce, G. J. Davies, A. J. Fairbanks, J. W. Moir and E. J. Taylor, *J. Mol. Biol.*, 2009, **389**, 1–9.
128. D. W. Abbott, M. S. Macauley, D. J. Vocadlo and A. B. Boraston, *J. Biol. Chem.*, 2009, **284**, 11676–11689.

129. B. Trastoy, J. V. Lomino, B. G. Pierce, L. G. Carter, S. Gunther, J. P. Giddens, G. A. Snyder, T. M. Weiss, Z. Weng, L. X. Wang and E. J. Sundberg, *Proc. Natl. Acad. Sci. U. S. A.*, 2014, **111**, 6714–6719.
130. Y. Wei, C. Li, W. Huang, B. Li, S. Strome and L. X. Wang, *Biochemistry*, 2008, **47**, 10294–10304.
131. G. Zou, H. Ochiai, W. Huang, Q. Yang, C. Li and L. X. Wang, *J. Am. Chem. Soc.*, 2011, **133**, 18975–18991.
132. C. W. Lin, M. H. Tsai, S. T. Li, T. I. Tsai, K. C. Chu, Y. C. Liu, M. Y. Lai, C. Y. Wu, Y. C. Tseng, S. S. Shivatare, C. H. Wang, P. Chao, S. Y. Wang, H. W. Shih, Y. F. Zeng, T. H. You, J. Y. Liao, Y. C. Tu, Y. S. Lin, H. Y. Chuang, C. L. Chen, C. S. Tsai, C. C. Huang, N. H. Lin, C. Ma, C. Y. Wu and C. H. Wong, *Proc. Natl. Acad. Sci. U. S. A.*, 2015, **112**, 10611–10616.
133. M. Kurogochi, M. Mori, K. Osumi, M. Tojino, S. Sugawara, S. Takashima, Y. Hirose, W. Tsukimura, M. Mizuno, J. Amano, A. Matsuda, M. Tomita, A. Takayanagi, S. Shoda and T. Shirai, *PLoS One*, 2015, **10**, e0132848.
134. J. P. Giddens and L. X. Wang, *Methods Mol. Biol.*, 2015, **1321**, 375–387.
135. I. Quast, C. W. Keller, M. A. Maurer, J. P. Giddens, B. Tackenberg, L. X. Wang, C. Munz, F. Nimmerjahn, M. C. Dalakas and J. D. Lunemann, *J. Clin. Invest.*, 2015, **125**, 4160–4170.
136. F. Schwarz, W. Huang, C. Li, B. L. Schulz, C. Lizak, A. Palumbo, S. Numao, D. Neri, M. Aebi and L. X. Wang, *Nat. Chem. Biol.*, 2010, **6**, 264–266.
137. M. Kowarik, S. Numao, M. F. Feldman, B. L. Schulz, N. Callewaert, E. Kiermaier, I. Catrein and M. Aebi, *Science*, 2006, **314**, 1148–1150.
138. J. D. Valderrama-Rincon, A. C. Fisher, J. H. Merritt, Y. Y. Fan, C. A. Reading, K. Chhiba, C. Heiss, P. Azadi, M. Aebi and M. P. DeLisa, *Nat. Chem. Biol.*, 2012, **8**, 434–436.

CHAPTER 3

Chemical Biology of Protein O-Glycosylation

PATRICK K. CHAFFEY[a], LIANLI CHI[b] AND ZHONGPING TAN*[a]

[a]Department of Chemistry and Biochemistry and BioFrontiers Institute, University of Colorado, Boulder, CO 80303, USA; [b]National Glycoengineering Research Center and the State Key Laboratory of Microbial Technology, Shandong University, Jinan, Shandong 250100, China
*E-mail: zhongping.tan@colorado.edu

3.1 Introduction

Glycosylation is an extremely common and complex post-translational modification of many biomolecules.[1] While this chapter will focus on the glycosylation of proteins, other types of biomolecules are known to carry glycans as well, most notably lipids. Protein glycosylation is commonly divided into three different categories based on the atom linking the glycan to the protein: *N*-type, *O*-type and the somewhat rarer *C*-type. For protein *N*-glycosylation, the carbohydrate chain is attached to the side chain of an asparagine (Asn) residue through the nitrogen atom of the terminal amide. *O*-Glycosylation, the subject of this chapter review, is the term for glycans attached to the side chain of a serine (Ser) or threonine (Thr) residue *via* the oxygen of the hydroxyl group. In *C*-glycosylation, the anomeric carbon of a sugar residue is attached directly to an aromatic carbon on the side chain of a tryptophan (Trp) residue through a carbon–carbon bond. Naturally occurring *C*-glycosylation always involves the sugar mannose (Man) and

Chemical Biology No. 2
Chemical Biology of Glycoproteins
Edited by Zhongping Tan and Lai-Xi Wang

Published by the Royal Society of Chemistry, www.rsc.org

has been observed almost exclusively within thrombospondin repeat domains.[2] *N*-Glycosylation, while present on a much wider variety of proteins, has a similarly restrictive consensus sequence for its incorporation into a protein; and it occurs at the Asn in the consensus sequence Asn–X–Ser/Thr where X is any amino acid except proline.[1] The fact that these types of glycosylation follow such easily identifiable, and thus controllable, patterns has greatly facilitated their study in biological systems. Unlike *N*- and *C*-glycosylation, however, most *O*-glycosylation sites have no obvious consensus sequence for their occurrence. This fact has hindered many traditional approaches to studying the phenomenon based on molecular biology, such as mutating in or out glycosylation sites in proteins of interest.

Nevertheless, a major goal of glycoscience is to understand the structure–function and the composition–function relationships of protein glycosylation (Figure 3.1). A better understanding of these relationships will lead to the development of better industrial and therapeutic proteins, and better diagnostic tools for human disease. While the complex nature of protein glycosylation makes such studies difficult, recent years have seen many advances from an interdisciplinary approach to the problem. New technologies and the chemical biology mindset have inspired studies that combine chemical synthesis of complex biomolecules and biologically relevant assays to reveal robust and important discoveries. The field of proteomics, or when applied to the study of glycoproteins glycoproteomics, has led to extremely useful tools for characterizing naturally occurring glycosylation and is also helping to advance our understanding in this area.

Protein *O*-glycosylation is a very complex type of post-translational modification (PTM) that covers a wide variety of structures.[3,4] It is usually divided into several different categories based on the identity of initial carbohydrate residue attached to the protein. In mammals, there are seven commonly acknowledged categories of *O*-glycosylation, each started by the following seven monosaccharides : *N*-acetylglucosamine (GalNAc), mannose (Man), fucose (Fuc), glucose (Glc), *N*-acetylglucosamine (GlcNAc), galactose (Gal) and xylose (Xyl) (Figure 3.2). By far the most common *O*-glycan is the *O*-GalNAc- or mucin-type glycan. This structure is extremely common on secreted proteins of many varieties and on most extracellular, membrane-bound proteins. *O*-Man was thought for many years to only occur on the protein α-dystroglycan, which is critical for cell adhesion, but has more recently become appreciated as the second most common type of *O*-glycan in the nervous system. *O*-Fuc glycans occur on many different extracellular proteins, but always in the context of specific peptide domains: epidermal growth factor-like repeats (EGF repeats) and thrombospondin-like repeats (TSRs). These domains are commonly involved in cellular communication. *O*-Glc glycans also appear on EGF repeats on a variety of membrane proteins. *O*-GlcNAc is most well known as a cytoplasmic modification of proteins somewhat akin to phosphorylation. Since its initial discovery, this type of glycosylation has become appreciated as an important aspect of signal transduction within the cell. Since this work focuses mostly on extracellular glycosylation, this unique

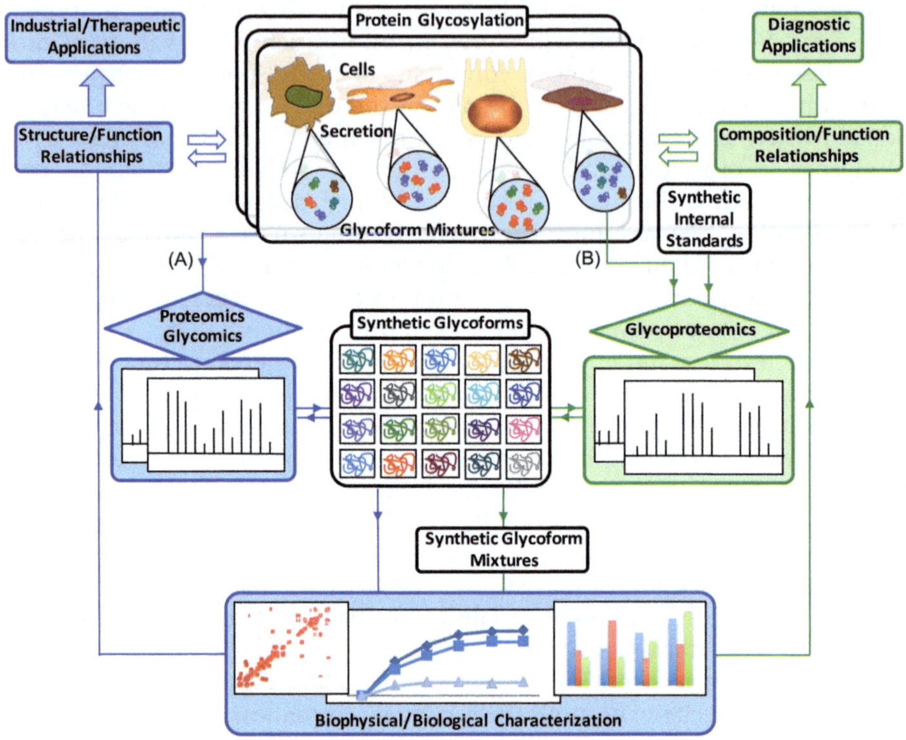

Figure 3.1 The overall strategy of chemical glycobiology research. (A) Identification of the naturally occurring glycans and glycosylation sites by proteomics and glycomics will inform the construction of a library of synthetic glycoforms, which will be individually characterized to determine the correlations between glycosylation patterns and functional properties of glycoproteins. (B) Synthetic glycoforms and internal glycopeptide standards will be used to enable the quantitative analysis of the composition of naturally-secreted mixtures of glycoforms under different conditions. Assays of reconstructed mixtures will lead to correlations between glycoform composition and mixture function. This type of knowledge is expected to greatly advance the understanding of protein glycosylation and promote its applications in the areas of enzyme and therapeutic protein engineering and disease diagnosis.

form of glycosylation will not be discussed in detail here. It is covered in Chapter 4. Recently, however, the *O*-GlcNAc modification was identified on EGF repeats of several important proteins. This extracellular *O*-GlcNAc appears to be controlled independently of the intracellular variety and is relevant to this work. *O*-Gal glycans occur mostly on collagen and are attached to the special, non-proteogenic amino acids hydroxylysine and hydroxyproline. Finally, *O*-Xyl glycans include both heparan sulfate and chondroitin sulfate, which are important members of the proteoglycan polysaccharide family.[1] These long and unbranched, poly-anionic carbohydrate chains are critical for cell adhesion and cellular communication events. Proteoglycans are not the focus of this review and are covered in Chapters 7 and 8.

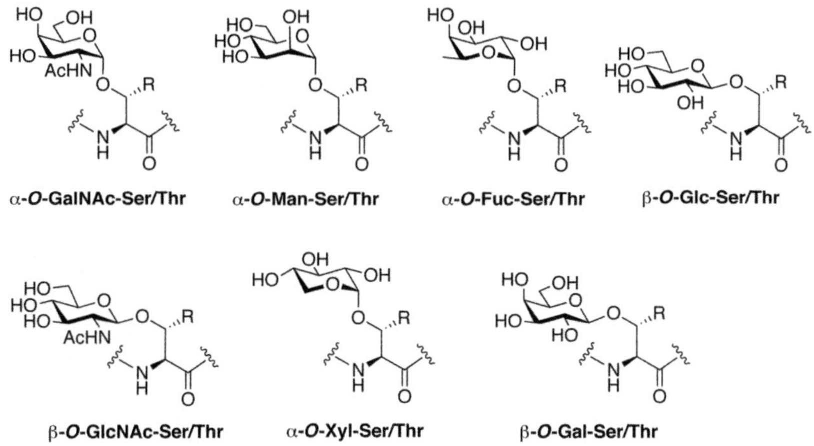

Figure 3.2 The seven types of *O*-glycosylation in mammals. R = hydrogen or methyl group.

This chapter will review the current literature discussing the five most important types of mammalian *O*-glycans, including α-*O*-GalNAc, α-*O*-Man, α-*O*-Fuc, β-*O*-Glc, and β-*O*-GlcNAc, which are naturally found outside the cell on secreted and membrane-bound proteins. Although much important work has been done using synthetic structures that closely mimic naturally occurring glycoproteins, such as several recent studies by the Bertozzi group,[5] these glycoprotein mimetics will not be covered and the discussion will be limited to studies of only actual glycopeptides and glycoproteins. The focus will be on their biosynthesis pathways, what is known of their *in vivo* functions, and how chemical biology has helped to advance the understanding of this important post-translational protein modification.

3.2 Biosynthesis of *O*-Glycoproteins

Protein *O*-glycosylation is an enzyme-catalyzed process that occurs in the endoplasmic reticulum (ER) and Golgi compartments (Figure 3.3). Each type of *O*-glycosylation has a unique biosynthetic pathway and uses a different range of enzymes and substrates.[6,7]

3.2.1 α-*O*-GalNAc

GalNAc is added to Ser or Thr residues in an α-linkage by a large family of over 20 different polypeptide *N*-acetylgalactosamine transferases (ppGalNAcTs). Unlike most glycosylation, which happens in the ER, GalNAc *O*-glycosylation occurs in the Golgi after protein folding.[8] Each ppGalNAcT has two domains: a catalytic glycosyltransferase domain and ricin-homology lectin domain. This lectin domain has been implicated in governing the selectivity and activity of several members of the family.[9,10] Unlike *N*-glycosylation and several other types of *O*-glycosylation, to date no consensus sequence

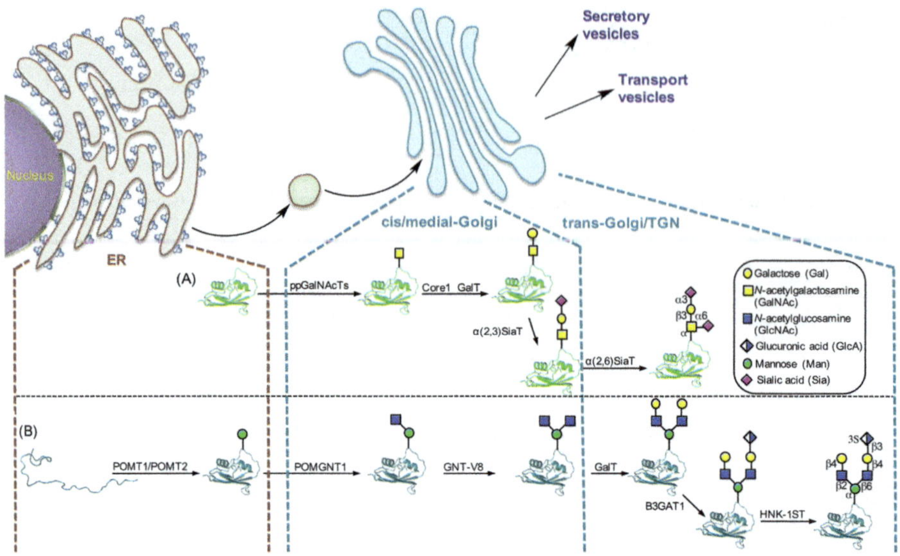

Figure 3.3 Biosynthesis of representative *O*-glycoproteins in the ER and Golgi apparatus. (A) GalNAc-type *O*-glycosylation is initiated in the Golgi after protein folding is completed. The first GalNAc residue is added by a large family of glycosyltransferases, of which at least 20 unique isoforms exist in mammals. (B) Man-type *O*-glycosylation is initiated in the ER by a pair of glycosyltransferases which appear to act as a functional heterodimer *in vivo*.

has been identified for the addition of *O*-GalNAc; it has even been suggested that no such sequence exists and that ppGalNAcTs recognize and modify protein secondary structures, like the β-turn, instead of specific primary sequences.[11] What is known, as demonstrated by several studies discussed in greater detail later on, is that each ppGalNAcT homolog has a slightly different activity profile for peptide and/or glycopeptide substrates. Additionally, the expression of each homolog varies spatially across tissues and temporally during development.[8,12] Such a high-level regulation during cell differentiation and organismal development is indicative of critical roles for these enzymes and the *O*-GalNAc modifications they impart. Interestingly, the deletion of the ppGalNAcT-1, T-4, T-5, or T-13 genes in mice fails to result in any observable phenotype.[13] This might indicate a high level of redundancy and overlap between the functions of each family member. However this situation does not hold for all systems studied, as the deletion of any one of five different *Drosophila melanogaster* *O*-GalNAc transferases is lethal[12] and inactivating mutations of ppGalNAcT-3 in humans cause the rare autosomal recessive metabolic disorder familial tumoral calcinosis.[14,15] Clearly, in certain cellular contexts, a single *O*-GalNAc transferase can have unique and critical functions. After the initial *O*-GalNAc transfer, further elaboration is accomplished in the Golgi apparatus by a suite of up to 30

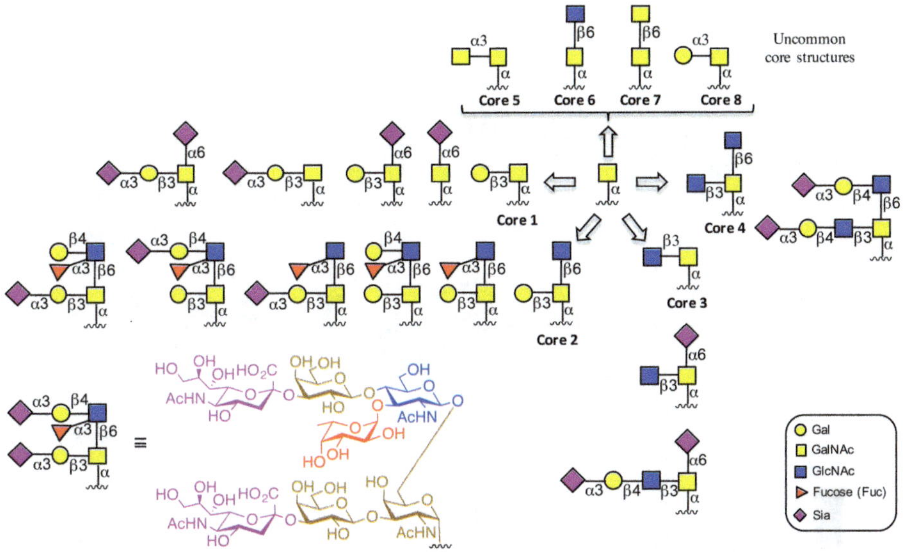

Figure 3.4 Core structures and their elaboration as seen in mucin-type *O*-glycans.

glycosyltransferases to yield a huge variety of mature "mucin-type" *O*-GalNAc glycans (see Figure 3.4).

3.2.2 α-*O*-Man

It is now recognized that *O*-Man glycans, like *O*-GalNAc glycans, can have a variety of branching architectures and end groups. It has been proposed that the naturally occurring mammalian *O*-Man glycans can be divided into three groups, each having a different core structure, similar to how mucin type glycans are classified (see Figure 3.5). All *O*-Man glycans are initiated by addition of mannose by a pair of mammalian protein *O*-mannosyl-transferases: POMT1 and POMT2, which appear to act as a functional heterodimer *in vivo*.[16] These enzymes have different expression patterns across tissues in humans, and different tissue-specific isoforms formed through alternative gene splicing.[17,18]

Like most *O*-glycan-transferases, both POMT1 and POMT2 are found in the ER, but interestingly they utilize a unique sugar donor: dolichol phosphate mannose (Dol–P–Man), unlike other glycosyltransferases which use nucleotide sugars.[19] After the attachment of the first mannosyl residue, glycoproteins are transported to the Golgi apparatus. Both Core M1 and Core M2 structures are then extended by addition of GlcNAc through a β1–2 linkage by the UDP–GlcNAc protein *O*-mannose β1-2 *N*-acetylglucosaminyltransferase 1 (POMGNT1).[20] POMGNT1 acts in the Golgi, and appears to be completely independent of the enzymes that catalyze formation of the GlcNAcβ1–2Man linkages in *N*-glycans, since those two enzymes (GnT-I and GnT-II) show no

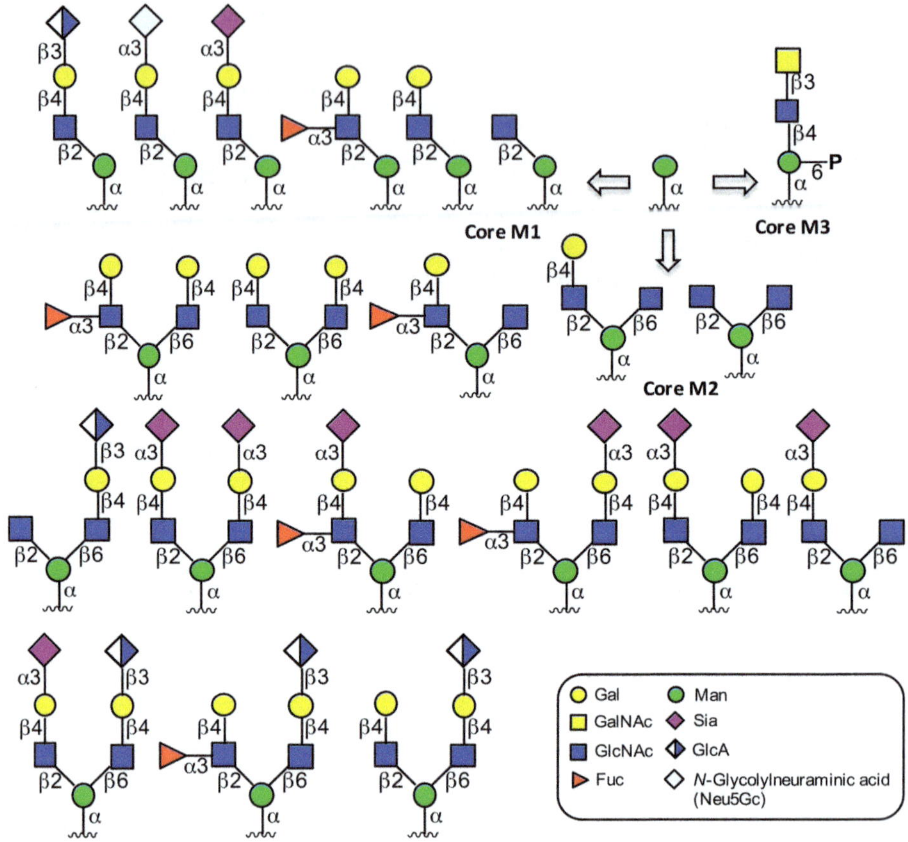

Figure 3.5 *O*-Mannose glycans found in mammals.

reactivity towards *O*-mannosyl peptides *in vitro*.[21] Core M1 structures are then further elongated by Golgi-resident glycosyltransferases, including galactosyltransferases (GALTs), sialyltransferases (SIATs), glucuronyltransferases (GLCATs), sulfotransferase HNK-1ST and α1-3 fucosyltransferase 9 (FUT9), to give a variety of end structures, but detailed mechanisms of how these steps are regulated are not well-understood.[19] Core M2 glycans, thought to exist only in the brain, have a branched trisaccharide core: GlcNAcβ1–6-(GlcNAcβ1–2)Man–Ser/Thr. After synthesis of the Core M1 dimer, a branching GlcNAcβ1–6 can be added to form Core M2 structures. This is done by the brain-specific *N*-acetylglucosaminyltransferase IX (GnT-IX) enzyme,[22] which is also involved in synthesis of highly branched complex-type *N*-glycans.[23]

Core M3 is a uniquely branched phosphodiester-containing trisaccharide: GalNAcβ1–3GlcNAcβ1–4(phosphate-6)Man–Ser/Thr. The phosphoglycan core structure was first identified and structurally characterized in 2010.[24] Unlike other complex *O*-glycans, this core structure is entirely assembled in the ER.[25] So far, this structure has only been found on a few specific sites on α-dystroglycan, a critical component of the dystroglycan complex that links the cytoskeleton of muscle cells to the extracellular matrix.[19] After initial mannosyltransfer to the peptide, glycosyltransferase-like domain-containing 2 (GTDC2/POMGNT2) adds GlcNAc, β1-3 *N*-acetylgalactosaminyltransferase 2 (B3GALNT2) adds GalNAc and the final trisaccharide is phosphorylated by protein kinase-like protein SGK196/POMK to produce the core M3 structure.[25] Notably, POMK is missing key catalytic residues found in other known kinases, and it has been suggested that it could be the first of a new class of kinases with a potentially novel catalytic mechanism.[26] After phosphorylation, the glycan is extended by a polymer of repeating (-3Xylα1–3GlcAβ1-) units synthesized by the enzyme LARGE.[27] This polymeric oligosaccharide has many similarities to glycosaminoglycans (GAGs), like heparin, and has even been proposed to be better thought of as a member of that glycan class.[28] However, LARGE cannot modify the phosphor-trisaccharide product of POMK. Recently, the β1-4 glucuronyltransferase B4GAT1 was found to add a glucuronic acid to xylose (most probably a β-Xyl) and the resulting dimer is capable of serving as the "primer" for (-3GlcAβ1-3Xylα1-) polymer synthesis by LARGE.[29] It is not yet known what the xylose that B4GAT1 modifies is attached to and so the exact details of how the LARGE glycan is attached to the Core M3 structure are not known. It has been confirmed that most, if not all, post-phosphorylation modification of Core M3 *O*-Man glycans occurs in the Golgi, and that neither sialic acid, galactose or fucose are involved in those transformations.[30] LARGE2, a paralog of LARGE, appears to have identical enzymatic activity, but a different optimal pH range and different tissue-expression patterns.[31,32] The significance of LARGE2 is not yet known. Further complicating the picture, sulfotransferase HNK-1ST has been shown to add a sulfate group to the post-phosphoryl carbohydrate moiety and block the activity of LARGE,[33] a probable negative regulatory mechanism for controlling the length of LARGE glycan polymers.

3.2.3 α-*O*-Fuc

O-Linked fucose appears on two distinct kinds of protein domain, both cysteine knots, present in many different proteins: EGF repeats and thrombospondin-like repeats (TSRs). Each one has its own protein *O*-fucosyl transferase: POFUT1 for EGF repeats and POFUT2 for TSRs.[34] Like most other protein *O*-glycosyltransferases, POFUT1 and POFUT2 are localized to the ER.[35,36] After initial protein *O*-fucosylation, each domain then follows a unique extension pattern. Interestingly, there appears to be no cross-talk between the two pathways, enzymes that build the *O*-Fuc glycans on TSRs have no activity towards EGF repeats and *vice versa*.[34] POFUT1 adds fucose to EGF repeats with the consensus sequence $C^2X_{4-5}(S/T)C^3$, where C^2 and C^3 are the second and third cysteines in the conserved EGF repeat sequence.[37] Under certain conditions, the initial fucose can be elongated with a β-GlcNAc at the 3-OH of fucose by the Fringe family of enzymes.[38] In mammals, Gal can be added to GlcNAcβ1–3Fuc by β1-4 galactosyltransferase 1(β4GalT1),[39] but in flies no extension past the disaccharide has been observed. Further elongation by sialic acid to a tetrasaccharide (Siaα2–3Galβ1–4GlcNAcβ1-3Fucα1-Ser) is also seen in mammals.[40] The consensus sequence for fucosylation of TSRs is $C^1X_{2-3}(S/T)C^2X_2G$, where C^1 and C^2 are the first and second Cys of a TSR sequence.[41] TSR *O*-Fuc glycans can be extended by the β1,3-glucosyltransferase B3GLCT to a disaccharide in the ER.[42,43]

3.2.4 β-*O*-Glc

Like *O*-Fuc, *O*-linked glucose is most commonly found on EGF repeats. In *Drosophila*, the protein Rumi was identified as the protein *O*-glucosyltransferase responsible for the modification[44] and the mammalian ortholog, protein *O*-glucosyltransferase 1 (POGLUT1), has also been identified and confirmed to serve the same function.[45,46] These enzymes add glucose to the serine within the $C^1XSX(P/A)C^2$ consensus sequence.[40,47] On several blood factors in both cows and humans as well as on the Notch receptor, the glucose-initiated glycan structure was found to be extended to a trisaccharide Xylα1–3Xylα1–3Glcβ1-Ser.[48,49] In humans the first xylose is added by either glucoside xylosyltransferase 1 (GXYLT1) or GXYLT2 and the second is added through action of xyloside xylosyltransferase 1 (XXYLT1). All three of these glycosyltransferases are active in the ER.[50,51] In flies, there is only one GXYLT, also known as Shams,[52] and extension to the trisaccharide has not been identified. It is not currently known why humans have two functional glucoside xylosyltransferases.[53]

3.2.5 β-*O*-GlcNAc

Recent years have established internal β-*O*-GlcNAc as an important and widespread modification catalyzed by the ubiquitously expressed protein *O*-glucosaminyltransferase (OGT). Advances in understanding the biology of that unique form of glycosylation have been reviewed previously. Relevant to this work, however, is the discovery that β-*O*-GlcNAc also exists extracellularly on the EGF repeats of several proteins. The ER-resident glycosyltransferase for this glycosylation has been identified in flies,[54] mice,[55] and humans,[56] and is known as EGF-specific protein *O*-glucosaminyltransferase (EOGT). Unexpectedly, this enzyme was found to share almost no sequence similarity to OGT, and instead is most closely related to a xylosyltransferase from plants.[55] Characterization of the known carriers of this glycan has revealed the consensus sequence for EOGT activity to be $C^5X_2GX(S/T)GX_2C^6$, where C^5 and C^6 are the fifth and sixth cysteines of the EGF sequence.[57] This consensus sequence shows up in many proteins in both fly and mammalian genomes, although most of these potential sites have not been experimentally validated.

3.3 Chemical Biology in Studying the Structural and Functional Consequences of Protein *O*-Glycosylation

While the unique features of *O*-glycosylation make its study difficult, recently the power of modern chemical biology has made it increasingly possible to investigate this important post-translational modification.[58] This is due, in large part, to the fact that chemical synthesis has advanced to the point where many homogenous glycoforms are available and now commonly used in studies of the phenomenon. Here, we will review the progress in using these new techniques and methods to study the chemical biology of protein *O*-glycosylation.

3.3.1 α-*O*-GalNAc

O-GalNAc glycosylation is the most common form of *O*-glycosylation in mammals, occurring in an estimated 10% of all mammalian proteins and half of all proteins passing though the secretory system.[59] The most well-known examples of proteins carrying this type of glycosylation are the mucins (Figure 3.6), and in fact *O*-GalNAc glycosylation is referred to as "mucin-type" glycosylation in the literature due to the strength of this association.[4,11,60] Mucins can be divided into two categories: secreted mucins that form extensive

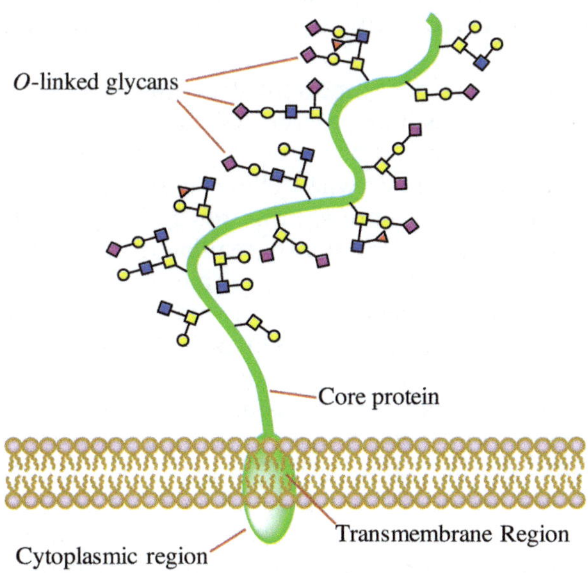

Figure 3.6 Schematic representation of the mucin structure. Mucins are a diverse family of proteins that are heavily *O*-glycosylated. Their protein core domains are often rich in serine, theronine, proline, alanine and glycine residues and adopt an extended "bottle brush" conformation.

oligomers and result in a viscous, mucosal layer around tissues; and membrane-bound mucins that are monomeric and form a significant amount of the glycocalyx that surrounds and protects many cells.[61] The most relevant feature of mucin proteins is the so-called "mucin-domain" which consists of a large number of repeats of a mucin-subtype-specific sequence.[62,63] These repeated sequences, often referred to as "tandem repeats", are heavily glycosylated with densely clustered *O*-GalNAc glycans.[11] Additionally, there are a number of non-mucin proteins contain domains that are structurally very similar to the heavily glycosylated, tandem-repeat regions of mucins, and these are also referred to as "mucin domains".[62,64]

O-GalNAc glycans in these mucin domains are frequently acknowledged for their ability to prevent proteolytic degradation.[65,66] Early light microscopy data indicated that mucin domains adopt a rigid, extended conformation in solution. A multitude of studies since then have established that this structure is a general feature of mucin domains and that this rod-like structure is dependent on the presence of *O*-GalNAc glycan clusters throughout the sequence.[67,68] Therefore, the function of *O*-GalNAc glycosylation in these domains has traditionally been thought of as mainly structural: producing a stable, extended structure on the cell surface for display of other, more functionally important, protein domains, with little importance placed on the actual glycosylation pattern(s) present.[9] For example, glycans on extracellular proteins are known to inhibit oligomerization through what is probably the nonspecific charge–charge repulsion of negatively charged end groups.[69,70] The fact that the sequence of mucins and mucin domains has been only lightly conserved during evolution has also been used as evidence for this assumption.[71]

While there is undeniably an important structural aspect to mucin glycosylation, numerous examples have emerged of glycans acting in a much more specific manner, for example, it has been suggested that they are integral to the molecular code used in everyday cellular communication and recognition.[72] As evidence of these more specific roles, the glycosylation patterns of many proteins appears to be tightly regulated during cellular maturation,[73] and both temporally and spatially during development.[74] Glycan patterns on mucins have also been shown to vary in a region-specific way in several different body systems.[61,75] The intestinal tract is a prime example, where different core structures are specific to each region and sialic acids appear in a well-defined gradient increasing from the ileum to the colon.[76,77] Mucin-type glycosylation is also known to be important in governing selectivity in cell–cell communication and signaling in several different systems including T cell maturation,[78,79] Notch signaling,[80] leukocyte migration[81] and mucin signaling.[82] This type of glycosylation is critical to proper processing of many proteins from the proprotein form to maturity[83] and important in regulating cleavage by a disintegrin and metalloprotease (ADAM) proteases and subsequent shedding of the extracellular domains of many diverse membrane proteins including Notch[80] and TNF-α.[84] Interestingly, this regulation has been observed to be both positive and negative, so it is not a simple steric-interference mechanism and may involve specific binding interactions.[84] Cytokines, secreted soluble proteins that are involved in cell communication, are also known to be regulated through several types of glycosylation, including *O*-glycosylation with mucin-type glycans.[85,86]

Chemical biology has clearly aided in the studies of *O*-GalNAc glycosylation. As discussed below, much of the early work on structural and functional effects of *O*-GalNAc glycosylation focused on small glycopeptides. The readily available and pure homogenous glycoforms made it possible to draw strong conclusions about the different substrate specificities of ppGalNAcTs and the ability of *O*-GalNAc glycans to affect the structural and functional properties of proteins.

3.3.1.1 Substrate Preferences of ppGalNAcTs

One of the most difficult aspects of studying *O*-glycosylation as compared with *N*-glycosylation is the lack of a defined consensus motif for the former. This is despite the efforts of many to characterize the sequence specificities of the large family of ppGalNAcTs over the years. Nevertheless, the ready availability of homogenous glycopeptide substrates that chemical synthesis allows has made strides in this area possible in recent years. In a seminal work by the Bertozzi group, for example, the specificities of many individual ppGalNAcT enzymes were characterized using an impressively large synthetic glycopeptide library.[13] The glycopeptide sequence was taken from a fragment of rat submandibular mucin and contains six potential glycosylation sites, all threonines. The 56-member library based on this sequence was composed of every possible combination of mono-, di-, tri-, and tetra-glycosylated peptide and all of the glycans were single GalNAc residues. Each member of the synthetic library was exposed to the eight pp-GalNAc-Ts studied and the activity of the transferase was subsequently quantified. The authors found distinct preferences for each of the transferases studied and were able to divide them into three groups: those that prefer unglycosylated or monoglycosylated substrates (ppGalNAcT-1, T-2 and T-5), those that prefer diglycosylated peptides (ppGalNAcT-3 and T-4), and those that prefer substrates with three or more glycans (ppGalNAcT-10). Interestingly, ppGalNAcT-7 and T-11 had unique and highly specific selectivities, suggesting more specialized roles for these two glycosyltransferases. From these results, the authors proposed a model of mucin glycosylation in the Golgi where the three groups of glycosyltransferases act in sequence to generate the highly clustered GalNAc glycans observed *in vivo*.[13]

It is well known that members of the ppGalNAcT family have two domains, a catalytic transferase domain and a carbohydrate-binding lectin domain, separated by a flexible linker.[8] In order to more explicitly investigate the influence of this lectin domain on the site-specificity of the ppGalNAcT family, the Gerken group used a series of randomized glycopeptides bearing a single GalNAc residue and a possible acceptor site between 6 and 16 amino acids away.[9] They then compared the ability of several ppGalNAcTs to act on these chemically synthesized substrates, and found distinct N- or C-terminal preferences for many of the enzymes. In particular, ppGalNAcT-1, T-2, and T-14 were found to prefer sites on the N-terminal side of pre-existing glycosylation and ppGalNAcT-3 and T-6 preferred the opposite orientation. Other transferases studied displayed no obvious preference. They also found that the optimal distance for new glycosylation varied across the enzymes from 10 residues for ppGalNAcT-1 and T-2 to 16 residues for ppGalNAcT-3, reflecting differences in the linker connecting the catalytic and lectin domains of the various family members.[9] Follow up work by the same group used a similar strategy to characterize the role of the catalytic domain in selectivity of the glycopeptide-preferring transferases. A set of random glycopeptides, each containing a single *O*-GalNAc–Thr residue and a possible glycosyltransfer acceptor site were tested as substrates for the ppGalNAcTs known to prefer glycopeptide substrates: ppGalNAcT-4, T-7, T-10 and T-12 and the fly ppGalNAcT-7 ortholog PGANT7.[10] It was found that all transferases tested except ppGalNAcT-4 significantly prefer acceptor sites near previous sites of glycosylation, and in addition, each transferase was highly specific in which site was modified. The authors propose that this is indicative of GalNAc binding sites within the catalytic domain of these transferases, and that the binding of these sites is much more important for catalytic activity than similar sites found in the lectin domains of the same enzymes. With their previous work, the authors propose that three distinct kinds of binding are used to various degrees by each member of the ppGalNAcT family: distal glycan recognition by the lectin domain, neighboring glycan recognition by the catalytic domain, and peptide sequence recognition by the catalytic domain. Together these three binding modes allow for glycosyltransfer to a diverse range of substrates with varying degrees of glycosylation and could explain the unusually large number of isoforms for this family.[10]

3.3.1.2 Structural Effects of O*-GalNAc Glycosylation*

Due to the repetitive sequence in the mucin domain with the presence of only a few different amino acids, it is often very difficult to obtain high-quality information from nuclear magnetic resonance (NMR) about the structure of the whole domain.[87] Therefore, most structural studies have been carried out on small glycopeptides. Since the extended structure of mucins ensures minimal interaction between regions of the glycopeptide more than a few amino acid residues apart, it is thought that even a very small model system could be relevant to much larger constructs.[68] This assumption has been repeatedly shown to be valid in heavily glycosylated systems.

Glycophorin A, which is an important glycoprotein found on erythrocytes that carries the epitope determining the MN blood type, is one of the earliest model systems used to study the structure of *O*-glycosylated peptides. It was recognized early on that the immunogenic nature of this glycoprotein was due to a small region of the N-terminus that is heavily glycosylated.[88] Since how the glycosylation affected the structure, and thus immunogenicity of the peptide, was of interest it was commonly studied as a model system for heavily glycosylated peptide domains. Synthesis of mono-glycosylated pentapeptides matching the N-terminal sequence of Glycophorin A^M, each with an *O*-GalNAc at a different one of four Ser/Thr residues, was carried out by the Dill group, and natural abundance ^{13}C-NMR was used to characterize them.[89] It was concluded that the placement of the glycan had noticeable effects on the properties and NMR signals of the pentapeptide, and in general the greatest changes were at the location of the glyco-amino acid in the sequence. Follow up work by the same lab with a higher field instrument, was able to show that glycosylated threonine residues in the sequence were more conformationally constrained than glycosylated serines.[90] More complicated constructs based on glycophorin were also noteworthy contributors. In an important 1999 study, a decapeptide containing six consecutive *O*-GalNAc residues was synthesized and studied.[91] The authors were able to obtain an impressive number of nuclear Overhauser effect (NOE)-distance and backbone angle restrictions, which were then used in a hydrated molecular modeling simulation of the structure. This simulation resulted in an extended "wave-like" structure for the glycosylated peptide, showing that the extended structure of densely glycosylated peptides is most-strongly dependent on the first carbohydrate unit.[91] The authors were also able to confirm that glycosylated Thr is a more rigid unit than glycosylated Ser.

Around this time, structural studies on mucin-derived glycopeptides were also reported. One such study looked at NMR structures of glycopeptides corresponding to fragments of MUC7.[92] Much like the glycophorin constructs an extended, rigid random-coil similar to a polyproline type II (PPII) coil was found for the peptide backbone. Also identified were NOE contacts between the backbone and the *O*-GalNAc of Thr-linked glycans but not Ser-linked glycans, indicating that the GalNAc–Thr linkage is more rigid than that of serine, which is similar to the results found in glycophorin systems.[90,91] The stability differences between Thr and Ser glycosylation are a general theme of *O*-GalNAc-type glycosylation. More detailed study of the structures of glycosylated Thr and Ser confirmed that the GalNAc–Thr is a much more rigid structure, and attributed this to a significant change in the way water is structured around each of the glyco-amino acids in solution.[93] The same group was able to show later that these structural differences are reflected in the binding preferences of several lectins, where a significant amount of selectivity was observed for GalNAc–Thr or GalNAc–Ser over the other in identical peptide sequences.[94] This shows that lectin recognition motifs can depend on the underlying peptide sequence as well as the glycan epitope.

There is also the very impressive work of the Danieshefsky and Live groups on structure of the mucin domain of CD43. The authors first prepared, through chemical synthesis, four glycosylated pentapeptides, each carrying three glycans on consecutive amino acid side chains.[87] The glycans were varied in size from single *O*-GalNAc to a sialylated trisaccharide. From the detailed NMR analysis of the synthetic glycopeptides, it was revealed that the glycans induce an extended and extremely stable structure, highly unusual for a peptide of the size.[87] Additionally, through synthesis of the β-linked stereoisomer of one of the glycopeptides, it was shown that the native α-linkage between Ser/Thr and GalNAc is critical to any stabilizing effect. By comparing the structures of glycopeptides bearing different sized glycans, it was revealed that the distal carbohydrate units played a very minor role in stabilizing the observed structure, confirming the conclusions of others. Together, these observations led to a general model for glycosylation in mucin domains where the peptide backbone and initial α-*O*-GalNAc carbohydrate residues form a very stable scaffold upon which important end-group carbohydrate epitopes can be mounted and act as signals in intercellular communications.[87] The authors followed up this study with more detailed work using even higher field-strength magnets a few years later.[95] In addition to the substrates previously examined, they also synthesized and characterized a slightly longer heptamer sequence containing three consecutive hexasaccharides. Much like the previous structures reported, they found an extremely rigid structure for the glycopeptides, and the length or size of the glycans did not have a significant effect on the stability of the molecules, even with hexasaccharides. With the more detailed NMR results, the authors could identify a highly organized network of hydrogen bonds linking the first α-*O*-GalNAc to the surrounding peptide backbone, which helps to explain the remarkable stabilization observed in the systems.[95]

MUC1, which is implicated in tumor progression and is thus a potential therapeutic target in several types of cancers,[68] has also been widely studied in terms of its structure. The 20-residue tandem repeat sequence of MUC1 contains two interesting structural features: a PDTRP motif that is responsible for immunogenicity of MUC1 and a GVTSAP motif that is a target for ppGalNAcT-1 and T-3. In cancers, MUC1 is often under-glycosylated and this leads to the revealing of the immunogenic PDTRP motif, which has important implications in cancer diagnosis and treatment.[96] To study possible influences on the immunogenic motif of glycosylation of the sequence, a synthetic 15-mer peptide and its glycosylated analog were synthesized and characterized with NMR.[97] Although the most populated NMR structures of the area immediately surrounding the glycosylation site were common to both glycosylated and unglycosylated constructs, there was also a unique conformation for each, showing some differences in conformational space available to the glycosylated peptide compared with the unglycosylated one and *vice versa*. Additionally, the authors were able to identify hydrogen bonding between the amide proton of the GalNAc moiety and the peptide backbone, much like in studies of other systems as discussed above.[95,97] This hints at a general mechanism for stabilization. The calculated structure closely matched a crystal structure of an anti-MUC1 antibody in complex with a 13-mer sequence containing the PDTRP motif.[98] A study of glycosylation directly on the PDTRP motif showed that the region adopted a β-I-type turn in the absence of glycosylation, but a much more extended conformation upon glycosylation by α-*O*-GalNAc at the Thr.[99] The Kunz group was able to synthesize and study the complete 20-mer tandem repeat sequence containing a GalNAc within the

GVTSAP motif.[100] They found that the entire sequence from the glycosylation site through the immunogenic motif (nine amino acids) was very rigid and stable. The glycosylation site adopted an extended, wave-like conformation very similar to those observed in the glycophorin and CD43 mucin domain systems discussed previously,[87,91] while the immunogenic domain was found in a β-turn structure.[100] Further work on the MUC1 repeat has also been carried out by the Nishimura group, who were able to chemically synthesize a series of glycopeptides of the full 20-mer repeat domain of MUC1 each containing one core-2-based tetrameric structure at one of the five potential glycosylation sites.[101] Enzymatic extension on both arms of the core oligosaccharide gave a series of glycopeptides displaying di-sialylhexasaccharides. NMR analysis and structural modelling of these glycopeptides showed that sialylation did indeed induce small changes to the backbone conformation, but only within the PDTRP immunogenic motif. The authors conclude that the underlying sequence is a noticeable factor in how distal sialylic acids alter the conformation of a glycopeptide. Together, these studies on MUC1 indicate that the immunogenicity of certain sequences can be modulated by conformational switches that are, in turn, controlled by the glycosylation state of the peptide. Since aberrant glycosylation is an extremely common feature of most cancers; this has the potential to inform new understanding of cancer biology and new therapies.[102] This is discussed in great detail in Chapter 12.

3.3.1.3 Functional Effects of O-GalNAc Glycosylation

Chemical biology has also contributed to the study of functional aspects of mucin-type glycans in biological systems. As mentioned earlier, glycosylation patterns on extracellular membrane proteins are very important for cellular communication and adhesion. For example, during inflammation, leukocytes adhere to endothelial cells near the site of injury before migration towards the injury site.[103] This adherence is the result of P-selectin on the endothelial cells binding specific ligands on leukocytes, the most important of which is known as P-selectin glycoprotein ligand-1 (PSGL-1). P-Selectin binds to the extreme terminus of PSGL-1, and both tyrosine sulfation and *O*-glycosylation of PSGL-1 in this region have been shown to be critical for high-affinity binding in studies of genetic knockout models.[103] To identify the particular modifications required for binding in this context, the Cummings group chemo-enzymatically synthesized a 23-mer glycosulfopeptide based on the N-terminal sequence of PSGL-1 that contained three sulfotyrosines and a sialyl Lewis x (sLe^x) hexasaccharide.[104,105] To avoid problems with site-specificity of ppGalNAcTs, the peptide was synthesized on solid phase with the first GalNAc at position Thr 57 in place. Enzymatic transformations by a series of five glycosyltransferases constructed the full hexasaccharide, and lastly addition of the sulfate groups was also accomplished enzymatically. It was found that the sulfotransferase was able to add sulfates to all three tyrosines in the glycopeptide simultaneously.[104,105] In addition, the authors synthesized a glycosulfopeptide with a core-1 sLe^X-containing hexasaccharide and the same three sulfotyrosines through a very similar chemoenzymatic approach. The authors then tested the two final glycosulfopeptides for binding affinity to P-selectin and found that only the construct with the core-2 glycan structure was bound to a measurable degree.[104,105] In addition, when the terminal sialic acid was removed from the glycan, binding was abolished. Moreover, they found that the synthetic intermediate carrying the full-length glycan, but not the sulfates, did not bind, nor did the isolated hexasaccharide bind. The results of this study indicate that both modifications, sulfation and glycosylation, are critical to high-affinity binding between PSGL-1 and P-selectin. However, because the authors were unable to chemo-enzymatically prepare any partially-sulfated glycopeptide products, they could not address the question of whether or not specific sulfo-tyrosines were more important to binding than others. This question was solved in a follow-up study by the same group that used chemical synthesis to introduce the sulfo-tyrosine residues in a completely controlled manner.[101] In this study they synthesized a large library of glycosulfopeptides containing each possible combination of 0–3 sulfate groups and several different glycan structures. Characterization of each construct and comparison of the results revealed that the most important determinant of binding was the presence of a fucose in the sLe^X epitope on the peptide, although the number and pattern of sulfate groups as well as the terminal sialic acid had strong effects on binding. Also noteworthy is the site-specific nature of this glycan dependence. Moving the full-length glycan from the natural glycosylation site to a nearby site (both threonines) resulted in complete loss of binding.[101]

3.3.2 α-*O*-Man

O-Mannosylation, originally thought to exist only in yeast and fungi, is now a well-known modification of mammalian systems. First discovered on proteoglycan core proteins in the brain and assumed to be relatively rare in mammals,[106] more recent estimates estimate mannose-initiated *O*-glycan structures to account for 30% of the total *O*-glycans in the brain.[107] Significantly, in contrast to yeast *O*-mannosyl glycans, which are mostly linear chains,[108] the *O*-mannosyl glycans observed in mammals have complicated structures with a wide variety of monosaccharides, branching, and charged terminal structures like Le^X and HNK-1 epitopes.[109–112] Over the last two decades, *O*-Man-linked glycans have been confirmed on a wide variety of proteins in both the brain and muscle cells, including the cadherin and plexin families of cell-membrane receptors,[113,114] IgG2,[115] CD24,[116] neurofascin 186,[117] receptor tyrosine phosphatase β (RPTPβ),[118] and the brain-derived lectican proteins aggrecan, versican, neurocan, and brevican.[119] Many of these proteins are important for their role in cell adhesion and the list includes almost every known member of the perineuronal net, which is responsible for stabilizing neuronal synapses in mature brain tissue.[119] Along with their widespread expression throughout the nervous system, the severe phenotypes that result from mutations in their biosynthetic machinery confirms that *O*-mannosyl glycans are functionally important in mammals. While many of the known carriers of this modification remain largely uncharacterized, there are a few examples of studies examining the detailed functional consequences of *O*-mannosylation. For example, recent evidence has pointed to the expression of HNK-1 epitopes on 2,6-branched *O*-mannosylglycans, particularly on RPTPβ in the brain, as being critical to proper myelination and brain function.[118,120] *O*-Man-linked glycans have also been implicated in critical stages of early neuronal development by serving as the most important scaffolds for both HNK-1 and Le^X epitopes in the developing brain.[121] But by far the best characterized *O*-mannosylated protein in humans is α-dystroglycan (α-DG).[122]

Dystroglycan is the product of a single gene, *DAG1*, which is cleaved into two mature proteins: α-and β-DG. β-DG is a transmembrane protein that binds intracellular dystrophin, which binds the cytoskeleton protein F-actin.[28] α-DG is secreted and binds the extracellular portion of β-DG at one end and several extracellular matrix components, most notably laminin, on the other end.[28] Together these proteins form the dystroglycan complex (DGC), which physically links the cytoskeleton of a cell to the surrounding basement membrane.[19] α-DG has a dumbbell shape with two globular domains separated by a conserved mucin-like core, which has over 40 potential *O*-glycosylation sites and is known to harbor both classic mucin-type (*O*-GalNAc) and *O*-Man glycans in close proximity to one another (Figure 3.7).[123] This glycosylation, and in particular the *O*-mannosylation, of α-DG is critical to the function of the complex.[122,124] Thanks to the many studies employing the powerful tools of chemical biology, over the past 20 years, our knowledge and understanding of this important system has been substantially enriched.

3.3.2.1 Biosynthetic Pathway of O-Man Glycans

As in the study of *O*-GalNAc glycosylation, several groups have attempted to settle the question of whether or not *O*-mannosylation has a consensus motif in recent years through the use of synthetic peptides. The Endo group, in a particularly relevant study, synthesized a series of 20-mer peptides covering the entire mucin-like domain of human α-DG and assayed each for ability to accept mannose transfer catalyzed by recombinant POMT1 and POMT2.[125] From this assay, they were able to identify two regions that were particularly prone to mannosylation corresponding to residues 336–355 and 401–420. Comparisons of these two regions of α-DG to one another and to regions of mucin proteins that are not able to be mannosylated in the same assay indicated that the consensus sequence for *O*-mannoylation is: IXPT(P/X)TXPXXXXPTX(T/X)XX.[125] However, this sequence has since been only mildly predictive of the observed mannosylation patterns in several recent mapping studies of α-DG.[123,126–128] Later work has pointed to a *cis*-regulatory 41-residue sequence located adjacent to the α-DG mucin that, along with the peptide primary sequence around glycosylation sites, influences *O*-mannosylation *in vivo*.[129]

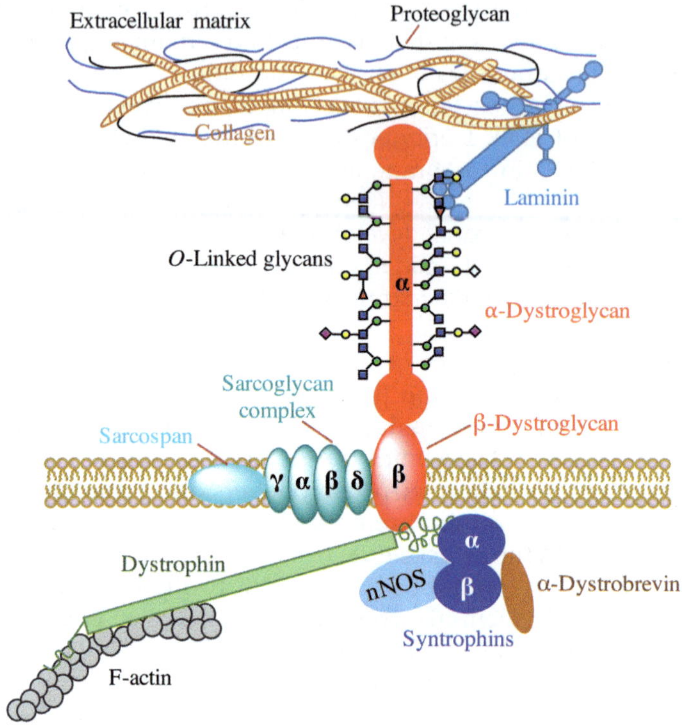

Figure 3.7 Schematic representation of the interactions between dystroglycan, dystrophin and the extracellular matrix. β-Dystroglycan binds several internal cellular components including the cytoskeletal protein dystrophin, which in turn binds actin filaments. α-Dystroglycan strongly binds extracellular matrix components, in particular laminin. Together these interactions are responsible for physically linking the cytoskeleton to the basement membrane of the surrounding tissue.

After initial mannose transfer, POMGNT1 is responsible for extending the glycan by attachment of a GlcNAc through a β1–2 linkage.[20] Study of this step in the biosynthetic pathway of both core M1 and M2 mannose-type glycans has also been advanced through the use of chemical biology. Initially, the presence of β1–2-GlcNAc transferase activity in brain extracts was confirmed using synthetic peptides and glycopeptides as probes for enzymatic activity.[21] That same study then made extensive use of synthetic carbohydrates as authentic chromatography standards to characterize the structure of the resulting disaccharide and classify the glycosyltransferase as a β1-2 *N*-acetylglucosaminyltransferase. The stereochemistry and connectivity of the linkage was later confirmed by NMR studies on glycopeptides synthesized chemoenzymatically with recombinant POMGNT1.[130] Mutations in the *POMGNT1* gene cause a severe form of muscular dystrophy, muscle–eye–brain (MEB) disease, and in MEB patients, α-DG is significantly hypoglycosylated and lacks laminin binding affinity.[20,131] These facts have prompted a deeper look at the functioning of the POMGNT1 protein. One such study looked at the effect of known MEB-causing mutations on the enzymatic activity by comparing the ability of mutant enzymes to incorporate GlcNAc into synthetic mannosyl-glycoproteins that mimic the natural sequence of α-DG.[132] Although they had difficulty correlating the structural effects of mutations on the enzyme with severity of the resulting phenotype, they did conclude that a complete loss of enzyme function is not necessary for an observable phenotype since several mutants studied still had appreciable activity towards synthetic glycopeptides *in vitro*. Also significantly, the authors of the study found that only small subset of the synthetic glycopeptides used as enzymatic probes were actually modified by POMGNT1, showing that, at least in this *in vitro* assay, only specific regions of α-DG are modified with extended mannose-type glycans.[132] The clear sequence dependence of POMGNT1 activity was confirmed in a very similar study that assayed for enzymatic activity using a series of glycopeptides of varying lengths.[133] The authors of that study concluded not only that the sequence surrounding the *O*-mannose affects glycan elongation, but also that a minimum of eight residues is required for efficient binding of POMGNT1.

Curiously, despite the fact that *POMGNT1* mutations result in MEB phenotypes,[20] the actual binding of α-DG to the basement membrane is mediated by a glycan structure that does not require POMGNT1 for assembly: the core M3 *O*-mannose glycan.[24] Using synthetic glycoconjugate substrates and recombinant enzymes, the entire biosynthetic pathway for the core M3 phosphotrisaccharide was recently laid out in more detail.[25] The authors started by examining the activity of GTDC2 (POMGNT2) towards a synthetic mannosyl-glycopeptide corresponding to residues 316–329 of human α-DG's mucin-like domain, one of the select few regions of the domain where this structure occurs.[124] They found that GTDC2 was able to incorporate UDP–GlcNAc but not UDP–GalNAc into the synthetic glycopeptide substrate.[25] To more rigorously characterize the linkage and stereochemistry of the disaccharide, the synthetic glycoconjugate 4-methylumbelliferyl-α-D-mannoside was used as the enzyme substrate and the product was analyzed with NMR. The authors were able to clearly identify the disaccharide glycosyl transfer product as GlcNAcβ1-4Man, proving that GTDC2 has protein *O*-mannose β-1,4-*N*-acetylglucosaminyltransferase activity.[25] Next, the authors used this chemoenzymatically produced GlcNAcβ1-4Man glycoconjugate to test if B3GALNT2 possessed its predicted β-1,3-*N*-acetylgalactosaminyltransferase activity and found that it did. The authors thus concluded that POMGNT2 (GTDC2) and B3GALNT2 act in series to construct the trisaccharide moiety of M3 core glycans on α-DG, and turned their attention to the phosphorylation of the glycan which had previously been shown to be necessary for laminin binding.[24,124] This trisaccharide was then mixed with recombinant SGK196 (POMK) and ATP, which verified that SGK196 was, in fact, the kinase responsible for *O*-Man glycan phosphorylation.[25] The phosphorylation site was confirmed as being on the 6-position of the mannose by NMR. Interestingly, SGK196 was only able to phosphorylate the full-length trisaccharide, which helps to explain how mutations in *POMGNT2* (*GTDC2*) and *B3GALNT2* cause dystroglycanopathy phenotypes.

The enzyme LARGE has been known to be important in the functioning of α-DG for several years.[124,131] In recent years, the understanding of the function of this enzyme has also been vastly improved by use of synthetic glycoconjugates and other tools common to chemical biology. The specific transferase activity of LARGE was confirmed by Inamori and colleagues in 2012 through a series of chemical-biology-based experiments.[134] They took recombinant LARGE enzyme and mixed it with either an α-xyloside glycoconjugate in the presence of UDP–GlcA or a β-glucuronide glycoconjugate in the presence of UDP–Xyl. In both cases disaccharides were formed. Incubation of these dimer products with LARGE, UDP–Xyl and UDP–GlcA resulted in a polymer of repeating disaccharides (-3Xylα1-3GlcAβ1-). Using NMR, the authors were able to confirm from the anomeric stereochemistry and regioselectivity of each carbohydrate linkage, that LARGE alone is capable of forming carbohydrate polymers and that it possesses bifunctional glycosyltransferase activity.[134] They continued this study a year later to examine a LARGE paralog in mammals, LARGE2. In that follow up study, they found LARGE2 to have the same biufunctional glycosyltransferase activity, but a slightly different optimum pH range.[32] While this did answer the questions surrounding the nature of LARGE's enzymatic activity, it did not address several remaining questions concerning the relationship between the structures of LARGE-synthesized glycans on α-DG and binding affinity towards laminin. Work by the same group began to answer these later types of questions with the study of synthetically produced LARGE glycans.[27] In this study, chemoenzyamtically synthesized LARGE-glycan repeats were immobilized on ELISA assay plates and laminin binding affinity was measured. They found that the length of the repeating glycan was directly proportional to the binding affinity towards ECM components like laminin.[27] They went on to show that the expression of LARGE, and as a result the length of the LARGE-glycan repeats on α-DG, was temporally regulated during muscle regeneration after injury. Together, these two pieces of evidence were used by the authors to propose a model where the binding affinity of α-DG towards the basement is regulated by the length of LARGE-glycan repeats through variations in the level of expression of LARGE.[27] This has potential functional implications for muscular dystrophies, where mutations in LARGE inhibit its ability to synthesize glycans of the proper length. However, LARGE cannot modify the phosphotrisaccharide core of M3 glycans on α-DG, which indicates that further modifications on the glycan are necessary before LARGE polymerization can take place. A recent paper found that the misnamed β1-3 *N*-acetylglucosaminyltransferase 1 (B3GNT1) is actually a β1-4 glucuronyltransferase with a preference for xylose acceptors.[29] Through a combination of synthetic substrates and recombinant enzymes, the authors showed that B3GNT1 transfers a GlcA with a β-linkage to the 4-position of a xylose acceptor, and they propose a new name, β1-4 glucuronyltransferase 1 (B4GAT1), for this enzyme. B4GAT1 was not found to be capable of forming polymeric carbohydrates, but it did have a significantly higher activity towards monosaccharide acceptors than LARGE. Because of these activity differences, it was proposed by the authors that this newly characterized enzyme (B4GAT1) acts as a "priming" enzyme by transferring the first GlcA to a Xyl acceptor to form a dimer structure that is easily recognized and elongated by LARGE.[29] They did not report, however, on the nature of the connection between the M3 core glycan structure and the potential Xyl residue that B4GAT1 requires for activity, and so mysteries remain regarding the biosynthesis of *O*-mannose type glycans.

While understanding the biosynthetic pathways and regulation of those pathways behind each individual glycan is important, there are also interactions and possible cooperative regulatory effects between different glycans to consider. For example, it is well known that the core M3 glycan structure responsible for laminin binding is only present on a few specific Thr residues of the α-DG sequence, and that those residues are in close proximity to sites modified by the more common core M1 glycans.[123,126–128] Furthermore, mutations present in the enzyme POMGNT1, which is only involved in the synthesis of M1 glycans, are known to have a strong effect on α-DG laminin binding in patients with MEB.[20,131] These pieces of evidence together are indicative of significant cross talk between different types of *O*-mannosylation at different sites. Here again, the unique approaches and mindset of chemical biology can achieve significant advances. After the initial discovery and characterization of the unique M3 core structure,[24] the synthesis of the novel phosphoglycan component and its incorporation into glycopeptides was seen as a top priority in order to allow future *in vitro* studies to be carried out. The synthesis of the M3 core structure was successfully completed and the resulting phosphotrisaccharide as well as glycopeptides corresponding to the natural α-DG sequence known to bear the M3 structure were studied.[135] Using mono-mannosylated glycopeptides corresponding to the α-DG sequence surrounding the known M3 glycan site, it was found that POMGNT1 was able to extend a mannose at the known M3 site (T379), but not at a site two residues away (T381).[135] This, along with the studies discussed previously,[132,133] illustrates the sequence dependence of POMGNT1. In addition, synthetic glycopeptides containing single mannose units at both the M3 site and the nearby site were found to be modified by POMGNT1 at both mannose residues, although not at both simultaneously. This indicates that POMGNT1 activity and selectivity can be significantly altered by clustered presentations of glycans on a peptide backbone.[135] The fact that POMGNT1 can modify the site of M3 glycans *in vitro* and yet no M1-type glycans have been observed at this site *in vivo* was later explained when the biosynthetic pathway for the M3 core-glycan was found to occur entirely in the ER, before possible exposure to POMGNT1 in the Golgi.[25]

The probable interplay between classical mucin-type *O*-GalNAc-initiated glycans and *O*-Man-initiated glycans in the mucin region of α-DG was more explicitly investigated in a later study.[136] In that study, the authors examined the efficiency and selectivity of several different ppGalNAcTs towards glycopeptides already containing one or more *O*-Man glycans. This mimics the natural biosynthetic pathways of the two glycans where *O*-Man is added in the ER[19] and *O*-GalNAc is added later in the Golgi.[8] They found that the presence of *O*-Man glycans on the sequence has a significant effect on the activity of ppGalNAcTs, and furthermore that such effects are site-specific in nature.[136] These conclusions were drawn from studying synthetic glycopeptides derived from two regions of the α-DG sequence, each sequence containing four consecutive sites for potential glycosylation. The first sequence was from a region of the protein shown to only contain *O*-Man type glycans *in vivo*, and the second from a region thought to contain both *O*-Man and *O*-GalNAc type glycans. Both of the sequences were shown to be able to accept GalNAc transfer, unless specific *O*-mannosylation patterns were present before exposure to ppGalNAcTs.[136] This highlights the importance of specific glycosylation patterns in regulating subsequent post-translational modification and, eventually, protein function. Also noteworthy is the fact that the location of *O*-GalNAc incorporation depended on the *O*-mannosylation pattern, showing that *O*-Man glycans have the ability to not only up- or down-regulate the activity of subsequent glycosyltransferase enzymes, but can also strongly affect the sites at which those enzymes act.[136]

3.3.2.2 *Biophysical and Biological Effects of* O*-Mannosylation*

Understanding the effects of *O*-mannosylation on the biophysical and biological properties of a protein or peptide is also an important research area. Given the well documented structural effects of *O*-GalNAc type glycans on mucin domains in other systems, it was of significant interest to investigate how the presence of both *O*-Man and *O*-GalNAc glycans on the mucin domain of α-DG might affect the structure of the protein. As with *O*-GalNAc glycosylation, biophysical and structural studies of synthetic model systems allowed for quick and robust conclusions to be drawn. For example, by synthesizing a series of glycopeptides derived from the α-DG mucin region and containing up to four consecutive *O*-Man residues, it was possible to conclude that the glycopeptides existed in a mostly unordered structure.[137] Interestingly, when the glycans were switched to *O*-GalNAc, the same sequence was considerably more ordered, as would be expected given the previous work on mucin-domain structures.[95,99] This comparison held across three different biophysical characterization methods: circular dichroism (CD), NMR and hydrogen–deuterium exchange (HDX).[137] Combined with the well-documented importance of *O*-Man glycans in the functioning of α-DG,[19,26] these observations led to a proposed model where the *O*-GalNAc glycans provide the mucin domain with a defined, rigid structure that supports optimal display of the functionally important end-groups of the *O*-Man glycans.[137,138] The results of structural studies of the glycopeptide bearing a negatively charged 6-*O*-phosphomannose residue also supported this assertion. As with previous work on the structural effects of *O*-mannosylation, the authors found little structural difference between the unglycosylated peptide and the phosphomannose-containing glycopeptide.[135] Replacing the mannose derivative with an *O*-GalNAc, however, resulted in significant changes in the NMR spectra of the glycopeptide.

In addition to the studies of the effects of *O*-glycosylation on unstructured mucin domains, many groups, including our own, have undertaken research programs to get this knowledge with respect to folded peptides and proteins. One example of this type of research is the work done by the Tan group on a Family 1 carbohydrate-binding module (CBM) of a fungal cellulase. The CBM is a small, 36-residue peptide domain and is responsible for recognition and binding of cellulose (Figure 3.8). It is naturally glycosylated by several *O*-glycans, which have been implicated as critical in the functions of this domain.[139] In order to characterize the specific effects of *O*-mannosylation on the CBM, a library of CBM glycoforms bearing mono- and oligo-mannose chains at each of the three possible glycosylation sites were chemically synthesized. By comparing the individual members of this glycoform library, the Tan group was able to show that *O*-mannosylated CBM was more stable and bound tighter to cellulose.[140] Particularly intriguingly, the authors found that glycosylation at Ser3, more so than at either of the other two sites, was the single most important factor in the thermostability of the glycoforms (Figure 3.8). This finding shows that in certain contexts,

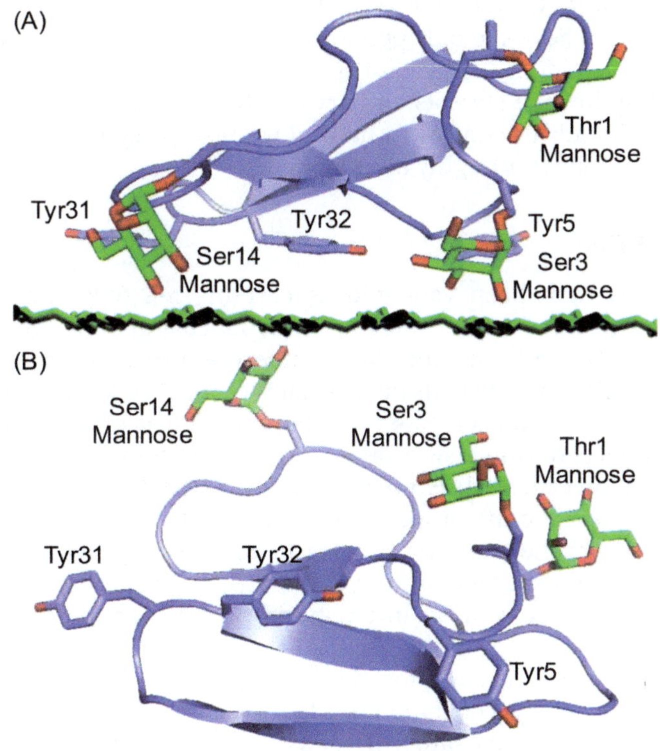

Figure 3.8 The NMR structure of the *O*-mannosylated Family 1 CBM. (A) Side view. (B) Top view. The top layer of the cellulose crystal is shown in green. The tyrosine residues that form the binding face to cellulose and the three mannose monosaccharides are shown as sticks.

glycosylation can site-specifically control the physical properties of glycoproteins. It also hinted at as-yet-underappreciated interplay between glycan structures and the underlying peptide sequence or structure that can lead to a specific outcome in the context of one glycosylation site but not another. This hypothesis warranted further investigation, and a follow-up study by the same group looked at the possible cooperative effects of amino acid side chains and glycan structures on the physical and functional properties of the CBM.[141] As with the previous study, a library approach was chosen and the authors synthesized 31 new CBM isoforms with amino acid mutations and a wide range of glycan structures varying in size, linkage stereochemistry, branch structure and charge state and each library member was rigorous characterized through a panel of biophysical and functional assays. By systematically investigating each structural feature, the authors determined that planar polar (Gln) and aromatic (Tyr) side-chains near the Ser3 site were critical to the increased stabilization observed upon glycosylation. Additionally, the stereochemistry of the linkage between glycan and peptide had to be in the α configuration. Thus, by comparing biophysical characteristics across library members, this study provided new insights into the molecular basis for the effects of *O*-glycosylation at Ser3 of the CBM. The strong conclusions made in these studies were only possible because of the ability to controllably synthesize a large variety of glycoforms. This ensured that any structural feature, including linkage stereochemistry and branching structure, which are notoriously difficult to control in natural expression systems or enzymatic syntheses, could be thoroughly investigated.

3.3.3 α-*O*-Fuc

Many proteins have been validated as carrying α-*O*-fucosylation, including plasminogen activator proteins,[142] blood coagulation factors,[37,143,144] Cripto,[140] and Notch.[145] In addition, numerous proteins containing EGF repeats and TSRs have the consensus sequence for *O*-fucosylation, although many of these have not been experimentally verified as glycosylation sites *in vivo*.[146] The effects of *O*-fucosylation on different proteins have also not been systematically investigated. However, the facts that POFUT1 and POFUT2 only recognize and modify properly folded substrates[147] and that *O*-fucosylation is essential for the secretion of at least two a disintegrin and metalloprotease with thrombospondin motifs (ADAMTS) superfamily proteases[148,149] indicated that *O*-fucosylation might be important for folding or quality control as well.[150]

Specific roles for *O*-fucosylation have also been identified, although they have not been extensively characterized. For example, defucosylated urokinase-type plasminogen activator (uPA) was shown to be able to bind its receptor as normal, but did not activate mitogenic activity in the bound cells, so it does not appear to stimulate the normal signal upon binding.[151] Recently, the presence of an *O*-Fuc on the EGF repeat of Cripto was identified and was initially thought to be essential for Cripto to mediate Nodal-dependent signaling since POFUT1 knockouts displayed a lethal phenotype related to Nodal signaling.[144,152] Later investigations showed that while POFUT1 does indeed fucosylate Cripto and that the Thr that gets fucosylated is essential, the carbohydrate itself is not. Even substituting the Fuc–Thr for Fuc–Ser resulted in the same lethal phenotype as an alanine mutation.[153] So far, the roles of *O*-fucosylation and the POFUT1 enzyme in the Cripto/Nodal signaling system remain cryptic.

The best characterized and most extensively investigated *O*-fucosylation substrate is Notch. Notch is an intercellular juxtacrine receptor critical for cell-fate decision making, development, and homeostatsis.[154] Notch structure and signaling is well conserved across metazoans, and all Notch receptors are large single-pass type I transmembrane proteins with a long extracellular domain containing between 29 and 36 EGF repeats in tandem.[155] These EGF repeats mediate interactions between Notch and its ligands and many of these repeats are post-translationally glycosylated with a variety of structures including Ser/Thr-linked *O*-Fuc and *O*-Glc glycans.[156] These glycans were first identified on mammalian Notch1 in 2000.[40] *O*-Fuc is added to Notch by a single glycosylfucosyl transferase: protein *O*-fucosyltransferase-1, POFUT1 in mammals and OFUT1 in *Drosophila*.[154] Effects of mutations which destroy the transferase activity of POFUT1 or OFUT1 imply that *O*-fucosylation is necessary for correct Notch signaling.[154] In *Drosophila*, the glycosyltransferase Fringe extends the glycan to a disaccharide: GlcNAcβ1-3Fuc.[38,157] In mammals, there is some evidence that this disaccharide can be further extended to a tetrasaccharide (Siaα2/3-6Galβ1-4GlcNAcβ1-3Fuc) by the galactosyl transferase B4GalT1 [37,39] and a sialic acid transferase.[37] In flies, it is well established that glycan extension by Fringe makes Notch more sensitive to the ligand Delta and simultaneously less sensitive to the competing ligand named Serrate.[158] However, in mammals, there are numerous isoforms of Notch, Fringe and each of the ligands, making the mammalian system significantly more complicated. Despite significant efforts, a clear picture of how glycosylation affects Notch signaling in mammals has not emerged. It is known that EGF repeats 11 and 12 of Notch are responsible for interactions with the ligand Delta.[159] Since EGF repeat 12 is known to be both *O*-fucosylated and *O*-glucosylated, this repeat has been the subject of several studies in this area. Recent work has revealed that Fringe-catalyzed extension of *O*-Fuc on EGF 12 results in an increased binding affinity of human Notch1 towards both Jagged-1 and Delta-like-1 (DLL1) ligands.[160] There is also evidence that the extension of *O*-Fuc affects the interaction of Notch1 with the different isoforms of Fringe found in mammals and the DLL1 ligand.[161] In another study, structural and computational analysis of a Notch1 fragment in complex with a DLL4 fragment showed that extension of the *O*-Fuc on EGF 12 adds a significant amount of contact area between the Fringe-added GlcNAc and both DLL4 and Notch1 surfaces.[162] Further extension of the glycan to a trisaccharide or tetrasaccharide has also been investigated as a potential regulatory event. Elongation of the glycan to a trisaccharide by B4GalT1 is necessary for Lunatic fringe-dependent inhibition of Jagged1 activation of Notch in mammals.[39]

Along with traditional biochemical approaches, the recently-reported chemical synthesis of homogenously glycosylated EGF repeat 12 has been instrumental in shedding light on the specific structural interactions of the glycopeptide domain. Both mouse and human EGF repeat 12 have been chemically synthesized by the Nishimura group in recent years, following a very similar synthetic strategy (Figure 3.9).[163] The mouse-derived sequence was first synthesized in 2010 by the group with a single *O*-Fuc-initiated disaccharide.[163,164] They were also able to enzymatically elongate the glycan after synthesis and folding to the full-length, naturally occurring tetrasaccharide. NMR studies of the synthetic glycoprotein domains revealed that the *O*-Fuc glycan is involved in several key contacts with amino acid residues opposite it on the anti-parallel β-sheet strand. The authors suggest that this is indicative of the glycan's stabilizing role as a sort of bridge across strands of the β-sheet and that it shows the importance of *O*-fucosylation for the structural stability of EGF repeat 12.[163] X-Ray structures of glycosylated human Notch a few years later confirmed many of these contacts.[160] Most recently, the Nishimura group has completed the synthesis of human EGF repeat 12 with both an *O*-fucose-initiated disacchride at Thr466 and an *O*-glucose-initiated Xylα1-3Xylα1-3Glcβ1-Ser trisaccharide at Ser458 (Figure 3.9).[164] They modified the previous folding protocol to include calcium ions in the folding buffer, which are known to be necessary for the final structure of EGF repeat 12.[160] As with the previous work, enzymatic elongation of the *O*-Fuc disaccharide yielded the full-length, sialylated tetrasaccharide at Thr466.[164] NMR analysis of the resulting glycopeptide domains confirmed many of the same contacts identified previously with glycosylated EGF repeat 12 from mouse Notch1. Additionally, they found that there were significant contacts between the terminal xylose of the *O*-Glc-initiated glycan and amino acid side chains in the structurally critical β-sheet of the EGF repeat.[164] By including calcium ions in the folding buffer for this synthesis, the authors were also able to investigate the influence of divalent metal ions on the structure. Most notably, they were able to show for the first time that calcium ions not only helped

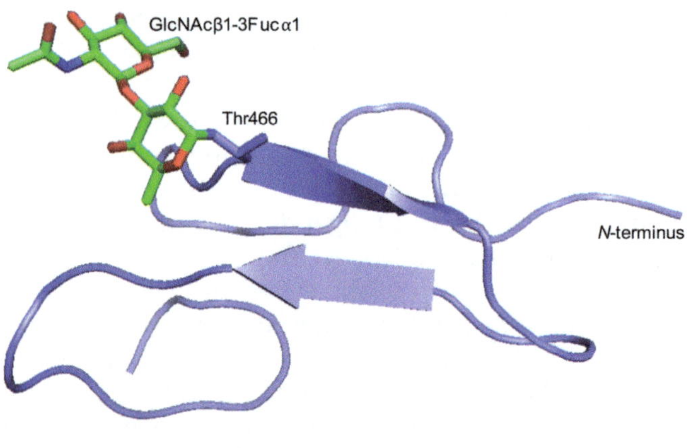

Figure 3.9 Three-dimensional structure of synthetic human Notch 1 EGF repeat 12. The *O*-fucosyl-linked Thr466 is shown as sticks.

to stabilize the final structure, but actually accelerated the folding reaction and decreased the presence of misfolded glycopeptide structures. Finally, by comparing the NMR structures obtained for both mouse and human EGF repeat 12, as well as X-ray structures, the authors were able to identify structural contributions by both types of *O*-glycans as well as the coordinated calcium ion to EGF repeat 12.[160,163,164] This led to the authors' conclusion that all three factors are important for maintaining the correct folding of the EGF domain, and thus most probably all three act as modulators of Notch signaling in concert.

In addition to EGF repeat 12, *O*-fucosylation of EGF repeats 26 and 27 is known to significantly affect the binding of Notch1 to its ligands, even though this region is not implicated in the binding event.[165] Recently it has also become appreciated that the Notch ligands themselves are also *O*-fucosylated by POFUT1[166] and that these modifications are likewise necessary for proper signaling,[167,168] but the molecular details of these effects are not known.

To date, almost all the functional work has been done on EGF domains and the role of fucosylation in the function of TSRs remains largely uncharacterized. TSRs and EGF repeats share many characteristics and both are involved in numerous physiologically-relevant protein–protein interactions.[169] Thus it is possible that fucosylation of TSRs has some functions analogous to those observed for fucosylation of EGF repeats. It is noteworthy that several important interactions between TSRs and ligands are thought to occur in sequences containing proposed *O*-fucosylation sites; but characterizations of how glycosylation affects these interactions have not been carried out.[150] Glucose can be added to Fuc by B3GlcT. The exact role of extension of *O*-Fuc on TSRs by Glc is not understood currently, but it is known that mutations in the glucosyltransferase responsible for the modification (B3GlcT) cause the disease Peter's plus syndrome.[170]

3.3.4 β-*O*-Glc

As with *O*-fucosylation, *O*-glucosylation is found on EGF repeats in several proteins, including blood coagulation factors,[171] but so far, only Notch *O*-Glc modification has been heavily studied.[40] The fact that POGLUT1 only recognizes and modifies properly folded substrates indicates that *O*-glucosylation may play some role in protein quality control.[172] In flies, half of the EGF repeats on Notch contain the consensus sequence for *O*-glucosylation, and these 18 potential sites seem to act redundantly to both promote correct folding of the Notch protein and to enable cleavage after ligand binding.[53] Despite the fact that the ligand binding site of Notch (EGF repeats 11–13) is *O*-glucosylated, so far *O*-Glc has not been shown to directly influence the binding of any ligands tested.[160,173] A recently solved crystal structure of Notch EGF repeats 11–13 in complex with DLL4 showed two *O*-glucose residues covering hydrophobic patches away from the site of ligand binding, which raises the interesting possibility that these glycans function by preventing aggregation of Notch receptors on the cell surface and thus promote proteolytic cleavage and signaling.[162] Further indirect effects on Notch signaling are evident in mammals, where EGF repeat 28 *O*-glucosylation appears to regulate DLL1 but not Jagged1 binding even though it is far from the known binding site for either ligand.[47] It is also worth noting that most Notch ligands in mammals contain several *O*-glucosylation consensus sequences, although validation of these sites as being glycosylated *in vivo* or functional characterization of the consequences of any glycosylation on these proteins has not been performed.[174] Finally, investigation of the extension of protein *O*-glucose by xylose has not yet been carried out in detail in mammalian cells. Studies of flies with mutated, non-functional Shams, however, seem to indicate that such an extension negatively regulates activity of Notch in flies.[52]

3.3.5 β-*O*-GlcNAc

The β-*O*-GlcNAc glycosylation of the extracellular domain of *Drosophila* Notch in EGF repeat 20 was first identified accidently while the authors were looking for the well-known *O*-Fuc and *O*-Glc modifications.[175] Subsequent investigations have confirmed that extracellular GlcNAc exists in humans and mice, as well as flies, on several EGF-repeat-containing membrane proteins in the brain and even on a secreted cytokine, aminoacyl tRNA synthetase complex interacting multifunctional protein 1 (AIMP1).[57] In addition, the glycosyltransferase that catalyzes the glycosylation appears to be ubiquitously expressed in mice.[55]

The *Drosophila* protein Dumpy is perhaps the best characterized example of a protein with this type of glycosylation, although it has no obvious mammalian ortholog.[54] Dumpy, which is thought to be critical to interactions between epithelial cells and cuticle cells in the fly, appears to require EOGT-catalyzed *O*-GlcNAcylation to function properly and mutation of *EOGT* in flies is lethal.[54] Curiously, even though Notch in *Drosophila* is known to be *O*-GlcNAcylated, the glycan does not appear to be important for Notch signaling.[54] As with other EGF-repeat-specific glycosyltransferases, EOGT appears to recognize exclusively fully folded EGF repeats, which indicates possible roles in protein quality control.[55] In humans, mutations in the *EOGT* gene have been found in individuals with Adams–Oliver Syndrome,[176] and the mutant EOGT proteins found in those individuals have been shown to impair the transferase activity of the enzyme.[177] So far no definitive role of β-*O*-GlcNAc has emerged in either *Drosophila* or mammals.

3.4 Chemical Biology in Studying the Composition of Mixtures of *O*-Glycoproteins

Natural glycoproteins are often secreted as a complex mixture of many glycoforms. To better understand protein glycosylation, it is also important to appreciate the biological consequences of forming such mixtures. To achieve this goal, it is essential to know the composition of glycoforms that are secreted by different cells and under different physiological conditions. The consequences can be determined by analyzing the functions of mixtures with known compositions.

Many different methods, including non-mass spectrometric (MS) and MS methods, have been developed for the analysis of protein glycosylation. MS methods are generally more sensitive and rapid, and have thus become more commonly used. The generally accepted workflow for MS methods typically involves four steps, all of which should be optimized for each particular analyte for best results: initial purification, degradation, on-line analytical separation and mass analysis (Figure 3.10).

3.4.1 Glycoprotein Purification and Enrichment

The first step is necessary to separate the target glycoprotein(s) from other molecules in the sample; for example, isolating a single protein and its various glycoforms from a blood-serum sample. This can involve target-specific affinity purifications or less specific separations.[178] Gel electrophoresis, either 1D-SDS-polyacrylamide gel electrophoresis (PAGE) or 2D-PAGE is a very common initial purification step in glycoform analysis.[179] Reverse-phase liquid chromatography is also very common for purification early on in the process.[178]

Intact protein molecules with relatively simple glycosylation, for example, most monoclonal antibodies, can be analyzed directly by electrospray ionization (ESI)–MS, and the ratios of unglycosylated to glycosylated protein can be calculated based on the intensities of deconvoluted MS peaks corresponding to each glycoform.[180] However, it is much more common for glycoproteins to carry a variety of glycans at multiple sites, and the mass spectra usually become too complicated to assign in such circumstances. It is thus common practice in many laboratories to degrade the glycoproteins using proteases and to analyze the glycoforms at the peptide level.[181]

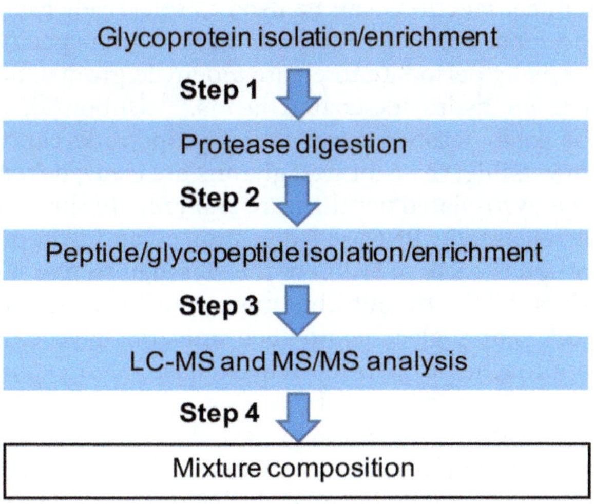

Figure 3.10 Experimental workflow for glycoprotein mixture analysis. Several steps of purification, enrichment, and processing are necessary for accurate characterization, and all of these steps must be individually optimized for each particular analyte studied.

The specific recognition of antibodies can be exploited to enrich glycoproteins. For example, *O*-GlcNAcylated proteins have been enriched and subsequently analyzed from rat brain samples using immunoaffinity chromatography that relied on an *O*-GlcNAc-specific antibody immobilized on a solid support.[182] Since antibody recognition is also reliant on the peptide backbone, particularly for small glycans, using multiple antibodies simultaneously significantly increases the coverage of enrichment.[183]

Lectins are another class of specific-epitope-binding molecules that recognize carbohydrates and there are a large variety of lectins available for use in enriching glycoproteins.[184] Often these lectins are immobilized on a solid support and packed in columns.[185] As with antibodies, combined use of multiple lectins significantly increases the coverage towards the full spectrum of glycan structures.[186] This is particularly relevant in discovery applications where the target glycan structure is perhaps unknown. While highly effective for enriching full-length glycoproteins, lectins tend to recognize clusters of glycans across multiple binding sites; which can make their use in purifying or enriching small, singly glycosylated peptides challenging. To solve this problem, long columns packed with immobilized lectin beads and isocratic elution conditions have been used in what is known as lectin weak affinity chromatography (LWAC) to enrich *O*-GlcNAcylated glycopeptides[187] or *O*-mannosylated peptides.[113] For *O*-GalNAcylated glycopeptides, jacalin, peanut agglutinin (PNA), and *Vicia villosa* lectin (VVA) are widely used, as jacalin and PNA prefer the T-antigen ligand motif and VVA prefers the Tn-antigen ligand motif. For example, the truncated GalNAc-type *O*-glycoproteome of "SimpleCell" cultures was mapped by using LWAC packed with VVA to enrich the glycoforms at the both glycoprotein stage and glycopeptide stage.[188]

Due to the unique concentration of vicinal-diol moieties in carbohydrates, orthogonal chemical reactions can be used to selectively capture glycans as well. One of the most common methods here is non-specific oxidation of *cis*-diols in glycans by periodate to create aldehyde groups, and subsequent covalent capture on hydrazide-coated beads.[189] Unbound, unglycosylated peptides can be easily washed away after glycopeptide capture. Since this procedure is irreversible, the captured glycans are cleaved from the peptides and the formerly-glycosylated peptides are analyzed. In the case of *N*-glycans, peptides can be released by PNGase F digestion,[190] and isotopic labels can be introduced through the use of $H_2{}^{18}O$ in the cleavage step to label the former glycosylation sites,[191,192] although this might not be as selective an isotope label as desired.[193] For *O*-glycans, no such universal glycosidase exists, but several other approaches to peptide release have been explored, including β-elimination and direct hydrazine bond cleavage by mildly acidic hydroxylamine treatment.[194] Milder oxidation by periodate can selectively target sialic acids on both *N*- and *O*-glycoproteins. This is advantageous because sialic acids are often the most vulnerable to acidic hydrolysis, and so the de-sialyated glycopeptides can be easily released after capture and most of the glycan structure as well as site information can be characterized.[195] Obviously, though, all information on the sialic acid content is lost with this approach.

Along similar lines, boronic acids can react selectively with the diols of glycans in the context of biological samples, and thus solid-support-immobilized boronic acids have also been used to enrich glycoproteins.[196] Usefully, this reaction is reversible and so captured glycoproteins can be released intact under mild acidolysis.[197]

3.4.2 Glycoprotein Digestion and Glycopeptide Separation

Once purified and, if necessary enriched, glycoproteins are often degraded by proteases before analysis. Breaking full length glycoproteins into glycopeptides and peptides allows for much more complete analysis. The protease digestion step can be performed in-gel or in-solution before desalting and MS or liquid chromatography (LC)–MS analysis of the resulting glycopeptide fragments.[179] The digestion can also be performed following hydrazide capture, which can be done while the glycoproteins are still bound to the beads for convenience.[195]

Most often, the protease trypsin is used for glycoprotein digestion.[198] In some cases, other enzymes are necessary, for example due to a lack of trypsin cleavage sites in a particular sample, and many other proteases have been used successfully.[185] In recent years, proteinase enzymes immobilized on solid supports and packed into columns have become commercially available.[199] These columns have the advantage of quicker, milder reaction conditions and minimal purification after digestion as compared with the traditional wet chemistry methods. Non-specific proteases have also been used for peptide digestion.[198] In particular, they are effective in combination with more specific proteases (like trypsin) for the analysis of glycoproteins that are difficult to digest and analyze, such as densely *O*-glycosylated mucin domains, which also consist of peptide sequences with few cleavage sites for specific proteases.[200]

Finally, often immediately before mass analysis, it is usually advantageous to separate the protease digestion product glycopeptides from one another. It is very helpful to temporally separate the analytes for clearest analysis and ideally, this would allow a single species into the mass analysis at a time. Many methods have been examined to separate and enrich small glycopeptides. Hydrophilic interaction chromatography (HILIC) is one such method. HILIC combines the highly polar stationary phases of normal-phase chromatography (sepharose, cellulose and silica) and the moderately polar mobile phases of reverse-phase chromatography (acetonitrile and water).[201] In this context, glycopeptides can be retained in the column while the comparatively more hydrophobic peptides are washed out; increasing the polarity of the mobile phase elutes the glycopeptides. Since its development, HILIC has become a standard part of glycoprotein and glycopeptide analysis.[202] Porous graphitized carbon is also commonly used in the analysis and separation of glycans and glycopeptides.[203] It has the advantage of separating not only on the basis of polarity but also of size and shape, so isomorphic glycans or glycopeptides containing isomorphic glycans can be separated from one another.[204] Ion-mobility can also separate molecules on the basis of size and shape, but in the gas-phase.[205] This has been used, for example, to separate two glycopeptides with identical sequence and glycan composition but different sites of glycosylation.[206]

Typically, this type of analytical separation is performed on-line, right before induction into the MS for mass analysis. Reverse-phase liquid chromatography (RPLC) is a very common method here. In RPLC the peptide backbone is the biggest determinant of the retention time, and so all the variously glycosylated isoforms of a given peptide sequence often elute very close to one another.[207] RPLC is commonly combined with other separation methods to achieve the best results, for example in combination with HILIC to isolate and characterize *N*-linked glycoproteins from the rat brain.[208] Multi-dimensional separations, done by combining three types of chromatography: electrostatic repulsion HILIC, jacalin affinity chromatography, and RPLC, resulted in more comprehensive coverage of mucin-type core 1 glycopeptides from bovine serum.[209]

3.4.3 Glycopeptide Analysis

Almost all mass analysis of glycopeptides is done using tandem mass spectroscopy, also known as MS/MS or MS^2.[184] Within MS/MS techniques, several methods exist for fragmenting the precursor ions with various levels of success. Collision-induced dissociation (CID) is one of the earliest and most widely available fragmentation methods for tandem MS. In a CID experiment, high-energy collisions with an inert gas in the fragmentation chamber cause ions to fragment, generating smaller ions that are then analyzed.[184] This almost exclusively results in the cleavage of glycosidic bonds, which makes CID very useful for characterization of glycans.[210] Unfortunately, information on the peptide sequence is almost absent from most CID experiments, and it is often impossible to located the glycosylation site. Since CID reliably generates easily-visible oxonium ions from glycosidic bond cleavage, it can be used as an easy way to identify chromatographic peaks that contain glycopeptides in an unknown sample. Other ionization methods can then be used to analyze the peptide sequence and get site information either in parallel or in subsequent runs. This technique was applied to great effect in a study of haptoglobin, even when the authors employed a novel nano-LC set-up that only used femtomolar quantities of analyte.[211] In certain circumstances, peptide bond cleavage can be achieved, and CID has been successfully used to sequence glycopeptides, identify glycosylation sites and characterize glycans. For example, CID paired with nanospray ESI-MS/MS and time of flight (TOF) detection has been successfully used to characterize sites of *O*-fucosylation.[212] Collision-based fragmentation has also seen use in the context of relatively new linear ion-trap orbitrap (LTQ Orbitrap) instruments, where high energy collisions (HCD) can take place before analysis in the orbitrap portion.[213] Because these collisions are higher in energy than standard CID, peptide backbone fragmentation is observed as well as glycosydic bond cleavage.

HCD has even been combined with traditional CID for a more complete analysis of glycoproteins in the rat brain.[208]

In some respects, electron-based fragmentation methods have an advantage over collision-based methods in the analysis of glycopeptides since electron-based fragmentation tends to favor backbone cleavage, leaving glycans attached to the resulting fragments intact.[185] This can simplify the assignment of glycosylation sites and peptide sequences. Electron capture dissociation (ECD) is a common feature of Fourier transform ion-cyclotron resonance (FT–ICR) mass spectroscopy that involves bombardment of multiply-charged ions with low-energy electrons to generate odd-electron radical cations, which readily dissociate across the N–Cα bonds of the peptide backbone.[214] ECD is a powerful technique that can allow for complete glycoform analysis, including differentiating glycoforms carrying identical glycans on different sites and glycoforms carrying different glycans on identical sites, as shown by a recent analysis of IgA *O*-glycosylation.[215] Electron transfer dissociation (ETD) uses a carrier gas to transfer the electron to the multiply charged precursor ions, and can be used in a much wider variety of MS instruments.[198] Much like ECD, this electron transfer results in the formation of radical cations that fragment preferentially across N–Cα bonds, leading to fragments of the peptide backbone with intact glycans.[185] A complication of both ETD and ECD is that fragmentation efficiency is heavily dependent on highly charged precursor ions. One solution to this problem is the use of "super-charging" reagents that favor formation of highly charged ions, which was found to significantly improve analysis and site-specific glycan assignment of glycopeptides derived from both erythropoietin and a monoclonal antibody, trastuzumab.[216] ETD has also been shown to be highly efficient at identifying glycosylation sites and peptide sequences for densely-glycosylated *O*-glycopeptides, such as those from the mucin family.[217]

3.4.4 Importance of Synthetic Glycopeptides in Protein Glycosylation Analysis

Each of the steps outlined above has been the subject of numerous efforts at optimization and improvement to get to the current state of the art. Throughout this process of advancement, synthetic glycopeptides have served as one of the most useful tools for validation of newer methods, and analysis of their relative strengths and weaknesses.[198] For example, novel high-affinity, pan-specific antibodies were raised against an epitope derived from casein kinase II (CKII) in a study that made use of synthetic glycopeptides in several ways.[183] The authors first used synthetic glycopeptides and native chemical ligation to create a novel vaccine composed of three parts: an *O*-GlcNAcylated glycopeptide taken from CKII, a helper T-cell epitope, and a Toll-like receptor 2 agonist as an adjuvant. This vaccine was then used for the successful production of many highly selective antibodies against the *O*-GlcNAcylated CKII peptide epitope. Synthetic glycopeptides and peptides were then used to validate each of the antibodies and show conclusively that many were selective for only the *O*-GlcNAc-containing sequence.[183] Such clear validation of a method is often easiest with well-characterized synthetic standards.

Synthetic glycopeptides' usefulness is also reflected in the development of negative-mode CID paired with ECD as a powerful method to sequence both glycan and peptide moieties of *O*-glycopeptides containing both fucose and sialic acids.[218] The authors were able to show that negative-mode CID favors a clean break between the peptide and glycan, without shedding of the normally labile fucose or sialic acid residues. Triple-stage mass spectrometry (MS^3) spectra of the cleaved glycan precursor ion and comparison with known fragmentation patterns of commercial standards allowed easy characterization of the complete *O*-glycans. Additionally, the peptide portion of the synthetic model molecule was sequenced through positive-ion ECD MS/MS in parallel. This study shows that using synthetic standards of known structures for validation greatly expedites the process of developing novel analytical methods.[218] This study also highlights the value and convenience that synthetic standards can provide when examining detailed fragmentation patterns in MS^n spectra.

Synthetic glycopeptides were especially convenient for validating ion-mobility methods as a way to separate and analyze isomorphic glycopeptides, such as positional isomers. In one such study, the authors used two synthetic standards which shared common peptide sequences and *O*-glycans, but differed in the glycosylation site.[206] They were able to show that, although the pair of positional isomers co-eluted on RPLC, they had different drift times under the experimental conditions, and could thus be separated by ion-mobility in the gas phase. This experiment would have been much more difficult had the two standards not been so readily available and well characterized.

Quantifying specific glycopeptides in complex mixtures is an important goal for the determination of the composition of mixtures of glycoproteins. Validating methods for quantification is another area where synthetic standards have proven invaluable, such as one study that aimed to quantify *O*-GlcNAc glycopeptides.[219] With the ability to control the exact amount of a previously characterized synthetic standard, the authors were able to carry-out detailed proof-of-concept experiments that would have been impossible otherwise. Thanks to such rigorous validation of their method, the authors convincingly showed that extremely targeted ion measurements using multiple-reaction monitoring in tandem MS spectra is a great way to quantify low-abundance *O*-glycosylated peptides amongst a complex background. Using this newly developed method, they went on to characterize a full-length glycoprotein: glycogen synthase kinase 3 beta (GSK-3β). Thanks to the in-depth studies carried out with model molecules, the authors confidently identified three new *O*-GlcNAcylation sites on the protein and quantified the increase in *O*-glycosylation upon inhibition of a glycosylhydrolase.[219]

Isotopically labelled synthetic peptides are commonly used to quantify proteins in complex biological samples.[220] Although similarly labeled standards for glycopeptides are not as widely available, they have also proven to be extremely helpful in quantification.[221]

3.5 Conclusion

Recent advances in the fields of chemical biology and glycoproteomics have combined to reveal an unprecedented amount of information regarding protein *O*-glycosylation. New technologies for the chemical synthesis of complex glycopeptides have allowed scientists to study many new glycan structures in a systematic and meaningful way. This is leading to a new understanding of the role of *O*-glycans in many biological systems and a renewed appreciation for the complicated mechanisms governing their introduction by glycosyltransferases. Thanks to the hard work of many in the glycoproteomics field, we know more about the natural glycan structures and glycosylation patterns of human proteins than ever before and this knowledge is helping researchers to make significant strides towards universal disease biomarkers and their use in diagnosis and treatment of human disease. Going forward, a concerted effort will be required to conquer the remaining challenges in the field. Two important goals for the future are quantifying individual glycoforms even in the presence of complex biological backgrounds and characterizing the composition of naturally produced glycoform mixtures under a variety of culture conditions and cellular contexts. Achieving this deep level of understanding will open new avenues for both the study of the biology of glycoproteins and their use in improving human health.

References

1. A. Varki, *Essentials of Glycobiology*, Cold Spring Harbor Laboratory Press, Cold Spring Harbor, N.Y., 2009.
2. Y. Ihara, Y. Inai, M. Ikezaki, I.-S. L. Matsui, S. Manabe and Y. Ito, in *Glycoscience: Biology and Medicine*, ed. N. Taniguchi, T. Endo, W. G. Hart, H. P. Seeberger and C.-H. Wong, Springer, Japan, Tokyo, 2015, pp. 1091–1099, DOI: 10.1007/978-4-431-54841-6_67.
3. P. V. d. Steen, P. M. Rudd, R. A. Dwek and G. Opdenakker, *Crit. Rev. Biochem. Mol. Biol.*, 1998, **33**, 151–208.
4. R. G. Spiro, *Glycobiology*, 2002, **12**, 43R–56R.
5. M. J. Paszek, C. C. DuFort, O. Rossier, R. Bainer, J. K. Mouw, K. Godula, J. E. Hudak, J. N. Lakins, A. C. Wijekoon, L. Cassereau, M. G. Rubashkin, M. J. Magbanua, K. S. Thorn, M. W. Davidson, H. S. Rugo, J. W. Park, D. A. Hammer, G. Giannone, C. R. Bertozzi and V. M. Weaver, *Nature*, 2014, **511**, 319–325.
6. J. B. McArthur and X. Chen, *Biochem. Soc. Trans.*, 2016, **44**, 129–142.
7. C. Chen, Y. Zhang, M. Xue, X. W. Liu, Y. Li, X. Chen, P. G. Wang, F. Wang and H. Cao, *Chem. Commun. (Cambridge, U. K.)*, 2015, **51**, 7689–7692.
8. E. P. Bennett, U. Mandel, H. Clausen, T. A. Gerken, T. A. Fritz and L. A. Tabak, *Glycobiology*, 2012, **22**, 736–756.
9. T. A. Gerken, L. Revoredo, J. J. C. Thome, L. A. Tabak, M. B. Vester-Christensen, H. Clausen, G. K. Gahlay, D. L. Jarvis, R. W. Johnson, H. A. Moniz and K. Moremen, *J. Biol. Chem.*, 2013, **288**, 19900–19914.

10. L. Revoredo, S. Wang, E. P. Bennett, H. Clausen, K. W. Moremen, D. L. Jarvis, K. G. Ten Hagen, L. A. Tabak and T. A. Gerken, *Glycobiology*, 2016, **26**, 360–376.
11. H. C. Hang and C. R. Bertozzi, *Bioorg. Med. Chem.*, 2005, **13**, 5021–5034.
12. D. T. Tran, L. Zhang, Y. Zhang, E. Tian, L. A. Earl and K. G. Ten Hagen, *J. Biol. Chem.*, 2012, **287**, 5243–5252.
13. M. R. Pratt, H. C. Hang, K. G. Ten Hagen, J. Rarick, T. A. Gerken, L. A. Tabak and C. R. Bertozzi, *Chem. Biol.*, 2004, **11**, 1009–1016.
14. O. Topaz, D. L. Shurman, R. Bergman, M. Indelman, P. Ratajczak, M. Mizrachi, Z. Khamaysi, D. Behar, D. Petronius, V. Friedman, I. Zelikovic, S. Raimer, A. Metzker, G. Richard and E. Sprecher, *Nat. Genet.*, 2004, **36**, 579–581.
15. K. Kato, C. Jeanneau, M. A. Tarp, A. Benet-Pagès, B. Lorenz-Depiereux, E. P. Bennett, U. Mandel, T. M. Strom and H. Clausen, *J. Biol. Chem.*, 2006, **281**, 18370–18377.
16. K. Akasaka-Manya, H. Manya, A. Nakajima, M. Kawakita and T. Endo, *J. Biol. Chem.*, 2006, **281**, 19339–19345.
17. L. A. P. Jurado, A. Coloma and J. Cruces, *Genomics*, 1999, **58**, 171–180.
18. T. Willer, W. Amselgruber, R. Deutzmann and S. Strahl, *Glycobiology*, 2002, **12**, 771–783.
19. T. Endo, *J. Biochem.*, 2015, **157**, 1–12.
20. A. Yoshida, K. Kobayashi, H. Manya, K. Taniguchi, H. Kano, M. Mizuno, T. Inazu, H. Mitsuhashi, S. Takahashi, M. Takeuchi, R. Herrmann, V. Straub, B. Talim, T. Voit, H. Topaloglu, T. Toda and T. Endo, *Dev. Cell*, 2001, **1**, 717–724.
21. S. Takahashi, T. Sasaki, H. Manya, Y. Chiba, A. Yoshida, M. Mizuno, H.-K. Ishida, F. Ito, T. Inazu, N. Kotani, S. Takasaki, M. Takeuchi and T. Endo, *Glycobiology*, 2001, **11**, 37–45.
22. K.-i. Inamori, T. Endo, Y. Ide, S. Fujii, J. Gu, K. Honke and N. Taniguchi, *J. Biol. Chem.*, 2003, **278**, 43102–43109.
23. K.-i. Inamori, T. Endo, J. Gu, I. Matsuo, Y. Ito, S. Fujii, H. Iwasaki, H. Narimatsu, E. Miyoshi, K. Honke and N. Taniguchi, *J. Biol. Chem.*, 2004, **279**, 2337–2340.
24. T. Yoshida-Moriguchi, L. P. Yu, S. H. Stalnaker, S. Davis, S. Kunz, M. Madson, M. B. A. Oldstone, H. Schachter, L. Wells and K. P. Campbell, *Science*, 2010, **327**, 88–92.
25. T. Yoshida-Moriguchi, T. Willer, M. E. Anderson, D. Venzke, T. Whyte, F. Muntoni, H. Lee, S. F. Nelson, L. Yu and K. P. Campbell, *Science*, 2013, **341**, 896–899.
26. J. L. Praissman and L. Wells, *Biochemistry*, 2014, **53**, 3066–3078.
27. M. M. Goddeeris, B. Wu, D. Venzke, T. Yoshida-Moriguchi, F. Saito, K. Matsumura, S. A. Moore and K. P. Campbell, *Nature*, 2013, **503**, 136–140.
28. T. Yoshida-Moriguchi and K. P. Campbell, *Glycobiology*, 2015, **25**, 702–713.
29. J. L. Praissman, D. H. Live, S. Wang, A. Ramiah, Z. S. Chinoy, G.-J. Boons, K. W. Moremen and L. Wells, *eLife*, 2014, **3**, e03943.

30. T. Willer, K. Inamori, D. Venzke, C. Harvey, G. Morgensen, Y. Hara, D. B. V. de Bernabe, L. P. Yu, K. M. Wright and K. P. Campbell, *elife*, 2014, **3**, e03941.
31. A. Ashikov, F. F. Buettner, B. Tiemann, R. Gerardy-Schahn and H. Bakker, *Glycobiology*, 2013, **23**, 303–309.
32. K.-i. Inamori, Y. Hara, T. Willer, M. E. Anderson, Z. Zhu, T. Yoshida-Moriguchi and K. P. Campbell, *Glycobiology*, 2013, **23**, 295–302.
33. N. Nakagawa, H. Takematsu and S. Oka, *Glycobiology*, 2013, **23**, 1066–1074.
34. Y. Luo, A. Nita-Lazar and R. S. Haltiwanger, *J. Biol. Chem.*, 2006, **281**, 9385–9392.
35. Y. Luo and R. S. Haltiwanger, *J. Biol. Chem.*, 2005, **280**, 11289–11294.
36. Y. Luo, K. Koles, W. Vorndam, R. S. Haltiwanger and V. M. Panin, *J. Biol. Chem.*, 2006, **281**, 9393–9399.
37. R. J. Harris and M. W. Spellman, *Glycobiology*, 1993, **3**, 219–224.
38. D. J. Moloney, V. M. Panin, S. H. Johnston, J. Chen, L. Shao, R. Wilson, Y. Wang, P. Stanley, K. D. Irvine, R. S. Haltiwanger and T. F. Vogt, *Nature*, 2000, **406**, 369–375.
39. J. Chen, D. J. Moloney and P. Stanley, *Proc. Natl. Acad. Sci. U. S. A.*, 2001, **98**, 13716–13721.
40. D. J. Moloney, L. H. Shair, F. M. Lu, J. Xia, R. Locke, K. L. Matta and R. S. Haltiwanger, *J. Biol. Chem.*, 2000, **275**, 9604–9611.
41. A. Gonzalez de Peredo, D. Klein, B. Macek, D. Hess, J. Peter-Katalinic and J. Hofsteenge, *Mol. Cell. Proteomics*, 2002, **1**, 11–18.
42. K. Kozma, J. J. Keusch, B. Hegemann, K. B. Luther, D. Klein, D. Hess, R. S. Haltiwanger and J. Hofsteenge, *J. Biol. Chem.*, 2006, **281**, 36742–36751.
43. T. Sato, M. Sato, K. Kiyohara, M. Sogabe, T. Shikanai, N. Kikuchi, A. Togayachi, H. Ishida, H. Ito, A. Kameyama, M. Gotoh and H. Narimatsu, *Glycobiology*, 2006, **16**, 1194–1206.
44. M. Acar, H. Jafar-Nejad, H. Takeuchi, A. Rajan, D. Ibrani, N. A. Rana, H. Pan, R. S. Haltiwanger and H. J. Bellen, *Cell*, 2008, **132**, 247–258.
45. R. Fernandez-Valdivia, H. Takeuchi, A. Samarghandi, M. Lopez, J. Leonardi, R. S. Haltiwanger and H. Jafar-Nejad, *Development*, 2011, **138**, 1925–1934.
46. H. Takeuchi, R. C. Fernández-Valdivia, D. S. Caswell, A. Nita-Lazar, N. A. Rana, T. P. Garner, T. K. Weldeghiorghis, M. A. Macnaughtan, H. Jafar-Nejad and R. S. Haltiwanger, *Proc. Natl. Acad. Sci. U. S. A.*, 2011, **108**, 16600–16605.
47. N. A. Rana, A. Nita-Lazar, H. Takeuchi, S. Kakuda, K. B. Luther and R. S. Haltiwanger, *J. Biol. Chem.*, 2011, **286**, 31623–31637.
48. H. Nishimura, S. Kawabata, W. Kisiel, S. Hase, T. Ikenaka, T. Takao, Y. Shimonishi and S. Iwanaga, *J. Biol. Chem.*, 1989, **264**, 20320–20325.
49. S. Hase, H. Nishimura, S. Kawabata, S. Iwanaga and T. Ikenaka, *J. Biol. Chem.*, 1990, **265**, 1858–1861.
50. M. K. Sethi, F. F. R. Buettner, V. B. Krylov, H. Takeuchi, N. E. Nifantiev, R. S. Haltiwanger, R. Gerardy-Schahn and H. Bakker, *J. Biol. Chem.*, 2010, **285**, 1582–1586.

51. M. K. Sethi, F. F. R. Buettner, A. Ashikov, V. B. Krylov, H. Takeuchi, N. E. Nifantiev, R. S. Haltiwanger, R. Gerardy-Schahn and H. Bakker, *J. Biol. Chem.*, 2012, **287**, 2739–2748.
52. T. V. Lee, M. K. Sethi, J. Leonardi, N. A. Rana, F. F. R. Buettner, R. S. Haltiwanger, H. Bakker and H. Jafar-Nejad, *PLoS Genet.*, 2013, **9**, e1003547.
53. A. R. Haltom and H. Jafar-Nejad, *Glycobiology*, 2015, **25**, 1027–1042.
54. Y. Sakaidani, T. Nomura, A. Matsuura, M. Ito, E. Suzuki, K. Murakami, D. Nadano, T. Matsuda, K. Furukawa and T. Okajima, *Nat. Commun.*, 2011, **2**, 9.
55. Y. Sakaidani, N. Ichiyanagi, C. Saito, T. Nomura, M. Ito, Y. Nishio, D. Nadano, T. Matsuda, K. Furukawa and T. Okajima, *Biochem. Biophys. Res. Commun.*, 2012, **419**, 14–19.
56. R. Muller, A. Jenny and P. Stanley, *PLoS One*, 2013, **8**, 17.
57. J. F. Alfaro, C. X. Gong, M. E. Monroe, J. T. Aldrich, T. R. W. Clauss, S. O. Purvine, Z. H. Wang, D. G. Camp, J. Shabanowitz, P. Stanley, G. W. Hart, D. F. Hunt, F. Yang and R. D. Smith, *Proc. Natl. Acad. Sci. U. S. A.*, 2012, **109**, 7280–7285.
58. K. N. Chuh, A. R. Batt and M. R. Pratt, *Cell Chem. Biol.*, 2016, **23**, 86–107.
59. D. J. Gill, H. Clausen and F. Bard, *Trends Cell Biol.*, 2011, **21**, 149–158.
60. K. W. Moremen, M. Tiemeyer and A. V. Nairn, *Nat. Rev. Mol. Cell Biol.*, 2012, **13**, 448–462.
61. A. P. Corfield, *Biochim. Biophys. Acta*, 2015, **1850**, 236–252.
62. J. Perez-Vilar and R. L. Hill, *J. Biol. Chem.*, 1999, **274**, 31751–31754.
63. J.-L. Desseyn, D. Tetaert and V. Gouyer, *Gene*, 2008, **410**, 215–222.
64. A. P. Corfield and M. Berry, *Trends Biochem. Sci.*, 2015, **40**, 351–359.
65. K. Kozarsky, D. Kingsley and M. Krieger, *Proc. Natl. Acad. Sci. U. S. A.*, 1988, **85**, 4335–4339.
66. E. B. Maryon, S. A. Molloy and J. H. Kaplan, *J. Biol. Chem.*, 2007, **282**, 20376–20387.
67. R. Shogren, T. A. Gerken and N. Jentoft, *Biochemistry*, 1989, **28**, 5525–5536.
68. J. J. Barchi, *Biopolymers*, 2013, **99**, 713–723.
69. Z. Xu and A. Weiss, *Nat. Immunol.*, 2002, **3**, 764–771.
70. K. W. Wagner, E. A. Punnoose, T. Januario, D. A. Lawrence, R. M. Pitti, K. Lancaster, D. Lee, M. von Goetz, S. F. Yee, K. Totpal, L. Huw, V. Katta, G. Cavet, S. G. Hymowitz, L. Amler and A. Ashkenazi, *Nat. Med.*, 2007, **13**, 1070–1077.
71. T. Lang, G. C. Hansson and T. Samuelsson, *Proc. Natl. Acad. Sci. U. S. A.*, 2007, **104**, 16209–16214.
72. H. J. Gabius, *The Sugar Code: Fundamentals of Glycosciences*, Wiley, 2009.
73. M. Amado, Q. Yan, E. M. Comelli, B. E. Collins and J. C. Paulson, *J. Biol. Chem.*, 2004, **279**, 36689–36697.
74. D. T. Tran and K. G. Ten Hagen, *J. Biol. Chem.*, 2013, **288**, 6921–6929.
75. M. Kilcoyne, J. Q. Gerlach, R. Gough, M. E. Gallagher, M. Kane, S. D. Carrington and L. Joshi, *Anal. Chem.*, 2012, **84**, 3330–3338.

76. C. Robbe, C. Capon, E. Maes, M. Rousset, A. Zweibaum, J.-P. Zanetta and J.-C. Michalski, *J. Biol. Chem.*, 2003, **278**, 46337–46348.
77. C. Robbe, C. Capon, B. Coddeville and J.-C. Michalski, *Biochem. J.*, 2004, **384**, 307–316.
78. L. A. Earl, S. G. Bi and L. G. Baum, *J. Biol. Chem.*, 2010, **285**, 2242–2254.
79. M. C. Clark, L. G. Baum and N. Y. A. S. Annals, *Glycobiology of the Immune Response*, 2012, vol. 1253, pp. 58–67.
80. M. T. Boskovski, S. Yuan, N. B. Pedersen, C. K. Goth, S. Makova, H. Clausen, M. Brueckner and M. K. Khokha, *Nature*, 2013, **504**, 456–459.
81. J. C. Brazil, R. P. Liu, R. Sumagin, K. N. Kolegraff, A. Nusrat, R. D. Cummings, C. A. Parkos and N. A. Louis, *J. Immunol.*, 2013, **191**, 4804–4817.
82. Y. Mori, K. Akita, M. Yashiro, T. Sawada, K. Hirakawa, T. Murata and H. Nakada, *J. Biol. Chem.*, 2015, **290**, 26125–26140.
83. K. T. B. G. Schjoldager and H. Clausen, *Biochim. Biophys. Acta, Gen. Subj.*, 2012, **1820**, 2079–2094.
84. C. K. Goth, A. Halim, S. A. Khetarpal, D. J. Rader, H. Clausen and K. T.-B. G. Schjoldager, *Proc. Natl. Acad. Sci. U. S. A.*, 2015, **112**, 14623–14628.
85. G. Opdenakker, P. M. Rudd, M. Wormald, R. A. Dwek and J. Van Damme, *FASEB J.*, 1995, **9**, 453–457.
86. C. Dong, A. Chua, S. Ganguly, A. M. Krensky and C. Clayberger, *J. Immunol. Methods*, 2005, **302**, 136–144.
87. D. H. Live, L. J. Williams, S. D. Kuduk, J. B. Schwarz, P. W. Glunz, X.-T. Chen, D. Sames, R. A. Kumar and S. J. Danishefsky, *Proc. Natl. Acad. Sci. U. S. A.*, 1999, **96**, 3489–3493.
88. W. L. Bigbee, R. G. Langlois, M. Vanderlaan and R. H. Jensen, *J. Immunol.*, 1984, **133**, 3149–3155.
89. K. Dill, R. D. Carter, J. M. Lacombe and A. A. Pavia, *Carbohydr. Res.*, 1986, **152**, 217–228.
90. K. Dill, S. H. Hu, E. Berman, A. A. Pavia and J. M. Lacombe, *J. Protein Chem.*, 1990, **9**, 129–136.
91. O. Schuster, G. Klich, V. Sinnwell, H. Kränz, H. Paulsen and B. Meyer, *J. Biomol. NMR*, 1999, **14**, 33–45.
92. G. A. Naganagowda, T. E. L. Gururaja, J. Satyanarayana and M. J. Levine, *J. Pept. Res.*, 1999, **54**, 290–310.
93. F. Corzana, J. H. Busto, G. Jiménez-Osés, M. García de Luis, J. L. Asensio, J. Jiménez-Barbero, J. M. Peregrina and A. Avenoza, *J. Am. Chem. Soc.*, 2007, **129**, 9458–9467.
94. D. Madariaga, N. Martinez-Saez, V. J. Somovilla, L. Garcia-Garcia, M. A. Berbis, J. Valero-Gonzalez, S. Martin-Santamaria, R. Hurtado-Guerrero, J. L. Asensio, J. Jimenez-Barbero, A. Avenoza, J. H. Busto, F. Corzana and J. M. Peregrina, *Chem.–Eur. J.*, 2014, **20**, 12616–12627.
95. D. M. Coltart, A. K. Royyuru, L. J. Williams, P. W. Glunz, D. Sames, S. D. Kuduk, J. B. Schwarz, X.-T. Chen, S. J. Danishefsky and D. H. Live, *J. Am. Chem. Soc.*, 2002, **124**, 9833–9844.
96. V. Apostolopoulos and I. F. McKenzie, *Crit. Rev. Immunol.*, 1994, **14**, 293–309.

97. L. Kirnarsky, O. Prakash, S. M. Vogen, M. Nomoto, M. A. Hollingsworth and S. Sherman, *Biochemistry*, 2000, **39**, 12076–12082.
98. P. Dokurno, P. A. Bates, H. A. Band, L. M. D. Stewart, J. M. Lally, J. M. Burchell, J. Taylor-Papadimitriou, D. Snary, M. J. E. Sternberg and P. S. Freemont, *J. Mol. Biol.*, 1998, **284**, 713–728.
99. J. Schuman, A. P. Campbell, R. R. Koganty and B. M. Longenecker, *J. Pept. Res.*, 2003, **61**, 91–108.
100. S. Dziadek, C. Griesinger, H. Kunz and U. M. Reinscheid, *Chem.–Eur. J.*, 2006, **12**, 4981–4993.
101. T. Matsushita, N. Ohyabu, N. Fujitani, K. Naruchi, H. Shimizu, H. Hinou and S. I. Nishimura, *Biochemistry*, 2013, **52**, 402–414.
102. S. Nath and P. Mukherjee, *Trends Mol. Med.*, 2014, **20**, 332–342.
103. R. D. Wright and D. Cooper, *Glycobiology*, 2014, **24**, 1242–1251.
104. A. Leppänen, P. Mehta, Y.-B. Ouyang, T. Ju, J. Helin, K. L. Moore, I. van Die, W. M. Canfield, R. P. McEver and R. D. Cummings, *J. Biol. Chem.*, 1999, **274**, 24838–24848.
105. A. Leppänen, S. P. White, J. Helin, R. P. McEver and R. D. Cummings, *J. Biol. Chem.*, 2000, **275**, 39569–39578.
106. J. Finne, T. Krusius, R. K. Margolis and R. U. Margolis, *J. Biol. Chem.*, 1979, **254**, 10295–10300.
107. W. Chai, C.-T. Yuen, H. Kogelberg, R. A. Carruthers, R. U. Margolis, T. Feizi and A. M. Lawson, *Eur. J. Biochem.*, 1999, **263**, 879–888.
108. M. Lommel and S. Strahl, *Glycobiology*, 2009, **19**, 816–828.
109. A. Chiba, K. Matsumura, H. Yamada, T. Inazu, T. Shimizu, S. Kusunoki, I. Kanazawa, A. Kobata and T. Endo, *J. Biol. Chem.*, 1997, **272**, 2156–2162.
110. C.-T. Yuen, W. Chai, R. W. Loveless, A. M. Lawson, R. U. Margolis and T. Feizi, *J. Biol. Chem.*, 1997, **272**, 8924–8931.
111. T. Sasaki, H. Yamada, K. Matsumura, T. Shimizu, A. Kobata and T. Endo, *Biochim. Biophys. Acta, Gen. Subj.*, 1998, **1425**, 599–606.
112. N. R. Smalheiser, S. M. Haslam, M. Sutton-Smith, H. R. Morris and A. Dell, *J. Biol. Chem.*, 1998, **273**, 23698–23703.
113. M. B. Vester-Christensen, A. Halim, H. J. Joshi, C. Steentoft, E. P. Bennett, S. B. Levery, S. Y. Vakhrushev and H. Clausen, *Proc. Natl. Acad. Sci. U. S. A.*, 2013, **110**, 21018–21023.
114. M. Lommel, P. R. Winterhalter, T. Willer, M. Dahlhoff, M. R. Schneider, M. F. Bartels, I. Renner-Müller, T. Ruppert, E. Wolf and S. Strahl, *Proc. Natl. Acad. Sci. U. S. A.*, 2013, **110**, 21024–21029.
115. T. Martinez, D. Pace, L. Brady, M. Gerhart and A. Balland, *J. Chromatogr. A*, 2007, **1156**, 183–187.
116. C. Bleckmann, H. Geyer, A. Lieberoth, F. Splittstoesser, Y. Liu, T. Feizi, M. Schachner, R. Kleene, V. Reinhold and R. Geyer, *Biol. Chem.*, 2009, **390**, 627–645.
117. S. Pacharra, F.-G. Hanisch and I. Breloy, *J. Proteome Res.*, 2012, **11**, 3955–3964.
118. K. L. Abbott, R. T. Matthews and M. Pierce, *J. Biol. Chem.*, 2008, **283**, 33026–33035.

119. S. Pacharra, F.-G. Hanisch, M. Mühlenhoff, A. Faissner, U. Rauch and I. Breloy, *J. Proteome Res.*, 2013, **12**, 1764–1771.
120. K. Kanekiyo, K.-i. Inamori, S. Kitazume, K. Sato, J. Maeda, M. Higuchi, Y. Kizuka, H. Korekane, I. Matsuo, K. Honke and N. Taniguchi, *J. Neurosci.*, 2013, **33**, 10037–10047.
121. S. Yaji, H. Manya, N. Nakagawa, H. Takematsu, T. Endo, R. Kannagi, T. Yoshihara, M. Asano and S. Oka, *Glycobiology*, 2015, **25**, 376–385.
122. L. Wells, *J. Biol. Chem.*, 2013, **288**, 6930–6935.
123. A. G. Toledo, M. Raducu, J. Cruces, J. Nilsson, A. Halim, G. Larson, U. Ruetschi and A. Grahn, *Glycobiology*, 2012, **22**, 1413–1423.
124. Y. Hara, M. Kanagawa, S. Kunz, T. Yoshida-Moriguchi, J. S. Satz, Y. M. Kobayashi, Z. Zhu, S. J. Burden, M. B. A. Oldstone and K. P. Campbell, *Proc. Natl. Acad. Sci. U. S. A.*, 2011, **108**, 17426–17431.
125. H. Manya, T. Suzuki, K. Akasaka-Manya, H.-K. Ishida, M. Mizuno, Y. Suzuki, T. Inazu, N. Dohmae and T. Endo, *J. Biol. Chem.*, 2007, **282**, 20200–20206.
126. S. H. Stalnaker, S. Hashmi, J.-M. Lim, K. Aoki, M. Porterfield, G. Gutierrez-Sanchez, J. Wheeler, J. M. Ervasti, C. Bergmann, M. Tiemeyer and L. Wells, *J. Biol. Chem.*, 2010, **285**, 24882–24891.
127. J. Nilsson, J. Nilsson, G. Larson and A. Grahn, *Glycobiology*, 2010, **20**, 1160–1169.
128. R. Harrison, P. G. Hitchen, M. Panico, H. R. Morris, D. Mekhaiel, R. J. Pleass, A. Dell, J. E. Hewitt and S. M. Haslam, *Glycobiology*, 2012, **22**, 662–675.
129. I. Breloy, T. Schwientek, B. Gries, H. Razawi, M. Macht, C. Albers and F.-G. Hanisch, *J. Biol. Chem.*, 2008, **283**, 18832–18840.
130. S. H. Stalnaker, K. Aoki, J.-M. Lim, M. Porterfield, M. Liu, J. S. Satz, S. Buskirk, Y. Xiong, P. Zhang, K. P. Campbell, H. Hu, D. Live, M. Tiemeyer and L. Wells, *J. Biol. Chem.*, 2011, **286**, 21180–21190.
131. D. E. Michele, R. Barresi, M. Kanagawa, F. Saito, R. D. Cohn, J. S. Satz, J. Dollar, I. Nishino, R. I. Kelley, H. Somer, V. Straub, K. D. Mathews, S. A. Moore and K. P. Campbell, *Nature*, 2002, **418**, 417–421.
132. J. Voglmeir, S. Kaloo, N. Laurent, M. M. Meloni, L. Bohlmann, I. B. H. Wilson and S. L. Flitsch, *Biochem. J.*, 2011, **436**, 447–455.
133. K. Akasaka-Manya, H. Manya, M. Mizuno, T. Inazu and T. Endo, *Biochem. Biophys. Res. Commun.*, 2011, **410**, 632–636.
134. K. Inamori, T. Yoshida-Moriguchi, Y. Hara, M. E. Anderson, L. P. Yu and K. P. Campbell, *Science*, 2012, **335**, 93–96.
135. K. F. Mo, T. Fang, S. H. Stalnaker, P. S. Kirby, M. Liu, L. Wells, M. Pierce, D. H. Live and G. J. Boons, *J. Am. Chem. Soc.*, 2011, **133**, 14418–14430.
136. D. T. Tran, J. M. Lim, M. Liu, S. H. Stalnaker, L. Wells, K. G. Ten Hagen and D. Live, *J. Biol. Chem.*, 2012, **287**, 20967–20974.
137. M. A. Liu, A. Borgert, G. Barany and D. Live, *Biopolymers*, 2008, **90**, 358–368.
138. A. W. Barb, A. J. Borgert, M. Liu, G. Barany and D. Live, *Methods Enzymol.*, 2010, **478**, 365–388.

139. C. B. Taylor, M. F. Talib, C. McCabe, L. Bu, W. S. Adney, M. E. Himmel, M. F. Crowley and G. T. Beckham, *J. Biol. Chem.*, 2012, **287**, 3147–3155.
140. L. Chen, M. R. Drake, M. G. Resch, E. R. Greene, M. E. Himmel, P. K. Chaffey, G. T. Beckham and Z. Tan, *Proc. Natl. Acad. Sci. U. S. A.*, 2014, **111**, 7612–7617.
141. X. Guan, P. K. Chaffey, C. Zeng, E. R. Greene, L. Chen, M. R. Drake, C. Chen, A. Groobman, M. G. Resch, M. E. Himmel, G. T. Beckham and Z. Tan, *Chem. Sci.*, 2015, **6**, 7185–7189.
142. E. J. Kentzer, A. Buko, G. Menon and V. K. Sarin, *Biochem. Biophys. Res. Commun.*, 1990, **171**, 401–406.
143. H. Nishimura, T. Takao, S. Hase, Y. Shimonishi and S. Iwanaga, *J. Biol. Chem.*, 1992, **267**, 17520–17525.
144. S. G. Schiffer, S. Foley, A. Kaffashan, X. Hronowski, A. E. Zichittella, C.-Y. Yeo, K. Miatkowski, H. B. Adkins, B. Damon, M. Whitman, D. Salomon, M. Sanicola and K. P. Williams, *J. Biol. Chem.*, 2001, **276**, 37769–37778.
145. T. Okajima and K. D. Irvine, *Cell*, 2002, **111**, 893–904.
146. R. Rampal, K. B. Luther and R. S. Haltiwanger, *Curr. Mol. Med.*, 2007, **7**, 427–445.
147. C. I. Chen, J. J. Keusch, D. Klein, D. Hess, J. Hofsteenge and H. Gut, *EMBO J.*, 2012, **31**, 3183–3197.
148. L. W. Wang, M. Dlugosz, R. P. T. Somerville, M. Raed, R. S. Haltiwanger and S. S. Apte, *J. Biol. Chem.*, 2007, **282**, 17024–17031.
149. L. M. Ricketts, M. Dlugosz, K. B. Luther, R. S. Haltiwanger and E. M. Majerus, *J. Biol. Chem.*, 2007, **282**, 17014–17023.
150. K. B. Luther and R. S. Haltiwanger, *Int. J. Biochem. Cell Biol.*, 2009, **41**, 1011–1024.
151. S. A. Rabbani, A. P. Mazar, S. M. Bernier, M. Haq, I. Bolivar, J. Henkin and D. Goltzman, *J. Biol. Chem.*, 1992, **267**, 14151–14156.
152. Y.-T. Yan, J.-J. Liu, Y. Luo, C. E, R. S. Haltiwanger, C. Abate-Shen and M. M. Shen, *Mol. Cell. Biol.*, 2002, **22**, 4439–4449.
153. S. Shi, C. Ge, Y. Luo, X. Hou, R. S. Haltiwanger and P. Stanley, *J. Biol. Chem.*, 2007, **282**, 20133–20141.
154. H. Takeuchi and R. S. Haltiwanger, *Biochem. Biophys. Res. Commun.*, 2014, **453**, 235–242.
155. R. Kopan and M. X. G. Ilagan, *Cell*, 2009, **137**, 216–233.
156. P. Stanley and T. Okajima, in *Notch Signaling*, ed. R. Kopan, Elsevier Academic Press Inc, San Diego, 2010, vol. 92, pp. 131–164.
157. K. Bruckner, L. Perez, H. Clausen and S. Cohen, *Nature*, 2000, **406**, 411–415.
158. A. Xu, N. Haines, M. Dlugosz, N. A. Rana, H. Takeuchi, R. S. Haltiwanger and K. D. Irvine, *J. Biol. Chem.*, 2007, **282**, 35153–35162.
159. I. Rebay, R. J. Fleming, R. G. Fehon, L. Cherbas, P. Cherbas and S. Artavanis-Tsakonas, *Cell*, 1991, **67**, 687–699.
160. P. Taylor, H. Takeuchi, D. Sheppard, C. Chillakuri, S. M. Lea, R. S. Haltiwanger and P. A. Handford, *Proc. Natl. Acad. Sci. U. S. A.*, 2014, **111**, 7290–7295.

161. X. Hou, Y. Tashima and P. Stanley, *J. Biol. Chem.*, 2012, **287**, 474–483.
162. V. C. Luca, K. M. Jude, N. W. Pierce, M. V. Nachury, S. Fischer and K. C. Garcia, *Science*, 2015, **347**, 847–853.
163. K. Hiruma-Shimizu, K. Hosoguchi, Y. Liu, N. Fujitani, T. Ohta, H. Hinou, T. Matsushita, H. Shimizu, T. Feizi and S.-I. Nishimura, *J. Am. Chem. Soc.*, 2010, **132**, 14857–14865.
164. S. Hayakawa, R. Koide, H. Hinou and S.-I. Nishimura, *Biochemistry*, 2016, **55**, 776–787.
165. R. Rampal, J. F. Arboleda-Velasquez, A. Nita-Lazar, K. S. Kosik and R. S. Haltiwanger, *J. Biol. Chem.*, 2005, **280**, 32133–32140.
166. V. M. Panin, L. Shao, L. Lei, D. J. Moloney, K. D. Irvine and R. S. Haltiwanger, *J. Biol. Chem.*, 2002, **277**, 29945–29952.
167. J. Muller, N. A. Rana, K. Serth, S. Kakuda, R. S. Haltiwanger and A. Gossler, *PLoS One*, 2014, **9**, 9.
168. K. Serth, K. Schuster-Gossler, E. Kremmer, B. Hansen, B. Marohn-Kohn and A. Gossler, *PLoS One*, 2015, **10**, 21.
169. J. C. Adams and R. P. Tucker, *Dev. Dyn.*, 2000, **218**, 280–299.
170. S. A. J. Lesnik Oberstein, M. Kriek, S. J. White, M. E. Kalf, K. Szuhai, J. T. den Dunnen, M. H. Breuning and R. C. M. Hennekam, *Am. J. Hum. Genet.*, 2006, **79**, 562–566.
171. S. Hase, S. Kawabata, H. Nishimura, H. Takeya, T. Sueyoshi, T. Miyata, S. Iwanaga, T. Takao, Y. Shimonishi and T. Ikenaka, *J. Biochem.*, 1988, **104**, 867–868.
172. H. Takeuchi, J. Kantharia, M. K. Sethi, H. Bakker and R. S. Haltiwanger, *J. Biol. Chem.*, 2012, **287**, 33934–33944.
173. J. Leonardi, R. Fernandez-Valdivia, Y.-D. Li, A. A. Simcox and H. Jafar-Nejad, *Development*, 2011, **138**, 3569–3578.
174. H. Jafar-Nejad, J. Leonardi and R. Fernandez-Valdivia, *Glycobiology*, 2010, **20**, 931–949.
175. A. Matsuura, M. Ito, Y. Sakaidani, T. Kondo, K. Murakami, K. Furukawa, D. Nadano, T. Matsuda and T. Okajima, *J. Biol. Chem.*, 2008, **283**, 35486–35495.
176. R. Shaheen, M. Aglan, K. Keppler-Noreuil, E. Faqeih, S. Ansari, K. Horton, A. Ashour, M. S. Zaki, F. Al-Zahrani, A. M. Cueto-González, G. Abdel-Salam, S. Temtamy and F. S. Alkuraya, *Am. J. Hum. Genet.*, 2013, **92**, 598–604.
177. M. Ogawa, S. Sawaguchi, T. Kawai, D. Nadano, T. Matsuda, H. Yagi, K. Kato, K. Furukawa and T. Okajima, *J. Biol. Chem.*, 2015, **290**, 2137–2149.
178. P. Pompach, Z. Brnakova, M. Sanda, J. Wu, N. Edwards and R. Goldman, *Mol. Cell. Proteomics*, 2013, **12**, 1281–1293.
179. D. Kolarich, P. H. Jensen, F. Altmann and N. H. Packer, *Nat. Protoc.*, 2012, **7**, 1285–1298.
180. J. P. Skinner, L. Chi, P. F. Ozeata, C. S. Ramsay, R. L. O'Hara, B. B. Calfin and S. Y. Tetin, *Anal. Chem.*, 2012, **84**, 1172–1177.
181. S. Ongay, A. Boichenko, N. Govorukhina and R. Bischoff, *J. Sep. Sci.*, 2012, **35**, 2341–2372.

182. L. Wells, K. Vosseller, R. N. Cole, J. M. Cronshaw, M. J. Matunis and G. W. Hart, *Mol. Cell. Proteomics*, 2002, **1**, 791–804.
183. C. F. Teo, S. Ingale, M. A. Wolfert, G. A. Elsayed, L. G. Nöt, J. C. Chatham, L. Wells and G.-J. Boons, *Nat. Chem. Biol.*, 2010, **6**, 338–343.
184. W. R. Alley, B. F. Mann and M. V. Novotny, *Chem. Rev.*, 2013, **113**, 2668–2732.
185. Z. Zhu and H. Desaire, *Annu. Rev. Anal. Chem.*, 2015, **8**, 463–483.
186. E. Ruiz-May, S. Hucko, K. J. Howe, S. Zhang, R. W. Sherwood, T. W. Thannhauser and J. K. C. Rose, *Mol. Cell. Proteomics*, 2014, **13**, 566–579.
187. K. Vosseller, J. C. Trinidad, R. J. Chalkley, C. G. Specht, A. Thalhammer, A. J. Lynn, J. O. Snedecor, S. Guan, K. F. Medzihradszky, D. A. Maltby, R. Schoepfer and A. L. Burlingame, *Mol. Cell. Proteomics*, 2006, **5**, 923–934.
188. S. Y. Vakhrushev, C. Steentoft, M. B. Vester-Christensen, E. P. Bennett, H. Clausen and S. B. Levery, *Mol. Cell. Proteomics*, 2013, **12**, 932–944.
189. H. Zhang, X.-j. Li, D. B. Martin and R. Aebersold, *Nat. Biotechnol.*, 2003, **21**, 660–666.
190. R. Chen, X. Jiang, D. Sun, G. Han, F. Wang, M. Ye, L. Wang and H. Zou, *J. Proteome Res.*, 2009, **8**, 651–661.
191. J. Gonzalez, T. Takao, H. Hori, V. Besada, R. Rodriguez, G. Padron and Y. Shimonishi, *Anal. Chem.*, 1992, **205**, 151–158.
192. D. F. Zielinska, F. Gnad, J. R. Wiśniewski and M. Mann, *Cell*, 2010, **141**, 897–907.
193. G. Palmisano, M. N. Melo-Braga, K. Engholm-Keller, B. L. Parker and M. R. Larsen, *J. Proteome Res.*, 2012, **11**, 1949–1957.
194. E. Klement, Z. Lipinszki, Z. Kupihár, A. Udvardy and K. F. Medzihradszky, *J. Proteome Res.*, 2010, **9**, 2200–2206.
195. J. Nilsson, U. Ruetschi, A. Halim, C. Hesse, E. Carlsohn, G. Brinkmalm and G. Larson, *Nat. Methods*, 2009, **6**, 809–811.
196. Y. Wang, M. Liu, L. Xie, C. Fang, H. Xiong and H. Lu, *Anal. Chem.*, 2014, **86**, 2057–2064.
197. M. Chen, Y. Lu, Q. Ma, L. Guo and Y.-Q. Feng, *Analyst*, 2009, **134**, 2158–2164.
198. G. Zauner, R. P. Kozak, R. A. Gardner, D. L. Fernandes, A. M. Deelder and M. Wuhrer, *Biol. Chem.*, 2012, **393**, 687–708.
199. J. Ma, L. Zhang, Z. Liang, Y. Shan and Y. Zhang, *TrAC, Trends Anal. Chem.*, 2011, **30**, 691–702.
200. M. N. Christiansen, D. Kolarich, H. Nevalainen, N. H. Packer and P. H. Jensen, *Anal. Chem.*, 2010, **82**, 3500–3509.
201. B. Buszewski and S. Noga, *Anal. Bioanal. Chem.*, 2011, **402**, 231–247.
202. G. Zauner, A. M. Deelder and M. Wuhrer, *Electrophoresis*, 2011, **32**, 3456–3466.
203. C. West, C. Elfakir and M. Lafosse, *J. Chromatogr. A*, 2010, **1217**, 3201–3216.
204. S. Hua, C. C. Nwosu, J. S. Strum, R. R. Seipert, H. J. An, A. M. Zivkovic, J. B. German and C. B. Lebrilla, *Anal. Bioanal. Chem.*, 2011, **403**, 1291–1302.

205. R. Guevremont, *J. Chromatogr. A*, 2004, **1058**, 3–19.
206. A. J. Creese and H. J. Cooper, *Anal. Chem.*, 2012, **84**, 2597–2601.
207. H. Desaire, *Mol. Cell. Proteomics*, 2013, **12**, 893–901.
208. B. L. Parker, M. Thaysen-Andersen, N. Solis, N. E. Scott, M. R. Larsen, M. E. Graham, N. H. Packer and S. J. Cordwell, *J. Proteome Res.*, 2013, **12**, 5791–5800.
209. Z. Darula, J. Sherman and K. F. Medzihradszky, *Mol. Cell. Proteomics*, 2012, **11**, O111 016774.
210. R. Goldman and M. Sanda, *Proteomics: Clin. Appl.*, 2015, **9**, 17–32.
211. D. Wang, M. Hincapie, T. Rejtar and B. L. Karger, *Anal. Chem.*, 2011, **83**, 2029–2037.
212. B. Maček, J. Hofsteenge and J. Peter-Katalinić, *Rapid Commun. Mass Spectrom.*, 2001, **15**, 771–777.
213. J. V. Olsen, B. Macek, O. Lange, A. Makarov, S. Horning and M. Mann, *Nat. Methods*, 2007, **4**, 709–712.
214. M. Mormann and J. Peter-Katalinić, *Rapid Commun. Mass Spectrom.*, 2003, **17**, 2208–2214.
215. K. Takahashi, A. D. Smith, K. Poulsen, M. Kilian, B. A. Julian, J. Mestecky, J. Novak and M. B. Renfrow, *J. Proteome Res.*, 2012, **11**, 692–702.
216. J. P. Williams, S. Pringle, K. Richardson, L. Gethings, J. P. C. Vissers, M. De Cecco, S. Houel, A. B. Chakraborty, Y. Q. Yu, W. Chen and J. M. Brown, *Rapid Commun. Mass Spectrom.*, 2013, **27**, 2383–2390.
217. M. Thaysen-Andersen, B. L. Wilkinson, R. J. Payne and N. H. Packer, *Electrophoresis*, 2011, **32**, 3536–3545.
218. K. Deguchi, H. Ito, T. Baba, A. Hirabayashi, H. Nakagawa, M. Fumoto, H. Hinou and S.-I. Nishimura, *Rapid Commun. Mass Spectrom.*, 2007, **21**, 691–698.
219. J. J. P. Maury, D. Ng, X. Bi, M. Bardor and A. B.-H. Choo, *Anal. Chem.*, 2014, **86**, 395–402.
220. M. A. Kuzyk, D. Smith, J. Yang, T. J. Cross, A. M. Jackson, D. B. Hardie, N. L. Anderson and C. H. Borchers, *Mol. Cell. Proteomics*, 2009, **8**, 1860–1877.
221. N. Leymarie, P. J. Griffin, K. Jonscher, D. Kolarich, R. Orlando, M. McComb, J. Zaia, J. Aguilan, W. R. Alley, F. Altmann, L. E. Ball, L. Basumallick, C. R. Bazemore-Walker, H. Behnken, M. A. Blank, K. J. Brown, S.-C. Bunz, C. W. Cairo, J. F. Cipollo, R. Daneshfar, H. Desaire, R. R. Drake, E. P. Go, R. Goldman, C. Gruber, A. Halim, Y. Hathout, P. J. Hensbergen, D. M. Horn, D. Hurum, W. Jabs, G. Larson, M. Ly, B. F. Mann, K. Marx, Y. Mechref, B. Meyer, U. Möginger, C. Neusüβ, J. Nilsson, M. V. Novotny, J. O. Nyalwidhe, N. H. Packer, P. Pompach, B. Reiz, A. Resemann, J. S. Rohrer, A. Ruthenbeck, M. Sanda, J. M. Schulz, U. Schweiger-Hufnagel, C. Sihlbom, E. Song, G. O. Staples, D. Suckau, H. Tang, M. Thaysen-Andersen, R. I. Viner, Y. An, L. Valmu, Y. Wada, M. Watson, M. Windwarder, R. Whittal, M. Wuhrer, Y. Zhu and C. Zou, *Mol. Cell. Proteomics*, 2013, **12**, 2935–2951.

CHAPTER 4

Chemical Biology of O-*GlcNAc Glycosylation*

NATHAN J. COX[a,b], THOMAS R. MEISTER[a] AND MICHAEL BOYCE*[a,b]

[a]Department of Biochemistry, Duke University School of Medicine, Durham, NC 27710, USA; [b]Pharmacological Sciences Training Program, Duke University School of Medicine, Durham, NC 27710, USA
*E-mail: michael.boyce@duke.edu

4.1 Introduction

Post-translational modifications (PTMs) are used by all cells to increase the complexity and functionality of their proteomes. Intracellular PTMs often consist of the covalent linkage of a small moiety to side chains of amino acids within a protein, and can exert myriad effects on protein localization, activity, function, stability or other properties.[1] *O*-Linked β-*N*-acetylglucosamine (*O*-GlcNAc), a form of intracellular glycosylation, was discovered only in the early 1980s.[2] *O*-GlcNAcylation consists of the dynamic addition and removal of a single GlcNAc monosaccharide to the hydroxyl side-chains of serine and threonine residues (Figure 4.1), distinguishing it from the more familiar *N*- and *O*-linked glycans of the secretory pathway, which are generally long polysaccharides irreversibly added to extracellular proteins. At the levels of cycling and regulation, *O*-GlcNAc more closely resembles signaling PTMs, such as phosphorylation, than it does other forms of glycosylation,

Chemical Biology No. 2
Chemical Biology of Glycoproteins
Edited by Zhongping Tan and Lai-Xi Wang

Published by the Royal Society of Chemistry, www.rsc.org

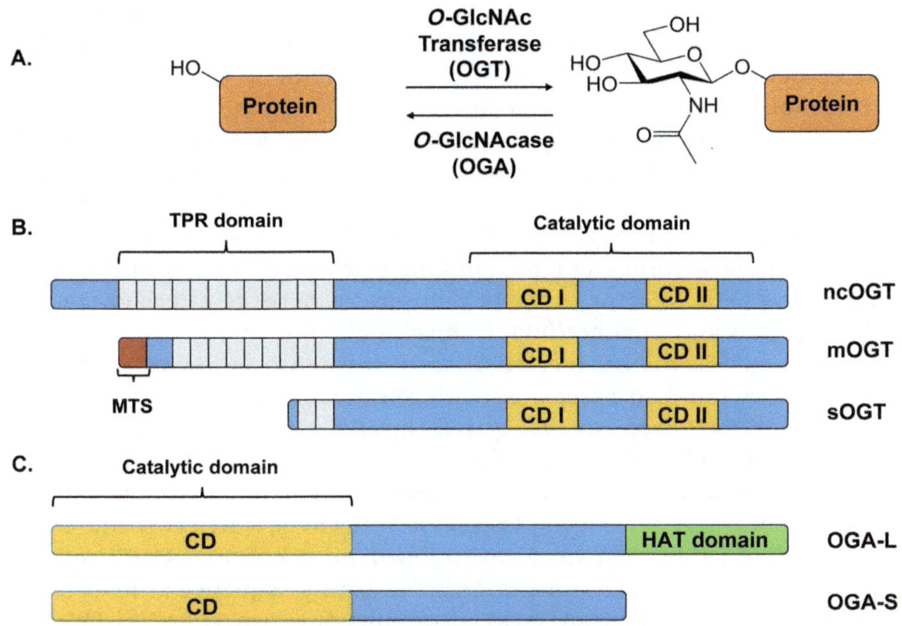

Figure 4.1 (A) *O*-GlcNAcylation consists of the addition of a single β-*N*-acetylglucosamine reversibly attached to the hydroxyl side-chains of serines and threonines of intracellular proteins. *O*-GlcNAc is dynamically regulated by *O*-GlcNAc transferase (OGT) and *O*-GlcNAcase (OGA). (B) The three isoforms of OGT. Nucleocytoplasmic OGT (ncOGT) contains twelve tetratricopeptide repeats (TPR); mitochondrial OGT (mOGT) contains nine TPRs and a mitochondrial targeting sequence (MTS); short OGT (sOGT) contains two TPRs. The C-terminal split glycosyltransferase catalytic domain (CD) is conserved among all three isoforms. (C) The two isoforms of OGA, long and short (OGA-L and OGA-S). The glycoside hydrolase catalytic domain (CD) is conserved in both. OGA-L also contains a C-terminal domain that resembles a histone acetyl transferase (HAT) but is probably catalytically inactive.

insofar as *O*-GlcNAc is used to control cellular processes on a relatively short time scale (minutes).

O-GlcNAc was discovered by Hart and coworkers while using bovine milk galactosyltransferase and uridine diphosphate (UDP)–[^{3}H]galactose to radiolabel terminal GlcNAc on mammalian lymphocyte proteins.[2] Surprisingly, the authors found that the bulk of the labeled GlcNAc residues were intracellular.[2] Shortly after, it was determined that the majority of *O*-GlcNAc-modified proteins resides in the nuclear and cytoplasmic compartments, and that its cycling is rapid and variable on specific substrates within these organelles, reminiscent of other intracellular PTMs.[3,4]

In mammals, *O*-GlcNAc is controlled by two enzymes, *O*-GlcNAc transferase (OGT) and *O*-GlcNAcase (OGA) (Figure 4.1). OGT, a member of the carbohydrate active enzymes (CAZy) glycosyltransferase 41 family (GT41), is the sole enzyme responsible for intracellular *O*-GlcNAc addition. [We note that another recently discovered glycosyltransferase, termed EOGT, *O*-GlcNAcylates epidermal growth factor (EGF)-repeat proteins in the secretory pathway.[5] Although OGT and EOGT create the same glycan, they are thought to be evolutionarily and functionally unrelated, and so we will not discuss EOGT further here.] While no clear consensus sequence has been identified among OGT substrates, a large minority share a proline–valine–serine/threonine motif.[6,7] *In vivo*, it is thought that OGT gains further specificity *via* peptide sequences or structural elements flanking glycosylation sites, and through associations with cofactor proteins that govern both its direct interaction with substrates and its subcellular localization[8,9] (see below).

The human OGT gene is located on the X chromosome near the *Xist* locus (Xq13.1), a region also associated with Parkinson's disease.[10,11] There are three human OGT isoforms generated through alternative splicing, all of which contain an identical C-terminal catalytic domain, a linker region and N-terminal tetratricopeptide repeats (TPRs).[12,13] The predominant and longest OGT isoform is nucleocytoplasmic OGT (ncOGT), which contains twelve TPRs, while an intermediate form, dubbed mitochondrial OGT (mOGT), contains nine TPRs and a mitochondrial targeting sequence at its N-terminus.[12,13] The shortest isoform (sOGT) contains only two TPRs.[12,13] TPRs are found throughout nature and consist of a single module of two antiparallel α-helices.[14] When these are arranged serially, as in OGT, an amphipathic channel is formed, aiding in the binding of target proteins.[14] In the case of OGT, the TPRs are thought to control cofactor and substrate–protein interactions.[8]

OGT activity is highly dependent on the concentration of its nucleotide-sugar donor substrate, UDP–GlcNAc.[8,9,15] UDP–GlcNAc is the terminal product of the hexosamine biosynthetic pathway (HBP), which integrates flux through glucose, amino acid, fatty acid and nucleotide metabolic pathways.[16] Therefore, OGT serves in part as a sentinel for nutrient availability in the cell.[17–20] An inverting glycosyltransferase, OGT converts the α-linked GlcNAc of its donor to a β-linked PTM on its substrates.[21–23] Structural analysis of human OGT indicates an ordered bi–bi mechanism, in which UDP–GlcNAc binds the catalytic site first, followed by the polypeptide acceptor.[24,25] Remarkably, OGT also functions as a co-protease for host cell factor-1 (HCF-1), a transcriptional co-regulator of human cell-cycle progression that undergoes cleavage as a maturation step.[26,27] HCF-1 proteolysis by OGT requires UDP–GlcNAc as a co-substrate, and is the first demonstration of a cleavage reaction catalyzed in the active site of a glycosyltransferase.[26,27] Proteolysis occurs between a cysteine and glutamate within one of six repeated HCF-1 sequences, and results in the formation of a pyroglutamate product.[26] Interestingly, mutating the glutamate at the scissile bond to a serine prevents cleavage but results in its *O*-GlcNAcylation.[26] At present, HCF-1 is the only reported proteolytic substrate of OGT; it will be interesting to learn in the future whether others exist.

Analogous to OGT, OGA is the only glycoside hydrolase known to remove *O*-GlcNAc from intracellular substrates in mammals. The human OGA gene lies on chromosome 10q24.1, which is associated with increased Alzheimer's disease (AD) risk.[28–30] Interestingly, a single nucleotide polymorphism (SNP) in the OGA gene also correlates with diabetes susceptibility among Mexican Americans,[31] though it remains unclear whether OGA dysregulation is a general feature of diabetes across other populations. OGA belongs to the CAZy glycoside hydrolase family 84 (GH84) and is a retaining enzyme, preserving the anomeric configuration of the β-GlcNAc after hydrolysis.[21,22,32] Unlike the majority of retaining glycoside hydrolases, which perform catalysis through the formation of a covalent glycosyl–enzyme intermediate, OGA uses a substrate-assisted mechanism, in which catalysis occurs through an oxazoline intermediate[22] (see below).

Like OGT, the function and substrate specificity of OGA are thought to be governed by interactions with various cofactor proteins, often forming stable multiprotein complexes with each other.[33] There are two major splice isoforms of mammalian OGA, long and short. The long OGA isoform, found throughout the cytoplasm and nucleus, contains a catalytic glycoside hydrolase domain and a domain resembling histone acetyltransferases (HATs).[34,35] The OGA HAT domain lacks key residues thought to be required for an active acetyltransferase, and most studies have failed to detect HAT activity by OGA,[36–38] leaving the function of this domain an open question. The short isoform of OGA contains only the glycoside hydrolase domain and has been shown to interact with lipid droplets, possibly controlling their assembly and mobilization by remodeling the droplet surfaces through local proteasome activation.[34]

O-GlcNAc is broadly conserved across animals, plants, some fungi and other clades. For example, it has been shown that *Trichoplax adhaerens*, among the most basal extant animals, has functional OGT, OGA and nucleocytoplasmic *O*-GlcNAcylation.[39] Remarkably, the *Trichoplax* OGT gene rescued *Drosophila* OGT mutants that would normally die as pharate adults, demonstrating a conservation of function even across distant animal groups.[39] Beyond animals, OGT homologs from the protozoans *Giardia lamblia* and *Cryptosporidium parvum* have been shown to have glycosyltransferase activity *in vitro*,[40] and OGT-like genes have been found in a large number of Gram-negative and Gram-positive bacteria (though not in Archaea).[41] Alignments of OGT-like catalytic domain sequences indicate a high level of conservation within the functional amino acids, implying that OGT activity may be conserved across these predicted proteins, though this hypothesis has not been rigorously tested.[41] OGA has also been described in invertebrates and prokaryotes, and putative GH84 enzymes are encoded in numerous bacterial genomes.[41,42] Many of these enzymes are distinct from eukaryotic OGA in that they are predicted to contain additional domains or secretion signals, leaving their physiological functions unclear and unstudied.[42] However, results from *in vitro* studies on several bacterial OGA homologs have demonstrated efficient removal of *O*-GlcNAc from eukaryotic glycoproteins, and these prokaryotic enzymes have served as extremely useful model systems for human OGA[43,44] (see below).

O-GlcNAc cycling regulates many cellular processes, including transcription, cell metabolism, cell cycle progression and cell death.[17,45] Proper *O*-GlcNAcylation is essential in mammals, as knockout of the OGT gene in mice is lethal at the one-cell stage, whereas OGA knockout mice die as neonates with developmental delays.[10,46] Due to the requirement for OGT and OGA in mammalian systems, other model organisms have been useful in the study of genetic alterations of *O*-GlcNAc cycling. Both *ogt* and *oga* deletions in the nematode *Caenorhabditis elegans* alter macronutrient storage and entry into the dauer larval state, a facultative form of diapause that allows the worm to resist unfavorable conditions, including stress or nutrient deprivation.[47,48] Interestingly, these signaling changes due to perturbed *O*-GlcNAc regulation are reminiscent of the reported role of *O*-GlcNAc cycling in mammalian insulin resistance.[47,48] A similar phenomenon is observed in *Drosophila*, where *Ogt* function is required for viability, and the mutant phenotypes associated with loss of *Ogt* and *Oga* function echo the pancreatic β-cell dysfunction, insulin resistance and glucose toxicity observed in mammalian systems with dysregulated *O*-GlcNAc.[49,50] In the mustard weed *Arabidopsis thaliana*, there are two OGT homologs, SECRET AGENT (SEC) and SPINDLY (SPY).[41,51–55] SPY functions as a gibberellin signaling (GA) repressor, and loss of SPY function results in a dwarf phenotype.[41,51–55] In contrast, SEC mutants have a mild phenotype, though SEC has been shown to O-GlcNAcylate proteins containing a DELLA domain, which mediate crosstalk between GA and other signaling pathways.[41,51–55] There is probably some overlap in function between SEC and SPY, because a SEC and SPY double mutation is embryonic-lethal.[41,52] Interestingly, no OGA homolog has yet been identified in *Arabidopsis*.[41]

As noted above, OGA and OGT gain substrate specificity, at least in part, from interacting partner proteins, either through changes in localization or by altering the binding affinity for substrates. Three examples of such binding partners are TET2, Milton and PP1.[56–58] Ten–eleven translocation (TET) enzymes are regulators of gene transcription, and it has been demonstrated that TET2 forms a DNA-bound complex with OGT, which may be responsible for chromatin and gene expression alterations.[57] The formation of this TET2–OGT complex promotes the *O*-GlcNAcylation of histone 2B (H2B) on Ser112.[57] *O*-GlcNAcylation at this residue has also been shown to promote Lys120 monoubiquitination by serving as an anchor for Bre1A, an H2B E3 ubiquitin ligase, underscoring the role of *O*-GlcNAc in epigenetic regulation.[59] In another example, Milton, a mitochondrial adaptor protein, has been shown to form a stable complex with OGT both *in vitro* and *in vivo*.[58,60–62] This OGT–Milton complex is thought to mediate mitochondrial transport in neurons and other cells.[58,60–62] Indeed, Milton is required for targeting OGT to the mitochondrial surface, and *O*-GlcNAcylation of Milton on key residues is required to arrest mitochondrial motility in response to physiological cues, such as high local glucose concentrations.[58] As a final example, protein phosphatase-1 (PP1) removes phosphoryl groups from serines and threonines of thousands of intracellular proteins and is found in stable and active complexes with OGT.[56] A transient complex consisting of PP1, OGT, OGA and the kinase Aurora B forms during mitosis and alters the PTM status of target proteins, such as vimentin.[63] This observation implies a critical role for the concerted action of these four enzymes in mitotic progression, and provides evidence for the highly dynamic interplay between phosphorylation and *O*-GlcNAcylation within cells.[63]

Improper cycling of *O*-GlcNAc has been implicated in myriad human diseases, including cancer,[64–67] diabetes,[20,68,69] cardiac dysfunction[70–73] and neurodegeneration.[74–77] For example, phosphofructokinase-1 (PFK1), an important glycolytic enzyme, becomes *O*-GlcNAcylated at Ser529 under hypoxic conditions, a feature common to many tumors.[78,79] *O*-GlcNAc modification of PFK1 inhibits its activity and redirects metabolic flux to the oxidative pentose phosphate pathway, providing a selective growth advantage to cancer cells through increased availability of biosynthetic precursors and NADPH to maintain redox homeostasis.[79] Blocking *O*-GlcNAcylation at Ser529 inhibits cancer cell proliferation *in vitro* and reduces tumor mass in a mouse xenograft model, demonstrating a novel mechanism for the regulation of metabolic flux, and possibly suggesting a target for future therapeutics.[79]

Hyperglycemia, a hallmark of diabetes, increases flux through the HBP, leading to increases in UDP–GlcNAc.[80] This, in turn, boosts global *O*-GlcNAc levels and can alter proper cycling of the modification on native substrates.[80,81] Ca^{2+}/calmodulin-dependent protein kinase II (CaMKII) is one substrate that becomes improperly regulated during hyperglycemia.[81] Under normal conditions, CaMKII has important regulatory functions in the heart and brain, but chronic activation can be pathological. During hyperglycemia, Ser279 is *O*-GlcNAcylated, which autonomously activates CaMKIIδ in the heart, contributing to cardiac dysfunction and arrhythmias.[81]

As these examples illustrate, *O*-GlcNAc cycling is an essential feature of nearly all mammalian cell and tissue types, and dysregulated *O*-GlcNAcylation probably contributes to myriad human pathologies. Therefore, there is great interest in understanding the mechanisms and effects of *O*-GlcNAcylation in both health and disease. However, like other PTMs, *O*-GlcNAc can be challenging to study with traditional molecular biology or genetic methods alone, because it is transient, rapidly cycling, and not encoded directly by the genome, as other central biological processes are (*e.g.*, transcription, translation). Fortunately, chemical biology tools are uniquely suited to addressing many of these challenges, thanks to their titratability, excellent control over time resolution, and broad applicability across biological systems (unlike, for example, mutations, which by definition remain within the genomes in which they were made). Many creative researchers have developed chemical approaches for studying *O*-GlcNAc and used them to understand its role in pathophysiological processes. Here we review the major contributions that chemical biology has made to this field, and provide an outlook on prospective future directions.

4.2 Chemical Blockade of *O*-GlcNAc Addition

4.2.1 Inhibitors of UDP–GlcNAc Biosynthesis

Small-molecule inhibitors have been valuable tools for studying protein *O*-GlcNAcylation since its discovery, and they have been applied in a wide variety of ways. Early efforts involved blocking the biosynthesis of UDP–GlcNAc, thereby depriving OGT of its obligate sugar-nucleotide donor and reducing global *O*-GlcNAcylation. The two most commonly used inhibitors of UDP–GlcNAc biosynthesis are azaserine and 6-diazo-5-oxo-norleucine (DON), which were independently described in the mid-1950's as antitumor compounds[82,83] (Figure 4.2). Subsequently, Marshall and colleagues proposed the hypothesis that, since azaserine and DON both act as "glutamate antagonists," they might inhibit glutamine:fructose-6-phosphate amidotransferase (GFAT), which produces glucosamine-6-phosphate and is the rate-limiting enzyme in the HBP.[84] Indeed, the authors showed that both azaserine and DON inhibit GFAT (and therefore UDP–GlcNAc production), and that they affect the cellular response to insulin signaling.[84]

Azaserine and DON have served as useful tool compounds in a variety of studies. For example, Erickson and colleagues used DON, in the above mentioned work, to show that glucose-induced calcium spikes in isolated rat cardiomyocytes depend on UDP–GlcNAc biosynthesis, implicating *O*-GlcNAcylation in this phenomenon.[81] Indeed, the authors showed that DON treatment reduces glucose-triggered premature ventricular complexes in

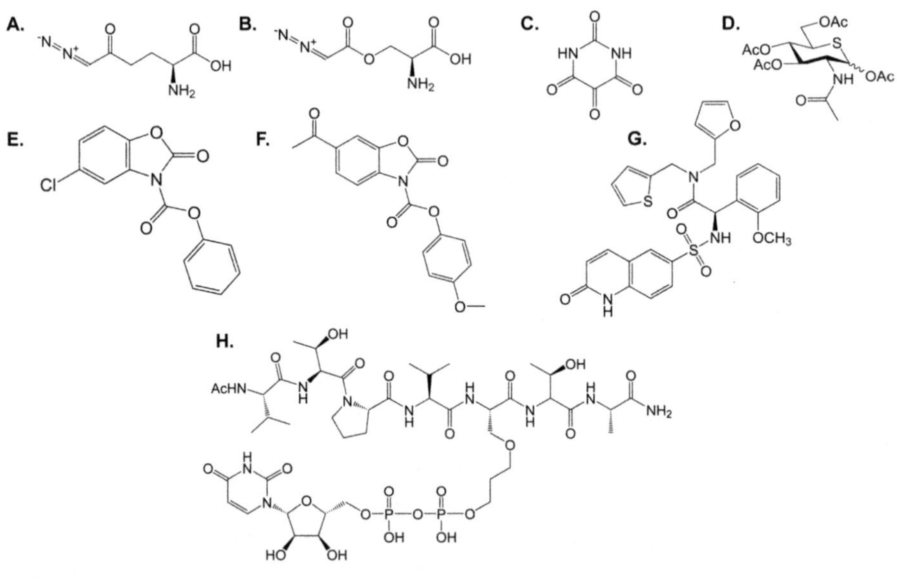

Figure 4.2 Inhibitors of *O*-GlcNAc addition. (A) DON (B) Azaserine (C) Alloxan (D) 5SGlcNAc (E) BZX (F) BZX derivative (G) OSMI-1 (H) Goblin-1. See text for details and references.

perfused rat hearts, and DON pre-treatment suppresses caffeine- and dobutamine-induced arrhythmias in diabetic animals.[81] Remarkably, the authors pinpointed the *O*-GlcNAcylation of Ser279 of the cardiac enzyme CaMKIIδ as the single glycosylation event most responsible for the effects of DON in this system.[81] These results may have implications for *O*-GlcNAc signaling in the heart arrhythmias that afflict some diabetic patients. In another example, the Yang lab showed that azaserine treatment of cultured human cells increases circadian period length, boosts the amplitude of the mRNA oscillations of the clock regulator Bmal1 and suppresses the high-glucose-induced amplitude increases in the mRNAs of Bmal1 and a second clock component, Cry1.[85] The authors later demonstrated that these effects are due, at least in part, to the rhythmic *O*-GlcNAcylation of the BMAL1 and CLOCK regulatory proteins, providing intriguing evidence that *O*-GlcNAcylation modulates peripheral circadian clock oscillations in mammals.[85]

Although azaserine and DON have been useful when applied judiciously, they also have two important liabilities in studies of *O*-GlcNAcylation. First, both compounds are relatively nonspecific and exhibit effects beyond GFAT inhibition. For example, it has been known since 1980 that azaserine is a mutagen, as judged by the classic Ames test, and induces hyperplasia and tumorigenesis in rats, implying a variety of undesired cellular effects.[86–88] Likewise, DON is known to produce a dramatic loss of mitochondrial function, a disruption of organelle morphology and apoptosis.[89] Second, because azaserine and DON inhibit the HBP, their on-target effects are not limited to reducing *O*-GlcNAcylation, but instead affect a wide variety of UDP–GlcNAc-dependent processes, including the production of other nucleotide-sugars, *N*- and *O*-linked glycoproteins in the secretory pathway, glycosaminoglycans, and glycolipids. Because of these caveats, the biological effects of azaserine and DON must always be interpreted with caution, and additional controls (ideally both genetic and pharmacological) are necessary in order to assign any results specifically to *O*-GlcNAcylation.

4.2.2 Product Inhibition of OGT: UDP

Since its discovery, OGT has been known to be inhibited potently by its product, UDP. Hart and colleagues used an *in vitro* glycosyltransferase assay with UDP–[^{3}H]GlcNAc and model substrate peptides to show that OGT lost more than half its observed activity in the presence of 500 nM UDP.[90] Subsequent work in a similar system revealed an inhibition constant (K_i) of 1.8 μM for UDP against OGT, significantly lower than those of multiple unnatural inhibitors tested in parallel.[91] Other groups have taken advantage of these observations in later *in vitro* studies of OGT. For example, the Walker group reported the first atomic-resolution structure of the catalytic domain of human OGT bound to UDP in both the presence and absence of a casein kinase II (CKII) peptide substrate, exploiting UDP as a high-affinity ligand to facilitate crystallization and shed light on catalytic mechanisms.[92] More recently, the Vocadlo group used UDP as a convenient OGT inhibitor in a cell-free coupled transcription–translation assay to provide evidence that *O*-GlcNAcylation occurs co-translationally on nascent polypeptides, probably to help guard against premature destruction by the proteasome.[93] Of course, UDP is neither cell-permeable nor specific in its inhibition of OGT, limiting its utility to particular *in vitro* applications.

4.2.3 Alloxan

Given the disadvantages of azaserine, DON and UDP as inhibitors of *O*-GlcNAc addition, it is clear that potent and specific small-molecule inhibitors of OGT itself would be extremely valuable research tools. The first compound to be widely used in this capacity was alloxan, an oxygenated pyrimidine derivative (Figure 4.2). Interestingly, alloxan is one of the oldest named organic compounds, having been discovered by Wöhler and Liebig soon after their landmark 1828 synthesis of urea.[94] By the 1940s, it was clear that alloxan treatment selectively kills pancreatic β islet cells in rabbits, hinting that it may perturb glucose and/or hexosamine metabolism.[95] In 2002, the Kudlow group proposed the hypothesis that alloxan, because it is a diabetogenic uracil analog, might block OGT directly.[96] Indeed, the authors used a chemoenzymatic assay to show that alloxan inhibits OGT *in vitro*, with a concentration giving 50% inhibition (IC_{50}) of 100 μM.[96] Furthermore, in explanted rodent β islets, alloxan prevented the rise in global *O*-GlcNAc induced by glucosamine and OGA inhibition, indicating that alloxan could indeed antagonize OGT activity in live cells.[96]

Since then, alloxan has been used in numerous studies of *O*-GlcNAcylation. For example, the Herr lab used alloxan in an *in vitro* assay to demonstrate that OGT enzymatic activity was required for the proteolysis of HCF-1 in both purified systems and HeLa cell extracts.[97] In whole-animal experiments, the Chatham and Marchase groups found that treatment with either glucosamine or glutamine, both of which feed the HBP, protects rat cardiac function and contractile performance in an ischemia–reperfusion model of injury.[98,99] Alloxan reverses this effect, implicating *O*-GlcNAcylation in cardioprotection.[98,99] Finally, in work complementary to the above-mentioned azaserine studies, Kaasik and colleagues demonstrated that alloxan treatment shortens period rhythmicity in the circadian clock of cultured mammalian cells, again implying a key role for OGT activity in peripheral clock regulation.[100]

Despite these notable examples, alloxan has considerable drawbacks as an OGT inhibitor. For example, it has been known for at least 35 years that alloxan induces DNA strand breaks and consequent poly-ADP ribose polymerase (PARP) activation in β islet cells,[101] implying that the compound is toxic and has targets beyond OGT. Indeed, even more importantly, the Konrad group has shown that alloxan inhibits not only OGT but also human OGA *in vitro*.[102] Therefore, any cell-based experiments with alloxan must be interpreted with caution, even with respect to specific measurements of *O*-GlcNAc levels or kinetics. Alloxan may remain a useful tool in properly controlled experimental systems. For example, Kaasik and colleagues confirmed their alloxan results with complementary genetic knockdown of OGT *via* small interfering RNAs (siRNAs), bolstering confidence that the effects observed were indeed due to OGT inhibition in this case.[100] However, the liabilities of alloxan have motivated researchers to discover newer and more specific OGT inhibitors.

4.2.4 Screening Approaches to Discover OGT Inhibitors

In the last ten years, the Walker group has pioneered the use of untargeted small-molecule screens to discover useful new inhibitors of OGT. In initial work, Gross and colleagues developed a ligand displacement assay, wherein the binding of a fluorescein-functionalized UDP–GlcNAc analog to human OGT could be monitored by fluorescence polarization (FP).[103] This FP assay allowed for the screening of over 64 000 small molecules for chemicals that would displace the ligand, providing candidate competitive inhibitors of OGT.[103] The authors reported a 0.2% hit rate in their primary screen, and used an orthogonal radiometric OGT assay to show that 19 of the initial hits inhibited OGT more than 40% at 25 μM.[103] Of the confirmed hits, the authors measured IC_{50}'s of 10 μM, 53 μM and 100–150 μM for three model compounds against the ncOGT isoform, and demonstrated that at least two of these three were competitive with UDP–GlcNAc.[103] The most potent compound, a benzoxazolinone (BZX), has received the most attention in later research (see Figure 4.2 and below).

Building on their initial work, Gross and colleagues subsequently developed a distinct Förster resonance energy transfer (FRET)-based OGT assay, wherein fluorescein and QXL-520 fluorescence quencher moieties were appended to either terminus of a peptide, with an OGT site and adjacent proteinase K cleavage site in the middle.[104] Based on similar work with kinases,[105] the authors anticipated that OGT modification of the glycosylation site would inhibit proteolysis at the adjacent site, keeping the FRET reporter quenched. However, inhibition of OGT activity would allow for proteolysis, liberating the fluorescein from the quencher, resulting in a gain of fluorescence. Though somewhat indirect, this assay does have the advantage of weeding out certain classes of false-negative chemicals common in compound libraries, such as aggregators, because those agents would presumably block both proteinase K and OGT activities nonspecifically. The authors screened approximately 124 000 compounds with their assay, finding 84 that inhibited OGT more than 30% (0.065% hit rate).[104] Of these, 45% were confirmed in a radiometric assay, and nine of those exhibited IC_{50}'s ranging from 900 nM to 20 μM, making them among the most potent artificial OGT inhibitors known.[104]

Other researchers have eagerly capitalized on the Walker lab's inhibitors, reporting activity in living cells for at least some of them. For example, Ngoh and colleagues found that two inhibitors discovered by Gross and colleagues exacerbate the death of cardiac myocytes in a hypoxia injury model.[106]

Moreover, BZX treatment reduced the *O*-GlcNAcylation of voltage-dependent anion channel (VDAC) proteins in cardiac mitochondria and sensitized cardiomyocytes to calcium-induced swelling in mice.[106] These results imply that OGT activity is pro-survival during hypoxic cardiac injury, and indeed the authors show that BZX suppresses the recovery of mitochondrial membrane potential in reoxygenated cardiac myocytes.[106] Although the *in vivo* specificity and toxicity of BZX remain poorly characterized, these results add to the evidence that *O*-GlcNAc cycling is important for cardiac resilience in response to injury. In another study, Caldwell and colleagues found that BZX reduced *O*-GlcNAc levels, growth in soft agar and cell migration in human breast epithelial cells transformed by the oncogenic receptor tyrosine kinase ErbB2.[107] The authors demonstrated that OGT inhibition upregulates markers of cell-cycle arrest, providing one potential mechanism to explain their other observations.[107] In a separate study, Shi and colleagues used BZX to demonstrate that high-baseline *O*-GlcNAcylation in primary chronic lymphocytic leukemias (CLL) is associated with impaired responses to various stimuli, including the chemotherapeutic drug vincristine, and that the aggressiveness of primary CLL tumors is correlated with reduced *O*-GlcNAc levels.[108] Because global *O*-GlcNAcylation is increased in many human tumor types,[64–67] these (and other) observations indicate that rational manipulation of OGT activity, at least on certain substrates, may, in principle, be a future anti-cancer therapeutic strategy.

Since their initial untargeted screening efforts, the Walker group has continued to improve and characterize their novel OGT inhibitors. In 2011, Jiang and colleagues reported the structure of human OGT bound to a BZX-derived inhibitor, revealing that a five-heteroatom dicarbamate at its core acts as a neutral mimic of the diphosphate moiety of UDP–GlcNAc.[109] Interestingly, the compound is a suicide inhibitor, proposed to act through a double-displacement reaction. First, one carbamate reacts with an essential lysine (Lys842) previously shown to contact the diphosphate of UDP–GlcNAc.[109] Then, a cysteine (Cys917) reacts with the lysine adduct and forms an irreversible carbonyl crosslink in the active site.[109] Unfortunately, the authors report that the BZX compound is a poor OGT inhibitor in live cells, probably due to low solubility, side-reactions and/or decomposition in aqueous solution.[109] Nevertheless, the Walker group has continued to use BZX compounds in important *in vitro* studies. For example, the Herr and Walker labs used BZX (as well as other OGT inhibitors) in their above-mentioned work demonstrating that HCF-1 is cleaved in the OGT active site.[97,110] This innovative work, then, was enabled in part by novel small-molecule OGT inhibitors.

Most recently, the Walker group has continued to search for additional OGT inhibitors that are more useful in living systems. In 2015, Ortiz-Meoz and colleagues reexamined prior screening data and noted that several hit compounds contained a quinolinone-6-sulfonamide (Q6S) scaffold.[111] The authors assembled a biased library of 1280 Q6S compounds and screened it with their FP assay.[111] The best hit from this effort, termed OSMI-1

(Figure 4.2), exhibited an IC_{50} of approximately 2.7 μM in two different *in vitro* assays and showed clear activity in cultured mammalian cells, raising global *O*-GlcNAc levels without producing undesired changes in cell-surface staining by several lectins.[111] Although further studies are needed, these preliminary results indicate that OSMI-1 may be relatively specific for OGT *versus* other glycosyltransferases *in vivo*.[111] Indeed, the Knipe lab has recently used OSMI-1 in herpes simplex virus-infected cells to demonstrate that OGT activity is required for normal viral gene expression, capsid assembly and replication.[112] Interestingly, the Walker lab reported that OGT inhibition by OSMI-1 is not competitive with UDP–GlcNAc, which is surprising, given that the Q6S scaffold emerged from libraries screened with a UDP–GlcNAc ligand displacement assay.[111] It may be that OSMI-1 binds OGT at an allosteric site but effects conformational changes that alter the active site, an important question for future structural studies. Ongoing work in this area will probably lead to improved OGT inhibitors—based on BZX, Q6S or other scaffolds—with more desirable properties, including improved potency, specificity, aqueous stability and *in vivo* utility.

4.2.5 Substrate Mimicry Approach to Discover OGT Inhibitors

As noted above, UDP itself is a tight-binding small-molecule inhibitor of OGT. Motivated in part by this knowledge, the Vocadlo lab sought to design potent OGT inhibitors using substrate mimicry. In 2011, Gloster and colleagues reported 2-acetamido-2-deoxy-5-thio-D-glucopyranose (5SGlcNAc) as a prodrug version of an effective in-cell OGT inhibitor[113] (Figure 4.2). The authors showed that peracetylated 5SGlcNAc ($Ac_4$5SGlcNAc) diffused across the membranes of cultured mammalian cells and then was deacetylated and converted by endogenous enzymes to UDP–5SGlcNAc, a thio-analog of the natural OGT nucleotide-sugar donor.[113] Results of *in vitro* studies indicated that UDP–5SGlcNAc was a poor substrate for OGT but instead proved to be an inhibitor, exhibiting a K_i of 8 μM in an *in vitro* radiometric OGT assay with nucleoporin-62 as a model substrate.[113] $Ac_4$5SGlcNAc also inhibited OGT in COS7 cells, with an apparent half maximal effective concentration (EC_{50}) of 5 μM.[113] Although it may seem counterintuitive that the *in vitro* K_i of UDP–5SGlcNAc is higher than the apparent EC_{50} of $Ac_4$5SGlcNAcin cells, these observations are probably explained by an accumulation of UDP–5SGlcNAc in the cytoplasm, since 5SGlcNAc is trapped by deacetylation upon crossing the membrane. Therefore, the effective concentration of UDP–5SGlcNAc inside $Ac_4$5SGlcNAc-treated cells is likely to be much higher than the initial treatment dose of $Ac_4$5SGlcNAc itself.

UDP–5SGlcNAc has been very useful for *in vitro* studies of OGT structure and function. In 2012, Lazarus and colleagues solved the crystal structure of a ternary complex of human OGT, CKII substrate peptide and UDP–5SGlcNAc, a feat that had not been possible with UDP–GlcNAc, since glycosyl transfer occurs much faster than protein crystallization.[114] Importantly, this work provided evidence for a proposed electrophilic migration mechanism for OGT catalysis, in which an oxocarbenium-ion-like species (the anomeric GlcNAc carbon) moves away from the leaving group (the UDP–GlcNAc pyrophosphate) towards the acceptor (the hydroxyl side-chain of the modified serine on the peptide substrate).[114] Electrophilic migration is a well-established mechanism of glycoside hydrolases, but this work provided the first structural evidence for its performance by a glycosyltransferase. In complementary later work, the van Aalten group crystallized the catalytic domain of OGT with UDP–5SGlcNAc and each of four peptides that had been identified as substrates in a screening effort.[7] This work provided insight into the orientation and sequence constraints for OGT substrates.[7]

$Ac_4$5SGlcNAc has also emerged as a powerful OGT inhibitor in living systems. For example, Ma and colleagues showed that $Ac_4$5SGlcNAc treatment reduces proliferation and anchorage-independent growth of MiaPaCa pancreatic ductal adenocarcinoma cells and confirmed these observations with short hairpin RNA (shRNA)-mediated OGT knockdown.[115] These results complement those of earlier studies indicating that increased global *O*-GlcNAcylation promotes survival and growth in cancer cells.[65] In a separate study, $Ac_4$5SGlcNAc was found to inhibit the insulin-stimulated activation of phosphoinositide-3 kinase (PI3K) and mitogen-activated protein kinase (MAPK) pathways in a human hepatocellular carcinoma cell line, shedding light on the coordination between glycosylation and phosphorylation cascades in nutrient sensing and growth signaling.[116] More recently, the Vocadlo group used complementary $Ac_4$5SGlcNAc and genetic approaches to show that OGT inhibition decreases nucleoporin glycosylation and renders the proteins susceptible to ubiquitination and proteasome-mediated destruction.[117] As a result, the selectivity barrier of the nuclear pore complex was compromised when OGT was inhibited.[117] Prior work has shown that many nucleoporins are heavily glycosylated and are extremely long-lived proteins, particularly in post-mitotic cells.[118,119] Therefore, the Vocadlo lab's work indicates that OGT activity may be essential for maintaining nuclear pore architecture and function in mammals.[117]

Despite these successful applications, questions remain as to the specificity of UDP–5SGlcNAc in cells. In particular, it is not completely clear whether UDP–5SGlcNAc accesses the secretory compartment and/or inhibits non-OGT glycosyltransferases or other enzymes, such as the hexosamine biosynthesis machinery. Recent data indicate that $Ac_4$5SGlcNAc treatment affects at least some subsets of cell-surface glycans, whereas structurally unrelated inhibitors (*e.g.*, OSMI-I) do not,[111] though whether this is due to an off-target effect of UDP–5SGlcNAc on unknown glycosyltransferases, or an indirect effect of potent on-target OGT inhibition remains to be determined. Regardless, 5SGlcNAc and its derivatives are likely to remain crucial tools for studying OGT *in vitro* and *in vivo*, particularly when combined with complementary pharmacological and genetic tools.

4.2.6 New Directions in OGT Inhibitor Development

Clearly, great progress has been made in the in the development of OGT inhibitors, but there remains room for improvement. For example, all inhibitors thus far are predicted to inhibit OGT activity on all substrates. Agents that selectively inhibit the glycosylation of only a subset of OGT substrates would be extremely valuable research tools. Interestingly, the van Aalten lab recently took a potential first step in this direction when they described the "Goblins," a family of bisubstrate UDP–peptide-conjugate OGT inhibitors.[120] Inspired by previous work,[121] the authors fused UDP to a substrate peptide (VTPVSTA) in the hopes of combining the high affinity of the UDP moiety with the selectivity of the acceptor peptide, all in one inhibitory molecule.[120] Indeed, Goblin-1 (Figure 4.2) showed a promising 18 μM IC_{50} in an *in vitro* assay.[120] While peptidyl inhibitors are unlikely to be cell-permeable and their selectivity against other eukaryotic glycosyltransferases remains unknown, it may be possible to use the structural biology characterization of goblins or other bisubstrate inhibitors to design potent and cell-active peptidomimetic OGT inhibitors in the future.

4.3 Chemical Blockade of *O*-GlcNAc Removal: OGA Inhibitors

4.3.1 Streptozotocin

Just as OGT inhibitor development has received intense and sustained research interest, many groups have worked to discover and optimize inhibitors of OGA. The earliest compound to be used frequently as an OGA inhibitor is streptozotocin (STZ), a nitrogen urea-functionalized GlcNAc analog originally identified as an antibiotic produced by the soil microbe *Streptomyces achromogenes*[122] (Figure 4.3). Early work showed that STZ was a diabetogenic

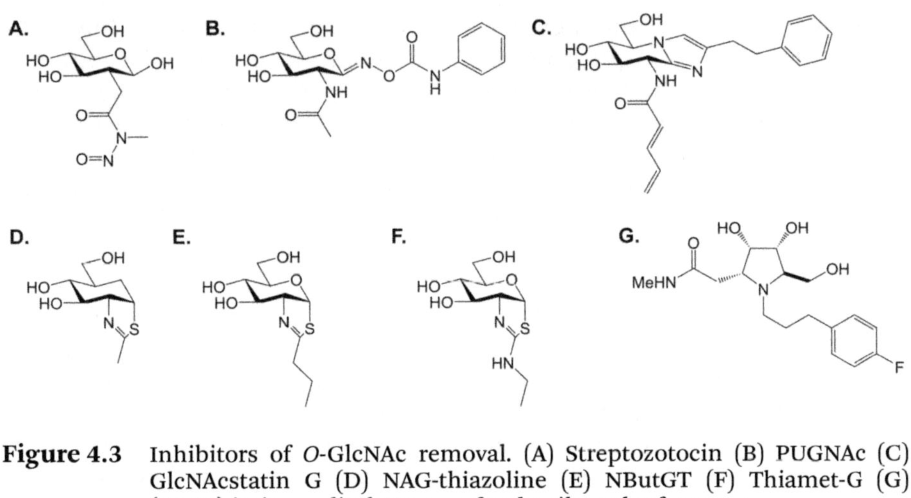

Figure 4.3 Inhibitors of *O*-GlcNAc removal. (A) Streptozotocin (B) PUGNAc (C) GlcNAcstatin G (D) NAG-thiazoline (E) NButGT (F) Thiamet-G (G) (2*R*,3*S*)-iminocyclitol. See text for details and references.

β islet toxin in mammals.[101] In the late 1990s, the Kudlow[123] and Hanover[124] groups demonstrated that STZ inhibited OGA *in vitro* and increased *O*-GlcNAc levels in β cells, apparently through OGA inhibition. Because β cells express high levels of OGT, the authors speculated that this cell type may be especially sensitive to OGA blockade and subsequent *O*-GlcNAc imbalance, triggering selective β cell death.[123,124]

Subsequent studies have used STZ as an OGA inhibitor in a variety of contexts. For example, Slawson and colleagues used STZ in an *in vitro* assay to examine dynamic *O*-GlcNAc changes in *Xenopus* oogenesis and progesterone-mediated oocyte maturation, well-established experimental systems of vertebrate cell-cycle progression and embryogenesis.[125] In addition, the Kudlow lab used STZ-enforced glycosylation of the transcription factor Sp1 in a biochemical characterization of the effects of *O*-GlcNAc on gene regulation.[126] Despite these examples, however, STZ has liabilities for studies of *O*-GlcNAcylation. Like alloxan, STZ is known to induce DNA damage and PARP activity in β cells, indicating targets beyond OGA.[101] Indeed, Turk and colleagues demonstrated that STZ treatment of rat islets resulted in nitrite generation, guanylyl cyclase stimulation, cyclic GMP accumulation and mitochondrial dysfunction, possibly due to nitric oxide liberated by STZ (and not OGA inhibition).[127] Similarly, Gao and colleagues showed that STZ, but not other small-molecule OGA inhibitors, triggered protein synthesis arrest, DNA fragmentation and cell death in Min6 insulinoma cells, indicating that STZ produces pleiotropic effects beyond OGA inhibition,[128] and the van Aalten lab demonstrated that the 4-epimer of STZ (*i.e.*, with galactose stereochemistry) induces insulinoma apoptosis but does not inhibit OGA.[129] Given these findings, it is clear that STZ produces toxic effects independent of *O*-GlcNAc cycling, and is no longer the reagent of choice for OGA inhibition.

4.3.2 PUGNAc

Since its initial purification, OGA has been known to be inhibited by *O*-(2-acetamido-2-deoxy-D-glucopyranosylidene)amino-*N*-phenylcarbamate (PUGNAc)[130] (Figure 4.3). PUGNAc was originally synthesized by the Vasella group in 1990 and investigated as a general inhibitor of *N*-acetylglucosaminidases, which are found throughout prokaryotic and eukaryotic clades.[131] When Dong and Hart reported the first purification of mammalian OGA from rat spleen homogenate, they observed that PUGNAc inhibited the enzyme with a K_i of 52 nM in an *in vitro* assay using *para*-nitrophenyl-*O*-GlcNAc as a substrate.[130]

OGA inhibition by PUGNAc has been extensively characterized at the chemical, biochemical and biophysical levels. The Hanover group used organic synthesis and both *in vitro* and cell-based assays to show that the *Z* isomer of PUGNAc is much more active against OGA than the *E* isomer.[132] Subsequently, the van Aalten lab solved the crystal structure of PUGNAc bound to the NagJ protein from *Clostridium perfringens*, a GH84 family glycoside hydrolase with significant homology to human OGA (CpOGA).[133] This work confirmed that

PUGNAc is a transition state analog and revealed that the PUGNAc oxime was in the *Z* orientation in the co-crystal, as predicted by the Hanover group's chemical analysis.[133] Importantly, the structure also provided support for the substrate-assisted catalysis model proposed previously for the OGA, wherein the 2-acetamido oxygen of the GlcNAc, and not an enzyme side-chain, is the catalytic nucleophile.[133] Indeed, the authors' structure showed that the 2-acetamido oxygen lay within 3 Å of the C1 (anomeric) carbon, demonstrating that its proposed role as a nucleophile was plausible.[133]

PUGNAc has also proven a useful tool in a variety of live-cell and -organism studies. For instance, the Jones group used both PUGNAc treatment and complementary genetic experiments to demonstrate that increased *O*-GlcNAcylation protects rat cardiomyocytes against endoplasmic reticulum (ER) stress-induced cell death, a common form of apoptosis in ischemia–reperfusion injury.[134] In another example, Ku and colleagues used PUGNAc treatment and genetic tools to show that keratin-18 *O*-GlcNAcylation is required for the recruitment and activation of the pro-survival kinase Akt1.[135] Moreover, this *O*-GlcNAc-dependent Akt1 activation is required in hepatocytes for resistance to apoptosis in two distinct liver injury models.[135] On a molecular level, the Lewis group used PUGNAc and genetic methods to show that *O*-GlcNAc cycles on and off specific residues of the long, unstructured C-terminal domain (CTD) of RNA polymerase II (pol II), the enzyme responsible for mRNA transcription.[136] The authors found that this glycosylation cycle occurs during pre-initiation complex formation to govern promoter occupancy and transcription, thereby establishing pol II CTD *O*-GlcNAcylation as an important regulator of global gene expression.[136] PUGNAc has also been applied to non-mammalian organisms with good success. For example, Lee and colleagues used PUGNAc and traditional genetic methods in *C. elegans* to show that increased *O*-GlcNAc levels promote entry into the dauer state.[137]

Though PUGNAc represents a significant improvement over STZ, it nevertheless suffers from particular drawbacks. Perhaps most importantly, PUGNAc is not specific to OGA but also inhibits hexosaminidase (Hex) A and B, two lysosomal human enzymes involved in degradation of endocytosed glycoconjugates.[138] In fact, results of *in vitro* assays indicate that PUGNAc inhibits OGA, HexA and HexB equally well, providing little opportunity to adjust dosing to discriminate among these enzymes in cells.[138] Indeed, Mehdy and colleagues demonstrated that PUGNAc treatment of Chinese hamster ovary cells leads to an unusual accumulation of free intracellular oligosaccharides, presumably by preventing HexA and HexB from catabolizing endocytosed *N*-glycans.[139] Beyond these off-target effects, PUGNAc's utility is also limited by its imperfect mimicry of the putative OGA transition state. The Vocadlo and Davies groups solved the structure of PUGNAc bound to a GH84 family enzyme from *Bacteroides thetaiotamicron* (BtGH84), another prokaryotic model protein for mammalian OGA, and found evidence of sub-optimal bond geometry in the co-complex.[140] These results implied that PUGNAc leaves room for improvement as an OGA inhibitor, and have motivated several research groups to continue to develop superior reagents.

4.3.3 NAG-Thiazoline

Vocadlo and coworkers have applied classic structural, enzymological and synthetic approaches to develop improved OGA small-molecule inhibitors. In an initial effort by the Withers group, Knapp and colleagues synthesized *N*-acetylglucosamine (NAG)-thiazoline as a rationally designed transition state (oxazolinium) mimic for β-hexosaminidases[141] (Figure 4.3). Although the aqueous stability of NAG-thiazoline is not optimal, it has served as a useful tool in *in vitro* studies. For example, Macauley and colleagues used NAG-thiazoline and its derivatives to provide early biochemical evidence for the above-mentioned retaining, 2-acetamido substrate-assisted catalysis model of OGA action.[142] The authors created a small library of inhibitors based on the NAG-thiazoline scaffold, some of which showed greater than 3000-fold selectivity for OGA over other β-hexosaminidases, in part because OGA more readily tolerates larger groups in its acetamido binding pocket than does HexA or HexB.[142] In subsequent work, the Vocadlo and Davies groups solved the structure of BtGH84 bound to NAG-thiazoline, revealing it to be a better transition-state mimic than PUGNAc, and providing additional biophysical support for the substrate-assisted-mechanism hypothesis.[140,143]

These studies lay the groundwork for the development of improved OGA inhibitors based on NAG-thiazoline. Notably, Macauley and colleagues reported the synthesis of a panel of NAG-thiazoline derivatives, including NButGT (1,2-dideoxy-2′-propyl-α-D-glucopyranoso-[2,1-D]-Δ2′-thiazoline), which exhibited a K_i of 230 nM for OGA and 1200-fold selectivity over β-hexosaminidase[142] (Figure 4.3). In related work, the same group reported that NButGT and PUGNAc both inhibit OGA and raise *O*-GlcNAc levels in cultured 3T3-L1 adipocytes to a similar extent, but that only PUGNAc, and not NButGT, induces insulin resistance and perturbed Akt signaling.[144] It may be that these additional effects of PUGNAc are due to its inhibition of HexA and HexB, highlighting again the importance of thoroughly characterizing the *in vivo* specificity of a small molecule before drawing mechanistic conclusions about its effects.[144] Other groups have also exploited the NAG–thiazoline scaffold to develop new OGA inhibitors. For example, the Hanover lab synthesized fluoro- and azido-thiazoline triols that are also selective inhibitors of human OGA *versus* placental hexosaminidase.[145]

Both NAG-thiazoline and NButGT have been used to study the role of *O*-GlcNAc cycling in a range of experimental contexts. For instance, the Chatham group used two different NAG-thiazoline derivatives to boost *O*-GlcNAc levels after ischemia–reperfusion injury in the rat heart, demonstrating improved recovery of contractile function and reduced tissue damage.[146] Though this study did not include complementary genetic controls, its results are broadly consistent with prior above-mentioned work on the protective role of *O*-GlcNAc in injured cardiomyocytes.[81,98,99,106,134] In another example, Ji and colleagues used NButGT to demonstrate that the transcriptional coactivator peroxisome proliferator-activated receptor gamma (PPARγ) is *O*-GlcNAcylated on Thr54 in 3T3-L1 adipocytes, decreasing its transactivation activity.[147] Correspondingly, NButGT impairs the differentiation of parental 3T3-L1 cells to adipocytes, at least in part by inhibiting PPARγ deglycosylation.[147] Finally, Shan and colleagues used NButGT to treat mutant superoxide dismutase (SOD)-overexpressing mice, a model of amyotrophic lateral sclerosis (ALS).[148] The authors found that NButGT increased *O*-GlcNAc levels in the spinal cords of wild type mice but not in those of SOD-overexpressors.[148] Though the mechanistic explanation for these observations remains unclear, the authors suggest that it may indicate an imbalance in *O*-GlcNAc cycling in this animal model of ALS, and, perhaps, in the human disease itself.

4.3.4 GlcNAcstatins

In parallel to the development of NAG-thiazoline and NButGT, the van Aalten group discovered and optimized a class of OGA inhibitors now termed the GlcNAcstatins (Figure 4.3). The origin of these efforts lies in the discovery and characterization of the natural product nagstatin by Takeuchi and colleagues.[149] Nagstatin was isolated from a fermentation broth of *Streptomyces amakusaensis*, and was determined to be an inhibitor of lysosomal hexosaminidases.[149] In later work, the authors reported that nagstatin was a tetrahydroimidazo[1,2-*a*]pyridine, providing a new scaffold for the development of additional hexosaminidase inhibitors.[150] Armed with this knowledge, the van Aalten group developed the synthesis of a gluco-configured derivative of nagstatin, rationally designed to be an OGA inhibitor *via* new N8 acyl substitutions (for selectivity) and C2 substituents (for increased affinity).[151] Remarkably, the most potent of these new GlcNAcstatins exhibited picomolar inhibition of CpOGA and approximately 100 000-fold selectivity over lysosomal hexosaminidases.[151] Although GlcNAcstatin inhibition of purified human OGA was not examined in this study, the compound did increase *O*-GlcNAc levels in cultured cells at concentrations similar to PUGNAc, indicating that it would be at least a useful starting point for developing tools for use in living systems.[151]

In subsequent work, the van Aalten group built on this success to improve GlcNAcstatins as research tools. For example, Dorfmueller and colleagues described a panel of GlcNAcstatin analogs, including GlcNAcstatin G, a penta-2,4-dienamide derivative with K_i of 4.1 nM against human OGA and over 900 000-fold selectivity against lysosomal hexosaminidases.[152] The authors solved the structure of CpOGA bound to GlcNAcstatin G, paving the way for future optimization, including the intriguing possibility of adding thiol-reactive moieties to exploit the Cys215 residue in the enzyme active site for the creation of potent and specific irreversible suicide inhibitors.[152] Finally, the authors demonstrated that GlcNAcstatin G increases *O*-GlcNAc levels in mammalian cells.[152] Though its low aqueous solubility and onerous synthesis are drawbacks of GlcNAcstatin, the van Aalten group recently reported a more succinct synthetic scheme for these compounds.[153]

GlcNAcstatins have found use in several living systems to date. For instance, Mariappa and colleagues used the compound in *Drosophila* to show that *O*-GlcNAc cycling is required for fibroblast-derived growth factor (FGF) signaling in that organism, probably because of the glycosylation of Downstream of FGF, an adaptor protein that mediates Rac and MAPK signaling in response to FGF receptor activation.[154] Genetic or chemical reduction of *O*-GlcNAcylation perturbs FGF-dependent tissue differentiation and morphogenesis.[154] In another example, Pathak and colleagues used GlcNAcstatin to investigate the signaling mediated by TAK1, a serine/threonine kinase downstream of pro-inflammatory cytokines, such as TGFβ, TNFα and Wnt ligands.[155] TAK1 binds to TAB1–3 as regulatory proteins. The authors used GlcNAcstatin to show that TAB1 is *O*-GlcNAcylated at Ser395, and this glycosylation is induced by known TAK1-activating stimuli, such as IL-1 and osmotic stress.[155] TAB1 glycosylation is required for full TAK1 activation and downstream signaling, including NF-κB activity and the production of IL-6 and TNFα.[155] These results underline the role of *O*-GlcNAc in inflammatory signaling and showcase the utility of OGA inhibitors in facilitating these biological discoveries.

4.3.5 Thiamet-G

Recently, the Vocadlo group has extended their OGA inhibitor program beyond NButGT. In 2008, Yuzwa and colleagues sought to design a new compound with a higher pK_a for the thiazolinium ion as compared with prior thiazolines. The authors reasoned from their earlier structural studies[140,143] that this change would produce a better match for the pK_a of the corresponding enzyme base, increasing electrostatic interaction and therefore binding.[156] The new inhibitor, termed Thiamet-G, exhibited a K_i of 21 nM against human OGA and 750 μM against human lysosomal hexosaminidase *in vitro*, with good stability in aqueous solution[156] (Figure 4.3). The authors solved the structure of BtGH84 bound to Thiamet-G, revealing a similar binding mode to NButGT, but with several small differences that may account for the increased potency of the former, including more optimal interactions with Asp242, a key catalytic residue for stabilizing the transition state.[156] Thiamet-G exhibited an EC_{50} of 30 nM in rat PC12 cells as judged by anti-*O*-GlcNAc Western blot, without apparent toxicity.[156]

As a result of its potency, stability and bioavailability, Thiamet-G has become a key reagent for studying the biology of OGA in live cells and organisms. For example, in a series of studies, the Vocadlo group has used Thiamet-G to explore the particular effects of OGA inhibition on the microtubule binding protein tau, a major constituent of the neurofibrillary tangles that are a classic pathological hallmark of AD. Tau hyperphosphorylation is correlated with aggregation and disease progression in some models. Interestingly, tau can also be *O*-GlcNAcylated, and this modification is often reciprocal with its phosphorylation.[76] Therefore, it is conceivable that enforcing tau glycosylation could reduce its phosphorylation and perhaps aggregation, possibly ameliorating its neurotoxic effects *in vivo*. Indeed, Yuzwa and colleagues showed in their initial work that Thiamet-G is orally bioavailable in rats, crosses the blood–brain barrier (BBB) and reduces the phosphorylation of tau in the brain.[156] The authors have used *in vitro* biochemical experiments to show that *O*-GlcNAcylation of tau reduces its aggregation, perhaps by increasing its monomer solubility and/or decreasing tau fibril or soluble aggregate stability.[157] Consistent with this, Thiamet-G mitigates the neuronal loss characteristic of AD in the JNPL3 tau transgenic mouse, which harbors a hemizygous mutant human tau allele (P301L) that is prone to aggregation.[74] Even more convincingly, Thiamet-G treatment also inhibited the production of Aβ42 peptides, which are a major constituent of the amyloid plaques characteristic of AD, in a double-transgenic mouse expressing both mutant tau and amyloid precursor protein (Swedish mutation, K670N/M671L).[158] Remarkably, Thiamet-G treatment also improved the scores of the double-transgenic mice in the Morris water maze behavioral test, implying that inhibiting OGA systemically may spare neurons and improve cognitive function in an AD model system.[158]

While these results are promising, important questions remain. For example, the mechanism of Thiamet-G action in these model systems is probably not as straightforward as had originally been anticipated. Indeed, Thiamet-G seems to exert effects on tau *in vivo* without consistently reducing its phosphorylation, calling into question the simple reciprocal model of *O*-GlcNAc/*O*-phosphate signaling.[74,158,159] Moreover, OGA probably has hundreds or thousands of substrates in neurons alone, and so it seems likely that long-term administration of a catalytic OGA inhibitor could produce adverse effects in human patients. Nevertheless, these important proof-of-principle experiments point to the power of Thiamet-G as a research tool, and may indicate that rational manipulation of *O*-GlcNAc cycling could show promise as a therapeutic strategy in AD.

Beyond neurodegeneration, Thiamet-G has become a useful compound in a variety of other contexts. For instance, Mi and colleagues used Thiamet-G to demonstrate that increased *O*-GlcNAcylation enhances anchorage-independent growth in soft agar in lung and colon cancer cell types and invasion by lung (though not colon) cancer lines in a transwell migration assay.[160] These results are consistent with other work indicating a correlation between elevated *O*-GlcNAc levels and cell behaviors associated with transformation and metastasis.[65] In another example, Goldberg and colleagues reported that high glucose levels stimulate p38 MAPK signaling in glomerular mesangial cells by increasing *O*-GlcNAcylation.[161] The authors proposed that this occurs when OGT inhibits Akt, leading to the activation of the kinase ASK1 and a consequent MAPK cascade.[161] Conversely, the authors showed that Thiamet-G treatment increased p38 activity, implicating increased *O*-GlcNAc levels in this effect.[161] In a final recent example, the Hart lab used Thiamet-G and HexA/HexB-specific inhibitors to provide evidence that OGA activity exists in mitochondria.[162] Indeed, Thiamet-G treatment of rat cardiomyocytes prevented the fluorogenic deglycosylation of a fluorescein–GlcNAc probe that co-localized with a mitochondrial marker.[162] Although *O*-GlcNAcylation of mitochondrial proteins had been observed before, this study provided the first evidence for OGA activity in the intact organelle, indicating that dynamic *O*-GlcNAc cycling may participate in some (or all) aspects of mitochondrial metabolism, an interesting area of future investigation.[162]

4.3.6 Other Approaches to OGA Inhibition

At present, potent and selective OGA inhibitors, including GlcNAcstatin and Thiamet-G, have empowered studies of *O*-GlcNAc cycling in a range of experimental systems. However, new inhibitors with improved pharmacokinetic and pharmacodynamic properties would be valuable additions to this scientific toolkit. Several groups have explored distinct scaffolds as rationally designed OGA inhibitors, including thiosulfonate- and selenazoline-functionalized GlcNAc analogs,[163,164] as well as pyrrolidine iminocylitols.[165] Indeed, some compounds in the latter category have proven to be single-digit nanomolar inhibitors of OGA, with good oral bioavailability and reported BBB permeability, though more work is needed to fully characterize their selectivity.[165] Other groups have expanded the effort to find non-carbohydrate OGA inhibitors through small-molecule screens, an approach that may be particularly useful for discovering new lead pharmacophores and/or allosteric inhibitors.[166] Finally, copper-catalyzed azide–alkyne cycloaddition (CuAAC), often known as "click chemistry," has been used recently to explore candidate OGA inhibitors.[167] Though the reported potencies were modest in this case,[167] similar studies may lay the groundwork for inhibitor discovery *via in situ* click chemistry, as has been pursued for other enzyme classes.[168] Overall, it seems likely that even more specific and powerful OGA inhibitors remain to be discovered, enhancing future investigations of *O*-GlcNAc biology.

4.4 Detection of *O*-GlcNAcylated Substrates: Lectins and Antibodies

Since its discovery, the sensitive and specific detection of *O*-GlcNAcylation on substrate proteins has been a perennial technical challenge. Here and in the following sections, we outline some of the commonly used methods for detecting *O*-GlcNAc on intracellular proteins.

One early and general strategy for *O*-GlcNAc detection was the leveraging of protein reagents, including lectins and antibodies, that bind selectively to *O*-GlcNAc moieties. For example, Hanover and colleagues reported as early as 1987 that the lectin wheat germ agglutinin (WGA) binds to pore components on both faces of the nuclear membrane, providing early evidence for nucleocytoplasmic *O*-GlcNAc.[169] Although WGA has been used widely to detect and purify *O*-GlcNAcylated proteins, it suffers from relatively low affinity and specificity, binding also to sialic acid residues on cell surface glycans, for example.[170] More recently, the Van Damme group has found that *Nicotiana tabacum* agglutinin (Nictaba lectin) is expressed in the tobacco cytoplasm and nucleus only after stress stimuli, including jasmonate signaling or insect attack.[171] Interestingly, Nictaba interacts with plant histone proteins in an *O*-GlcNAc-influenced manner, and also binds to mammalian histones.[171–173] While Nictaba may prove to be a second useful lectin for *O*-GlcNAc detection, its specificity and affinity remain to be fully characterized.

To overcome the liabilities of lectins, several groups have sought to generate pan-anti-*O*-GlcNAc antibodies. In one example, the Gerace group used the nuclear pore/lamina fraction of rat liver nuclei to immunize mice and generate monoclonal antibodies.[174] One of these clones, termed RL2, has been especially useful for detecting *O*-GlcNAc *via* Western blot, immunoprecipitation (IP) and electron microscopy (EM) *via* immunogold labeling.[174] Indeed, the Gerace group showed that RL2 labels discrete nuclear pores and that RL2 immunoreactivity disperses throughout the cell during mitosis, as the nuclear envelope is known to do.[174] In cooperation with the Hart lab, the Gerace group showed that RL2 immunoreactivity correlated with chemoenzymatic labeling of *O*-GlcNAc moieties (see below) and was ablated by hexosaminidase treatment or mild β-elimination of protein samples *in vitro*, providing further evidence that RL2 specifically recognizes *O*-GlcNAc.[175] In subsequent efforts, the Hart group generated a new anti-*O*-GlcNAc monoclonal antibody, termed CTD110.6, by immunizing mice with a synthetic glycopeptide from the RNA pol II CTD.[176] CTD110.6, an IgM, recognizes *O*-GlcNAc on serine and threonine residues, but not α-*O*-GlcNAc or *O*-GalNAc, confirming its specificity.[176]

Though RL2 and CTD110.6 are valuable research tools, evidence indicates that they detect only a subset of *O*-GlcNAc substrates, probably due in part to poorly characterized amino acid sequence preferences surrounding the *O*-GlcNAc moiety itself. Therefore, additional monoclonal antibodies would be useful for expanding the repertoire of detectable proteins. To address this need, the Boons lab recently devised a synthetic immunogen that combines, in a single molecule, a Toll-like receptor ligand, a helper T cell epitope (to enhance immunogenicity) and a serine-*O*-GlcNAcylated peptide derived from CKIIβ.[177] The authors inserted this immunogen into phospholipid unilamellar vesicles and immunized mice.[177] These efforts yielded several new monoclonal antibodies that recognize *O*-GlcNAc by IP and Western blotting, with apparent, and non-overlapping, peptide sequence preferences.[177] As a demonstration of the utility of these new antibodies, the authors IP-ed proteins from 293T human embryonic kidney epithelial cells and identified more than 200 putative *O*-GlcNAc substrates, including both well established and previously unreported glycoproteins.[177] Moreover, the authors used their new antibodies to detect and identify changes in *O*-GlcNAc substrates in the rat liver in response to a trauma–hemorrhage model.[177]

Because WGA and all the antibodies described above are commercially available, they often make a good starting point for pan-*O*-GlcNAc detection on substrates of interest. However, none of these reagents shows specificity for particular *O*-GlcNAcylated proteins or specific *O*-GlcNAc sites on those proteins. Therefore, while they are useful tools for profiling global *O*-GlcNAc, site-specific anti-*O*-GlcNAc antibodies would also be powerful assets for studying glycosylation-dependent signaling on particular proteins of interest, just as site-specific antibodies detecting phosphorylation or methylation, for example, have proven extremely useful in other experimental contexts. Though the development of site-specific anti-*O*-GlcNAc antibodies has lagged significantly, there have been a few recent success stories. For example, Klein and colleagues reported the creation of an antibody specific for glycosylated Ser1011 on the insulin receptor substrate-1 (IRS-1) protein, a residue that lies near to binding sites for PI3K and the tyrosine phosphatase SHP2, indicating that Ser1011-*O*-GlcNAc may regulate insulin signaling.[178] The authors created a synthetic *O*-GlcNAc peptide conjugated to keyhole limpet hemocyanin and immunized rabbits.[178] Signal from this antibody increased on IRS-1 from PUGNAc-treated cells, indicating that it may be a useful tool for future studies of insulin signaling.[178] In another example, mentioned above, Fujiki and colleagues reported that *O*-GlcNAcylation of Ser112 on histone H2B recruits the ubiquitin ligase Bre1A, leading to H2B ubiquitination, chromatin remodeling and broad transcriptional changes.[179] In this work, the authors created and characterized a Ser112-*O*-GlcNAc-specific antibody, showing that it stains euchromatin on *Drosophila* polytene chromosomes, indicating that it recognizes an active chromatin mark, as predicted.[179] Furthermore, chromatin IP followed by DNA sequencing (ChIP-seq) revealed that Ser112-*O*-GlcNAc is found frequently at transcription start sites and may regulate nutrient-dependent gene expression changes.[179] In other work cited above, the Vocadlo group reported the generation of an antibody that specifically recognizes Ser400-*O*-GlcNAc on tau and used it to monitor endogenous glycosylated tau in the rat brain.[180] Finally, the van Aalten lab generated a specific Ser395-*O*-GlcNAc antibody for TAB1 and used it in their work to dissect the role of glycosylation in TAK1 signaling.[155]

Beyond antibody and lectins, results of new research indicate that other classes of engineered protein reagents may prove useful for detecting *O*-GlcNAc. For example, the van Aalten group recently demonstrated that a catalytic-dead mutant of CpOGA (D298N mutation) binds to *O*-GlcNAc moieties with nanomolar affinity.[181] Indeed, the authors used a CpOGA(D298N)-glutathione-*S*-transferase fusion in a Far Western assay to show dynamic changes across the *Drosophila* proteome as a function of developmental stage.[181] In the future, additional natural or evolved proteins, including lectins, glycosidases or other *O*-GlcNAc-binding proteins, might be exploited for facile detection of the modification *in vitro* and in cells.

4.5 Detection of *O*-GlcNAcylated Substrates: Mass Spectrometry-Based Methods

In recent years, mass spectrometry (MS) has emerged as the most powerful tool for detecting, quantifying and localizing *O*-GlcNAc on specific substrates and residues of interest.[182] However, detection of *O*-GlcNAc by MS still presents considerable challenges. For example, like many PTMs, *O*-GlcNAcylation of individual substrates is often sub-stoichiometric, even when glycosylation exerts significant biological effects. This phenomenon alone creates a signal-to-noise problem for *O*-GlcNAc detection. In addition, the glycosidic bond of *O*-GlcNAc is labile under the standard collision-induced dissociation (CID) fragmentation modalities that are most commonly used to identify proteins by MS, meaning that *O*-GlcNAc moieties are routinely lost in conventional proteomics experiments.[182,183] Finally, ion suppression of *O*-GlcNAcylated peptides by their cognate unmodified peptides has been observed, further obscuring the signal from the modification.[182,183] In response to these challenges, a number of creative MS, enrichment and chemical derivatization methods have been developed to improve the sensitivity and utility of MS in studying *O*-GlcNAc. Because MS-based characterization of *O*-GlcNAc has been extensively reviewed recently,[182] here we simply highlight the major recent contributions of chemical biology in this area.

4.5.1 Electron Transfer Dissociation MS

The rise of electron transfer dissociation (ETD) technology has been a boon to the study of *O*-GlcNAc and other labile modifications. Unlike standard CID fragmentation, ETD uses small-molecule radical anions (*e.g.*, fluoranthene) to deliver electrons to cationic precursor peptides, resulting in peptide backbone fragmentation to produce sequence-specific c- and z-type product ions.[182] Importantly, ETD fragmentation largely spares glycosidic bonds and so analyte peptides retain *O*-GlcNAc, which can then be detected. An early application of this technology appeared in 2007, when the Hsieh-Wilson group used both ETD and CID methods to map glycosylation sites on mouse cortical neuronal proteins, including several novel sites on regulators of transcription and translation.[184]

Since then, other groups have capitalized on ETD to identify hundreds of new *O*-GlcNAcylation substrates and modification sites. For example, the Hart lab used ETD to map *O*-GlcNAc sites on the transcription factor FoxO1 and its transcriptional coactivator PGC1α.[185,186] In the liver, these glycosylation events are induced by high glucose, leading (somewhat counterintuitively) to the induction of gluconeogenic and stress-response genes.[185,186] These results may shed important light on the pathological nutrient signaling observed in the diabetic liver. In another example, the Burlingame group performed ETD analysis on *O*-GlcNAcylated peptides enriched by WGA affinity chromatography to map 58 glycosylation sites on proteins from murine cortical neurons, including 28 sites on the protein Bassoon alone.[187] Given Bassoon's putative role in scaffolding the cytomatrix of the pre-synaptic active zone, these data indicate that *O*-GlcNAcylation may play a critical role in synaptic transmission.[187] In a similar effort, the same lab identified over 60 *O*-GlcNAcylated nuclear proteins in mouse embryonic stem cells (mESCs), including several required for stem-cell maintenance.[188] The authors showed that polycomb repressive complex 2 (PRC2) is required for OGT expression in mESCs, implying that a PRC2–OGT–*O*-GlcNAc axis may govern mESC self-renewal.[188] In a third example, the Burlingame group also identified at least 19 glycosylation sites on HCF-1, demonstrating the power of ETD to illuminate *O*-GlcNAc modification of individual substrates of interest.[183] Finally, the Wells lab combined immunoaffinity purification of *O*-GlcNAcylated proteins with ETD and other fragmentation modalities to identify 83 sites on proteins from 293T cells.[189] As the results of these and other studies demonstrate, ETD analysis has vastly expanded our "parts list" of *O*-GlcNAcylated substrates and residues, particularly in the mouse and human proteomes. Future applications of ETD will undoubtedly discover many new *O*-GlcNAcylation events, both in other species and in mammalian cells or tissues exposed to specific signals, stresses, or pathological stimuli.

4.5.2 MS Approaches to Complement ETD

Although ETD is emerging as the prime method for mapping new *O*-GlcNAc sites in discovery experiments, other types of MS analysis may prove useful for distinct experimental goals. For example, multiple reaction monitoring (MRM) MS may hold promise for quantifying changes in *O*-GlcNAcylation at particular known sites of interest. MRM MS has been used extensively to study other PTMs (*e.g.*, phosphorylation) in this way, providing a label- and enrichment-free way of detecting sub-stoichiometric modifications in targeted proteomics experiments.[190] Indeed, in a study described above, the Vocadlo group used MRM MS to identify and monitor *O*-GlcNAcylation sites on tau.[180] Similarly, Maury and colleagues recently reported an MRM method to quantify *O*-GlcNAcylation of GSK3β in a complex mixture of human ESCs extracts, with no enrichment of target proteins required beyond simple sodium dodecyl sulfate–polyacrylamide gel electrophoresis (SDS–PAGE).[191] Though further work is needed, MRM MS may provide a helpful complement to ETD in targeted proteomic characterization of pre-selected *O*-GlcNAc substrates.

Interestingly, results of recent work also indicated that it may be possible to use improved bioinformatic methods to glean important information on *O*-GlcNAcylation from existing proteomics data. The Kuster lab developed Oscore, a software algorithm that searches existing datasets for characteristic masses from *O*-GlcNAc fragmentation, such as diagnostic neutral GlcNAc loss or the GlcNAc oxonium ion.[192] Though these species are rarely observed in non-ETD proteomics experiments, due to the aforementioned lability of the glycosidic bond, data from MS instruments with high resolution and mass accuracy may nevertheless contain these fragment ions buried in them, unrecognized. In a proof-of-principle experiment, the authors used Oscore to re-search three public MS datasets, spanning nearly 14 000 000 tandem mass spectra.[193] In this effort alone, Oscore found evidence for 158 putative *O*-GlcNAcylated peptides, with 194 sites from 628 spectra, and subsequent manual inspection of these data led to 26 unambiguously localized *O*-GlcNAc sites.[193] While ETD remains the preferred method for *O*-GlcNAc site-mapping, the authors' work indicates that the vast amount of existing proteomic data from traditional MS experiments could be mined productively for undiscovered information on OGT substrates, almost as a bonus. It will be interesting to see whether larger-scale efforts to organize and re-search these data succeed in adding new validated *O*-GlcNAc substrates to the growing list.

4.5.3 Chemical Derivatization of MS Samples for *O*-GlcNAc Enrichment and Detection

Because of the stoichiometry and ion suppression challenges mentioned above, it is often advantageous to enrich *O*-GlcNAcylated proteins or peptides prior to MS analysis. Natural *O*-GlcNAc moieties can be enriched by lectin- or immunoaffinity chromatography, but those approaches require expensive reagents and laborious purification and are subject to specificity problems arising from the lectins or antibodies themselves.[194] To address these difficulties, several types of chemical derivatization have been employed to label and enrich natural *O*-GlcNAc substrates. The Hart group reported an early example of this approach in the form of a combined β-elimination–Michael addition (BEMAD) workflow[195] (Figure 4.4). The authors used mild alkaline β-elimination to create dehydro-amino acids at sites of *O*-GlcNAc, followed by Michael addition with dithiothreitol (DTT) into the newly formed double bond.[195] As the authors showed, the resulting adduct is stable under CID conditions, allowing for MS-based identification of *O*-GlcNAc sites *via* searching for BEMAD product ions.[195] In addition, the authors demonstrated a version of BEMAD with biotin pentylamine, allowing for the subsequent affinity purification of *O*-GlcNAc substrates *via* stringent streptavidin affinity chromatography.[195] In a proof-of-principle experiment, the authors used BEMAD methods to identify *O*-GlcNAc sites on model peptides as well as glycoproteins from the rat brain, including Nup155, the lamin B receptor and synapsin I.[195]

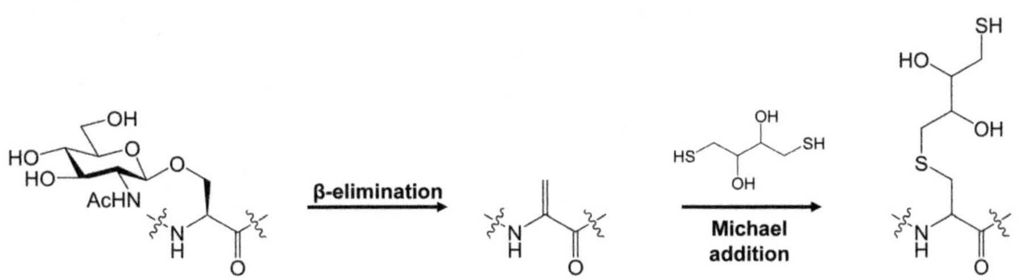

Figure 4.4 The β-elimination–Michael addition (BEMAD) workflow creates an MS-stable modification in place of *O*-GlcNAc. First, modified hydroxyl residues on proteins of interest are β-eliminated using mild base (*e.g.*, dilute NaOH, $NaBH_4$, triethylamine, *etc.*) to yield dehydro-amino acids at modification sites. This is followed by Michael addition at the resulting α,β-unsaturated carbonyl, often using dithiothreitol (shown) or biotin pentylamine.

Later, Vosseller and colleagues extended these results, introducing a deuterated DTT to create isotopic mass tags for comparing *O*-GlcNAc substrates across distinct protein samples.[196] In a related effort, the authors combined WGA affinity chromatography with BEMAD to identify 145 unique *O*-GlcNAc peptides in murine post-synaptic density preparations.[197] Interestingly, these results indicated that a large minority of *O*-GlcNAc substrates contained previously unrecognized amino acid motifs (PVST and/or TTA), providing new insight into OGT substrate selectivity *in vivo*.[197] In separate work, the Janek group combined the features of several BEMAD protocols in a version with isotopically labeled cystamine to enrich, quantify and site-map *O*-GlcNAcylated peptides from purified mouse proteasomes, revealing several sites of modification.[198] Finally, in recent work, the Hart group used a combination of BEMAD with isotopically labeled DTT, commercial tandem mass tag peptide labeling and thio-affinity chromatography to profile *O*-GlcNAcylation in vehicle- and Thiamet-G-treated rat heart mitochondria.[199] Using this elaborate MS workflow, the authors discovered that Thiamet-G increases the *O*-GlcNAcylation of many proteins in the electron transport chain and went on to show that mitochondrial *O*-GlcNAc governs both oxidative phosphorylation and calcium capacity in that organelle.[199]

Based on the results of these and other studies, it is clear that BEMAD can be a useful derivatization method for native *O*-GlcNAc enrichment and identification. However, caveats to the method exist. For example, BEMAD will label other labile *O*-linked PTMs, including phosphorylation, which may introduce confusion in determining the type of modification at any given site. Control experiments with phosphatase and/or hexosaminidase pre-treatment can improve the specificity of BEMAD results,[199,200] though this introduces additional sample-handling steps and potentially increased experimental variability. In the future, new chemical methods may emerge for tagging native *O*-GlcNAc substrates. Recently, for example, Klement and colleagues reported a method that exploits periodate oxidation of *O*-GlcNAc moieties, followed by hydrazide resin capture, on-resin trypsin digestion and the release of captured peptides *via* hydroxylamine treatment.[201] One potential benefit of this approach is that the oxime derivative that results from the hydroxylamine step (*i.e.*, the open carbohydrate ring) produces a characteristic fragmentation pattern in CID MS experiments, facilitating detection and site assignment without requiring expertise in ETD methods.[201] Indeed, the authors used their workflow to identify several *O*-GlcNAc substrates in the *Drosophila* proteome.[201] However, one drawback to this approach is that periodate will also oxidize secretory pathway *N*- or *O*-glycans, as well as tryptic peptides with N-terminal serine or threonine residues (which display vicinal amino alcohols), leading to their hydrazide capture and subsequent potential false-positive signal.[201] Performing the method on intact proteins treated with appropriate glycosidases to remove secretory pathway glycans may ameliorate these difficulties. Clearly, however, novel chemistries that permit the facile and specific tagging of *O*-GlcNAc moieties would be a significant achievement for future research.

4.6 Detection of *O*-GlcNAcylated Substrates: A Chemoenzymatic Approach

Chemoenzymatic methods to detect *O*-GlcNAcylation have been instrumental in the field since its inception. Indeed, the Hart lab harnessed existing enzyme tools to provide some of the very first evidence for the existence of *O*-GlcNAc. In 1984, Torres and colleagues demonstrated that bovine milk galactosyltransferase (GalT), which transfers galactose onto terminal GlcNAc acceptor glycans, labeled intracellular proteins from mammalian lymphocytes in a [^{3}H]-galactose transfer assay.[2] The results of these studies demonstrated differences in the abundance and patterns of the apparent *O*-GlcNAc modification among different subpopulations of lymphocytes, indicating that intracellular glycosylation may play broad functional roles in mammals.[2] In related work, the Hart lab used similar GalT experiments to show that *O*-GlcNAc is distributed among subcellular organelles and is especially abundant in the nuclear envelope, although it also occurs even in anucleate cells, such as erythrocytes.[202,203]

Early GalT studies of *O*-GlcNAc relied on wild-type enzyme and radiolabeled nucleotide-sugars for detection.[2,202,203] In a significant breakthrough, in 2002 the Qasba lab reported the rational design of a GalT mutant (Y289L) that relaxed the natural substrate specificity of GalT, allowing it to transfer GalNAc efficiently.[204] Soon thereafter, the Qasba and Hsieh-Wilson groups collaborated to exploit this increased GalT donor promiscuity, showing that GalT(Y289L) could transfer GalNAc analogs with *N*-acetyl substitutions, such as ketones, onto *O*-GlcNAc acceptors.[205] This permitted the subsequent chemical modification of the resulting *O*-GlcNAc–keto-Gal disaccharide with aminooxy-functionalized reagents, which react with the ketone to form a stable oxime adduct[205] (Figure 4.5). The authors used GalT(Y289L) and biotin-functionalized aminooxy probes to tag and purify mammalian *O*-GlcNAc substrates, including the transcriptional coactivator CREB and α-crystallin.[205] Interestingly, the glycosylation of the latter is of low enough stoichiometry that it is challenging to detect with older antibody-, lectin- or radiolabeling-based methods.[205] However, tagging with GalT(Y289L) and aminooxy-biotin followed by enhanced chemiluminescent detection of an anti-biotin Western blot showed an impressive gain in sensitivity, demonstrating the utility of this new chemoenzymatic approach for detecting low-abundance glycans.[205]

The GalT(Y289L)-mediated detection of *O*-GlcNAc offers several advantages, including the high specificity of GalT for a unique, native intracellular acceptor (*O*-GlcNAc) and the potential for signal amplification *via* secondary reagents recognizing the chemical handles (*e.g.*, biotin, fluorophores) enzymatically attached to *O*-GlcNAc sites.[204,205] In addition, GalT(Y289L) appears to recognize and accept a wider range of *O*-GlcNAc-modified proteins than do anti-*O*-GlcNAc antibodies, reducing the false-negative rate of the assay.[205] Because of these benefits, numerous reports have exploited and improved upon the initial GalT system reported by Hsieh-Wilson, Qasba and colleagues.

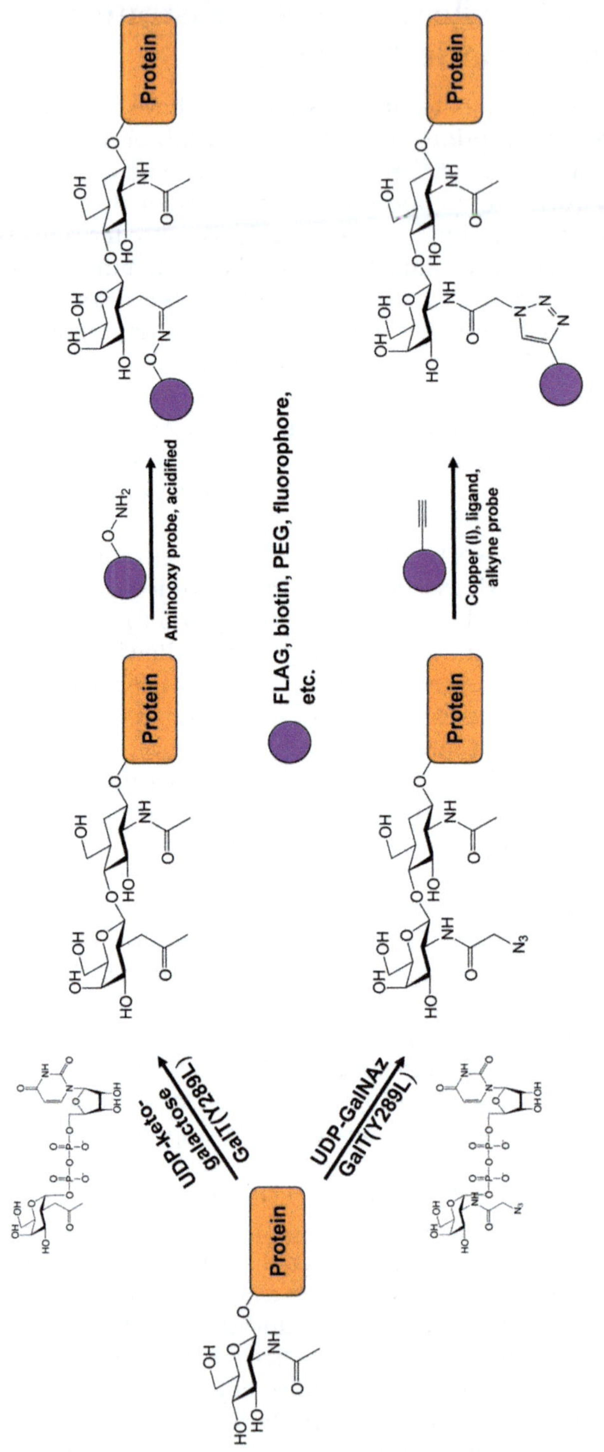

Figure 4.5 A chemoenzymatic approach to detect *O*-GlcNAcylation. A mutant galactosyltransferase (GalT) (Y289L) is used to transfer modified galactose units from a synthetic nucleotide-sugar donor onto native *O*-GlcNAc substrates. Commonly used GalT(Y289L) substrates include UDP-keto-galactose (top) and UDP-GalNAz (bottom), which are then reacted with aminooxy- or alkyne-functionalized probes, respectively. Probes often feature Flag epitopes, PEG, biotin, fluorophores, *etc.* (purple circles), for downstream detection and/or purification.

For example, Tai and colleagues used GalT(Y289L) and UDP–keto-Gal to profile many *O*-GlcNAcylated proteins in cell lysates and identify new substrates by affinity purification and MS, including CREB and the c-Fos transcription factor.[206] In related work, Khidekel and colleagues leveraged this GalT assay to create a "QUIC-Tag" workflow, wherein cortical neuron extracts were labeled with GalT(Y289L) and UDP–keto-Gal, tagged *via* an aminooxy-biotin probe, treated with either formaldehyde/$NaCNBH_3$ or deuterated formaldehyde/$NaCNBD_3$ to incorporate isotopic labels, combined, protease-digested and purified by streptavidin affinity chromatography to enrich for labeled glycopeptides.[184] The glycopeptides could then be identified and quantified by MS, thanks to the isotopic mass tags added during chemical derivatization.[184] Only two of the 25 proteins discovered in this effort had previously been identified as OGT substrates, underlining the power of GalT assays coupled with MS proteomics to discover new *O*-GlcNAcylated proteins.[184] The authors used these tools in combination with chemical inhibitors *in vitro* (PUGNAc) and *in vivo* (the glutamate receptor agonist kainic acid) to demonstrate that the *O*-GlcNAcylation they detected was specific and actively cycling and could be dynamically regulated by changes in brain activity.[184]

In another creative variation on the GalT(Y289L)/UDP–keto-Gal assay, the Hsieh-Wilson group used aminooxy-functionalized polyethylene glycol (PEG) molecules of 2 or 5 kDa as mass tags to investigate *O*-GlcNAc stoichiometry on specific proteins of interest.[207] When large PEGs are ligated onto an OGT substrate *via* the GalT(Y289L)/UDP–keto-Gal and subsequent oxime formation, the protein will gain 2 (or 5) kDa of mass per *O*-GlcNAc glycan, changing its migration in SDS–PAGE and allowing the stoichiometry of the modification to be read out from the banding pattern on a Western blot.[207] The authors used this method to dissect the stoichiometry of *O*-GlcNAcylation on CREB in the rodent brain and liver, showing intriguing organ-specific differences in *O*-GlcNAc occupancy.[207] The authors also combined their GalT(Y289L)/PEG method with existing phospho-specific antibodies to analyze the relationship between *O*-GlcNAc and *O*-phosphate on specific proteins.[207] Interestingly, they found that phosphorylation and glycosylation appeared independent on CREB (*i.e.*, the protein could be glycosylated, phosphorylated or both, without apparent crosstalk between the modifications).[207] On the other hand, the transcriptional repressor MeCP2 displayed what the authors dubbed a "reverse yin–yang" phenomenon, wherein increased *O*-GlcNAcylation resulted in increased phosphorylation at Ser80.[207] These results underscore both the complex interplay between these PTMs and the ability of chemical biology methods to elucidate their relationship.[207] In related follow-up work, the Hsieh-Wilson lab used their mass tagging strategy to demonstrate that CREB is mainly monoglycosylated in neurons at Ser40.[208] The authors showed that *O*-GlcNAcylation at this site is increased by membrane depolarization in a CaMK/MAPK-dependent manner and that it represses the transcriptional activity of CREB.[208] Taken together, these results illustrate the power of GalT(Y289L) labeling and mass tagging to dissect the interplay between *O*-linked PTMs.

Another recent innovation in the GalT(Y289L) labeling strategy has been the incorporation of clickable probes. Around the same time, the Hart[200] and Hsieh-Wilson[209] labs independently reported the transfer of *N*-azidoacetylgalactosamine (GalNAz) by GalT(Y289L) from UDP–GalNAz onto *O*-GlcNAc acceptor species. The resulting azido-disaccharide could then be "clicked" to an alkyne-functionalized probe *via* CuAAC[200,209] (Figure 4.5). Using this approach, combined with strong cation exchange and affinity chromatographics plus BEMAD, the Hart lab identified new *O*-GlcNAcylated sites on the intermediate filament protein vimentin.[200] Similarly, the Hsieh-Wilson group used in-gel fluorescence to detect the glycosylation of GalT(Y289L)/UDP–GalNAz-modified proteins that were then clicked to an alkyne-fluorophore.[209] After excision from the gel and MS analysis, the authors identified 67 known *O*-GlcNAcylated proteins and 146 novel ones, many of which were involved in metabolism, nutrient sensing and neural communication.[209] Similar approaches also allowed the bulk labeling of *O*-GlcNAc in fixed mammalian cells for imaging by fluorescence microscopy.[209]

Importantly, the Hsieh-Wilson lab reported that GalT(Y289L) labeling followed by detection through clickable probes resulted in significantly lower background signal than did detection *via* keto-Gal labeling and aminooxy probe conjugation, probably due to the relative promiscuity of aminooxy reagents as compared with CuAAC.[209] Thanks to these advantages, numerous other studies have exploited this "clickable" version of the GalT(Y289L) assay in myriad experimental contexts. For example, in separate studies, the Hart group used GalT(Y289L) and clickable probes to investigate *O*-GlcNAcylation sites in the tau-enriched fraction of rat brains, identifying eight new glycosylation sites,[210] and also reported the glycosylation and site-mapping of mammalian histones H2A, H2B and H4, providing interesting evidence that *O*-GlcNAc is a functional part of the epigenetic regulatory network of combinatorial PTMs known as the histone code.[211] In addition, the Hsieh-Wilson group used GalT(Y289L) and clickable, biotin-functionalized mass tags to confirm the presence and stoichiometry of *O*-GlcNAc on Ser529 of the glycolytic enzyme phosphofructokinase 1 (PFK1).[66] Importantly, the authors found that glycosylation inhibits PFK1 activity through an allosteric mechanism and is induced specifically in cancer cells, resulting in a redirection of glucose flux through the pentose phosphate shunt and a consequent growth advantage for tumor cells in mouse xenograft models.[66] Therefore, glycosylation of PFK1 may present a new therapeutic opportunity for targeting *O*-GlcNAcylation on a specific substrate.[66] More recently, Alfaro and colleagues used GalT(Y289L) and clickable biotin probes to identify 274 *O*-GlcNAc substrates in the mouse brain, 168 of which had not previously been reported.[212] The authors identified 448 modification sites on 195 proteins, many of which lie at or near known phosphorylation sites, implying a complex interplay between *O*-GlcNAc and *O*-phosphate in the brain.[212] Interestingly, the authors also identified six proteins with *O*-GlcNAc sites on extracellular domains, indicating that they are substrates of EOGT, rather than OGT.[212] This observation indicates that future cell-surface-targeted GalT(Y289L) assays could prove useful in identifying EOGT substrates. In another example, the Baltimore lab used GalT(Y289L) and clickable probes to show that the NF-κB protein c-Rel is *O*-GlcNAcylated on Ser350 and that this modification activates the protein, promoting its DNA binding and transactivation activities in lymphocytes.[213] Indeed, preventing c-Rel glycosylation inhibited cytokine gene expression in response to T cell receptor activation, identifying a new role for *O*-GlcNAc in immune signaling.[213] Finally, Xiao and colleagues provided an interesting recent example of the utility of the GalT(Y289L) assay in a non-animal system.[214] The authors used GalT(Y289L) labeling to confirm the *O*-GlcNAcylation of the gene expression regulator TaGRP2 in winter wheat.[214] TaGRP2 glycosylation promotes its binding to an inhibitory partner, VER2, causing it to dissociate from and de-repress the mRNA encoding critical proteins to mediate the vernalization response, through which plants sense and adjust to prolonged cold at the onset of winter.[214] Along with a previous study from the same group, these results demonstrate that *O*-GlcNAc governs the cold-induced seasonal response of winter wheat through specific gene expression circuits.[214,215] Taken together, these and other examples amply demonstrate the power of versatile GalT assays for studying the function of *O*-GlcNAc cycling in a wide range of tissues, organisms and experimental contexts.

4.7 Detection of *O*-GlcNAcylated Substrates: Metabolic Labeling Approaches

Chemoenzymatic methods have clearly made major contributions to our understanding of *O*-GlcNAc biology. Nevertheless, the GalT(Y289L) system is not well suited to monitoring certain aspects of *O*-GlcNAc cycling. For example, in a GalT assay, any *O*-GlcNAc moiety will be tagged and detected, whether it was added to its substrate protein three seconds, three hours or three days prior to the GalT reaction. Because *O*-GlcNAc is a dynamically cycling modification, the ability to discriminate between newer, signal-dependent glycans and older, "housekeeping" glycans is often essential for the analysis of its effects. While GalT labeling does not offer this time resolution, metabolic labeling strategies for tagging *O*-GlcNAc do.

4.7.1 *N*-Azidoacetylglucosamine (GlcNAz)

The first reagent specifically designed for metabolic labeling of *O*-GlcNAc was *N*-azidoacetylglucosamine (GlcNAz)[216] (Figure 4.6). In pioneering work with other classes of glycans, the Bertozzi group had previously demonstrated that azide-functionalized sugars ("azidosugars") are often accepted by endogenous nucleotide-sugar biosynthetic enzymes and glycosyltransferases.[217–219] Therefore, a cell-permeable form of an appropriate azidosugar can be provided to live cells or animals, resulting in its metabolic conversion to, first, a nucleotide-sugar and, second, a glycoconjugate.[217–219] These azidoglycans

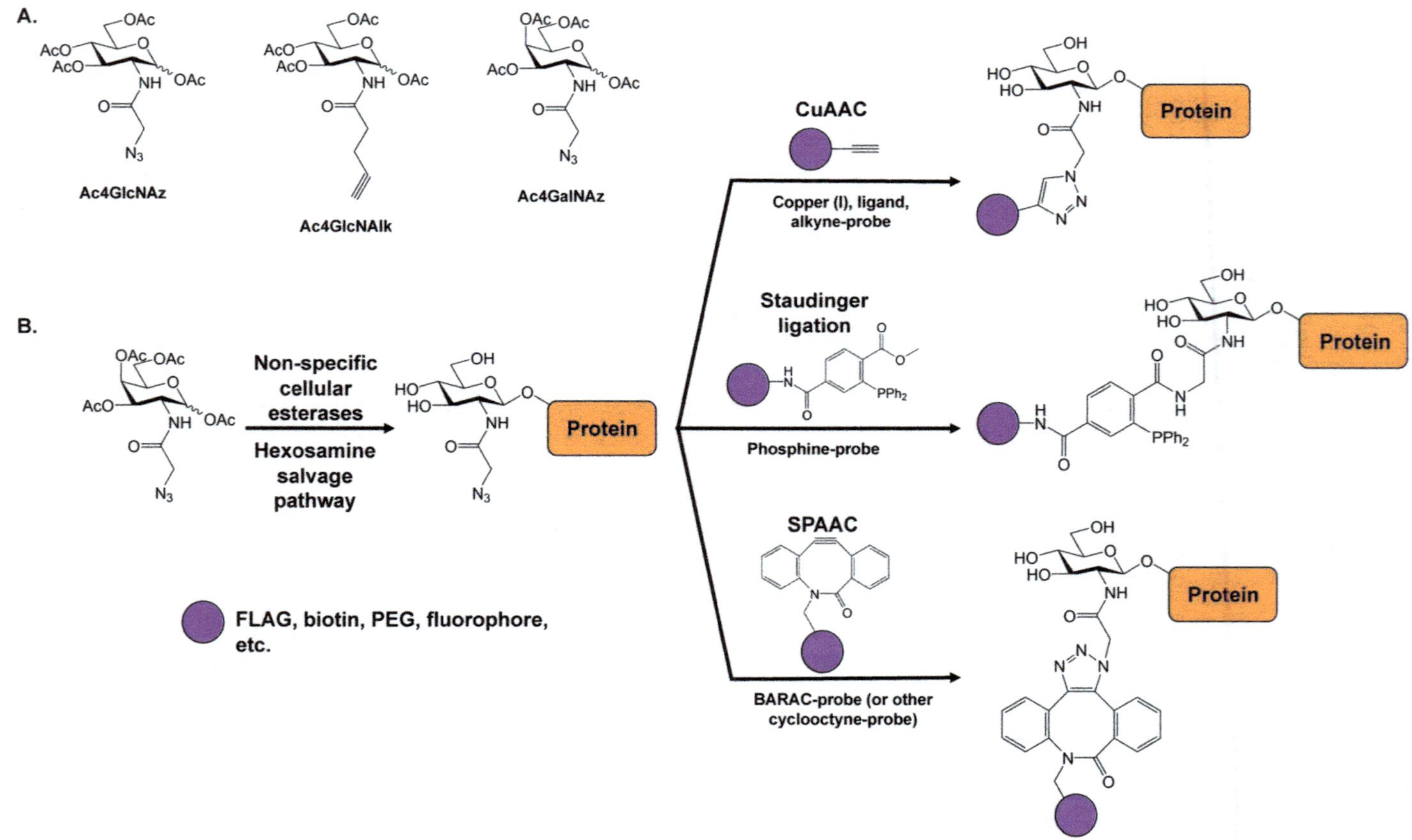

Figure 4.6 Metabolic-labeling reagents to detect *O*-GlcNAc within live cells or organisms. (A) Ac_4GlcNAz, Ac_4GlcNAlk and Ac_4GalNAz are commonly used to metabolically install bioorthogonal, reactive moieties onto endogenous *O*-GlcNAc substrates. (B) In cells treated with Ac_4GalNAz, non-specific esterases and the GalNAc salvage pathway (including galactose 4-epimerase) convert the precursor sugar to UDP–GlcNAz, which is utilized by OGT. Subsequently, various bioorthogonal reactions can be used to tag and detect the OGT substrate. Examples of three commonly used reactions are shown: i. Copper-catalyzed azide–alkyne cycloaddition (CuAAC); ii. The Staudinger ligation; iii. Strain-promoted azide–alkyne cycloaddition (SPAAC). Conversely, alkynyl sugars can be used with azide-functionalized probes (not shown). More recently, cyclopropene-functionalized sugars and tetrazine probes have been used in analogous experiments (not shown). See text for details and references.

can then be detected *via* reaction with a bioorthogonal probe, including an alkyne, as described above, or phosphine (the Staudinger ligation).[217–219] In 2003, Vocadlo and colleagues described GlcNAz as a specific reagent for the metabolic labeling of *O*-GlcNAc in mammalian cells.[216] Careful *in vitro* analysis showed that the enzymes of the GlcNAc salvage pathway could act on the appropriate GlcNAz-derived species and that OGT itself accepted UDP–GlcNAz as a substrate.[216] These results indicated that peracetylated GlcNAz (Ac_4GlcNAz), which is cell-permeable, could serve as a metabolic label for tagging endogenous *O*-GlcNAcylated proteins. Indeed, the authors used the Staudinger ligation to show that incubation of the human Jurkat T cell lymphoma line with Ac_4GlcNAz resulted in the labeling of intracellular proteins, including the well characterized and heavily glycosylated nucleoporin-62.[216]

Since this initial report, several groups have used Ac_4GlcNAz to profile and study *O*-GlcNAc substrates. For example, Sprung and colleagues cultured *Drosophila*, mouse, monkey and human cells with Ac_4GlcNAz and reacted azide-labeled proteins with a phosphine–biotin probe.[220] Anti-biotin chromatography followed by MS revealed a number of candidate *O*-GlcNAcylated proteins.[220] Similarly, other studies have used Ac_4GlcNAz and azide-reactive biotin probes or matrices to purify putative OGT substrates from human HeLa, MCF-7 and 293T cell lines.[221–223] While all studies showed clear evidence of GlcNAz-dependent labeling of intracellular proteins, in no case did the authors use GlcNAz-independent methods to confirm that these putative OGT substrates were natively *O*-GlcNAcylated.[220–223] Moreover, in later work, we noticed that Ac_4GlcNAz labeling of mammalian glycoproteins was substantially less robust than analogous results with other azidosugars.[224] In investigating possible explanations for this observation, we discovered that GlcNAz is not processed efficiently by the GlcNAc salvage pathway *in vivo*.[224] In particular, we found that the mammalian pyrophosphorylase enzymes AGX1/2, which convert GlcNAc-1-phosphate to UDP–GlcNAc in the GlcNAc salvage pathway, process only minimal levels of GlcNAz-1-phosphate in cells, resulting in essentially undetectable levels of UDP–GlcNAz under steady-state conditions.[224] Ac_4GlcNAz may prove useful for labeling *O*-GlcNAc substrates in certain contexts (*e.g.*, particular mammalian cell types, or non-mammalian organisms, where pyrophosphorylase activity may be higher or more tolerant of the *N*-azidoacetyl substituent). However, given these difficulties, Ac_4GlcNAz is not an ideal choice for labeling OGT substrates in mammalian cells, and newer, more robust, metabolic-labeling reagents were needed.

4.7.2 *N*-Azidoacetylgalactosamine (GalNAz)

In searching for alternatives to GlcNAz as a metabolic label for *O*-GlcNAc, we turned to *N*-azidoacetylgalactosamine (GalNAz), a GalNAc analog and the 4-epimer of GlcNAz (Figure 4.6).[224] In initial experiments, we found that peracetylated GalNAz (Ac_4GalNAz) transited the endogenous GalNAc salvage pathway and was converted to UDP–GalNAz by cultured mammalian cells.[224] Interestingly, most UDP–GalNAz was then converted to UDP–GlcNAz by endogenous UDP–galactose 4-epimerase (GALE), an enzyme that interconverts both UDP–Glc/UDP–Gal and UDP–GlcNAc/UDP–GalNAc *in vivo*.[224] After Ac_4GalNAz treatment, we observed an approximately 3 : 1 ratio of UDP–GlcNAz : UDP–GalNAz in cultured human cells, roughly in line with predictions for the relative respective thermodynamic stabilities of these compounds.[224] This observation indicated that GalNAz might be used to label OGT substrates by taking advantage of the GalNAc salvage and GALE pathways to efficiently form UDP–GlcNAz in cells. Indeed, we observed robust labeling of nuclear and cytoplasmic proteins from cells treated with Ac_4GalNAz, and used OGT and OGA overexpression experiments as well as *in vitro* deglycosylation assays with BtGH84 to confirm that Ac_4GalNAz-dependent signal resulted from an authentic *O*-GlcNAc mimic ("*O*-GlcNAz").[224] It is important to note that Ac_4GalNAz treatment also results in the labeling of secretory pathway glycans (*e.g.*, mucins) due to its minor metabolic conversion to UDP–GalNAz.[224] Therefore, when using Ac_4GalNAz to examine intracellular *O*-GlcNAc, it is advisable to use simple subcellular fractionation to separate the nuclear and cytoplasmic contents from the remainder of the cell. Indeed, in a proof-of-principle experiment, we used Ac_4GalNAz, cell fractionation and a phosphine-Flag–hexahistidine dual affinity reagent to purify many *O*-GlcNAc substrates, including nucleoporins, HCF-1 and OGT itself.[224] This work established GalNAz as a robust and faithful metabolic reporter for *O*-GlcNAc in mammalian cells.

Several groups have subsequently used GalNAz to study *O*-GlcNAc in diverse contexts. For example, we combined GalNAz labeling with novel alkyne-functionalized fluorescent probes and 2D difference gel electrophoresis (DIGE) to detect differences in *O*-GlcNAcylated proteins across two protein samples.[225] We used this method, termed "glyco-DIGE," to compare cytoplasmic and mitochondrial *O*-GlcNAcylated proteins and identify substrates exclusive to the mitochondria, an underexplored site of OGT action.[225] Through this effort, we identified and validated VDAC2 as a mitochondria-specific *O*-GlcNAcylated protein and demonstrated that the genetic ablation of VDAC2 protects cells from the mitochondrial dysfunction and cell death triggered by pharmacological perturbation of global *O*-GlcNAc cycling.[225] Though the 2D gel format of glyco-DIGE is labor-intensive, it offers the advantage of avoiding any affinity-purification steps for *O*-GlcNAc enrichment, which often bias proteomic results in favor of highly abundant and/or multiply glycosylated proteins, such as nucleoporins or HCF-1, at the expense of less abundant or less highly modified proteins of interest.[225] Therefore, glyco-DIGE can serve as a useful complement to affinity-based proteomics techniques in identifying signal-dependent changes in *O*-GlcNAcylated proteins.

Other groups have also used Ac_4GalNAz successfully to focus on *O*-GlcNAcylated proteins. For example, the Wells group combined Ac_4GalNAz labeling and an azide-reactive azadibenzylcyclooctyne probe linked to PEG to analyze the stoichiometry of *O*-GlcNAc on known substrates, such as OGA and Sp1,[226] similar to the above-mentioned PEG ligation strategy from the Hsieh-Wilson lab.[207,208] Along similar lines, Vocadlo and coworkers used

Ac_4GalNAz labeling of 293T cells to provide evidence that Sp1 is co-translationally *O*-GlcNAcylated and that this modification inhibits the premature proteasome-mediated degradation of Sp1 *in vivo*.[93] This work suggested the interesting new hypothesis that transient co-translational *O*-GlcNAcylation may be a key feature of protein quality control in mammals, preventing the inappropriate destruction of nascent polypeptides as they emerge from the ribosome and actively fold.[93] Even more recently, the Bertozzi group combined Ac_4GalNAz labeling of three different human cell lines with a new multifunctional "IsoTag" probe, which incorporates a terminal alkyne for CuAAC, a dibrominated isotopic recoding moiety, a cleavable silane linker for liberation of captured proteins and a biotin moiety for affinity purification.[227] The authors used targeted MS to identify and characterize the captured peptides, as detected *via* the isotopic mass tag.[227] This effort identified approximately 500 *O*-glycoproteins, many of which are known *O*-GlcNAc substrates, including HCF-1 and nucleoporins.[227] Although the current IsoTag probe may not be compatible with ETD analysis, since the authors report that the resulting halogenated peptides fragment poorly in this dissociation modality, future tag designs may ameliorate this drawback and facilitate site-mapping as well.[227] Finally, Ac_4GalNAz has served as a useful tool in non-mammalian systems. For example, Burnham-Marusich and colleagues reported the labeling of *C. elegans* primary embryonic cells in culture with Ac_4GalNAz and obtained evidence of glycosylation on the α subunit of the mitochondrial ATP synthase, presumably attributable to *O*-GlcNAc.[228] More recently, the Sun group, with assistance from our lab, labeled seedlings in liquid culture with Ac_4GalNAz to tag endogenous *O*-GlcNAcylated *Arabidopsis* proteins.[229] This work demonstrated that the *Arabidopsis* OGT homolog SECRET AGENT glycosylates REPRESSOR OF ga1–3 (RGA), a transcription factor that activates downstream signaling in response to such stimuli as light, jasmonate and brassinosteroids.[229] Therefore, GalNAz labeling, in combination with other biochemical, cell biological and genetic methods, identified a key *O*-GlcNAcylation event that regulates a range of stimulus responses in plants. Taken together, the results of these studies firmly establish GalNAz labeling as a powerful metabolic reporter for *O*-GlcNAc in a wide range of biological contexts.

4.7.3 *N*-Butynyl-Glucosamine (GlcNAlk) and Other Alkynylsugars

The above-mentioned studies with GlcNAz and GalNAz clearly illustrate the utility of clickable probes to metabolically label *O*-GlcNAc substrates. Of course, in principle it should be possible to swap the azide and alkyne moieties, using alkyne-functionalized monosaccharides to label glycoproteins and azide-functionalized probes for detection. Both alkyl azides and alkynes are small and nearly bioorthogonal (*i.e.*, not perturbative to living systems),[219] so it is reasonable to expect that alkynyl sugars may behave similarly to azido sugars in live cells. In addition, some evidence indicates that alkyne probes—both terminal alkynes for CuAAC and strained alkynes for copper-free reactions[230]—can react with thiols or other biotic nucleophiles, resulting in undesired, azide-independent conjugation to biological macromolecules.[231,232] If the alkyne resides on the probe molecule, this will result in a false-positive signal from an off-target reaction, whereas if the alkyne resides on the sugar, it will merely cause an unproductive but undetected reaction with that particular glycoconjugate. For these reasons, researchers have explored the utility of alkynyl sugars in studies of *O*-GlcNAc.

The Pratt group has led these efforts, first reporting in 2011 that a peracetylated form of *N*-butynyl-glucosamine (GlcNAlk) (Figure 4.6) labeled both *O*-GlcNAc and *N*-glycan glycoconjugates in mammalian cells.[233] The authors used a combination of Ac_4GlcNAlk labeling of NIH3T3 cells and a chemically cleavable azido-azo-biotin probe to purify GlcNAlk-tagged substrates, including neural precursor cell expressed developmentally down-regulated protein 4-1 (NEDD4-1).[233] NEDD4-1 is an E3 ubiquitin ligase that governs the proteolytic recycling of cell surface proteins in such diverse contexts as cancer cell signaling and HIV budding. Though it was not a previously identified OGT substrate, the authors confirmed that NEDD4-1 is modified by natural *O*-GlcNAc, providing important validation for this GlcNAlk result.[233] In subsequent work, the Pratt and Sheibani groups used GlcNAlk to label mouse retinal pericytes and enrich glycoproteins *via* an azido–biotin tag.[234] The authors identified 431 proteins, including 115 not previously reported as *O*-GlcNAc substrates.[234] Interestingly, results of follow-up experiments implicated the *O*-GlcNAcylation of pericyte proteins in p53-dependent apoptosis, revealing a potentially new functional role for *O*-GlcNAc cycling in the eye.[234]

Though alkynyl sugars are a promising alternative to GalNAz, their specificity and metabolism have not yet been fully characterized. In initial efforts in this direction, the Pratt group synthesized *N*-propargyloxycarbamate (Poc)-functionalized Glc, Gal and mannose (Man) analogs and showed that their peracetylated precursors are metabolically incorporated into mammalian proteins.[235] Curiously, however, both Ac_4GlcPoc and Ac_4ManPoc labeled NEDD4 in this system.[235] Since NEDD4 is a soluble cytoplasmic protein, it lacks secretory pathway glycans with Man, *N*-acetyl-D-mannosamine (ManNAc), or sialic acids (which are biosynthesized from ManNAc). Therefore, it is difficult to explain how NEDD4 could acquire alkynes *via* ManPoc labeling without invoking unexpected metabolic conversion of ManPoc to other, as-yet unidentified species. Similarly, the Pratt group also synthesized 1-deoxy-GlcNAlk and unexpectedly discovered that the peracetylated form of this molecule labeled proteins in a glycosylation-independent manner.[236] Because this labeling was competed away by sodium acetate or acetyltransferase inhibitors, the authors speculated that 1-deoxy-GlcNAlk may enter protein acetylation pathways, though this hypothesis has not yet been rigorously tested.[236] While GlcNAlk itself may not undergo unintended transformations in mammalian cells, these results indicate that it will be important to carefully characterize the cellular processing of alkynyl sugars before using them as metabolic labels.

4.7.4 Other Metabolic Labeling Reagents

Beyond GalNAz and GlcNAlk, additional metabolic labeling tools with increased specificity and/or orthogonal chemical reactivity would be valuable for future studies. As one step in this direction, Chuh and colleagues reported the synthesis of 6-azido-GlcNAc as a potentially more specific reagent for tagging *O*-GlcNAc substrates and found that its peracetylated form labeled intracellular proteins exclusively.[237] However, given that the GlcNAc salvage pathway begins with the 6-phosphorylation of the sugar, it is not clear how 6-azido-GlcNAc (which lacks the 6-hydroxyl) could be accepted by the UDP–GlcNAc biosynthetic enzymes as currently understood. More work is needed to confirm the specificity of this labeling.

In other studies, the Prescher lab reported the synthesis of monosaccharides bearing methylcyclopropene (CCp) substituents at the 2 position, including GlcNAc and GalNAc analogs (GlcCCp and GalCCp).[238] Cyclopropenes undergo an extremely rapid and bioorthogonal ligation with tetrazine species *via* an inverse electron-demand Diels–Alder reaction to form adducts stable in biological systems.[239] The authors demonstrated that both GlcCCp and GalCCp labeled cellular proteins in mammalian cells, as judged by tetrazine–biotin ligation and Western blotting, though the glycan classes and specificities of the reagents were not determined.[238] Very recent work by the Wittmann group, however, indicated that GlcCCp may be an extraordinarily useful metabolic label for *O*-GlcNAc.[240] First, the authors demonstrated that peracetylated GlcCCp labeled intracellular *O*-GlcNAc substrates, including p53, Akt, FoxO1, CaMK IV, vinculin and OGT itself, suggesting that GlcCCp may indeed be a faithful reporter of *O*-GlcNAcylation.[240] The authors also used a cell-permeable fluorescent tetrazine probe [Carboxytetramethylrhodamine (TAMRA)-tetrazine] to demonstrate the live-cell labeling of *O*-GlcNAcylated substrates.[240] Remarkably, these reagents also permitted detection of protein-specific glycosylation events.[240] The authors fused *O*-GlcNAc substrate proteins of interest to GFP and expressed them in GlcCCp-treated cells.[240] Because TAMRA and GFP constitute a FRET donor–acceptor pair, the authors were able to treat the cells with TAMRA-tetrazine and then use fluorescence lifetime imaging microscopy (FLIM) to assess specific glycosylation of the doubly tagged protein *via* FRET in living cells, without interference from GlcCCp-labeled proteins that lacked the GFP.[240] Previous work from the Chen lab had analyzed protein-specific glycosylation with a similar FLIM strategy, using Ac_4GalNAz labeling, but had required cell fixation because the alkyne detection probe was not cell-permeable.[241] In the case of the Wittmann study, more work is required to thoroughly assess the specificity and robustness of these cyclopropene sugars and FLIM–FRET methods. However, if they prove specific and generalizable, these reagents would not only complement azido sugars and alkynyl sugars in studies of *O*-GlcNAc, but may provide unprecedented chemical access to live-cell imaging of specific glycosylation events in a variety of systems.

4.8 Methods to Deduce the Biochemical and Cellular Functions of *O*-GlcNAc

As discussed above, many excellent studies have focused on the chemical manipulation of *O*-GlcNAc, *via* OGT and OGA inhibitors, or the detection and identification of *O*-GlcNAcylated proteins and residues, using a range of MS, lectin, antibody, chemoenzymatic and metabolic labeling tools. By comparison, relatively little is known about the specific biochemical and cell biological effects that *O*-GlcNAcylation exerts on individual substrate proteins. Recently, however, researchers have increasingly turned their attention to this problem and some have invented and deployed innovative chemical tools as part of the solution.

4.8.1 Live-Cell Assays of OGT or OGA Activity

As noted above, numerous groups have developed robust and sensitive *in vitro* assays for measuring the activity of OGA and, especially, OGT. Indeed, several of these assays have already been used to study the enzymes themselves, or to screen for and characterize useful inhibitors.[90,103,104,142,242–248] Future studies will undoubtedly harness these and new methods for additional *in vitro* purposes. Despite this progress, the measurement of OGT or OGA activity in live cells has proven a more elusive goal. Recently, however, the Mahal group described a suite of chimeric proteins that serve as live-cell reporters of *O*-GlcNAc cycling. In an initial report, Carrillo and colleagues created a fusion protein consisting of enhanced yellow and cyan fluorescent protein modules (Venus and eCFP) at either terminus, flanking a GafD module and CKII acceptor peptide for *O*-GlcNAc in the middle[249] (Figure 4.7). GafD is an *E. coli* protein that binds to GlcNAc moieties with good affinity.[250] The authors anticipated that when the CKII peptide was glycosylated (*i.e.*, when OGT activity was high and/or OGA activity was low), the adjacent GafD domain of the chimeric reporter would bind intramolecularly, thereby bringing the distant Venus and eCFP into close proximity.[249] Because Venus/eCFP form a FRET donor/acceptor pair, Carrillo and colleagues could monitor the extent of reporter glycosylation *via* fluorescence microscopy.[249] When the authors expressed the reporter in HeLa cells, they observed a 7–30% increase in FRET signal (depending on the individual cell) after treatment with PUGNAc and glucosamine to boost global *O*-GlcNAc, whereas a non-GlcNAc-binding control reporter showed less than 1% increase in FRET signal under the same conditions.[249] In a follow-up study, the same group adjusted the linker length between the Venus and CKII peptide, improving the FRET signal.[251] In addition, they created versions of the reporters that were localized specifically to the cytoplasm, nucleus or mitochondria, through the addition of organelle-specific signal peptides.[251] In this way, the authors were able to image real-time dynamics in reporter *O*-GlcNAcylation in response to serum stimulation and show that the resulting changes were dependent on PI3K signaling, because they were blocked by the specific inhibitor wortmannin.[251]

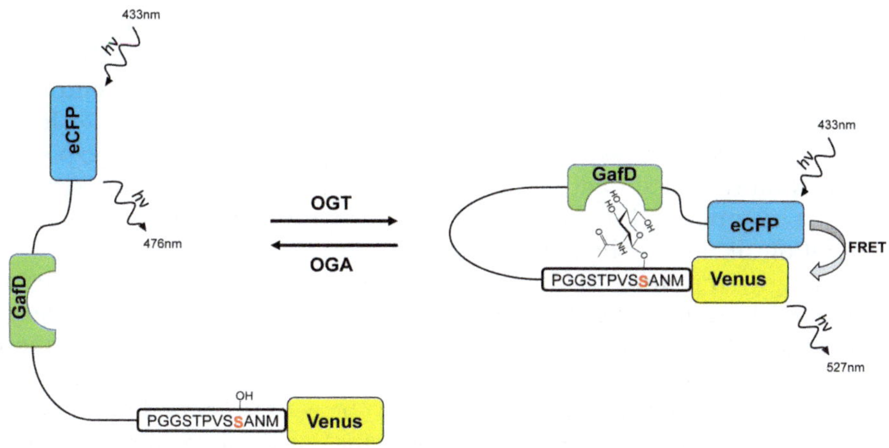

Figure 4.7 Live-cell reporter of *O*-GlcNAc cycling. This chimeric protein biosensor consists of enhanced cyan fluorescent protein (eCFP), GafD (an *E. coli* protein that binds *O*-GlcNAc), a CKII acceptor peptide for OGT and Venus (a modified yellow fluorescent protein). When the CKII acceptor peptide is *O*-GlcNAcylated within the cell, GafD binds the glycan on the neighboring domain and brings eCFP close to Venus, increasing the FRET signal between the fluorophores. In this way, FRET serves as a read-out of the extent of global *O*-GlcNAc levels in living cells.

The dynamic range of this *in vivo* reporter assay remains modest, leaving room for improvement.[251] However, this method may provide an elegant template for future live-cell approaches to assess localized *O*-GlcNAc cycling under a range of conditions.

4.8.2 Chemical Modification and Semi-Synthesis of Model Glycoproteins and Glycopeptides

Even when specific *O*-GlcNAc sites are mapped on a substrate, it is often unclear what biochemical function the glycosylation might exert on that protein. Unglycosylatable point mutants (Ser/Thr → Ala) can help to address this question, but this strategy also eliminates any other modifications at the same site (*e.g.*, phosphorylation) and only provides loss-of-function data. To fully understand the effects of specific *O*-GlcNAcylation events, it would also be valuable to assess phenotypes caused by a gain of glycosylation at a particular site. These goals are difficult to achieve using traditional cell biology or genetics alone, but chemical tools have provided new solutions to this longstanding problem.

Several groups have exploited chemical methods to modify or semi-synthesize homogeneous and defined glycoforms of proteins of interest, greatly aiding the biochemical characterization of *O*-GlcNAcylation. For example, the Davis group reported a bis-alkylation–elimination reaction of cysteine side-chains on intact proteins using a novel dibromide reagent.[252] Then, the resulting dehydroalanine can undergo simple conjugate addition by a thiol, all under mild aqueous conditions.[252] Using this approach, the authors successfully installed thioether-linked GlcNAc moieties (*S*-GlcNAc) onto prokaryotic and eukaryotic model proteins, thereby installing a Ser-*O*-GlcNAc mimetic at desired sites.[252] In follow-up work, the Davis group exploited this method to add similar *O*-GlcNAc-mimetic moieties to specific sites on multiple histones and incorporate them into synthetic nucleosomes, in an exploration of the biochemical role of *O*-GlcNAc in the histone code.[253,254] Interestingly, the authors found that incorporation of Thr101-*S*-GlcNAc histone H2A reduced the stability of the assembled nucleosomes by destabilizing the H3–H4 tetramer subcomplex.[254] The authors also used these nucleosomes and biotinylated DNA to affinity-purify proteins from HeLa nuclear extracts that bound specifically to the *S*-GlcNAc-modified nucleosomes.[254] They reported more than 20 proteins enriched greater than threefold by the *S*-GlcNAc modification, including DNA binding and mismatch repair proteins implicated in nucleosome remodeling.[254] Together, these data provide *in vitro* evidence that site-specific *O*-GlcNAcylation of histones may influence chromatin architecture and regulatory protein recruitment. In the future, it will be important to test the hypotheses generated from these elegant *in vitro* studies in live cells and model organisms.

In parallel, the Cole lab reported the semi-synthesis of CKIIα with a site-specific *O*-GlcNAc-mimetic moiety.[255] The authors found that CKIIα is glycosylated on Ser347, which lies near an activating phosphorylation site (Thr344), indicating that interplay between these PTMs may regulate CKIIα activity.[255] To address this question, the authors used expressed protein ligation (EPL)[256] to create a homogeneously glycosylated form of CKIIα.[255] First, they produced a peptide with an *S*-GlcNAcylated Ser at the 347 position *via* standard solid-phase peptide synthesis.[255] Then, they ligated this thioglycopeptide to an expressed N-terminal fragment of CKIIα using established EPL methods.[255] In cellular microinjection experiments, Ser347-*S*-GlcNAc antagonized Thr344 phosphorylation on CKIIα and led to destabilization of the protein, probably because the protein Pin1 recognizes phospho-Thr344 and binds to CKIIα *in vivo*, shielding it from proteasome-mediated degradation.[255] Interestingly, the authors also used *in vitro* protein microarrays to show that Ser347-*S*-GlcNAc affects CKIIα substrate specificity [but not catalytic constant/Michaelis constant (k_{cat}/K_m)].[255] Indeed, about 10% of CKIIα substrates on a 17 000-protein array exhibited differential phosphorylation depending on the PTM state of the semi-synthetic CKIIα used.[255] The results of these studies provided strong biochemical evidence that specific glycosylation events regulate CKIIα substrate selection and stability.

Finally, the Pratt lab has used synthetic glycopeptides and glycoproteins to test the effects of *O*-GlcNAc on α-synuclein pathophysiology. Aggregated α-synuclein in the brain is a pathological hallmark of Lewy body dementia diseases, including Parkinson's disease. Therefore, it is of great interest to understand how α-synuclein aggregation occurs and how it might be prevented for therapeutic benefit. In an initial study along these lines, Marotta and colleagues demonstrated that an *O*-GlcNAc moiety at the native Thr72 position reduced the propensity of a synthetic α-synuclein peptide to aggregate *in vitro* and blocked its ability to accelerate the aggregation of full-length α-synuclein, as compared with the cognate unmodified peptide.[257] In subsequent work, the same authors used EPL to semi-synthesize full-length Thr72-*O*-GlcNAcylated α-synuclein and found that this glycoform showed reduced aggregation, but undiminished membrane binding and bending properties, thought to be two normal functions of the protein.[258] Interestingly, Thr72-*O*-GlcNAc also reduced the toxicity of α-synuclein exogenously added to cultured neuroblastoma cells, a model system for the emerging idea in the Parkinson's field that posits neuron-to-neuron spread of disease pathology *via* extracellular α-synuclein.[258] The authors' results are reminiscent of the work of the Vocadlo lab and others with tau *O*-GlcNAcylation[74,76,156–159,180,210] (see above). Taken together, these results indicate that *O*-GlcNAcylation of specific neuronal proteins may mitigate both pathological aggregation and cellular and organ dysfunction in a range of neurodegenerative disease models.

4.8.3 Photocrosslinking Tools to Capture *O*-GlcNAc-Mediated Protein–Protein Interactions

Recently it has become clear that one biochemical function of *O*-GlcNAc is to mediate protein–protein interactions in contexts as diverse as chromatin remodeling,[179] nutrient sensing[259] and circadian entrainment.[85] In order to fully understand how *O*-GlcNAc governs these processes, it will be necessary to characterize the protein–glycoprotein interactions involved. However, *O*-GlcNAc-mediated protein–protein interactions pose technical challenges for conventional cell biological or biochemical assays, because they are often sub-stoichiometric, transient and/or low-affinity. To address this challenge, the Kohler lab, with assistance from us, designed a creative photocrosslinking strategy to capture and analyze *O*-GlcNAc-mediated protein–protein interactions *in vivo*[260] (Figure 4.8).

In a first report, Yu and colleagues used previous crystal structures of pyrophosphorylase enzymes to rationally design a mutant AGX1/2 protein (F383G) that would accommodate GlcNAc-1-phosphate analogs with extended *N*-acetyl substituents.[260] In particular, the authors showed that AGX1/2(F383G) could accept a diazirine-functionalized GlcNAc analog, termed GlcNDAz, thereby incorporating a widely used photocrosslinking moiety into the sugar.[260] AGX1/2(F383G)-expressing human cells readily salvaged a cell-permeable precursor form of GlcNDAz, converting it to UDP–GlcNDAz.[260] Moreover, OGT accepted UDP–GlcNDAz as a donor and appended *O*-GlcNDAz moieties to native substrates.[260] Upon irradiation with 365 nm UV light, diazirines eliminate molecular nitrogen, forming a highly reactive carbene, which can insert into carbon–carbon or carbon–heteroatom bonds of macromolecules nearby. Because the lifetime of a carbene in cytoplasm was expected to be extremely short, the authors anticipated that

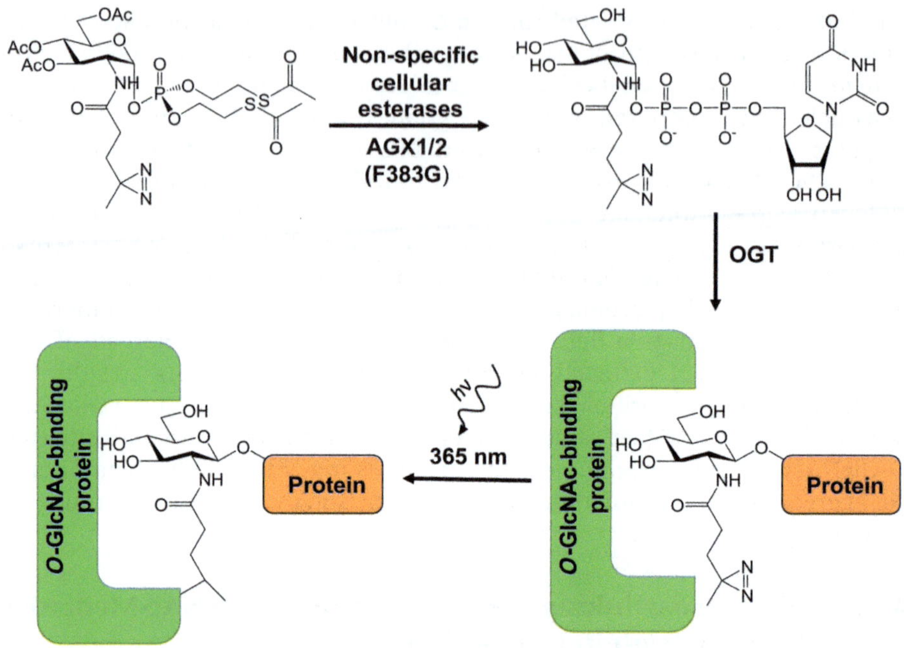

Figure 4.8 A metabolic labeling strategy to capture *O*-GlcNAc-mediated protein–protein interactions. Ac_3GlcNDAz-1-P(Ac-SATE)$_2$, a cell-permeable, 1-phosphorylated, protected precursor of the diazirine sugar GlcNDAz, is fed to cells expressing mutant AGX1/2(F383G). The Ac_3GlcNDAz-1-P(Ac-SATE)$_2$ is deprotected and converted to UDP–GlcNDAz by endogenous enzymes and AGX1/2(F383G). OGT accepts UDP–GlcNDAz and creates *O*-GlcNDAz moieties on its native substrates. When the cells are irradiated with 365 nm UV light, the diazirines eliminate molecular nitrogen and form highly reactive carbenes, which can irreversibly insert into the carbon–carbon or carbon–heteroatom bonds of proteins directly contacting the diazirine glycans. These covalently crosslinked protein–glycoprotein complexes can then be purified and analyzed.

O-GlcNDAz-mediated crosslinking inside cells would occur only between binding partner proteins that directly contacted the diazirine glycans. Indeed, when AGX1/2(F383G)-expressing, GlcNDAz-labeled human cells were exposed to a brief UV treatment, they demonstrated robust crosslinking among known OGT substrates, including several nucleoporins, as judged by mass shifts in an anti-nucleoporin Western blot, where crosslinked adducts were readily apparent.[260] Interestingly, the authors immunopurified these nucleoporins and identified the crosslinked binding partners by MS, finding several proteins known to regulate nuclear trafficking.[260] These results indicate that *O*-GlcNAc may participate in the nuclear pore selectivity filter, perhaps, in part, through direct contact with karyopherins and other components of the nuclear transport machinery.[117,260]

In related work, the Kohler lab again turned to structure-guided rational design to engineer a mutation in human OGT (C917A) that enlarged the

UDP–GlcNAc binding pocket, with the goal of better accommodating the diazirine of UDP–GlcNDAz.[261] Indeed, the authors showed that OGT(C917A) used UDP–GlcNDAz preferentially to natural UDP–GlcNAc itself, whereas wild-type OGT displayed the reverse preference.[261] Compared with the wild-type, OGT(C917A) promoted increased crosslinking of a glycosylated CKII peptide to the RL2 antibody *in vitro*, even in the presence of UDP–GlcNAc, and enhanced the photocrosslinking of endogenous nucleoporin-153 when expressed in human cells.[261] Dozens or hundreds of functionally important *O*-GlcNAc-mediated protein–protein interactions probably still await discovery. Photocrosslinking *via* GlcNDAz will probably provide a powerful tool for investigating this biochemical function of *O*-GlcNAc in living systems.

4.9 Conclusions and Outlook

Since its discovery over 30 years ago, *O*-GlcNAc has emerged as a critical PTM in a wide range of physiological and pathological processes. Despite tremendous strides in our understanding of *O*-GlcNAc biology, important fundamental questions remain: Which changes in *O*-GlcNAcylation are most important in the response to any given signal or stimulus of interest? How do OGT and OGA select substrates *in vivo*? What is the role of different OGT and OGA isoforms? What are the biochemical effects of *O*-GlcNAcylation on individual substrates? What is the nature of the interplay between *O*-GlcNAc and other PTMs, especially phosphorylation? Do OGT and OGA perform important cellular functions beyond glycosyl transfer and glycoside hydrolysis? How does *O*-GlcNAcylation become dysregulated in human disease and can it be rationally manipulated for therapeutic benefit? Discovering the answers to these questions will undoubtedly require contributions from many scientific disciplines. In particular, chemical biology is uniquely well positioned to address several aspects of *O*-GlcNAcylation, since it is a template-independent, rapidly cycling PTM. Indeed, as we have argued here, chemical methods have already been instrumental in our understanding of *O*-GlcNAc. We expect that continued close collaboration between chemists and biologists will empower many fruitful discoveries about *O*-GlcNAcylation in the decades to come.

References

1. C. T. Walsh, S. Garneau-Tsodikova and G. J. Gatto Jr, *Angew. Chem., Int. Ed.*, 2005, **44**, 7342–7372.
2. C. R. Torres and G. W. Hart, *J. Biol. Chem.*, 1984, **259**, 3308–3317.
3. G. D. Holt and G. W. Hart, *J. Biol. Chem.*, 1986, **261**, 8049–8057.
4. K. P. Kearse and G. W. Hart, *Proc. Natl. Acad. Sci.*, 1991, **88**, 1701–1705.
5. Y. Sakaidani, T. Nomura, A. Matsuura, M. Ito, E. Suzuki, K. Murakami, D. Nadano, T. Matsuda, K. Furukawa and T. Okajima, *Nat. Commun.*, 2011, **2**, 583.

6. J. Wang, M. Torii, H. Liu, G. W. Hart and Z.-Z. Z. Hu, *BMC Bioinf.*, 2011, **12**, 91.
7. S. Pathak, J. Alonso, M. Schimpl, K. Rafie, D. E. Blair, V. S. Borodkin, A. W. Schuttelkopf, O. Albarbarawi and D. M. van Aalten, *Nat. Struct. Mol. Biol.*, 2015, **22**, 744–750.
8. L. K. Kreppel and G. W. Hart, *J. Biol. Chem.*, 1999, **274**, 32015–32022.
9. W. A. Lubas and J. A. Hanover, *J. Biol. Chem.*, 2000, **275**, 10983–10988.
10. R. Shafi, S. N. Iyer, L. G. Ellies, N. O'Donnell, K. W. Marek, D. Chui, G. W. Hart and J. D. Marth, *Proc. Natl. Acad. Sci.*, 2000, **97**, 5735–5739.
11. S. à. p. Stichelen, L. K. Abramowitz and J. A. Hanover, *Biochem. Biophys. Res. Commun.*, 2014, **453**, 201–207.
12. D. C. Love, J. Kochan, R. L. Cathey, S.-H. H. Shin, J. A. Hanover and J. Kochran, *J. Cell Sci.*, 2003, **116**, 647–654.
13. B. D. Lazarus, D. C. Love and J. A. Hanover, *Glycobiology*, 2006, **16**, 415–421.
14. G. L. Blatch and M. Lässle, *BioEssays*, 1999, **21**, 932–939.
15. M.-D. D. Li, H.-B. B. Ruan, M. E. Hughes, J.-S. S. Lee, J. P. Singh, S. P. Jones, M. N. Nitabach and X. Yang, *Cell Metab.*, 2013, **17**, 303–310.
16. N. E. Zachara and G. W. Hart, *Biochim. Biophys. Acta, Gen. Subj.*, 2004, **1673**, 13–28.
17. J. A. Hanover, M. W. Krause and D. C. Love, *Biochim. Biophys. Acta*, 2010, **1800**, 80–95.
18. M. A. Mondoux, D. C. Love, S. K. Ghosh, T. Fukushige, M. Bond, G. R. Weerasinghe, J. A. Hanover and M. W. Krause, *Genetics*, 2011, **188**, 369–382.
19. M. R. Bond and J. A. Hanover, *Annu. Rev. Nutr.*, 2013, **33**, 205–229.
20. S. Hardiville and G. W. Hart, *Cell Metab.*, 2014, **20**, 208–213.
21. V. Lombard, H. Golaconda Ramulu, E. Drula, P. M. Coutinho and B. Henrissat, *Nucleic Acids Res.*, 2014, **42**, 5.
22. C. Martinez-Fleites, Y. He and G. J. Davies, *Biochim. Biophys. Acta*, 2010, **1800**, 122–133.
23. P. M. Coutinho, E. Deleury, G. J. Davies and B. Henrissat, *J. Mol. Biol.*, 2003, **328**, 307–317.
24. M. B. Lazarus, Y. Nam, J. Jiang, P. Sliz and S. Walker, *Nature*, 2011, **469**, 564–567.
25. M. B. Lazarus, J. Jiang, T. M. Gloster, W. F. Zandberg, G. E. Whitworth, D. J. Vocadlo and S. Walker, *Nat. Chem. Biol.*, 2012, **8**, 966–968.
26. M. B. Lazarus, J. Jiang, V. Kapuria, T. Bhuiyan, J. Janetzko, W. F. Zandberg, D. J. Vocadlo, W. Herr and S. Walker, *Science*, 2013, **342**, 1235–1239.
27. M. P. Kötzler and S. G. Withers, *J. Biol. Chem.*, 2016, **291**, 429–434.
28. D. Heckel, N. Comtesse, N. Brass, N. Blin, K. D. Zang and E. Meese, *Hum. Mol. Genet.*, 1998, **7**, 1859–1872.
29. L. Bertram, D. Blacker, K. Mullin, D. Keeney, J. Jones, S. Basu, S. Yhu, M. G. McInnis, R. C. Go, K. Vekrellis, D. J. Selkoe, A. J. Saunders and R. E. Tanzi, *Science*, 2000, **290**, 2302–2303.
30. N. A. Twine, K. Janitz, M. R. Wilkins and M. Janitz, *PLoS One*, 2011, **6**, e16266.

31. D. M. Lehman, D.-J. J. Fu, A. B. Freeman, K. J. Hunt, R. J. Leach, T. Johnson-Pais, J. Hamlington, T. D. Dyer, R. Arya, H. Abboud, H. H. H. Göring, R. Duggirala, J. Blangero, R. J. Konrad and M. P. Stern, *Diabetes*, 2005, **54**, 1214–1221.
32. W. L. Sheldon, M. S. Macauley, E. J. Taylor, C. E. Robinson, S. J. Charnock, G. J. Davies, D. J. Vocadlo and G. W. Black, *Biochem. J.*, 2006, **399**, 241–247.
33. C. Slawson, T. Lakshmanan, S. Knapp and G. W. Hart, *Mol. Biol. Cell*, 2008, **19**, 4130–4140.
34. C. N. Keembiyehetty, A. Krzeslak, D. C. Love and J. A. Hanover, *J. Cell Sci.*, 2011, **124**, 2851–2860.
35. C. Toleman, A. J. Paterson and T. R. Whisenhunt, *J. Biol. Chem.*, 2004, **279**, 53665–53673.
36. C. Butkinaree, W. D. Cheung, S. Park, K. Park, M. Barber and G. W. Hart, *J. Biol. Chem.*, 2008, **283**, 23557–23566.
37. F. V. Rao, A. W. Schüttelkopf, H. C. Dorfmueller, A. T. Ferenbach, I. Navratilova and D. M. van Aalten, *Open Biol.*, 2013, **3**, 130021.
38. Y. He, C. Roth, J. P. Turkenburg and G. J. Davies, *Acta Crystallogr., Sect. D: Biol. Crystallogr.*, 2014, **70**, 186–195.
39. N. Selvan, D. Mariappa, H. W. P. van den Toorn, A. J. R. Heck, A. T. Ferenbach and D. M. F. van Aalten, *J. Biol. Chem.*, 2015, **290**, 11969–11982.
40. S. Banerjee, P. W. Robbins and J. Samuelson, *Glycobiology*, 2009, **19**, 331–336.
41. N. E. Olszewski, C. M. West, S. O. Sassi and L. M. Hartweck, *Biochim. Biophys. Acta, Gen. Subj.*, 2010, **1800**, 49–56.
42. J. Alonso, M. Schimpl and D. M. van Aalten, *J. Biol. Chem.*, 2014, **289**, 34433–34439.
43. F. V. Rao, H. C. Dorfmueller, F. Villa, M. Allwood, I. M. Eggleston and D. M. van Aalten, *EMBO J.*, 2006, **25**, 1569–1578.
44. R. J. Dennis, E. J. Taylor, M. S. Macauley, K. A. Stubbs, J. P. Turkenburg, S. J. Hart, G. N. Black, D. J. Vocadlo and G. J. Davies, *Nat. Struct. Mol. Biol.*, 2006, **13**, 365–371.
45. G. W. Hart, C. Slawson, G. Ramirez-Correa and O. Lagerlof, *Biochemistry*, 2011, **80**, 825–858.
46. Y. R. Yang, M. Song, H. Lee, Y. Jeon, E.-J. J. Choi, H.-J. J. Jang, H. Y. Moon, H.-Y. Y. Byun, E.-K. K. Kim, D. H. Kim, M. N. Lee, A. Koh, J. Ghim, J. H. Choi, W. Lee-Kwon, K. T. Kim, S. H. Ryu and P.-G. G. Suh, *Aging Cell*, 2012, **11**, 439–448.
47. J. A. Hanover, M. E. Forsythe, P. T. Hennessey, T. M. Brodigan, D. C. Love, G. Ashwell and M. Krause, *Proc. Natl. Acad. Sci. U. S. A.*, 2005, **102**, 11266–11271.
48. M. E. Forsythe, D. C. Love, B. D. Lazarus, E. J. Kim, W. A. Prinz, G. Ashwell, M. W. Krause and J. A. Hanover, *Proc. Natl. Acad. Sci. U. S. A.*, 2006, **103**, 11952–11957.
49. O. Sekine, D. C. Love, D. S. Rubenstein and J. A. Hanover, *J. Biol. Chem.*, 2010, **285**, 38684–38691.

50. D. A. Sinclair, M. Syrzycka, M. S. Macauley, T. Rastgardani, I. Komljenovic, D. J. Vocadlo, H. W. Brock and B. M. Honda, *Proc. Natl. Acad. Sci. U. S. A.*, 2009, **106**, 13427–13432.
51. S. E. Jacobsen and N. E. Olszewski, *Plant Cell*, 1993, **5**, 887–896.
52. L. M. Hartweck, C. L. Scott and N. E. Olszewski, *Genetics*, 2002, **161**, 1279–1291.
53. A. Shimada, M. Ueguchi-Tanaka, T. Sakamoto, S. Fujioka, S. Takatsuto, S. Yoshida, T. Sazuka, M. Ashikari and M. Matsuoka, *Plant J.*, 2006, **48**, 390–402.
54. A. L. Silverstone, T.-S. S. Tseng, S. M. Swain, A. Dill, S. Y. Jeong, N. E. Olszewski and T.-P. P. Sun, *Plant Physiol.*, 2007, **143**, 987–1000.
55. R. Zentella, J. Hu, W.-P. P. Hsieh, P. A. Matsumoto, A. Dawdy, B. Barnhill, H. à. t. Oldenhof, L. M. Hartweck, S. Maitra, S. G. Thomas, S. Cockrell, M. Boyce, J. Shabanowitz, D. F. Hunt, N. E. Olszewski and T.-P. P. Sun, *Genes Dev.*, 2016, **30**, 164–176.
56. L. Wells, L. K. Kreppel, F. I. Comer, B. E. Wadzinski and G. W. Hart, *J. Biol. Chem.*, 2004, **279**, 38466–38470.
57. Q. Chen, Y. Chen, C. Bian, R. Fujiki and X. Yu, *Nature*, 2013, **493**, 561–564.
58. G. Pekkurnaz, J. C. Trinidad, X. Wang, D. Kong and T. L. Schwarz, *Cell*, 2014, **158**, 54–68.
59. R. Fujiki, W. Hashiba, H. Sekine, A. Yokoyama, T. Chikanishi, S. Ito, Y. Imai, J. Kim, H. H. He, K. Igarashi, J. Kanno, F. Ohtake, H. Kitagawa, R. G. Roeder, M. Brown and S. Kato, *Nature*, 2011, **480**, 557–560.
60. S. P. N. Iyer, Y. Akimoto and G. W. Hart, *J. Biol. Chem.*, 2003, **278**, 5399–5409.
61. S. P. N. Iyer and G. W. Hart, *J. Biol. Chem.*, 2003, **278**, 24608–24616.
62. K. Brickley, K. Pozo and F. A. Stephenson, *Biochim. Biophys. Acta*, 2011, **1813**, 269–281.
63. T. R. Whisenhunt, X. Yang, D. B. Bowe, A. J. Paterson, B. A. Tine and J. E. Kudlow, *Glycobiology*, 2006, **16**, 551–563.
64. G. W. Hart, C. Slawson, G. Ramirez-Correa and O. Lagerlof, *Annu. Rev. Biochem.*, 2011, **80**, 825–858.
65. Z. Ma and K. Vosseller, *Amino Acids*, 2013, **45**, 719–733.
66. W. Yi, P. M. Clark, D. E. Mason, M. C. Keenan, C. Hill, W. A. Goddard 3rd, E. C. Peters, E. M. Driggers and L. C. Hsieh-Wilson, *Science*, 2012, **337**, 975–980.
67. J. P. Singh, K. Zhang, J. Wu and X. Yang, *Cancer Lett.*, 2015, **356**, 244–250.
68. K. Vaidyanathan and L. Wells, *J. Biol. Chem.*, 2014, **289**, 34466–34471.
69. J. Ma and G. W. Hart, *Expert Rev. Proteomics*, 2013, **10**, 365–380.
70. J. R. Erickson, L. Pereira, L. Wang, G. Han, A. Ferguson, K. Dao, R. J. Copeland, F. Despa, G. W. Hart, C. M. Ripplinger and D. M. Bers, *Nature*, 2013, **502**, 372–376.
71. J. R. Erickson, *Front. Pharmacol.*, 2014, **5**, 59.
72. S. Dassanayaka and S. P. Jones, *Pharmacol. Ther.*, 2014, **142**, 62–71.

73. V. M. Darley-Usmar, L. E. Ball and J. C. Chatham, *J. Mol. Cell. Cardiol.*, 2012, **52**, 538–549.
74. S. A. Yuzwa, X. Shan, M. S. Macauley, T. Clark, Y. Skorobogatko, K. Vosseller and D. J. Vocadlo, *Nat. Chem. Biol.*, 2012, **8**, 393–399.
75. K. Vaidyanathan, S. Durning and L. Wells, *Crit. Rev. Biochem. Mol. Biol.*, 2014, **49**, 140–163.
76. S. A. Yuzwa and D. J. Vocadlo, *Chem. Soc. Rev.*, 2014, **43**, 6839–6858.
77. Y. Zhu, X. Shan, S. A. Yuzwa and D. J. Vocadlo, *J. Biol. Chem.*, 2014, **289**, 34472–34481.
78. W. R. Wilson and M. P. Hay, *Nat. Rev. Cancer*, 2011, **11**, 393–410.
79. W. Yi, P. M. Clark, D. E. Mason, M. C. Keenan, C. Hill, W. A. Goddard, E. C. Peters, E. M. Driggers and L. C. Hsieh-Wilson, *Science*, 2012, **337**, 975–980.
80. M. G. Buse, *Am. J. Physiol.: Endocrinol. Metab.*, 2006, **290**, E1–E8.
81. J. R. Erickson, L. Pereira, L. Wang, G. Han, A. Ferguson, K. Dao, R. J. Copeland, F. Despa, G. W. Hart, C. M. Ripplinger and D. M. Bers, *Nature*, 2013, **502**, 372–376.
82. C. C. Stock, H. C. Reilly, S. M. Buckley, D. A. Clarke and C. P. Rhoads, *Nature*, 1954, **173**, 71–72.
83. G. L. Coffey, J. Ehrlich, M. W. Fisher, A. B. Hillegas, D. L. Kohberger, H. E. Machamer, W. A. Rightsel and F. R. Roegner, *Antibiot. Chemother. (Northfield)*, 1956, **6**, 487–497.
84. S. Marshall, V. Bacote and R. R. Traxinger, *J. Biol. Chem.*, 1991, **266**, 4706–4712.
85. M. D. Li, H. B. Ruan, M. E. Hughes, J. S. Lee, J. P. Singh, S. P. Jones, M. N. Nitabach and X. Yang, *Cell Metab.*, 2013, **17**, 303–310.
86. N. Staiano, R. B. Everson, D. A. Cooney, D. S. Longnecker and S. S. Thorgeirsson, *Mutat. Res.*, 1980, **79**, 387–390.
87. D. S. Longnecker and T. J. Curphey, *Cancer Res.*, 1975, **35**, 2249–2258.
88. D. S. Longnecker and B. G. Crawford, *J. Natl. Cancer Inst.*, 1974, **53**, 573–577.
89. F. Wu, A. Lukinius, M. Bergstrom, B. Eriksson, Y. Watanabe and B. Langstrom, *Eur. J. Cancer*, 1999, **35**, 1155–1161.
90. R. S. Haltiwanger, G. D. Holt and G. W. Hart, *J. Biol. Chem.*, 1990, **265**, 2563–2568.
91. H. C. Dorfmueller, V. S. Borodkin, D. E. Blair, S. Pathak, I. Navratilova and D. M. van Aalten, *Amino Acids*, 2011, **40**, 781–792.
92. M. B. Lazarus, Y. Nam, J. Jiang, P. Sliz and S. Walker, *Nature*, 2011, **469**, 564–567.
93. Y. Zhu, T. W. Liu, S. Cecioni, R. Eskandari, W. F. Zandberg and D. J. Vocadlo, *Nat. Chem. Biol.*, 2015, **11**, 319–325.
94. N. G. McLetchie, *J. R. Coll. Physicians Edinb.*, 2002, **32**, 134–142.
95. J. Dunn, H. Sheehan and N. G. McLetchie, *Lancet*, 1943, **241**, 484.
96. R. J. Konrad, F. Zhang, J. E. Hale, M. D. Knierman, G. W. Becker and J. E. Kudlow, *Biochem. Biophys. Res. Commun.*, 2002, **293**, 207–212.

97. F. Capotosti, S. Guernier, F. Lammers, P. Waridel, Y. Cai, J. Jin, J. W. Conaway, R. C. Conaway and W. Herr, *Cell*, 2011, **144**, 376–388.
98. J. Liu, Y. Pang, T. Chang, P. Bounelis, J. C. Chatham and R. B. Marchase, *J. Mol. Cell. Cardiol.*, 2006, **40**, 303–312.
99. J. Liu, R. B. Marchase and J. C. Chatham, *J. Mol. Cell. Cardiol.*, 2007, **42**, 177–185.
100. K. Kaasik, S. Kivimae, J. J. Allen, R. J. Chalkley, Y. Huang, K. Baer, H. Kissel, A. L. Burlingame, K. M. Shokat, L. J. Ptacek and Y. H. Fu, *Cell Metab.*, 2013, **17**, 291–302.
101. H. Yamamoto, Y. Uchigata and H. Okamoto, *Nature*, 1981, **294**, 284–286.
102. T. N. Lee, W. E. Alborn, M. D. Knierman and R. J. Konrad, *Biochem. Biophys. Res. Commun.*, 2006, **350**, 1038–1043.
103. B. J. Gross, B. C. Kraybill and S. Walker, *J. Am. Chem. Soc.*, 2005, **127**, 14588–14589.
104. B. J. Gross, J. G. Swoboda and S. Walker, *J. Am. Chem. Soc.*, 2008, **130**, 440–441.
105. S. M. Rodems, B. D. Hamman, C. Lin, J. Zhao, S. Shah, D. Heidary, L. Makings, J. H. Stack and B. A. Pollok, *Assay Drug Dev. Technol.*, 2002, **1**, 9–19.
106. G. A. Ngoh, L. J. Watson, H. T. Facundo, W. Dillmann and S. P. Jones, *J. Mol. Cell. Cardiol.*, 2008, **45**, 313–325.
107. S. A. Caldwell, S. R. Jackson, K. S. Shahriari, T. P. Lynch, G. Sethi, S. Walker, K. Vosseller and M. J. Reginato, *Oncogene*, 2010, **29**, 2831–2842.
108. Y. Shi, J. Tomic, F. Wen, S. Shaha, A. Bahlo, R. Harrison, J. W. Dennis, R. Williams, B. J. Gross, S. Walker, J. Zuccolo, J. P. Deans, G. W. Hart and D. E. Spaner, *Leukemia*, 2010, **24**, 1588–1598.
109. J. Jiang, M. B. Lazarus, L. Pasquina, P. Sliz and S. Walker, *Nat. Chem. Biol.*, 2012, **8**, 72–77.
110. M. B. Lazarus, J. Jiang, V. Kapuria, T. Bhuiyan, J. Janetzko, W. F. Zandberg, D. J. Vocadlo, W. Herr and S. Walker, *Science*, 2013, **342**, 1235–1239.
111. R. F. Ortiz-Meoz, J. Jiang, M. B. Lazarus, M. Orman, J. Janetzko, C. Fan, D. Y. Duveau, Z. W. Tan, C. J. Thomas and S. Walker, *ACS Chem. Biol.*, 2015, **10**, 1392–1397.
112. M. Angelova, R. F. Ortiz-Meoz, S. Walker and D. M. Knipe, *J. Virol.*, 2015, **89**, 8474–8483.
113. T. M. Gloster, W. F. Zandberg, J. E. Heinonen, D. L. Shen, L. Deng and D. J. Vocadlo, *Nat. Chem. Biol.*, 2011, **7**, 174–181.
114. M. B. Lazarus, J. Jiang, T. M. Gloster, W. F. Zandberg, G. E. Whitworth, D. J. Vocadlo and S. Walker, *Nat. Chem. Biol.*, 2012, **8**, 966–968.
115. Z. Ma, D. J. Vocadlo and K. Vosseller, *J. Biol. Chem.*, 2013, **288**, 15121–15130.
116. Y. Perez-Cervera, V. Dehennaut, M. Aquino Gil, K. Guedri, C. J. Solorzano Mata, S. Olivier-Van Stichelen, J. C. Michalski, F. Foulquier and T. Lefebvre, *FASEB J.*, 2013, **27**, 3478–3486.
117. Y. Zhu, T. W. Liu, Z. Madden, S. A. Yuzwa, K. Murray, S. Cecioni, N. Zachara and D. J. Vocadlo, *J. Mol. Cell Biol.*, 2016, **8**, 2–16.

118. J. N. Savas, B. H. Toyama, T. Xu, J. R. Yates 3rd and M. W. Hetzer, *Science*, 2012, **335**, 942.
119. B. H. Toyama, J. N. Savas, S. K. Park, M. S. Harris, N. T. Ingolia, J. R. Yates 3rd and M. W. Hetzer, *Cell*, 2013, **154**, 971–982.
120. V. S. Borodkin, M. Schimpl, M. Gundogdu, K. Rafie, H. C. Dorfmueller, D. A. Robinson and D. M. van Aalten, *Biochem. J.*, 2014, **457**, 497–502.
121. M. Izumi, H. Yuasa and H. Hashimoto, *Curr. Top. Med. Chem.*, 2009, **9**, 87–105.
122. J. J. Vavra, C. Deboer, A. Dietz, L. J. Hanka and W. T. Sokolski, *Antibiot. Annu.*, 1959, 7, 230–235.
123. M. D. Roos, W. Xie, K. Su, J. A. Clark, X. Yang, E. Chin, A. J. Paterson and J. E. Kudlow, *Proc. Assoc. Am. Physicians*, 1998, **110**, 422–432.
124. J. A. Hanover, Z. Lai, G. Lee, W. A. Lubas and S. M. Sato, *Arch. Biochem. Biophys.*, 1999, **362**, 38–45.
125. C. Slawson, S. Shafii, J. Amburgey and R. Potter, *Biochim. Biophys. Acta*, 2002, **1573**, 121–129.
126. X. Yang, K. Su, M. D. Roos, Q. Chang, A. J. Paterson and J. E. Kudlow, *Proc. Natl. Acad. Sci. U. S. A.*, 2001, **98**, 6611–6616.
127. J. Turk, J. A. Corbett, S. Ramanadham, A. Bohrer and M. L. McDaniel, *Biochem. Biophys. Res. Commun.*, 1993, **197**, 1458–1464.
128. Y. Gao, G. J. Parker and G. W. Hart, *Arch. Biochem. Biophys.*, 2000, **383**, 296–302.
129. S. Pathak, H. C. Dorfmueller, V. S. Borodkin and D. M. van Aalten, *Chem. Biol.*, 2008, **15**, 799–807.
130. D. L. Dong and G. W. Hart, *J. Biol. Chem.*, 1994, **269**, 19321–19330.
131. D. Beer, J.-L. Maloisel, D. Rast and A. Vasella, *Helv. Chim. Acta*, 1990, **73**, 1918–1922.
132. M. Perreira, E. J. Kim, C. J. Thomas and J. A. Hanover, *Bioorg. Med. Chem.*, 2006, **14**, 837–846.
133. F. V. Rao, H. C. Dorfmueller, F. Villa, M. Allwood, I. M. Eggleston and D. M. van Aalten, *EMBO J.*, 2006, **25**, 1569–1578.
134. G. A. Ngoh, T. Hamid, S. D. Prabhu and S. P. Jones, *Am. J. Physiol.: Heart Circ. Physiol.*, 2009, **297**, H1711–H1719.
135. N. O. Ku, D. M. Toivola, P. Strnad and M. B. Omary, *Nat. Cell Biol.*, 2010, **12**, 876–885.
136. S. M. Ranuncolo, S. Ghosh, J. A. Hanover, G. W. Hart and B. A. Lewis, *J. Biol. Chem.*, 2012, **287**, 23549–23561.
137. J. Lee, K. Y. Kim, J. Lee and Y. K. Paik, *J. Biol. Chem.*, 2010, **285**, 2930–2939.
138. K. A. Stubbs, N. Zhang and D. J. Vocadlo, *Org. Biomol. Chem.*, 2006, **4**, 839–845.
139. A. Mehdy, W. Morelle, C. Rosnoblet, D. Legrand, T. Lefebvre, S. Duvet and F. Foulquier, *J. Biochem.*, 2012, **151**, 439–446.
140. G. E. Whitworth, M. S. Macauley, K. A. Stubbs, R. J. Dennis, E. J. Taylor, G. J. Davies, I. R. Greig and D. J. Vocadlo, *J. Am. Chem. Soc.*, 2007, **129**, 635–644.

141. S. Knapp, D. J. Vocadlo, Z. Gao, B. Kirk, J. Lou and S. G. Withers, *J. Am. Chem. Soc.*, 1996, **118**, 6804–6805.
142. M. S. Macauley, G. E. Whitworth, A. W. Debowski, D. Chin and D. J. Vocadlo, *J. Biol. Chem.*, 2005, **280**, 25313–25322.
143. R. J. Dennis, E. J. Taylor, M. S. Macauley, K. A. Stubbs, J. P. Turkenburg, S. J. Hart, G. N. Black, D. J. Vocadlo and G. J. Davies, *Nat. Struct. Mol. Biol.*, 2006, **13**, 365–371.
144. M. S. Macauley, A. K. Bubb, C. Martinez-Fleites, G. J. Davies and D. J. Vocadlo, *J. Biol. Chem.*, 2008, **283**, 34687–34695.
145. S. Knapp, M. Abdo, K. Ajayi, R. A. Huhn, T. J. Emge, E. J. Kim and J. A. Hanover, *Org. Lett.*, 2007, **9**, 2321–2324.
146. B. Laczy, S. A. Marsh, C. A. Brocks, I. Wittmann and J. C. Chatham, *Am. J. Physiol.: Heart Circ. Physiol.*, 2010, **299**, H1715–H1727.
147. S. Ji, S. Y. Park, J. Roth, H. S. Kim and J. W. Cho, *Biochem. Biophys. Res. Commun.*, 2012, **417**, 1158–1163.
148. X. Shan, D. J. Vocadlo and C. Krieger, *Neurosci. Lett.*, 2012, **516**, 296–301.
149. T. Aoyagi, H. Suda, K. Uotani, F. Kojima, T. Aoyama, K. Horiguchi, M. Hamada and T. Takeuchi, *J. Antibiot.*, 1992, **45**, 1404–1408.
150. T. Aoyama, H. Naganawa, H. Suda, K. Uotani, T. Aoyagi and T. Takeuchi, *J. Antibiot.*, 1992, **45**, 1557–1558.
151. H. C. Dorfmueller, V. S. Borodkin, M. Schimpl and D. M. van Aalten, *Biochem. J.*, 2009, **420**, 221–227.
152. H. C. Dorfmueller, V. S. Borodkin, M. Schimpl, X. Zheng, R. Kime, K. D. Read and D. M. van Aalten, *Chem. Biol.*, 2010, **17**, 1250–1255.
153. V. S. Borodkin and D. M. van Aalten, *Tetrahedron*, 2010, **66**, 7838–7849.
154. D. Mariappa, K. Sauert, K. Marino, D. Turnock, R. Webster, D. M. van Aalten, M. A. Ferguson and H. A. Muller, *Sci. Signaling*, 2011, **4**, ra89.
155. S. Pathak, V. S. Borodkin, O. Albarbarawi, D. G. Campbell, A. Ibrahim and D. M. van Aalten, *EMBO J.*, 2012, **31**, 1394–1404.
156. S. A. Yuzwa, M. S. Macauley, J. E. Heinonen, X. Shan, R. J. Dennis, Y. He, G. E. Whitworth, K. A. Stubbs, E. J. McEachern, G. J. Davies and D. J. Vocadlo, *Nat. Chem. Biol.*, 2008, **4**, 483–490.
157. S. A. Yuzwa, A. H. Cheung, M. Okon, L. P. McIntosh and D. J. Vocadlo, *J. Mol. Biol.*, 2014, **426**, 1736–1752.
158. S. A. Yuzwa, X. Shan, B. A. Jones, G. Zhao, M. L. Woodward, X. Li, Y. Zhu, E. J. McEachern, M. A. Silverman, N. V. Watson, C. X. Gong and D. J. Vocadlo, *Mol. Neurodegener.*, 2014, **9**, 42.
159. Y. Yu, L. Zhang, X. Li, X. Run, Z. Liang, Y. Li, Y. Liu, M. H. Lee, I. Grundke-Iqbal, K. Iqbal, D. J. Vocadlo, F. Liu and C. X. Gong, *PLoS One*, 2012, 7, e35277.
160. W. Mi, Y. Gu, C. Han, H. Liu, Q. Fan, X. Zhang, Q. Cong and W. Yu, *Biochim. Biophys. Acta*, 2011, **1812**, 514–519.
161. H. Goldberg, C. Whiteside and I. G. Fantus, *Am. J. Physiol.*, 2011, **301**, E713–E726.
162. P. S. Banerjee, J. Ma and G. W. Hart, *Proc. Natl. Acad. Sci. U. S. A.*, 2015, **112**, 6050–6055.

163. E. J. Kim, B. Amorelli, M. Abdo, C. J. Thomas, D. C. Love, S. Knapp and J. A. Hanover, *J. Am. Chem. Soc.*, 2007, **129**, 14854–14855.
164. E. J. Kim, D. C. Love, E. Darout, M. Abdo, B. Rempel, S. G. Withers, P. R. Rablen, J. A. Hanover and S. Knapp, *Bioorg. Med. Chem.*, 2010, **18**, 7058–7064.
165. M. Bergeron-Brlek, J. Goodwin-Tindall, N. Cekic, C. Roth, W. F. Zandberg, X. Shan, V. Varghese, S. Chan, G. J. Davies, D. J. Vocadlo and R. Britton, *Angew. Chem., Int. Ed.*, 2015, **54**, 15429–15433.
166. H. C. Dorfmueller and D. M. van Aalten, *FEBS Lett.*, 2010, **584**, 694–700.
167. T. Li, L. Guo, Y. Zhang, J. Wang, Z. Li, L. Lin, Z. Zhang, L. Li, J. Lin, W. Zhao, J. Li and P. G. Wang, *Carbohydr. Res.*, 2011, **346**, 1083–1092.
168. K. B. Sharpless and R. Manetsch, *Expert Opin. Drug Discovery*, 2006, **1**, 525–538.
169. J. A. Hanover, C. K. Cohen, M. C. Willingham and M. K. Park, *J. Biol. Chem.*, 1987, **262**, 9887–9894.
170. N. E. Zachara, G. W. Hart, R. N. Cole and Y. Gao, *Current Protocols in Molecular Biology*, ed. F. M. Ausubel, 2002, Chapter 17, Unit 17 16.
171. D. Schouppe, B. Ghesquiere, G. Menschaert, W. H. De Vos, S. Bourque, G. Trooskens, P. Proost, K. Gevaert and E. J. Van Damme, *Plant Physiol.*, 2011, **155**, 1091–1102.
172. A. Delporte, W. H. De Vos and E. J. Van Damme, *J. Plant Physiol.*, 2014, **171**, 1149–1156.
173. A. Delporte, J. De Zaeytijd, N. De Storme, A. Azmi, D. Geelen, G. Smagghe, Y. Guisez and E. J. Van Damme, *Plant Physiol. Biochem.*, 2014, **83**, 151–158.
174. C. M. Snow, A. Senior and L. Gerace, *J. Cell Biol.*, 1987, **104**, 1143–1156.
175. G. D. Holt, C. M. Snow, A. Senior, R. S. Haltiwanger, L. Gerace and G. W. Hart, *J. Cell Biol.*, 1987, **104**, 1157–1164.
176. F. I. Comer, K. Vosseller, L. Wells, M. A. Accavitti and G. W. Hart, *Anal. Biochem.*, 2001, **293**, 169–177.
177. C. F. Teo, S. Ingale, M. A. Wolfert, G. A. Elsayed, L. G. Not, J. C. Chatham, L. Wells and G. J. Boons, *Nat. Chem. Biol.*, 2010, **6**, 338–343.
178. A. L. Klein, M. N. Berkaw, M. G. Buse and L. E. Ball, *Mol. Cell. Proteomics*, 2009, **8**, 2733–2745.
179. R. Fujiki, W. Hashiba, H. Sekine, A. Yokoyama, T. Chikanishi, S. Ito, Y. Imai, J. Kim, H. H. He, K. Igarashi, J. Kanno, F. Ohtake, H. Kitagawa, R. G. Roeder, M. Brown and S. Kato, *Nature*, 2011, **480**, 557–560.
180. S. A. Yuzwa, A. K. Yadav, Y. Skorobogatko, T. Clark, K. Vosseller and D. J. Vocadlo, *Amino Acids*, 2011, **40**, 857–868.
181. D. Mariappa, N. Selvan, V. S. Borodkin, J. Alonso, A. T. Ferenbach, C. Shepherd, I. H. Navratilova and D. M. van Aalten, *Biochem. J.*, 2015, **470**, 255–262.
182. A. S. Vercoutter-Edouart, I. E. Yazidi-Belkoura, C. Guinez, S. Baldini, M. Leturcq, M. Mortuaire, A. M. Mir, A. Steenackers, V. Dehennaut, A. Pierce and T. Lefebvre, *Proteomics*, 2015, **15**, 1039–1050.
183. S. A. Myers, S. Daou, B. Affar el and A. Burlingame, *Proteomics*, 2013, **13**, 982–991.

184. N. Khidekel, S. B. Ficarro, P. M. Clark, M. C. Bryan, D. L. Swaney, J. E. Rexach, Y. E. Sun, J. J. Coon, E. C. Peters and L. C. Hsieh-Wilson, *Nat. Chem. Biol.*, 2007, **3**, 339–348.
185. M. P. Housley, J. T. Rodgers, N. D. Udeshi, T. J. Kelly, J. Shabanowitz, D. F. Hunt, P. Puigserver and G. W. Hart, *J. Biol. Chem.*, 2008, **283**, 16283–16292.
186. M. P. Housley, N. D. Udeshi, J. T. Rodgers, J. Shabanowitz, P. Puigserver, D. F. Hunt and G. W. Hart, *J. Biol. Chem.*, 2009, **284**, 5148–5157.
187. R. J. Chalkley, A. Thalhammer, R. Schoepfer and A. L. Burlingame, *Proc. Natl. Acad. Sci. U. S. A.*, 2009, **106**, 8894–8899.
188. S. A. Myers, B. Panning and A. L. Burlingame, *Proc. Natl. Acad. Sci. U. S. A.*, 2011, **108**, 9490–9495.
189. P. Zhao, R. Viner, C. F. Teo, G. J. Boons, D. Horn and L. Wells, *J. Proteome Res.*, 2011, **10**, 4088–4104.
190. H. A. Ebhardt, A. Root, C. Sander and R. Aebersold, *Proteomics*, 2015, **15**, 3193–3208.
191. J. J. Maury, D. Ng, X. Bi, M. Bardor and A. B. Choo, *Anal. Chem.*, 2014, **86**, 395–402.
192. C. Shao, W. Sun, F. Li, R. Yang, L. Zhang and Y. Gao, *J. Mass Spectrom.*, 2009, **44**, 25–31.
193. H. Hahne, A. Moghaddas Gholami and B. Kuster, *Mol. Cell. Proteomics*, 2012, **11**, 843–850.
194. N. E. Zachara, K. Vosseller and G. W. Hart, *Current Protocols in Protein Science*, ed. J. E. Coligan, 2011, Chapter 12, Unit12 18.
195. L. Wells, K. Vosseller, R. N. Cole, J. M. Cronshaw, M. J. Matunis and G. W. Hart, *Mol. Cell. Proteomics*, 2002, **1**, 791–804.
196. K. Vosseller, K. C. Hansen, R. J. Chalkley, J. C. Trinidad, L. Wells, G. W. Hart and A. L. Burlingame, *Proteomics*, 2005, **5**, 388–398.
197. K. Vosseller, J. C. Trinidad, R. J. Chalkley, C. G. Specht, A. Thalhammer, A. J. Lynn, J. O. Snedecor, S. Guan, K. F. Medzihradszky, D. A. Maltby, R. Schoepfer and A. L. Burlingame, *Mol. Cell. Proteomics*, 2006, **5**, 923–934.
198. T. Overath, U. Kuckelkorn, P. Henklein, B. Strehl, D. Bonar, A. Kloss, D. Siele, P. M. Kloetzel and K. Janek, *Mol. Cell. Proteomics*, 2012, **11**, 467–477.
199. J. Ma, T. Liu, A. C. Wei, P. Banerjee, B. O'Rourke and G. W. Hart, *J. Biol. Chem.*, 2015, **290**, 29141–29153.
200. Z. Wang, A. Pandey and G. W. Hart, *Mol. Cell. Proteomics*, 2007, **6**, 1365–1379.
201. E. Klement, Z. Lipinszki, Z. Kupihar, A. Udvardy and K. F. Medzihradszky, *J. Proteome Res.*, 2010, **9**, 2200–2206.
202. G. D. Holt and G. W. Hart, *J. Biol. Chem.*, 1986, **261**, 8049–8057.
203. G. D. Holt, R. S. Haltiwanger, C. R. Torres and G. W. Hart, *J. Biol. Chem.*, 1987, **262**, 14847–14850.
204. B. Ramakrishnan and P. K. Qasba, *J. Biol. Chem.*, 2002, **277**, 20833–20839.

205. N. Khidekel, S. Arndt, N. Lamarre-Vincent, A. Lippert, K. G. Poulin-Kerstien, B. Ramakrishnan, P. K. Qasba and L. C. Hsieh-Wilson, *J. Am. Chem. Soc.*, 2003, **125**, 16162–16163.
206. H. C. Tai, N. Khidekel, S. B. Ficarro, E. C. Peters and L. C. Hsieh-Wilson, *J. Am. Chem. Soc.*, 2004, **126**, 10500–10501.
207. J. E. Rexach, C. J. Rogers, S. H. Yu, J. Tao, Y. E. Sun and L. C. Hsieh-Wilson, *Nat. Chem. Biol.*, 2010, **6**, 645–651.
208. J. E. Rexach, P. M. Clark, D. E. Mason, R. L. Neve, E. C. Peters and L. C. Hsieh-Wilson, *Nat. Chem. Biol.*, 2012, **8**, 253–261.
209. P. M. Clark, J. F. Dweck, D. E. Mason, C. R. Hart, S. B. Buck, E. C. Peters, B. J. Agnew and L. C. Hsieh-Wilson, *J. Am. Chem. Soc.*, 2008, **130**, 11576–11577.
210. Z. Wang, N. D. Udeshi, M. O'Malley, J. Shabanowitz, D. F. Hunt and G. W. Hart, *Mol. Cell. Proteomics*, 2010, **9**, 153–160.
211. K. Sakabe, Z. Wang and G. W. Hart, *Proc. Natl. Acad. Sci. U. S. A.*, 2010, **107**, 19915–19920.
212. J. F. Alfaro, C. X. Gong, M. E. Monroe, J. T. Aldrich, T. R. Clauss, S. O. Purvine, Z. Wang, D. G. Camp 2nd, J. Shabanowitz, P. Stanley, G. W. Hart, D. F. Hunt, F. Yang and R. D. Smith, *Proc. Natl. Acad. Sci. U. S. A.*, 2012, **109**, 7280–7285.
213. P. Ramakrishnan, P. M. Clark, D. E. Mason, E. C. Peters, L. C. Hsieh-Wilson and D. Baltimore, *Sci. Signaling*, 2013, **6**, ra75.
214. J. Xiao, S. Xu, C. Li, Y. Xu, L. Xing, Y. Niu, Q. Huan, Y. Tang, C. Zhao, D. Wagner, C. Gao and K. Chong, *Nat. Commun.*, 2014, **5**, 4572.
215. L. Xing, J. Li, Y. Xu, Z. Xu and K. Chong, *PLoS One*, 2009, **4**, e4854.
216. D. J. Vocadlo, H. C. Hang, E. J. Kim, J. A. Hanover and C. R. Bertozzi, *Proc. Natl. Acad. Sci. U. S. A.*, 2003, **100**, 9116–9121.
217. D. H. Dube and C. R. Bertozzi, *Curr. Opin. Chem. Biol.*, 2003, 7, 616–625.
218. J. A. Prescher and C. R. Bertozzi, *Cell*, 2006, **126**, 851–854.
219. M. Boyce and C. R. Bertozzi, *Nat. Methods*, 2011, **8**, 638–642.
220. R. Sprung, A. Nandi, Y. Chen, S. C. Kim, D. Barma, J. R. Falck and Y. Zhao, *J. Proteome Res.*, 2005, **4**, 950–957.
221. A. Nandi, R. Sprung, D. K. Barma, Y. Zhao, S. C. Kim, J. R. Falck and Y. Zhao, *Anal. Chem.*, 2006, **78**, 452–458.
222. C. Gurcel, A. S. Vercoutter-Edouart, C. Fonbonne, M. Mortuaire, A. Salvador, J. C. Michalski and J. Lemoine, *Anal. Bioanal. Chem.*, 2008, **390**, 2089–2097.
223. H. Hahne, N. Sobotzki, T. Nyberg, D. Helm, V. S. Borodkin, D. M. van Aalten, B. Agnew and B. Kuster, *J. Proteome Res.*, 2013, **12**, 927–936.
224. M. Boyce, I. S. Carrico, A. S. Ganguli, S. H. Yu, M. J. Hangauer, S. C. Hubbard, J. J. Kohler and C. R. Bertozzi, *Proc. Natl. Acad. Sci. U. S. A.*, 2011, **108**, 3141–3146.
225. K. K. Palaniappan, M. J. Hangauer, T. J. Smith, B. P. Smart, A. A. Pitcher, E. H. Cheng, C. R. Bertozzi and M. Boyce, *Cell Rep.*, 2013, **5**, 546–552.
226. C. F. Teo and L. Wells, *Anal. Biochem.*, 2014, **464**, 70–72.

227. C. M. Woo, A. T. Iavarone, D. R. Spiciarich, K. K. Palaniappan and C. R. Bertozzi, *Nat. Methods*, 2015, **12**, 561–567.
228. A. R. Burnham-Marusich, C. J. Snodgrass, A. M. Johnson, C. M. Kiyoshi, S. E. Buzby, M. R. Gruner and P. M. Berninsone, *PLoS One*, 2012, 7, e49020.
229. R. Zentella, J. Hu, W. P. Hsieh, P. A. Matsumoto, A. Dawdy, B. Barnhill, H. Oldenhof, L. M. Hartweck, S. Maitra, S. G. Thomas, S. Cockrell, M. Boyce, J. Shabanowitz, D. F. Hunt, N. E. Olszewski and T. P. Sun, *Genes Dev.*, 2016, **30**, 164–176.
230. J. C. Jewett, E. M. Sletten and C. R. Bertozzi, *J. Am. Chem. Soc.*, 2010, **132**, 3688–3690.
231. E. M. Sletten and C. R. Bertozzi, *Angew. Chem., Int. Ed.*, 2009, **48**, 6974–6998.
232. E. J. Kim, D. W. Kang, H. F. Leucke, M. R. Bond, S. Ghosh, D. C. Love, J. S. Ahn, D. O. Kang and J. A. Hanover, *Carbohydr. Res.*, 2013, **377**, 18–27.
233. B. W. Zaro, Y. Y. Yang, H. C. Hang and M. R. Pratt, *Proc. Natl. Acad. Sci. U. S. A.*, 2011, **108**, 8146–8151.
234. Z. Gurel, B. W. Zaro, M. R. Pratt and N. Sheibani, *PLoS One*, 2014, **9**, e95561.
235. L. A. Bateman, B. W. Zaro, K. N. Chuh and M. R. Pratt, *Chem. Commun. (Camb.)*, 2013, **49**, 4328–4330.
236. B. W. Zaro, K. N. Chuh and M. R. Pratt, *ACS Chem. Biol.*, 2014, **9**, 1991–1996.
237. K. N. Chuh, B. W. Zaro, F. Piller, V. Piller and M. R. Pratt, *J. Am. Chem. Soc.*, 2014, **136**, 12283–12295.
238. D. M. Patterson, K. A. Jones and J. A. Prescher, *Mol. BioSyst.*, 2014, **10**, 1693–1697.
239. D. M. Patterson and J. A. Prescher, *Curr. Opin. Chem. Biol.*, 2015, **28**, 141–149.
240. F. Doll, A. Buntz, A. K. Spate, V. F. Schart, A. Timper, W. Schrimpf, C. R. Hauck, A. Zumbusch and V. Wittmann, *Angew. Chem., Int. Ed.*, 2016, **55**, 2262–2266.
241. W. Lin, L. Gao and X. Chen, *ChemBioChem*, 2015, **16**, 2571–2575.
242. R. S. Haltiwanger, M. A. Blomberg and G. W. Hart, *J. Biol. Chem.*, 1992, **267**, 9005–9013.
243. L. K. Kreppel and G. W. Hart, *J. Biol. Chem.*, 1999, **274**, 32015–32022.
244. W. A. Lubas and J. A. Hanover, *J. Biol. Chem.*, 2000, **275**, 10983–10988.
245. E. J. Kim, D. O. Kang, D. C. Love and J. A. Hanover, *Carbohydr. Res.*, 2006, **341**, 971–982.
246. T. M. Leavy and C. R. Bertozzi, *Bioorg. Med. Chem. Lett.*, 2007, **17**, 3851–3854.
247. E. J. Kim, L. K. Abramowitz, M. R. Bond, D. C. Love, D. W. Kang, H. F. Leucke, D. W. Kang, J. S. Ahn and J. A. Hanover, *Bioconjugate Chem.*, 2014, **25**, 1025–1030.
248. R. F. Ortiz-Meoz, Y. Merbl, M. W. Kirschner and S. Walker, *J. Am. Chem. Soc.*, 2014, **136**, 4845–4848.

249. L. D. Carrillo, L. Krishnamoorthy and L. K. Mahal, *J. Am. Chem. Soc.*, 2006, **128**, 14768–14769.
250. J. Tanskanen, S. Saarela, S. Tankka, N. Kalkkinen, M. Rhen, T. K. Korhonen and B. Westerlund-Wikstrom, *J. Bacteriol.*, 2001, **183**, 512–519.
251. L. D. Carrillo, J. A. Froemming and L. K. Mahal, *J. Biol. Chem.*, 2011, **286**, 6650–6658.
252. J. M. Chalker, S. B. Gunnoo, O. Boutureira, S. C. Gerstberger, M. Fernandez-Gonzalez, G. J. L. Bernardes, L. Griffin, H. Hailu, C. J. Schofield and B. G. Davis, *Chem. Sci.*, 2011, **2**, 1666–1676.
253. J. M. Chalker, L. Lercher, N. R. Rose, C. J. Schofield and B. G. Davis, *Angew. Chem., Int. Ed.*, 2012, **51**, 1835–1839.
254. L. Lercher, R. Raj, N. A. Patel, J. Price, S. Mohammed, C. V. Robinson, C. J. Schofield and B. G. Davis, *Nat. Commun.*, 2015, **6**, 7978.
255. M. K. Tarrant, H. S. Rho, Z. Xie, Y. L. Jiang, C. Gross, J. C. Culhane, G. Yan, J. Qian, Y. Ichikawa, T. Matsuoka, N. Zachara, F. A. Etzkorn, G. W. Hart, J. S. Jeong, S. Blackshaw, H. Zhu and P. A. Cole, *Nat. Chem. Biol.*, 2012, **8**, 262–269.
256. M. Holt and T. Muir, *Annu. Rev. Biochem.*, 2015, **84**, 265–290.
257. N. P. Marotta, C. A. Cherwien, T. Abeywardana and M. R. Pratt, *ChemBioChem*, 2012, **13**, 2665–2670.
258. N. P. Marotta, Y. H. Lin, Y. E. Lewis, M. R. Ambroso, B. W. Zaro, M. T. Roth, D. B. Arnold, R. Langen and M. R. Pratt, *Nat. Chem.*, 2015, **7**, 913–920.
259. R. Dentin, S. Hedrick, J. Xie, J. Yates 3rd and M. Montminy, *Science*, 2008, **319**, 1402–1405.
260. S. H. Yu, M. Boyce, A. M. Wands, M. R. Bond, C. R. Bertozzi and J. J. Kohler, *Proc. Natl. Acad. Sci. U. S. A.*, 2012, **109**, 4834–4839.
261. A. C. Rodriguez, S. H. Yu, B. Li, H. Zegzouti and J. J. Kohler, *J. Biol. Chem.*, 2015, **290**, 22638–22648.

CHAPTER 5

Chemical Synthesis and Engineering of N-Linked Glycoproteins

HONGXING LI[a], YUANKUN DAO[a] AND SUWEI DONG*[a]

[a]Peking University, School of Pharmaceutical Sciences, State Key Laboratory of Natural and Biomimetic Drugs, 38 Xueyuan Road, Haidian District, Beijing, 100191, PR China
*E-mail: dongs@hsc.pku.edu.cn

5.1 Introduction

As a key component of the central dogma, proteins have dominated numerous research efforts in many areas of life science. In contrast, the functions of glycans are still quite far from being completely understood, despite the fact that many important proteins are heavily glycosylated. The major challenges involved in solving the mysteries of protein glycosylation originate, mostly, from the incredibly diverse and complex heterogeneity of glycans. For instance, while *N*-linked glycans are known to be covalently attached to asparagine residues *via* amide linkages within the corresponding conserved consensus sequence: Asn–Xaa–Ser/Thr (Scheme 5.1), individual glycan structures usually share only a pentasaccharide core (Man_3–$GlcNAc_2$) and present tremendous diversity in their outer appendages (Scheme 5.2).[1]

To better understand the molecular details of glycan function, many efforts to produce homogeneous glycoforms of proteins have been pursued.[2]

Chemical Biology No. 2
Chemical Biology of Glycoproteins
Edited by Zhongping Tan and Lai-Xi Wang

Published by the Royal Society of Chemistry, www.rsc.org

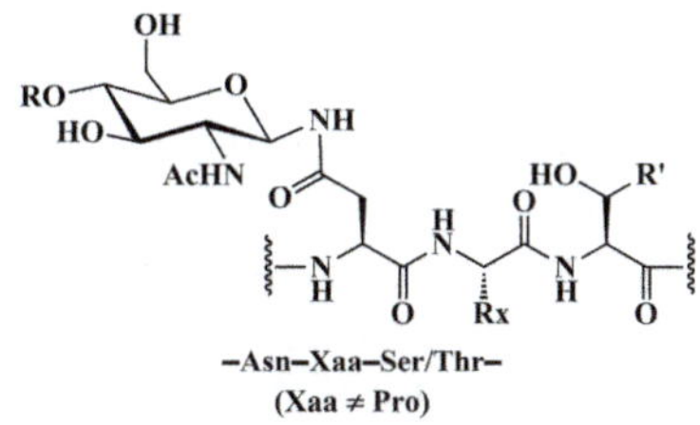

Scheme 5.1 Consensus sequon of *N*-linked glycosylation on proteins.

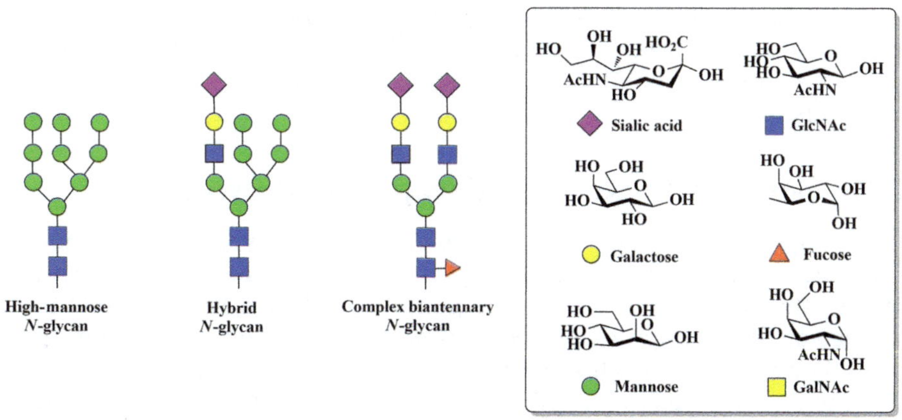

Scheme 5.2 Different types of protein *N*-glycans.

Among these efforts, chemical approaches have demonstrated wide applicability to the synthesis and engineering of certain glycoproteins.[3] Although the current synthetic techniques are the result of years of optimization and are quite powerful for producing complex small molecules, preparation of a biologically relevant molecule with the size and sophistication of a typical glycoprotein is still a formidable challenge. To address this issue, solutions to three key problems in preparing homogeneous glycoproteins are required: (1) strategies to obtain peptide segments and structurally well-defined pure carbohydrates; (2) methods to selectively introduce the glycans on peptide fragments or whole-protein peptide chains; (3) approaches to connect short peptide/glycopeptide fragments to form the complete protein sequence, and proper folding conditions to obtain the final protein with correct tertiary structure. Several of the advances in these directions have recently been reviewed and discussed, in particularly the ligation-desulfurization strategy common in full-length-protein chemical synthesis.[3,4] Thus, in this chapter, we intended to focus the discussion on techniques for preparing *N*-glycosylated proteins from a historical view, hoping to display the evolution of strategies applied in *de novo* synthesis of glycoproteins, starting from early glycosylated protein mimics to those containing fully natural glycosyl–amide linkages. Through the examination and analysis of representative cases, we hope also to highlight the key developments of technologies and methodologies in this research area.

To help better present these topics, we will focus our discussion on studies targeting proteins with naturally occurring *N*-glycan structures and those that synthesized glycoforms by introducing glycans through entirely chemical methods. However, studies targeting both glycoproteins where the glycan linkages are unnatural covalent linkers as well those with natural amide bond linkages will be included in the interest of completeness. Based on the different sources from which the peptide fragments and oligosaccharides used in the synthesis were obtained, all examples are further sub-categorized as "semisynthesis" (where any of the key building blocks were from natural sources, or recombinant methods), or "total synthesis" (where the whole protein was all built from the simplest commercially available starting materials).

5.2 Semisynthesis of *N*-Glycosylated Proteins

One of the long-standing challenges in engineering glycoproteins is how to introduce specific homogeneous glycans in a highly site-selective manner. Due to the size limitations of solid-phase peptide synthesis (SPPS),[5] early studies of chemical protein glycosylation were based on recombinant peptides. For instance, the Flitsch group reported *in vitro* experiments to attach one monosaccharide to recombinant erythropoietin (EPO).[6] As an essential glycoprotein hormone regulating red blood cell production,[7,8] as well as a broadly used therapeutic agent in treating anemia,[9] EPO has attracted significant attention in scientific research and drug development, and numerous studies have been conducted aiming to identify the functions of EPO's *N*-glycans (*vide infra*). Natural EPO consists of 166 amino acids, bearing three *N*-linked glycans at Asn24, 38, and 83, and one *O*-linked glycan at Ser126.[10] In Flitsch's study, the glycan was introduced on a mutated cysteine at one of the natural glycosylation sites through an unnatural acetamide linkage formed between the anomeric carbon of *N*-acetyl-glucosamine and the cysteine's thiol group. The Flitsch synthesis of an EPO mimic represents one of the earliest attempts at producing homogeneously glycosylated proteins, but it is clear from the work that many sites within the recombinant form of the whole protein sequence were significantly reactive towards the unprotected complex glycans, and many side-chain functional groups can interfere with the desired glycosylation. It is apparent that site-selective glycosylation on an unprotected peptide or whole protein using chemical methods was, and still is, a formidable challenge.

To circumvent any undesired reactions during the glycosylation reaction and introduce glycans in a more controllable fashion, an alternative strategy would be to focus initially on the construction of smaller glycopeptide fragments followed by connecting all necessary segments together. The success of such an approach relies on a highly chemoselective and powerful ligation method. While the revival of preparing glycoproteins using chemical methods has been based on the revolutionary technology of SPPS,[5] it was the combination of native chemical ligation (NCL)[11] and mild desulfurization[12] that overcame the size limit of SPPS, thus allowing for the production of native proteins from short peptide or glycopeptide segments (Scheme 5.3). However, the availability of structurally well-defined complex oligosaccharides still limits the synthesis and engineering of homogeneous glycoproteins. As precise as the current chemical protocols are, tedious synthetic routes are still required to prepare the many different types of carbohydrate building blocks required in a typical glycoprotein synthesis.[13] Taking advantage of isolated natural product or recombinant polypeptides, and using these materials in glycoprotein engineering, has thus emerged as an effective alternative strategy to total synthesis.

In contrast to the oligosaccharides obtained from time-consuming and labor-intensive chemical synthesis,[14] or costly chemoenzymatic manipulations,[15] a special complex oligosaccharide was found to be abundant in egg yolk (Scheme 5.4), which was first isolated by Koketsu and colleagues as a sialylglycopeptide (SGP).[16] The *N*-glycan moiety on SGP has since been identified as a biantennary complex-type undecasaccharide with two terminal $\alpha_{(2\rightarrow6)}$-linked sialic acids. This discovery has prompted numerous studies and advances since, in particularly from the Kajihara group, to derivatize SGP

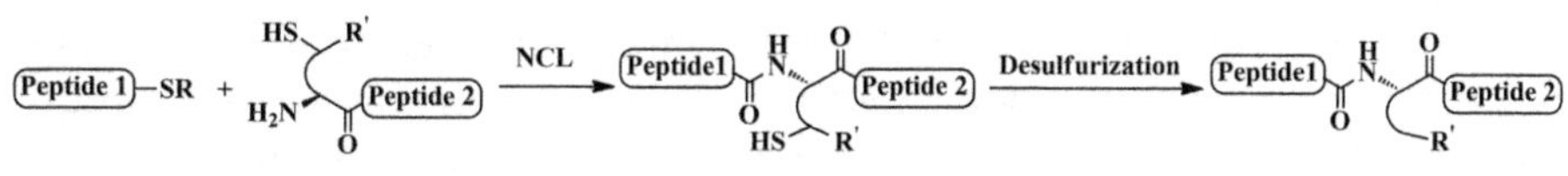

Scheme 5.3 Native chemical ligation and desulfurization in peptide synthesis.

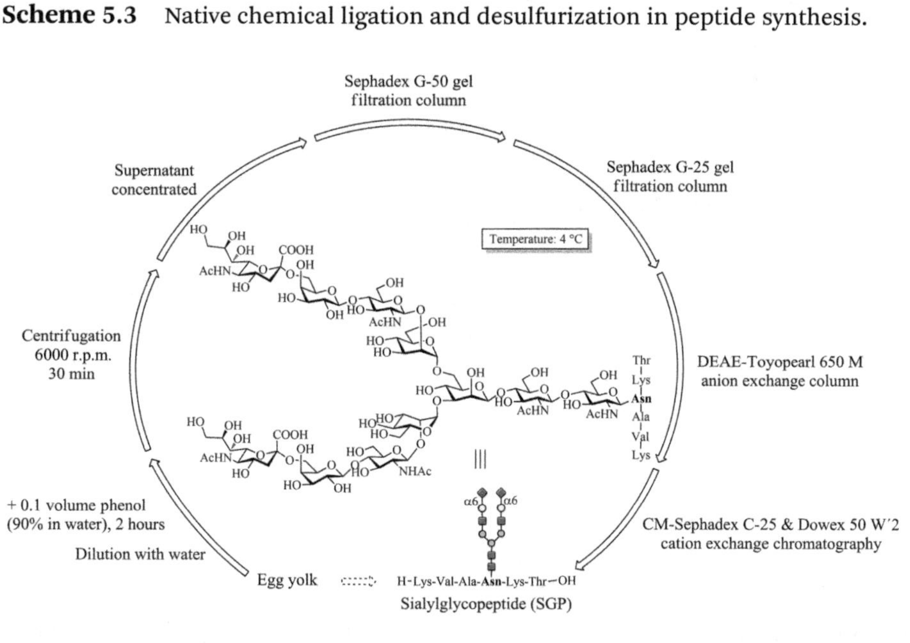

Scheme 5.4 Isolation and purification of sialylglycopeptide (SGP) from egg yolk.

using chemical or chemoenzymatic methods for use in subsequent glycoprotein synthesis.[17]

Another isolable pure *N*-glycan, a triantennary M9 high-mannose-type oligosaccharide, was also found to be available from soybean agglutinin (SBA),[18] albeit in much lower abundance.[19] However, Kajihara, Ito and colleagues managed to take the isolated carbohydrate from hen egg yolk on a preparative scale and elaborated it to the 9-fluorenylmethoxycarbonyl (Fmoc)–Asn($Man_9GlcNAc_2$)-OH building block (Scheme 5.5),[20] which was utilized further in glycoprotein synthesis (*cf.* Section 5.3).

These readily available complex oligosaccharides, along with recombinant methods for peptide preparation, have allowed for a number of semi-synthetic studies of production of homogeneous glycoproteins.

5.2.1 Semisynthesis of *N*-Glycoprotein Mimics with Unnatural Glycosyl Linkages at Cysteines

Utilizing the isolated SGP from egg yolk, Kajihara and colleagues designed and prepared a bromoacetamide derivative of the pure undecasaccharide,[17c] which was shown to selectively react with the thiol groups of free cysteines. With this important carbohydrate building block in hand, they synthesized a mimic of EPO bearing two complex-type biantennary undecasaccharides at glycosylation sites 24 and 30, both of which had been mutated to cysteines.[21] The glycosylation sites chosen in this case were based on the modified EPO therapeutic Darbepoetin alfa,[22] as the wild-type EPO contains three *N*-linked glycans at Asn residues 24, 38, and 83.[10] As shown in Scheme 5.6,

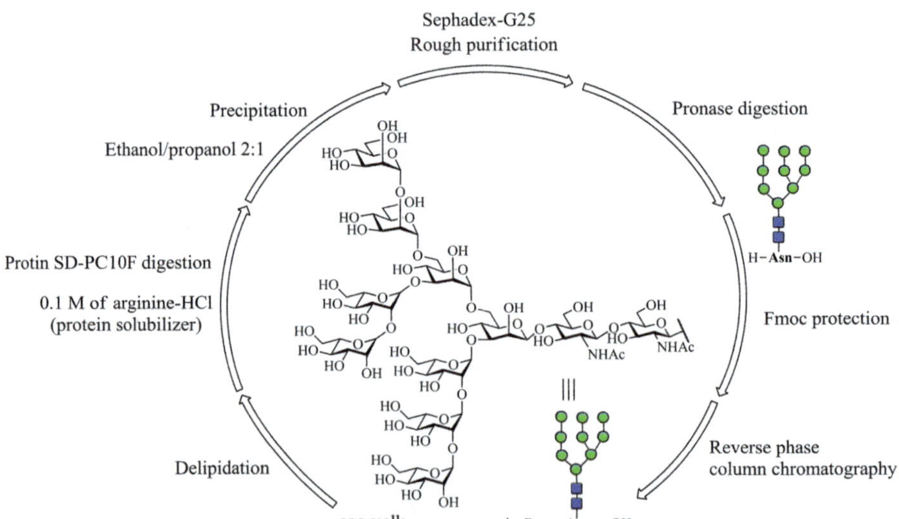

Scheme 5.5 Preparation of triantennary M9-oligosaccharide-derived building block from hen egg yolk.

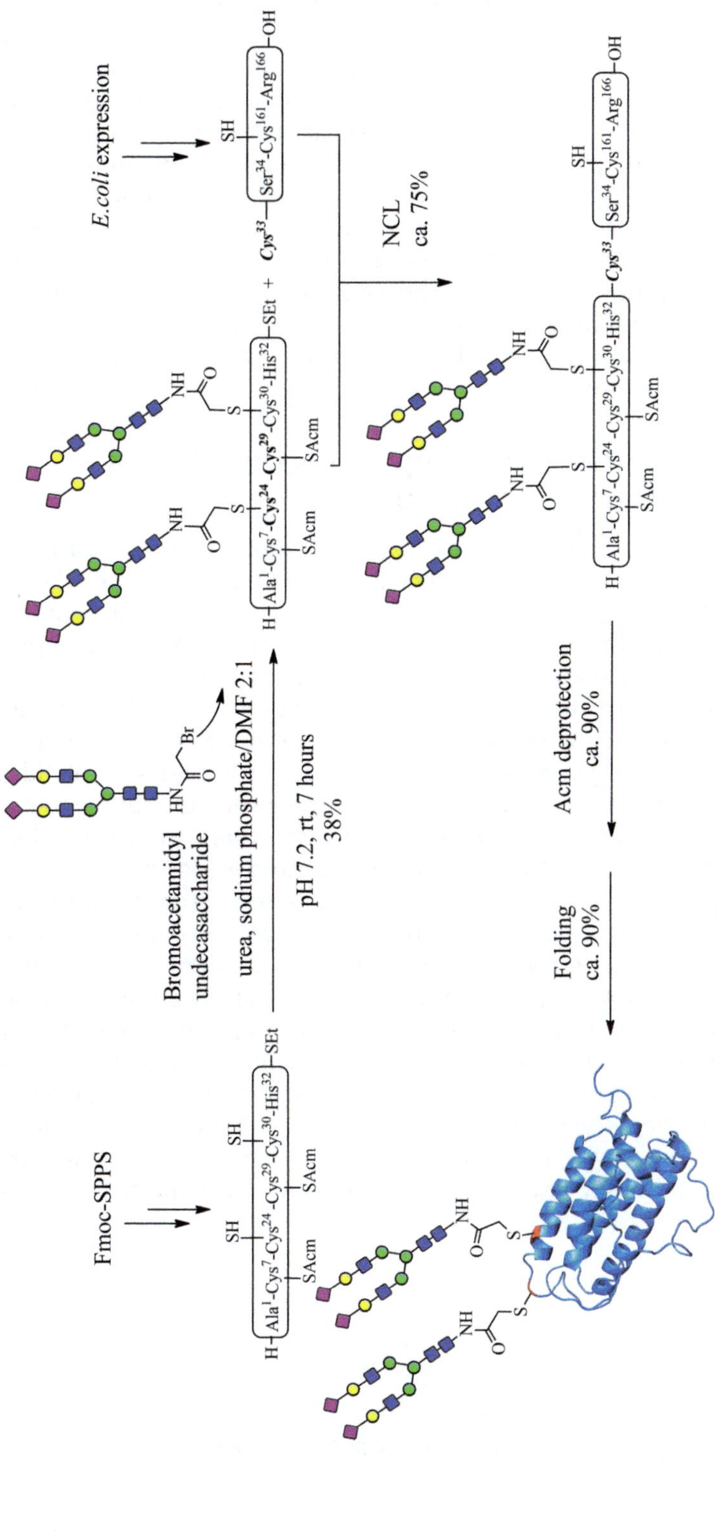

Scheme 5.6 Semisynthesis of an EPO mimic bearing two undecasaccharides at positions 24 and 30.

NCL between the chemically synthesized glycopeptide thioester EPO(1–32) and the recombinant polypeptide EPO(33–166), generated the entire protein sequence efficiently. After the removal of acetamidomethyl (Acm) protection on cysteines 7 and 29 and folding, an unnatural homogeneously disialyl-glycosylated EPO was obtained. In a similar manner, another homogeneous EPO analogue was also synthesized, which contained three biantennary sialyloligosaccharides at glycosylation sites 24, 28, and 32.[23] Although both these analogues were found to exhibit significantly reduced bioactivity *in vivo*, the successful synthetic attempts demonstrated the applicability of expressed peptides, and more importantly the isolated complex-type sialo-oligosaccharide, in semisynthesis of glycoproteins. Moreover, the cysteine-selective glycosylation strategy has provided an alternative method to engineer glycoproteins and generate homogeneously glycosylated protein analogues.

5.2.2 Semisynthesis of Glycoproteins with Natural *N*-Linkages on Asparagines

To introduce the *N*-glycan moiety on peptide chains and form a natural amide linkage selectively, early studies focused on using the preformed glycosyl asparagine building block directly in SPPS (Scheme 5.7), similar to the well-known "cassette" strategy used in the synthesis of *O*-linked glycopeptides.[24]

Since the terminal sialic acids (also known as *N*-acetylneuraminic acid, NeuAc) of these glycans are sensitive to strong acids, and the α-thioester is incompatible with base treatment during Fmoc-SPPS, the synthesis of peptide thioesters bearing complex-type oligosaccharides on solid phase presented a substantial challenge. To solve this problem, Kajihara and Dawson designed a minimal protection strategy,[25] where the hydroxyl groups of the carbohydrate and the side-chains of amino acids were free of protection during SPPS. As it had been previously demonstrated that undesired esterification involving carbohydrate hydroxyl groups could be avoided by simply reducing the concentration of substrates in Fmoc-SPPS,[26] only the carboxylate of NeuAc requires protection by a benzyl group (Scheme 5.8). After the successful synthesis of the first glycopeptide thioester using the corresponding Asn(Bn-carbohydrate)-OH [Asn(Bn-CHO)-OH] "cassette", the peptide segment was further elaborated to produce a single glycoform of monocyte chemotactic protein-3 (MCP-3). In this case, a repetitive NCL strategy was employed to connect peptide thioester (11–35) and peptide (36–76) first, followed by the ligation with sialylglycopeptide (1–10) to yield

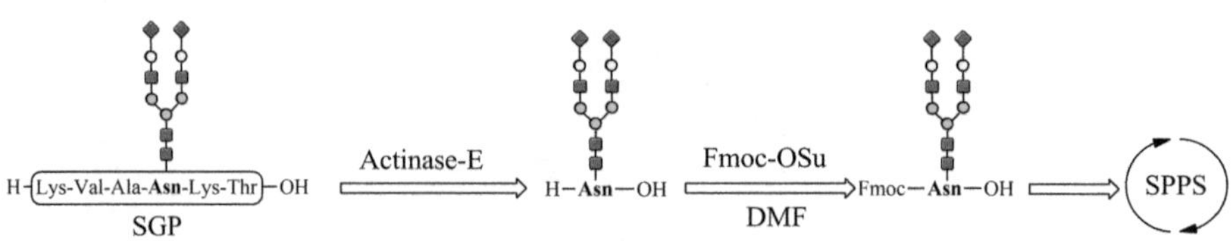

Scheme 5.7 Preparation of the glycosyl asparagine "cassette".

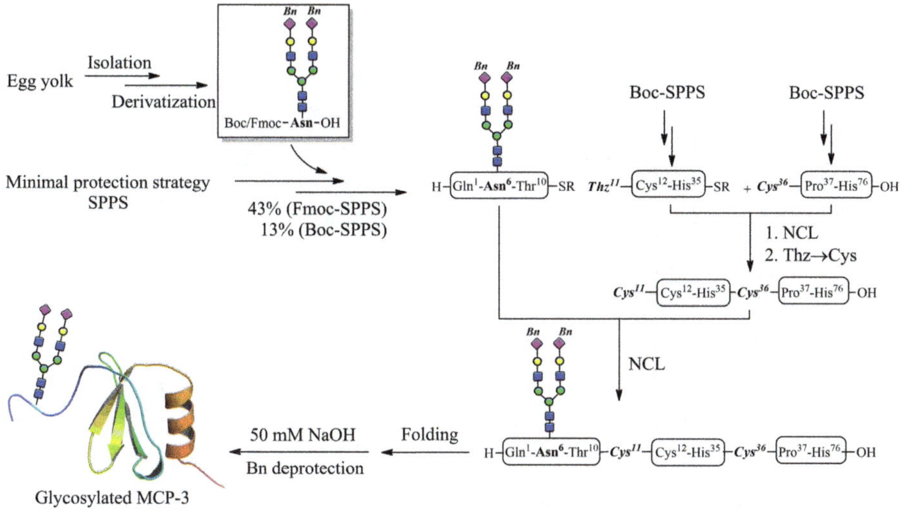

Scheme 5.8 "Cassette"-based SPPS using minimal protections and iterative NCL in the preparation of homogeneously glycosylated MCP-3.

the full sequence of MCP-3. Finally, the homogeneous glycoprotein MCP-3 was generated after folding and the removal of the benzyl protections on sialic acids. It is noteworthy that this work is the first example of chemically synthesized homogeneous glycoprotein with wild-type sialic-acid-containing *N*-glycans. It was also the first time the minimal side-chain protection strategy was employed in tert-butoxycarbonyl (Boc)–SPPS to prepare the glycopeptide thioester. However, one major disadvantage of this method is that the benzyl cation generated in the deprotection step may react with the peptide leading to undesired byproducts and reduced synthetic efficiency.[27]

Later in 2008, Unverzagt and colleagues accomplished the synthesis of ribonuclease C (Scheme 5.9).[28] Sequential NCL was employed to connect the synthetic glycopeptide thioester RNaseC(26–39), and the recombinant fragment RNaseC(40–124) with an *N*-terminal cysteine, followed by the ligation with peptide thioester RNaseC(1–25) to produce the entire protein sequence. After oxidative folding to generate the natural disulfide bonds, ribonuclease C was obtained as a single glycoform. In this work, facilitated by acetyl protecting groups on the carbohydrate hydroxyl groups, the authors used a complex-type biantennary nonasaccharide "cassette" in preparation of the glycopeptide thioester RNaseC(26–39) on solid phase.[29] Notably, the Fmoc–Asn(CHO)–OH building block was derived from the isolated SGP in a manner similar to that used in the work described above,[16] but the terminal *N*-acetylneuraminic acids were intentionally removed *via* acid hydrolysis.[30] This eliminated possible side reactions due to the presence of the carboxylic acids.

While the assembly of peptide segments using NCL is currently the most widely used method to overcome the size limitation of SPPS, a convergent fragment condensation method used in the synthesis of long peptide

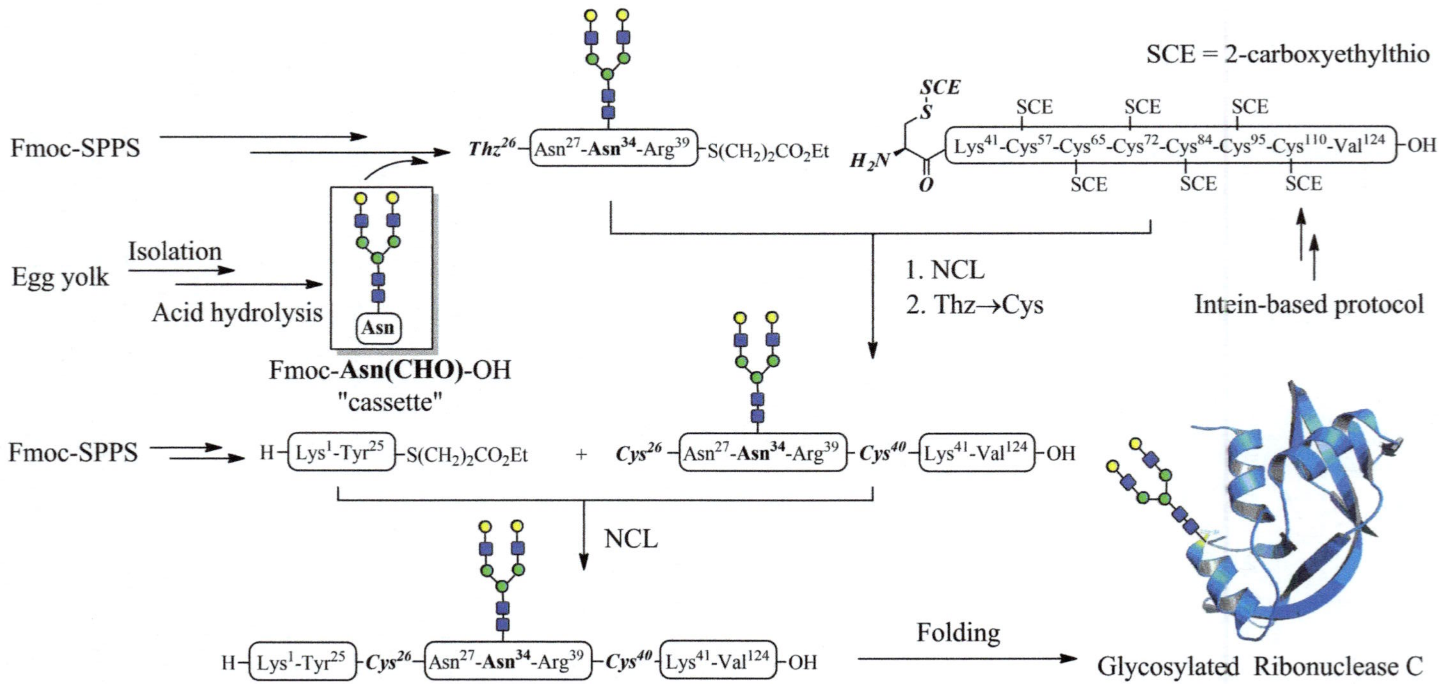

Scheme 5.9 Semisynthesis of ribonuclease C as performed by Unverzagt and colleagues.

sequences is also worth mentioning.[31] In a report from the Unverzagt group, three side-chain-protected segments, RNaseC(1–16), RNaseC(17–22), and RNaseC(23–32) each with pseudoproline serine($\Psi^{Me,Me}$Pro) residues at the C-termini, were prepared using Fmoc–SPPS followed by cleavage from the resin (Scheme 5.10). Glycopeptide RNaseC(33–39) bearing a complex-type biantennary asialyl nonasaccharide was synthesized using the "cassette" strategy as described previously.[29] After sequential condensation of the side-chain-protected peptide fragments to the glycopeptide on solid support in a C to N direction, thiolysis and global deprotection yielded the desired 39-mer glycopeptide thioester, RNaseC(1–39). This protocol is racemization-free due to the rigid five-member ring of pseudoproline, and thus offers potential access to the many glycoproteins which do not contain suitable ligation sites close to the glycosylation locus. However, further studies are required to optimize the Fmoc-SPPS conditions for preparing the peptide fragments with pseudoprolines at the C-termini, which are prone to form diketopiperazine structures, thus leading to diminished yield of the desired peptide fragment.

To overcome the disadvantages of using NeuAc benzyl ester-containing Asn(CHO)–OH "cassette" as the building block in glycoprotein synthesis, Kajihara and co-workers further investigated the utilization of phenacyl (Pac) ester-masked "cassette" in SPPS.[32] The Pac-ester-containing building block showed improved stability under peptide deprotection conditions using trifluoroacetic acid (TFA)/trifluoromethanesulfonic acid (TfOH)/dimethyl sulfide (DMS)/*m*-cresol cocktail, and the Pac ester was easy removed using piperidine after NCL. Utilizing this glycosyl amino acid derivative, they accomplished the synthesis of a single glycoform of EPO bearing one complex-type biantennary undecasaccharide at Asn83 (Scheme 5.11). With two mutation sites and four protected native cysteines, the full sequence was assembled from a total of six fragments. The first two segments, I(1–21) and II(22–49), were ligated using kinetically controlled ligation (KCL),[33] where the aryl thioester of I was preferentially activated and the alkyl thioester on II stayed intact under the reaction conditions. On the other hand, segments VI(128–166), V(98–127), IV(79–97), and III(50–78) were connected sequentially under iterative NCL conditions. After the subsequent ligation of the resulting fragments I–II and III–VI, desulfurization, cysteine deprotection and folding, the target glycoprotein EPO was obtained as a single glycoform. It is notable that during the synthesis, glycopeptide thioester IV was constructed *via* the "cassette" strategy, where a large excess of the oligosaccharyl asparagine building block (six equivalents) was required to ensure the synthetic efficiency during SPPS. The scalable separation and derivatization of SGP from egg yolk guarantees the satisfactory overall yield based on the peptide, but in cases involving more complex oligosaccharides with limited availability, a more efficient method for the incorporation of glycans would be required.

During these studies working towards preparing homogeneous glycoproteins, not only were several new strategies developed to incorporate glycans, but methods that aimed to improve the efficiency of peptide segment

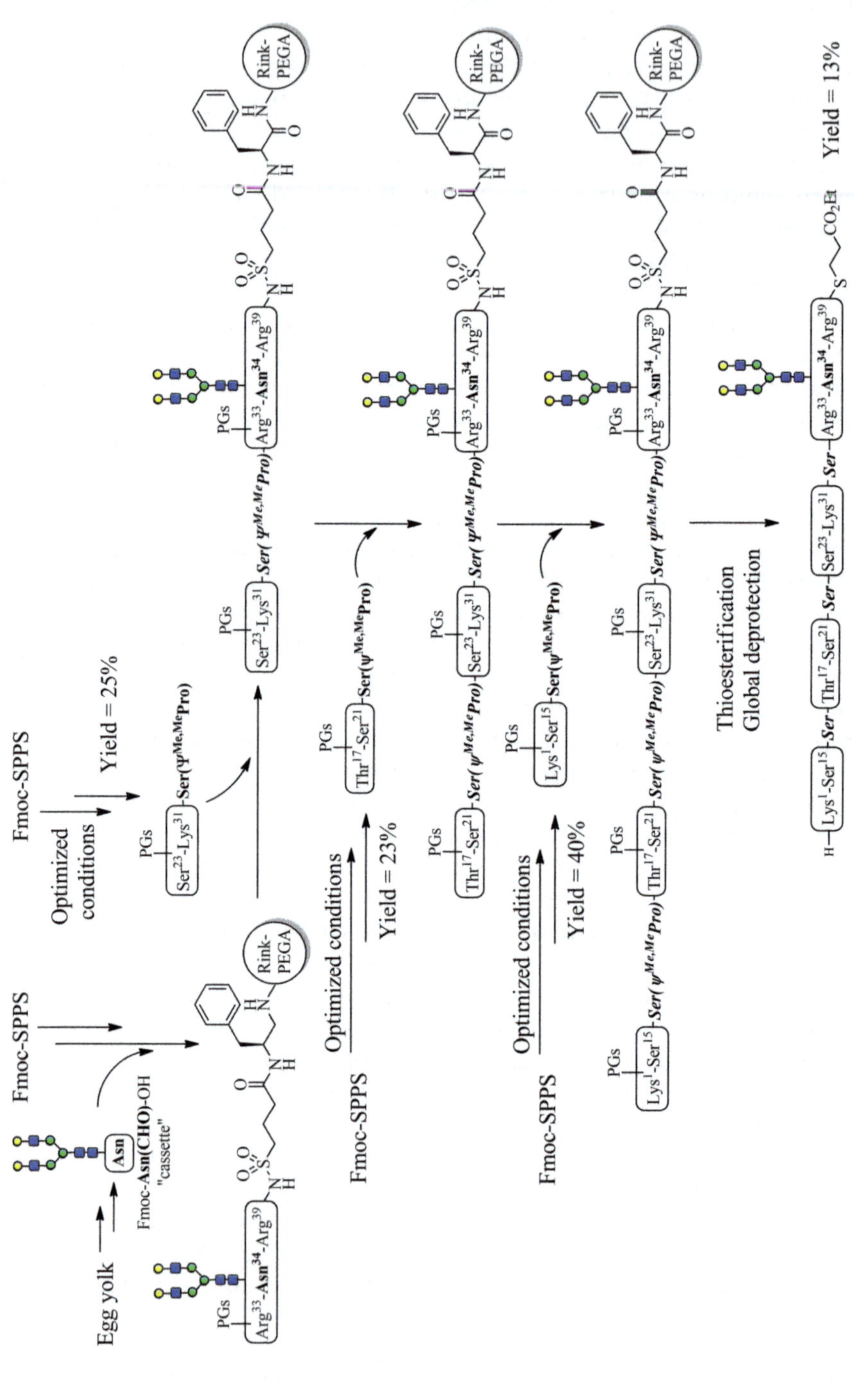

Scheme 5.10 Pseudoproline-based fragment condensation strategy for the synthesis of glycosylated RNaseC(1–39) thioester.

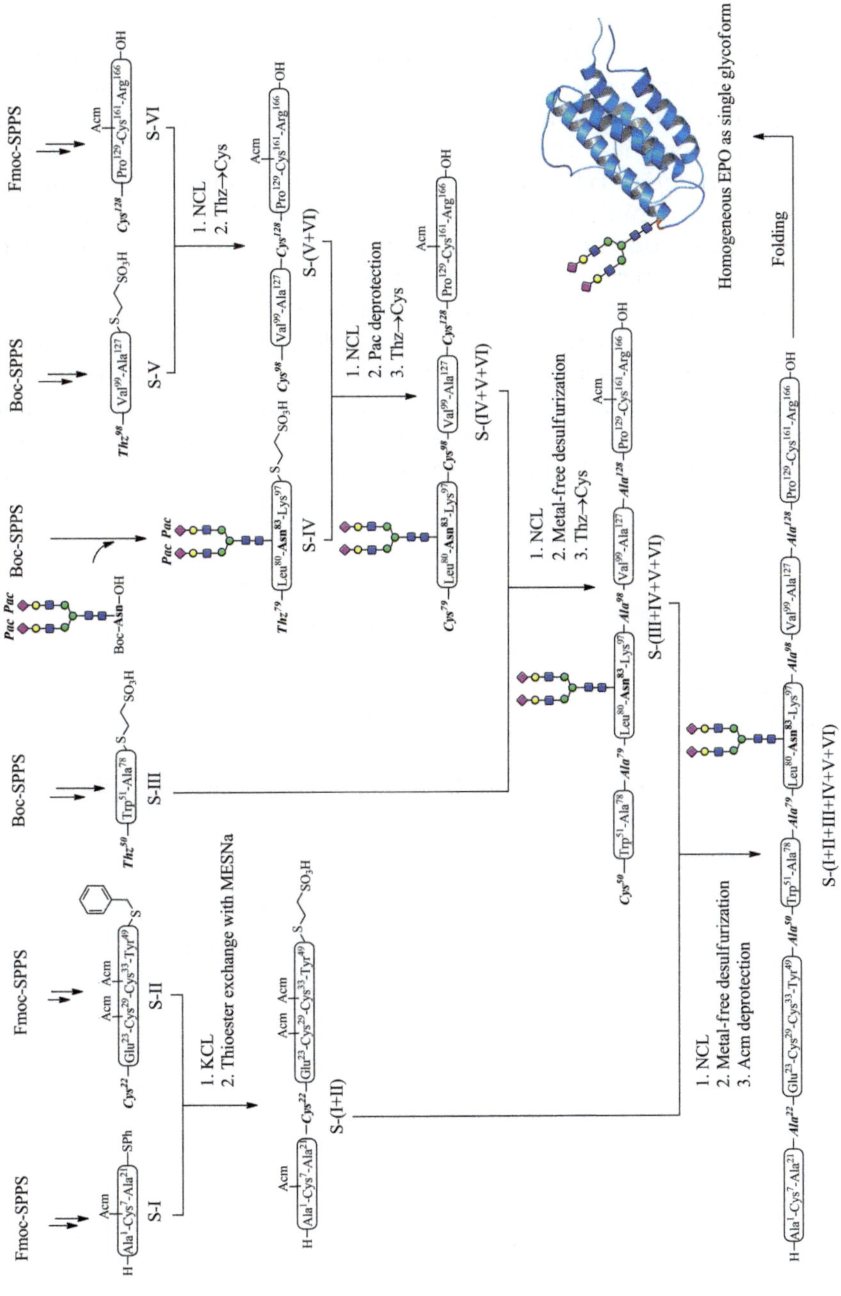

Scheme 5.11 Chemical synthesis of a monoglycosylated EPO analogue.

preparation were also investigated. For instance, a new system developed by the Kajihara group allows for the expression of a single recombinant polypeptide for the preparation of two peptide segments.[34] In a classical recombinant intein-based process, two different expression systems are required to separately prepare the N-terminal peptide thioester and the C-terminal segment bearing one free cysteine.[35] To showcase this protocol, interleukin-13 possessing one *N*-linked glycan at the middle region was synthesized from four fragments (Scheme 5.12). Both terminal fragments, peptide thioester I(1–27) and polypeptide IV(56–112), were generated from one recombinant fusion polypeptide simultaneously using a site-specific cyanogen bromide (CNBr) cleavage reaction and the "guanidine method".[36] Three iterative NCL reactions were then performed to sequentially assemble the fragments in the C to N direction and produce the full protein sequence. After folding, interleukin-13 was obtained efficiently as a single glycoform bearing a complex-type biantennary asialyl nonasaccharide. This efficient recombinant protocol utilized chemical derivatization instead of protein splicing, which may be applied to even denatured polypeptides from an *Escherichia coli* inclusion body, thus broadening the scope of its usage in protein chemical synthesis.

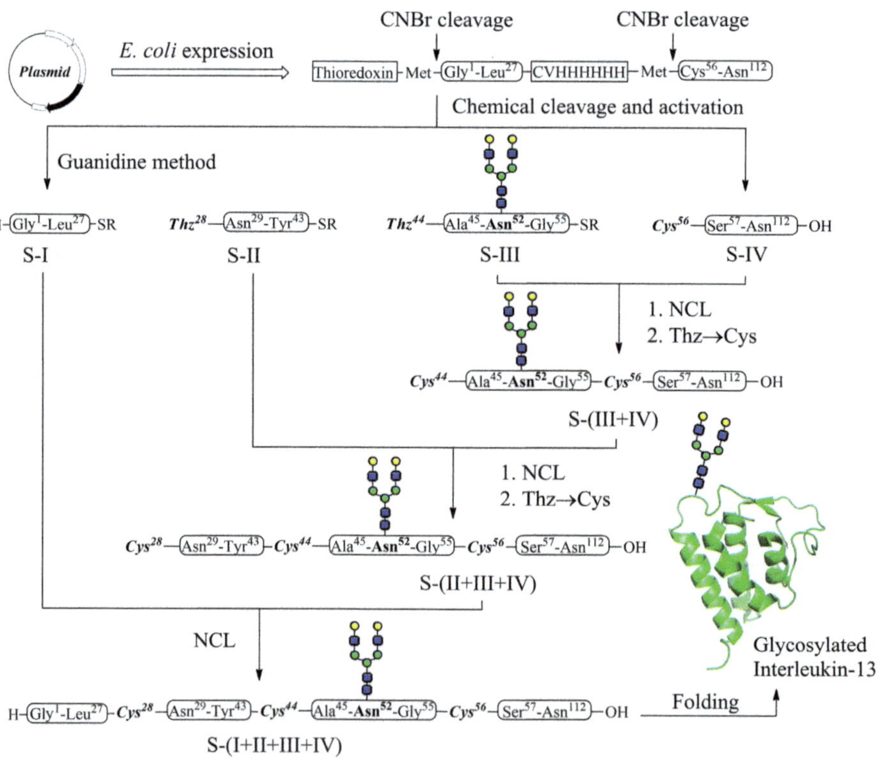

Scheme 5.12 Semisynthesis of interleukin-13 as a single glycoform.

Another recent example that utilized recombinant gene expression technology to prepare a large polypeptide resulted in the efficient semi-synthesis of homogeneously glycosylated interleukin-6.[37] In this work, Unverzagt and colleagues employed a sequential N to C NCL strategy, which began with the ligation of recombinant peptide thioester IL-6(1–42) and synthetic glycopeptide hydrazine IL-6(43–48) (Scheme 5.13). After the conversion of the peptidyl hydrazide to the corresponding thioester following Liu's protocol,[38] a subsequent NCL with recombinant fragment IL-6(50–183) produced the entire protein sequence. Finally, the IL-6 containing an asialyl nonasaccharide glycan was produced after folding. This work extended the use of a two-intein system, which was previously reported to produce cyclic peptides,[39] to prepare a linear peptide thioester for the first time, in which the desired peptide sequence was derivatized to the thioester after simultaneous cleavage of two co-expressed inteins. Another requisite fragment IL-6(50–183) was also prepared using recombinant methods, which highlights the capability of *E. coli* expression to prepare large polypeptides that are currently inaccessible using only SPPS. However, the success of any recombinant protocol greatly depends on the solubility of the expressed peptide sequences, and especially importantly in the case of glycosylated peptides, there is currently still a lack of easily-accessible and cost-effective recombinant methods to generate homogeneous glycoforms.

As we have shown in the above examples, based on Kajihara's work on the preparation of various Fmoc–Asn(CHO)–OH building blocks deriving from SGP, such a "cassette" strategy has been an effective method in the homogeneous semisynthesis of glycoproteins. Most of the time, the general stepwise SPPS protocol can efficiently produce a desired glycopeptide with a length of around 20 amino acids, but the efficiency decreases if the peptide becomes longer due to numerous side reactions. For glycopeptide fragments longer than 20 amino acids, and in particularly ones with glycans that are not readily available, a more efficient method was required. Recently, both the Unverzagt[40] and the Danishefsky groups[41] independently developed a convergent protocol to prepare *N*-glycosylated peptide segments using Lansbury aspartylation.[42] This technique allowed them to investigate glycan functions by synthesizing many homogeneous glycoforms of important proteins (Scheme 5.14).

The Unverzagt group utilized the pseudoproline-assisted Lansbury aspartylation to glycosylate the peptide moiety on solid phase, and successfully synthesize the glycopeptide segment RNaseC(1–39) (Scheme 5.15).[40] The nonasaccharide glycan, isolated from egg-derived SGP, was introduced after the peptide sequence was fully elaborated on resin by the previously developed pseudoproline fragment condensation method and selective removal of the allyl protection on aspartic acid at the glycosylation site. The glycosyl amide bond was formed by coupling the 1-amino glycan, prepared according to the Shoda protocol,[43] to the free-acid-containing peptide and the final global deprotection and cleavage from the resin generated glycosylated RNaseC(1–39) in a moderate yield. In a similar manner, a homogeneously

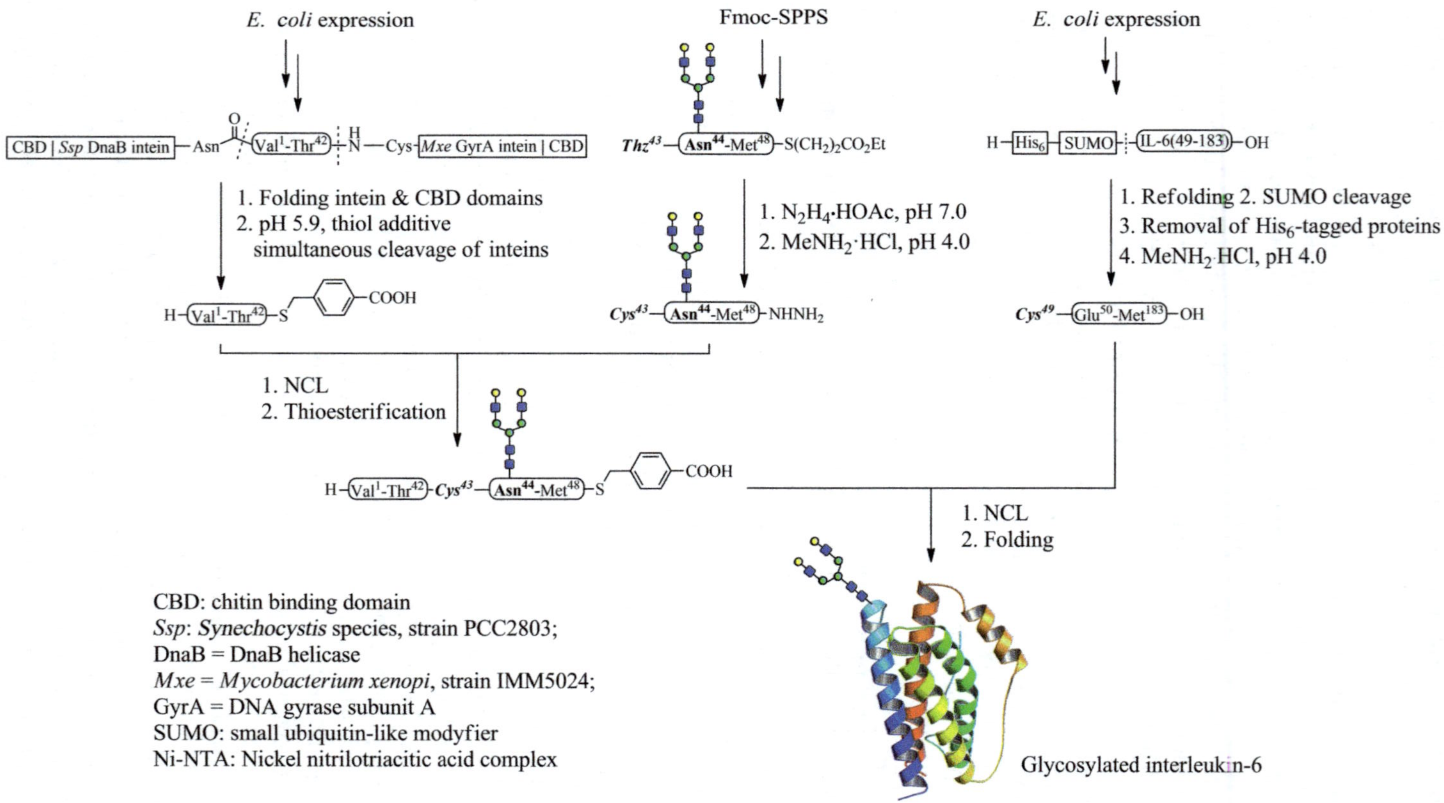

Scheme 5.13 Semisynthesis of interleukin-6 as a single glycoform.

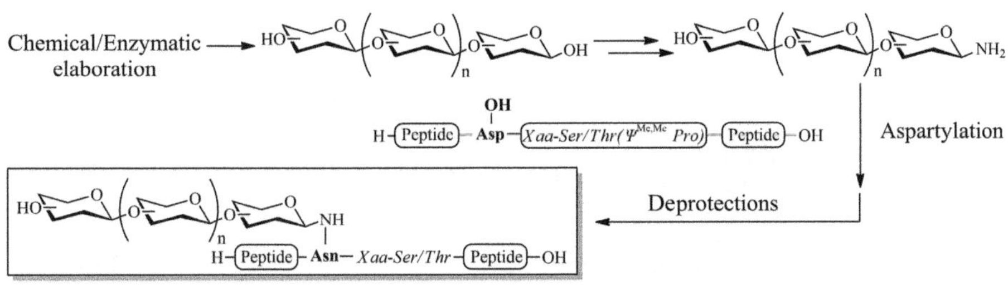

Scheme 5.14 Pseudoproline-assisted convergent aspartylation strategy for preparing glycopeptides.

glycosylated EPO(1–28) fragment was also engineered successfully. These examples, as well as the work from the Danishefsky group (*vide infra*), have demonstrated the utility of such convergent aspartylation methods, which have greatly expanded the diversity of glycopeptides and glycoproteins that can be prepared from small amounts of synthetically pure complex oligosaccharides.

5.3 Total Chemical Synthesis of *N*-Linked Glycoproteins

While an almost unlimited variety of sialyloligosaccharides exists in nature, so far those available in convenient homogeneous forms are extremely rare and only the undecasaccharide from egg yolk is suitable for large-scale use in the synthesis of homogeneous glycoproteins. The difficulty of discovering, isolating, and purifying suitable oligosaccharides from natural sources has significantly limited the diversity of glycoforms that could be obtained by such methods. In contrast, approaching glycoproteins through total synthesis offers high flexibility in a much more controllable way. Furthermore, many types of glycans can be prepared *via* chemical synthesis and subsequently installed on proteins, thus facilitating construction of a series of homogenous glycoforms. As attractive as this approach is, in practice it has been far from simple. The major issue currently being faced is a lack of effective ways to synthesize complex oligosaccharides *via* short routes and at large scales, and sialic-acid-containing *N*-glycans have been particularly problematic. The limited scales on which oligosaccharides can be synthetically prepared require extremely efficient methods for introducing glycans to peptides or proteins with high selectivity. Additionally, synthetic routes towards large glycoproteins are usually investigated and validated first using more readily available and simpler carbohydrates, such as chitobiose. Facing the formidable challenges inherent in the total synthesis of glycoproteins, much effort has been invested in advancing the synthetic methods used and several notable accomplishments have been achieved.

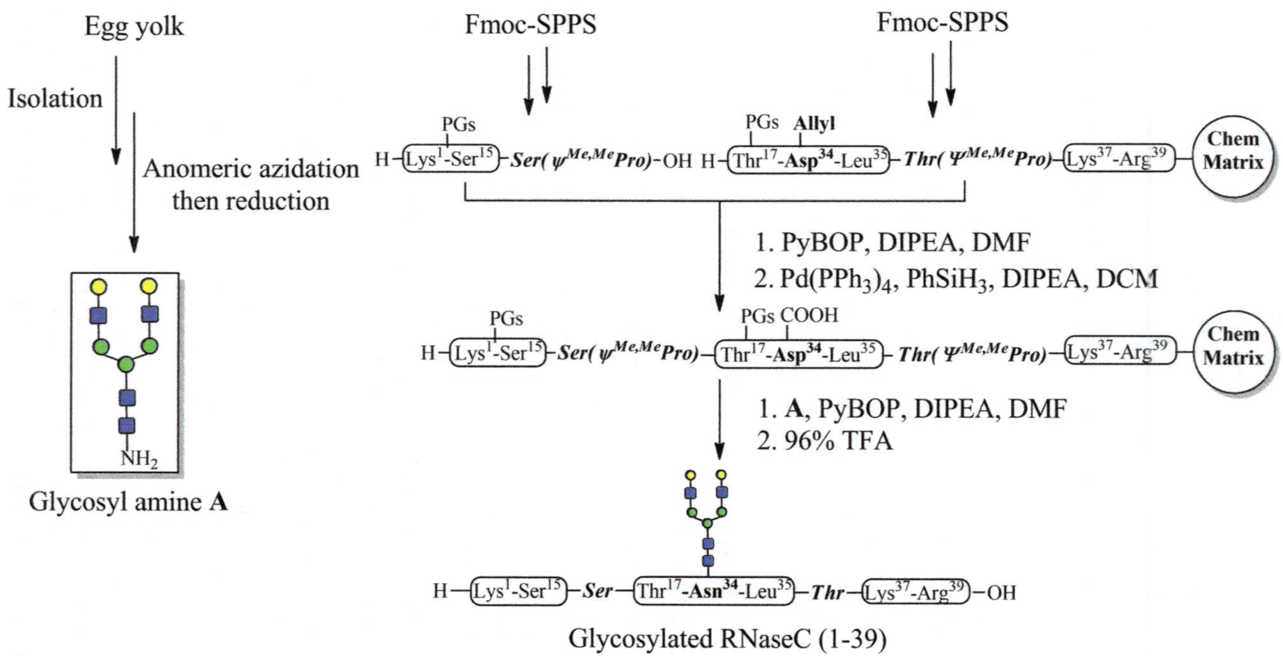

Scheme 5.15 Pseudoproline-assisted Lansbury aspartylation strategy in the synthesis of glycopeptide RNaseC(1–39).

5.3.1 Total Synthesis of *N*-Linked Glycoproteins Bearing *N*-Chitobioses

The Danishefsky group has made significant contributions in glycoprotein total synthesis, and their work in this research area has been recently reviewed.[44] One major synthetic target they have been extensively studied is EPO. In 2012, Danishefsky and colleagues finished the first total synthesis of homogenously glycosylated EPO.[45] Their strategy featured glycopeptidyl segments prepared using a one-flask convergent aspartylation and deprotection protocol,[46] as well as the combination of native chemical ligation and metal-free desulfurization to connect different segments. The total synthesis of full-length, homogeneously glycosylated EPO has been investigated through two routes (Scheme 5.16). The first one is a linear strategy, where segment IV was ligated sequentially with segment III at site 125, and the elongated peptide segment reacted with segment II at site 79. After the global desulfurization and Acm removal revealing the native Cys29 residues, the final NCL at Cys29 generated the full sequence of EPO. The second route involves a synthetically convergent strategy to assemble the EPO primary sequence, where a kinetically controlled ligation (KCL) approach was used to ligate the two *N*-terminal segments.[33] Final disulfide bond formation under oxidative folding conditions yielded homogenous EPO with three *N*-linked disaccharides and one *O*-linked sialic acid-containing tetrasaccharide. Importantly, the bioactivity of this EPO glycoform was confirmed in an *in vitro* assay.

Utilizing similar strategies, the Danishefsky group has also successfully synthesized the aglycon of granulocyte–macrophage colony-stimulating factor (GM-CSF) as well as an *N*-glycosylated glycoform (Scheme 5.17).[47] GM-CSF is an important glycoprotein cytokine that plays essential roles in cellular growth and proliferation,[48] and the recombinant form currently has multiple clinical indications.[49,50] Through the previously mentioned convergent aspartylation protocol and the ligation–desulfurization strategy, two *N*-chitobiose glycopeptide segments and one 47-mer peptide were combined to produce the full length GM-CSF. The folded protein bearing two disaccharides at the natural *N*-glycosylation sites was then shown to be bioactive in both *in vitro* and *in vivo* studies.

Chitobiose represents a much simplified *N*-glycan, which is useful in exploring the optimal synthetic route. However, compared with the large complex-type oligosaccharides, chitobiose may not be able to provide enough steric bulk or hydrophilicity to protect the protein from aggregation. Additionally, there have been some cases where an increase in glycan size has altered the synthetic route established using a smaller model disaccharide (*vide infra*).

5.3.2 Total Synthesis of *N*-Linked Glycoproteins Bearing "Wild-Type" *N*-Glycans

Since natural *N*-linked glycans come in several varieties, like the high-mannose, hydrid, or complex-types (Scheme 5.2), many groups have attempted to obtain glycoproteins containing oligosaccharides more complex than chitobiose to better simulate these "wild-type" *N*-glycans. To date, the only example

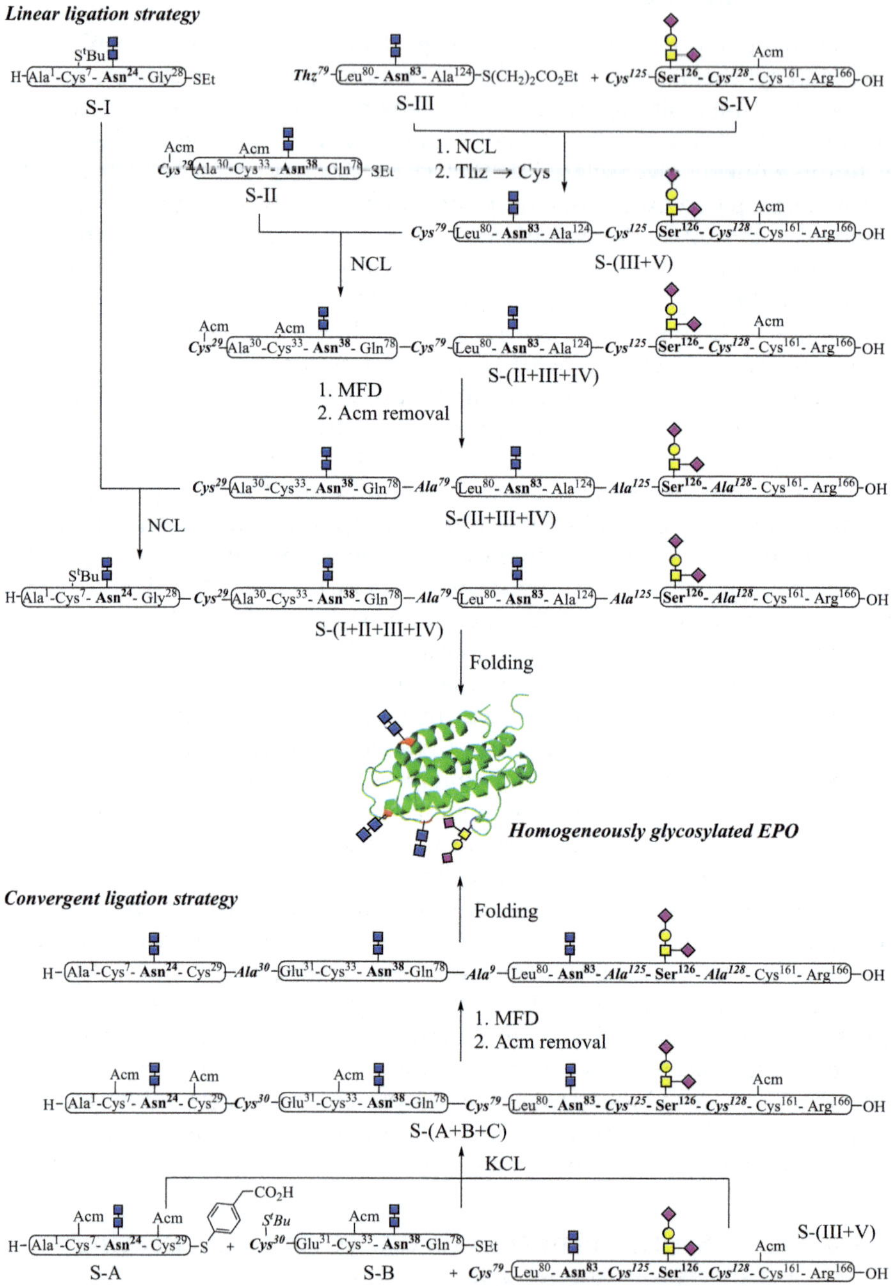

Scheme 5.16 Total synthesis of EPO bearing *N*-chitobiose and *O*-glycophorin.

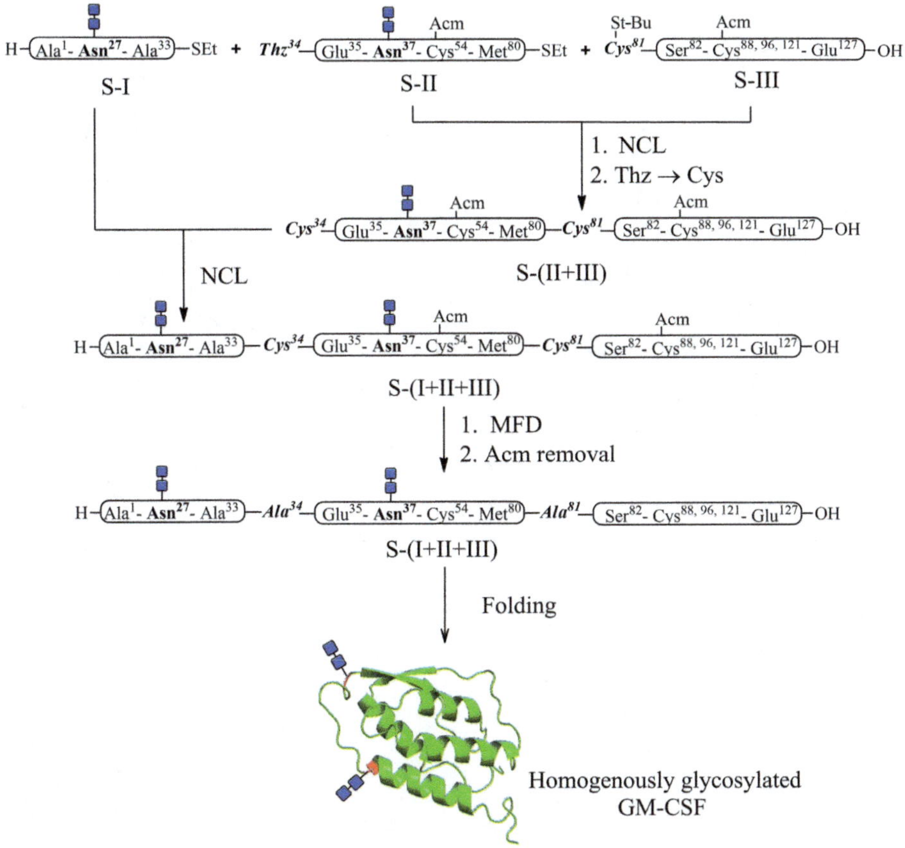

Scheme 5.17 Total synthesis of GM-CSF bearing two *N*-linked chitobiose.

of a glycoprotein bearing "wild-type" *N*-glycans constructed through chemical total synthesis is Danishefsky's work on EPO, which contained three *N*-linked complex dodecasaccharides at Asn residues 24, 38, and 83 and one *O*-linked tetrasaccharide at Ser 126. Considering the large protein backbone and complex carbohydrate domains, the first challenge in synthesizing such a giant molecule is to obtain appreciable amounts of the requisite oligosaccharides. Although carbohydrate synthesis is not our focus in this chapter, we would like to briefly review the synthetic strategy used to construct fucosylated biantennary dodecasaccharides and to highlight that synthesizing such a complex glycan itself is an impressive accomplishment.[13,51] As shown in Scheme 5.18, the final dodecasaccharide was assembled from two trisaccharides and a hexasaccharide *via* bis-glycosylation. The trisaccharides were synthesized from the corresponding glucosamine, galactose and sialic acid building blocks, while the hexasaccharide was derived from a mannose donor and a tetrasaccharide that was itself elaborated from a glycal unit by three separate glycosylations. Notably, this structure contains a branching fucose moiety, which is absent on the egg-yolk-derived undecasaccharide but has been identified in many natural *N*-glycoforms of proteins.[52]

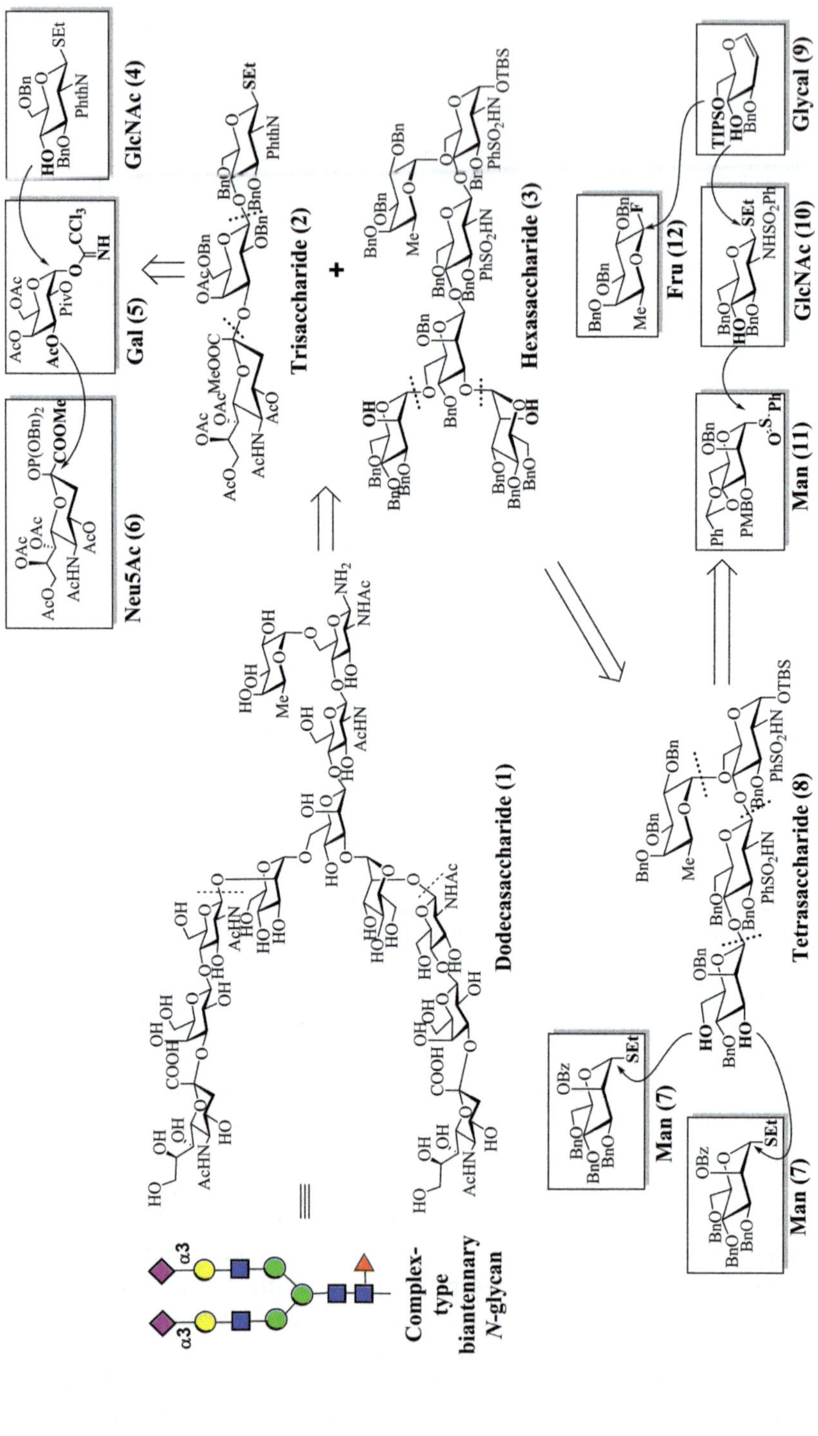

Scheme 5.18 Synthetic strategy for the preparation of a complex-type fucosylated dodecasaccharide.

The challenging and lengthy synthesis of the complex dodecasaccharide prompted Danishefsky and colleagues to seek out a more efficient method to prepare glycopeptide segments, as it is impossible to use the "cassette" strategy with such precious material. Accordingly, the pseudoproline-assisted aspartylation was developed and found to be an efficient protocol even on the complex dodecasaccharide substrate. It is notable that in the Danishefsky group's studies, both the $\alpha_{2,3}$-sialic acid and the fucose unit on the glycan were found to tolerate the TFA/triisopropylsilane (TIPS)/H_2O deprotection cocktail.[46]

With these advances in both carbohydrate and glycopepetide synthesis, the Danishefsky group was able to successfully accomplish the total synthesis of EPO with complex "wild-type" sialic acid-rich glycans (Scheme 5.19).[53]

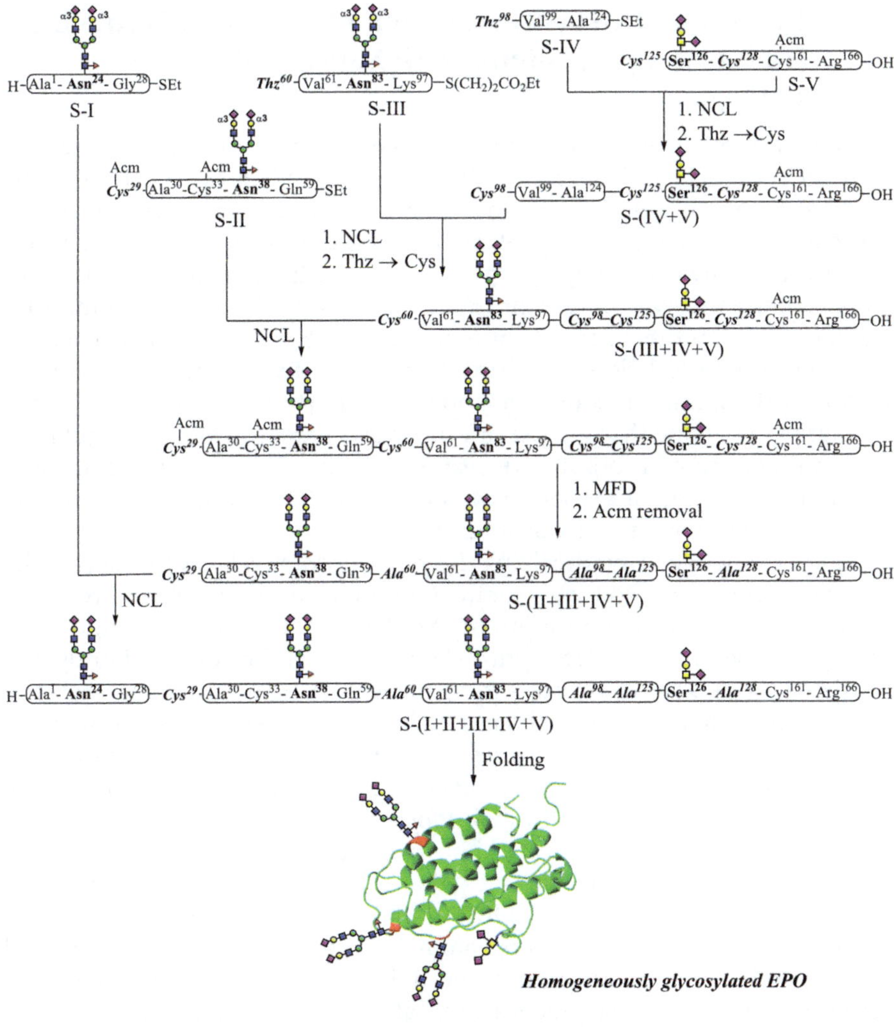

Scheme 5.19 Total synthesis of an EPO glycoform bearing three *N*-linked dodecasaccharides and one *O*-linked tetrasaccharide.

Unfortunately, due to the increased size of the *N*-linked glycan compared with the previous *N*-chitobiose model studies, the ligation site had to be adjusted. Ultimately, the homogeneously glycosylated EPO was evaluated *in vitro* and *in vivo* for hematopoietic activity, and it was found that the fully synthetic sample exhibited similar effects to the clinically used recombinant EPO (Procrit). The total synthesis of EPO bearing *N*-linked complex dodecasaccharide and *O*-linked tetrasaccharide has demonstrated the capability of chemical synthesis of large biologically relevant molecules, allowing for further interdisciplinary studies on chemistry and biology. This accomplishment has also painted a promising future for the development of synthetic organic chemistry.

5.3.3 Representative Synthetic Attempts Towards Producing *N*-Linked Glycoproteins with Multiple Disulfide Linkages

Although a number of glycoproteins have been successfully synthesized *via* chemical approaches, these molecules represent only a tiny portion of the extremely complex and diverse family of glycoproteins. At the current stage, synthesizing homogeneous glycoforms with the correct tertiary structures for most naturally occurring glycoproteins is still far beyond our reach. For instance, human glycoprotein hormones are one of the most important families of glycoproteins in the human body; the members of this family, including human follicle stimulating hormone (hFSH), human luteinizing hormone (hLH), human chorionic gonadotropin (hCG) and human thyroid stimulating hormone (hTSH), play a variety of critical roles in the regulation and maintenance of reproductive processes. These glycoproteins are heterodimers consisting of two non-covalently associated subunits: a common α-subunit and a characteristic β-subunit.[54]

The key challenge of synthesizing these glycoproteins is how to obtain the properly folded form, as they usually contain at least 12 cysteine residues that are required for the formation of the disulfide linkages. Although there still has not been a complete synthesis of any member of this family, the Danishefsky group has done extensive studies working towards this goal, and they have accumulated much experience in engineering of these protein targets that will be applicable to future studies. In most of the cases thus far disclosed, disaccharide chitobiose was used as a model *N*-linked glycan during preparation of the corresponding glycopeptide segments. Then, through a convergent aspartylation protocol that involved iterative segment ligations followed by desulfurization, the full sequence of several target proteins was assembled, including α-hGPH,[55] β-hFSH,[51] β-hCG, and β-hLH.[56]

In the case of human α-glycoprotein hormone subunit (α-hGPH) and human β-follicle stimulating hormone (β-hFSH),[57] the complex *N*-dodecasaccharide glycosylated sequences were also synthesized. A representative synthetic example is shown in Scheme 5.20. β-hFSH was dissected into four segments, where segments I and II contain *N*-linked dodecasaccharides.

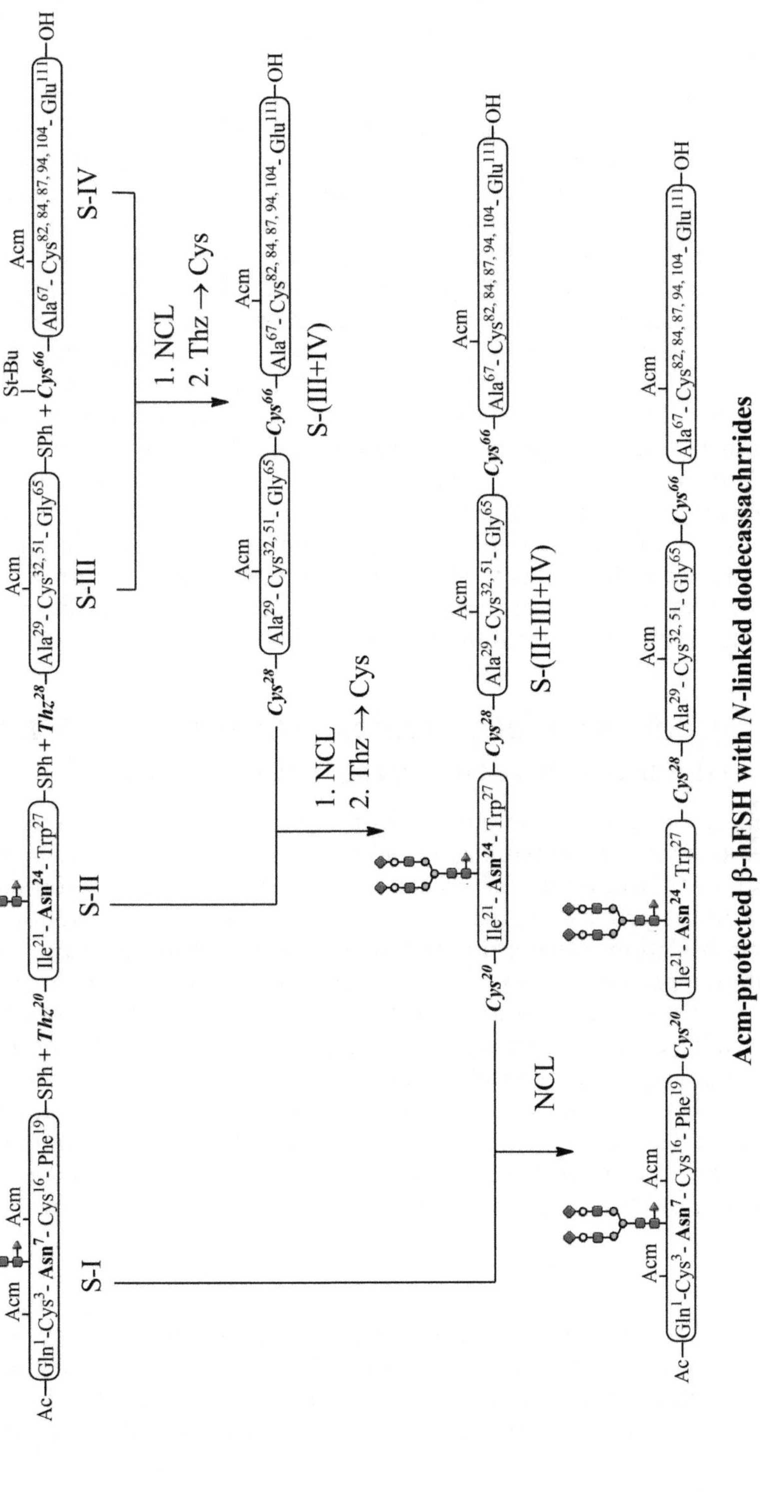

Scheme 5.20 Synthesis of Acm-protected β-hFSH sequence bearing two complex-type dodecasaccharides.

Taking advantage of three native cysteines, iterative NCL yielded the primary sequence of β-hFSH. Considering the multiple Cys residues, the folding step remains to be further investigated.

More recently, synthetic attempts working towards producing human β-chorionic gonadotropin (β-hCG) containing *O*-linked GalNAc type glycans have also been reported.[56] As shown in Scheme 5.21, the full sequence of β-hCG was dissected into four segments. Segment IV, which carried an *O*-linked GalNAc, was synthesized using the "cassette" strategy, while segment I, with two *N*-linked chitobiose glycans, was synthesized through convergent bis-aspartylation. Then three NCL reactions generated the primary sequence of β-hCG with the nine native cysteine residues still protected by Acm groups.

As we have shown through the above examples, although there has been significant progress toward the synthesis of human glycoprotein hormones, there are still two major challenges remaining. The first one is the difficulty of removing the Acm groups on cysteines, as the deprotection requires a large excess of silver acetate (AgOAc) and often results in either incomplete conversion or diminished yield. This is particularly problematic for proteins containing multiple cysteines. Secondly, identifying proper folding conditions that form the requisite disulfide linkages as well as the correct tertiary structure for each glycoprotein is a demanding task. These issues remain to be conquered and call for innovative solutions.

5.4 Application of Chemically Synthesized *N*-Linked Glycoproteins to Biological Processes

Chemical synthesis and engineering of glycoproteins is never practiced only to demonstrate the capability of scientific technology or manpower. The importance of this research field lies in the promise of utilizing homogeneous glycoforms to probe biological processes and to explore the functions of protein glycosylation. During the past several years, advances in the engineering of glycoproteins as homogeneous glycoforms through chemical means have allowed scientists to study problems that would be challenging, or even impossible, to answer using heterogeneous protein glycoforms obtained from traditional biochemical methods. For instance, to reveal the relationship between protein glycosylation and the bioactivity of individual glycoforms, a series of homogeneous glycoforms are required. The Kajihara group has done extensive work and made significant progress towards addressing this issue. Utilizing the "cassette" strategy and optimized Fmoc-SPPS conditions, Kajihara and colleagues efficiently completed the semisynthesis of two glycoforms of interferon-β containing either a neutral asialo-nonasaccharide, or an acidic sialo-undecasaccharide (Scheme 5.22).[58] By protecting the three native cysteines but "masking" two alanines as cysteines for the ligation sites, the three fragments I–III were assembled in the C to N direction and yielded the whole protein sequence. The synthesis was completed by "unmasking" the ligation-site alanines through desulfurization,

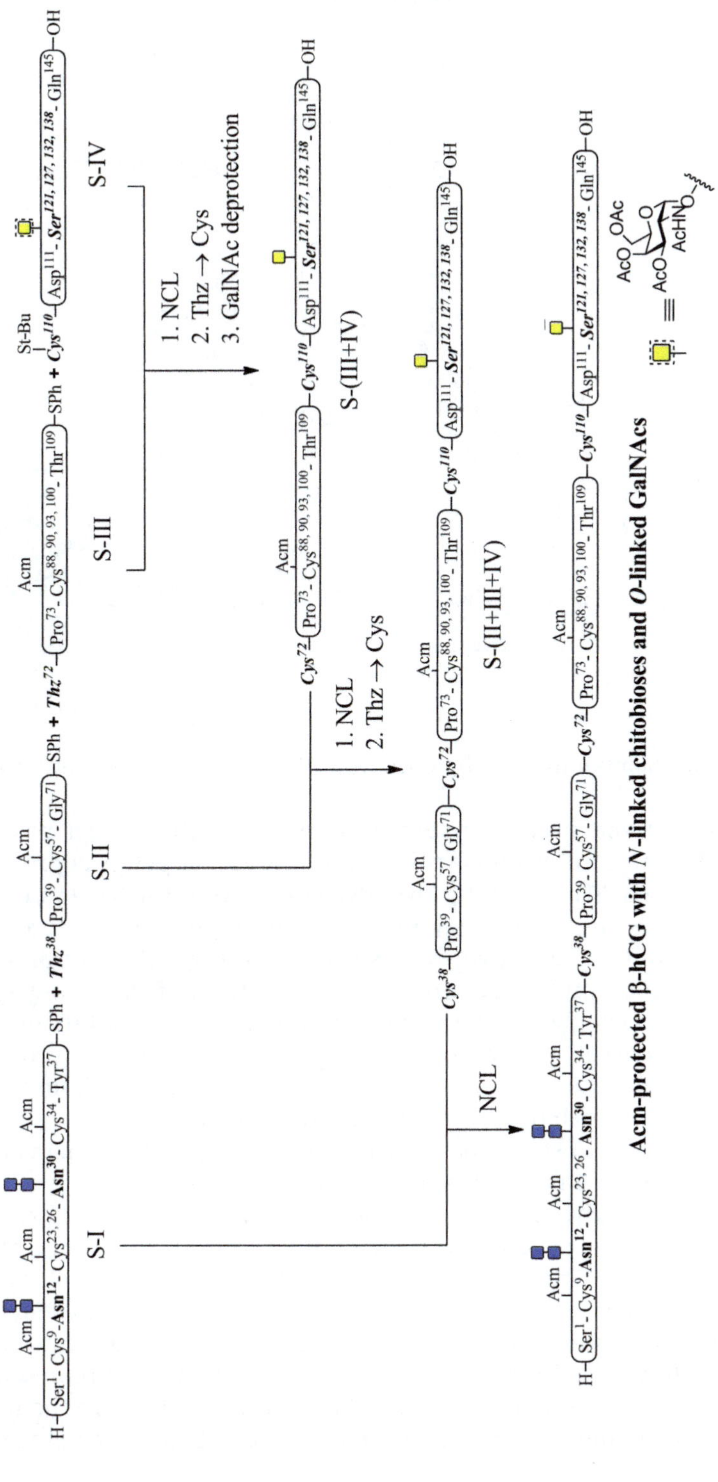

Scheme 5.21 Synthesis of Acm-protected β-hCG bearing two *N*-linked chitobioses and four *O*-linked GalNAcs.

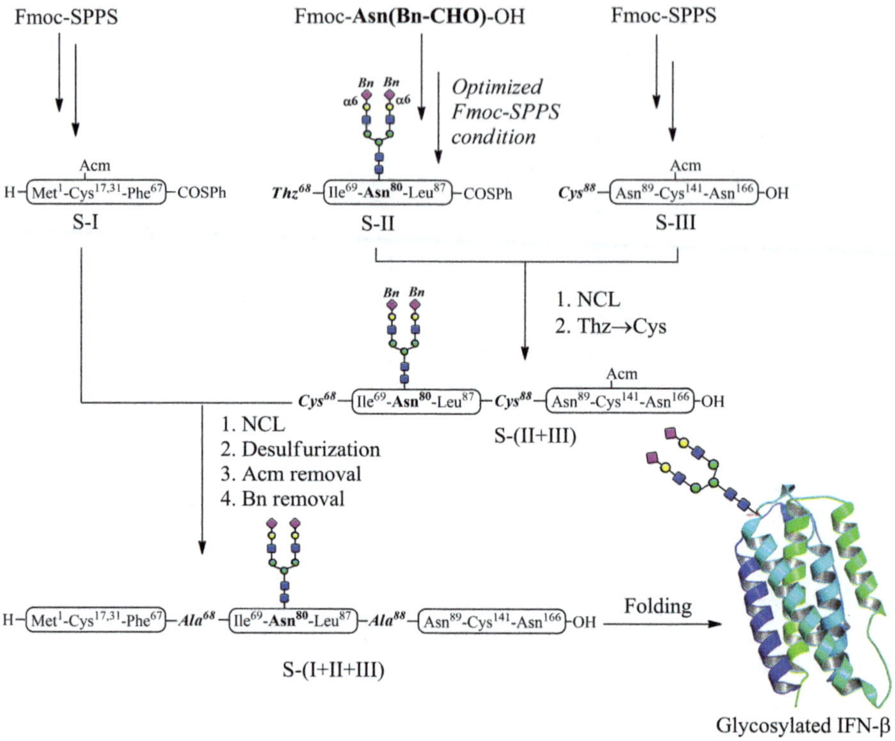

Scheme 5.22 Semisynthesis of homogeneous IFN-β using the "cassette" strategy.

deprotection of the natural cysteines, removal of the benzyl protecting groups on NeuAc, and oxidative folding, to give the target glycoproteins as single glycoforms. The resulting two glycoforms were further evaluated in biological studies, and it was demonstrated that the sialic acid residues were critical for the antitumor activity of IFN-β *in vivo*. It is worth mentioning that this was the first example of superior *in vivo* bioactivity of chemically synthesized homogeneous glycoproteins compared with the recombinant forms. This work is an excellent example of using homogeneous glycoproteins to probe glycan functions, as well as an illustrative instance of the importance of results from research on homogeneous glycoproteins.

Based on the previously discussed studies with EPO,[21,23,32] Kajihara and colleagues designed an efficient synthetic strategy and successfully synthesized five homogenously glycosylated EPO glycoforms which varied in glycan position and the number of terminal sialic acids.[59] From this work the authors were able to obtain many EPO glycoforms: one bearing three complex glycans at natural positions (Asn24, Asn38, and Asn83), three bis-glycosylated EPO glycoforms (Asn38 and Asn83), (Asn24 and Asn83), and (Asn24 and Asn38), and a monoglycosylated EPO with *N*-glylcosylation at Asn83. A representative synthetic route that led to the tri-glycosylated EPO is shown in Scheme 5.23. The synthesis combines both Boc- and Fmoc-based SPPS for the preparation

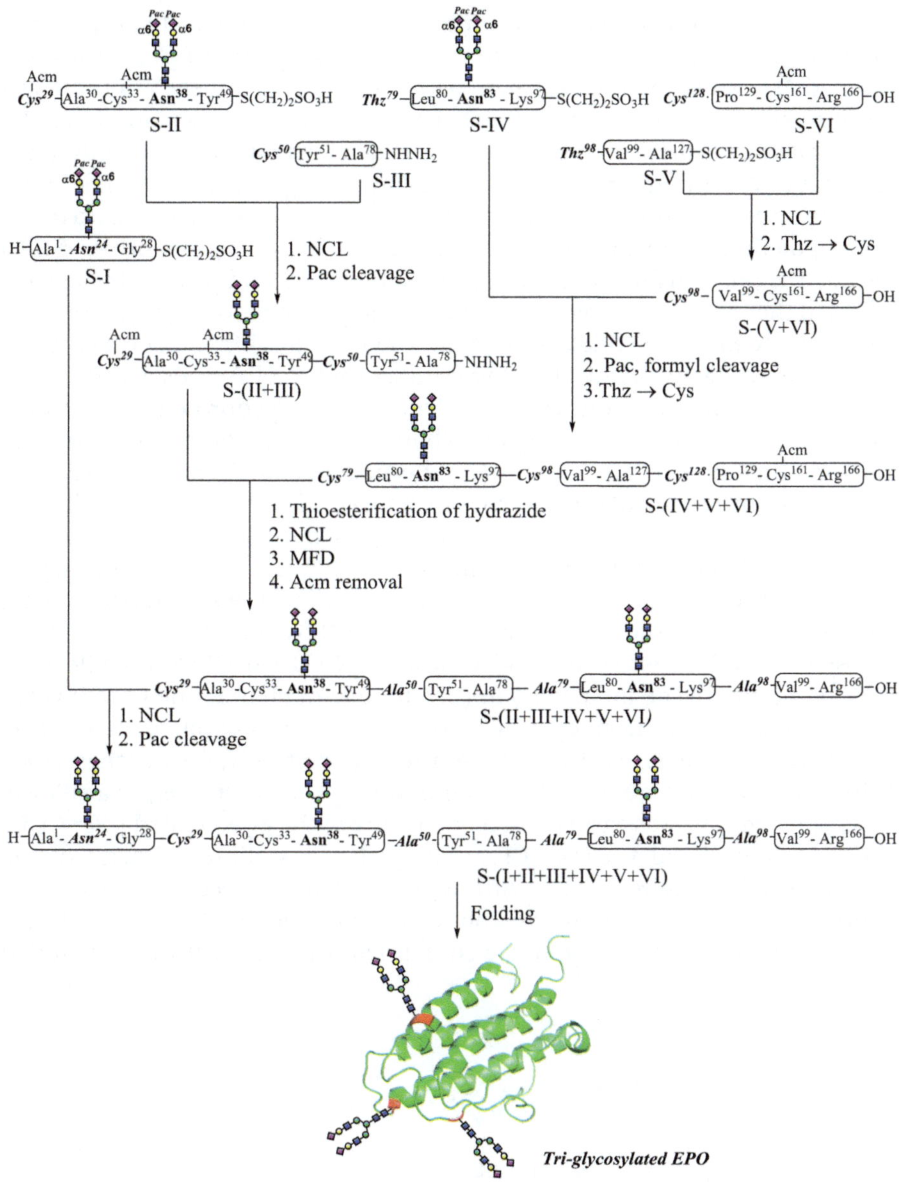

Scheme 5.23 Semisynthesis of a homogeneous EPO bearing three complex-type undecasaccharides.

of peptide segments, as well as a "cassette" strategy for the preparation of the sialylglycopeptide segments I, II and IV. The full sequence was assembled from these peptide segments after iterative native chemical ligations, and a peptidyl azide-based ligation was used to join together segments II + III and IV + V + VI.[60] After *in vitro* folding under redox conditions, the desired homogeneously glycosylated EPO was obtained. The other four glycoforms of EPO were prepared through similar means, and thus a small library of EPO glycoforms was obtained. By evaluating the *in vivo* hematopoietic activity of each of these glycoforms using recombinant EPO (EPOGEN) as reference, it was found that the triglycosylated synthetic EPO exhibited the best bioactivity, albeit slightly lower than that of recombinant EPOGEN. More interestingly, glycosylation sites 38 and 83 were demonstrated to be more important than Asn24 with regard to the hematopoietic activity. This systematic study on different glycoforms of EPO revealed the relationships between glycosylation sites and protein bioactivity, indicated that *N*-glycans may prevent protein aggregation or degradation thus extending the *in vivo* circulating time of proteins, and ultimately presented a text-book example of chemical biology studies on protein glycosylation.

While obtaining three-dimensional structures of glycoproteins using X-ray crystallography has been an ongoing challenge due to the heterogeneity of intact glycans, it is believed that using homogeneous glycoforms may be a potential solution to determine the structures of intact glycoproteins. In 2014, Okamoto, Kajihara, Kent and colleagues disclosed their studies on two *N*-glycosylated proteins, CCL-1 and Ser-CCL1.[61] The homogeneous glycoforms of CCL-1/Ser-CCL1 with an asialo-nonasaccharide were prepared using a strategy similar to the previously described IFN-β route. By comparing the glycosylated molecule to the nonglycosylated CCL-1/Ser-CCL1, the effects of *N*-glycosylation on the bioactivity of this important chemokine were investigated. The results of the study indicated that CCL-1 has stronger chemotaxis activity than that of Ser-CCL1, and also that, interestingly, the homogeneous *N*-glycan did not contribute to the chemotaxis activity. Moreover, (quasi-)racemic crystallography was carried out on a mixture of the glycosylated L-Ser-CCL1 and non-glycosylated D-Ser-CCL1 to study the crystalline behavior of the *N*-glycoprotein (Scheme 5.24).[62] Although this led to successful crystal formation, the glycan moiety was resolved in a disordered state and only the first *N*-linked GlcNAc residue was well-defined. Ultimately it was concluded

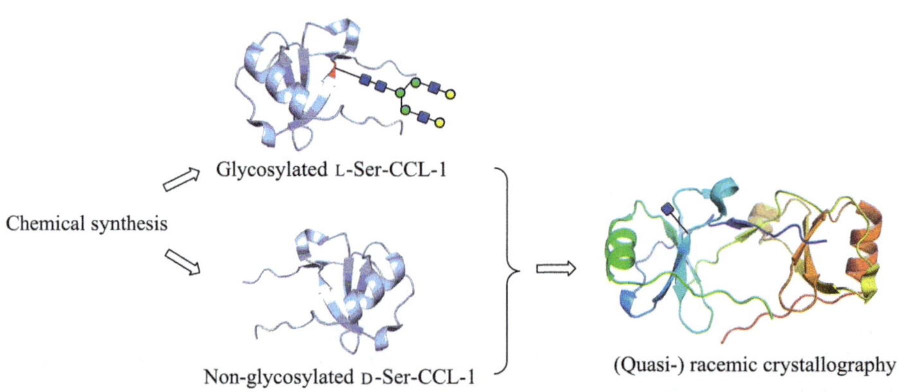

Scheme 5.24 (Quasi-)racemic crystallography of homogeneously *N*-glycosylated L-Ser-CCL-1 and non-glycosylated D-Ser-CCL-1.

that the *N*-glycosylation has no significant effect on the protein structure, but further investigations are still necessary to answer the question of how the carbohydrate affects the three-dimensional structure of the protein.

Besides the research on biological activities and X-ray crystallography of glycoproteins, studies have also investigated various intriguing biological processes using the synthesized homogeneous glycoforms, including the glycoprotein quality control (GQC) mechanisms in the endoplasmic reticulum (ER). This process involves the important glycoprotein folding sensor enzyme, UDP–glucose: glycoprotein glucosyltransferase (UGGT),[63] which facilitates correct glycoprotein folding by recognizing any misfolded glycoproteins and transferring one glucose unit to the misfolded substrate's glycan moiety.[64] However, since the glycoproteins used in the previous studies were not well-defined and existed as multiple isoforms of heterogeneous glycoforms, the details of this process still require further clarification. In 2012, Kajihara, Ito, and coworkers disclosed the results of their studies using a homogeneous glycoprotein model.[65] Utilizing the chemically synthesized interleukin-8 (IL-8) with a tri-antennary M9 high-mannose oligosaccharide at Asn36 (Scheme 5.25), they clearly demonstrated distinct reactivity of UGGT towards the correctly folded and misfolded IL-8. This evidence strongly indicates that the enzyme prefers misfolded substrates and the presence of recognition sites on UGGT for both the nonglucosylated glycan and misfolded region of the substrate protein.

Following up the previous work, they investigated another protein, crambin, as a substrate of UGGT.[66] Using a single glycoform of crambin appended with a tri-antennery M9 high-mannose oligosaccharide on Asn17 (Scheme 5.26), they designed an *in vitro* model system containing UGGT to simulate the folding in the ER environment. The correctly folded glycoprotein was mixed with a mutated form (Cys3 → Ser3, Cys4 → Ser4) that mimicked both the misfolded species and folding intermediates, and after liquid chromatography–mass spectrometry (LC-MS) analysis of the glycoprotein species in the folding mixture, it was concluded that UGGT recognized not only the final stable misfolded glycoprotein, but also the folding intermediates formed during the dynamic folding process.

In another study, Kajihara, Okamoto, and colleagues used crambin and ovomucoide as model *N*-glycoproteins to investigate the influence of intact carbohydrates on the folding, conformation and stability of glycoproteins.[67] SGP-derived asialylnonasaccharide was employed in the synthesis, and a three-step NCL strategy facilitated the generation of the non-glycosylated proteins, one homogeneously glycosylated ovomucoide, and two homogeneous glycoforms of crambin with *N*-glycans located at different sites (Scheme 5.27). All the unfolded glycoprotein precursors were found to smoothly undergo the folding step, while the loop-glycosylated isoform showed decreased folding velocity in comparison with the non-glycosylated protein. However, no significant conformational difference among the proteins was observed using either circular dichroism (CD) or NMR spectroscopy, which was explained by the authors as being a result of the presumably tight tertiary structure constrained by several disulfide bonds. The influence

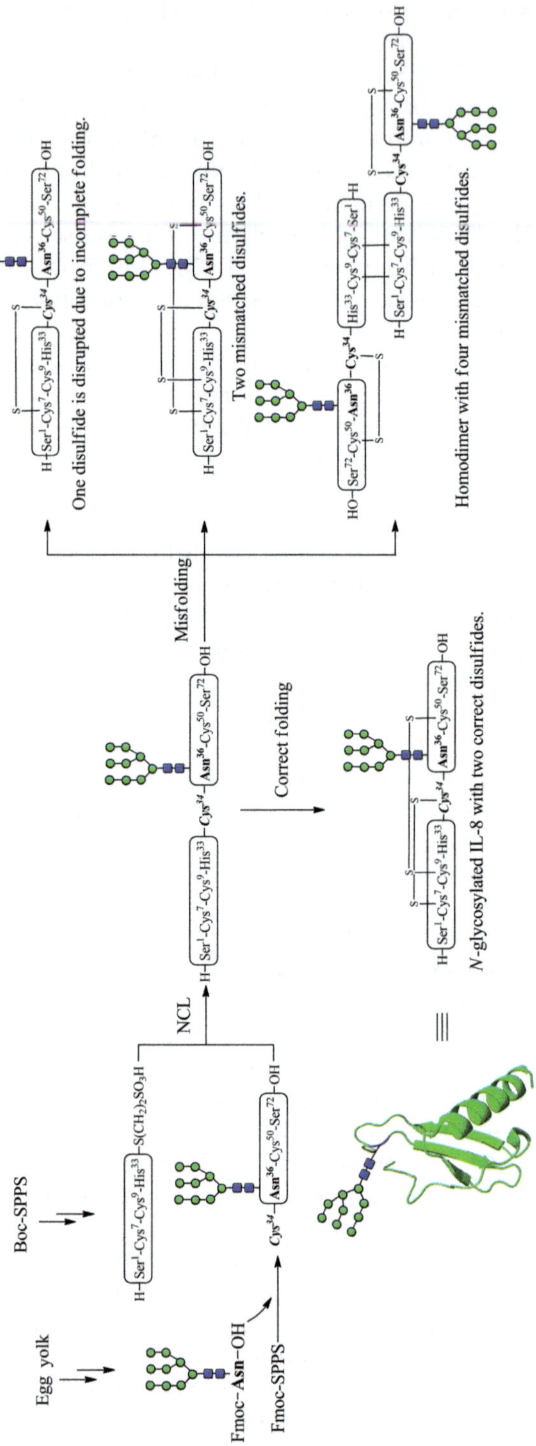

Scheme 5.25 Preparation of correctly folded and misfolded homogeneous IL-8.

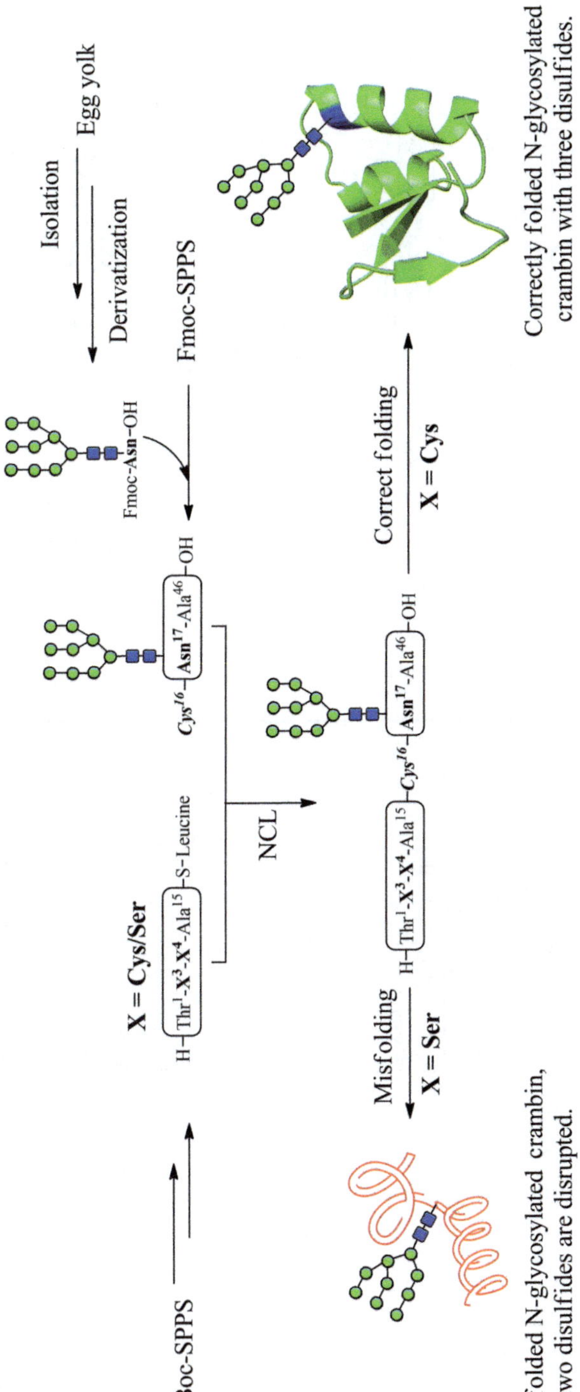

Scheme 5.26 Semisynthesis of homogeneous crambin and analogues.

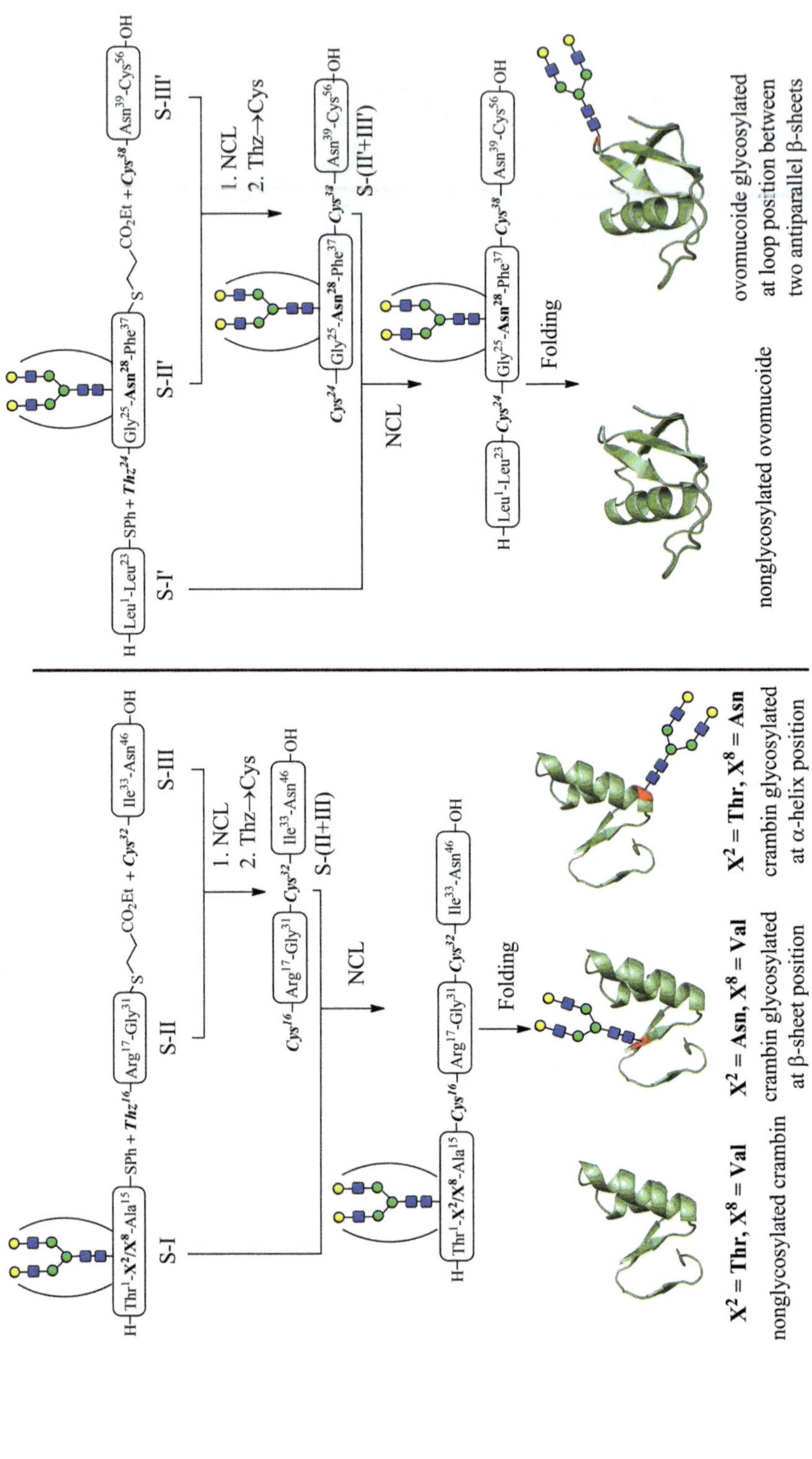

Scheme 5.27 Semisyntheses of crambin and ovomucoide in homogeneous glycoforms.

of fluctuating oligosaccharide chains on the conformational change still needs further exploration.

The examples described above have presented a picture of the current usage of homogeneous glycoproteins in research beyond chemical synthesis. It is clear that obtaining more diverse examples of well-defined protein glycoforms will require vast efforts and significant innovations, both of which will have huge effects on studies in the field of chemical biology.

5.5 Conclusion

Micro-heterogeneity of glycoproteins poses a huge challenge for scientists trying to identify the functions of these complex structures. The ultimate questions surrounding this field cannot be answered until glycoproteins of interest are readily available in a structurally defined and homogenous form. Although this remains an extremely challenging task, through the combined advancement of isolation/separation methods, recombinant techniques, and synthetic approaches, several groups have made significant strides in recent years. As summarized above, more than a few powerful strategies, in particular ligation–desulfurization, glycosyl asparagine "cassette"-based SPPS, and pseudoproline-assisted Lansbury aspartylation, have enabled the chemical syntheses and engineering of a number of *N*-linked glycoproteins of impressive size and complexity. More significantly, some examples have shown that the homogeneous glycoforms obtained as a result of these efforts can be effectively utilized in further studies aiming to solve challenging biological problems. These biological investigations have been of great value in helping us to understand living systems, which is the ultimate goal of pursuing the synthesis of these large molecules as homogeneous isoforms. The promising results and progress achieved thus far will no doubt encourage more extensive studies in this field, which calls for the development of yet more efficient methods and strategies for producing even more complicated proteins and larger libraries of glycoforms. In particular, solutions to the limited availability or diversity of homogeneous complex oligosaccharides still wait to be discovered.

References

1. (a) F. Schwarz and M. Aebi, *Curr. Opin. Struct. Biol.*, 2011, **21**, 576; (b) N. Lannoo and E. J. Van Damme, *Plant Sci.*, 2015, **239**, 67.
2. (a) R. M. Schmaltz, S. R. Hanson and C. H. Wong, *Chem. Rev.*, 2011, **111**, 4259; (b) J. L. Price, D. Shental-Bechor, A. Dhar, M. J. Turner, E. T. Powers, M. Gruebele, Y. Levy and J. W. Kelly, *J. Am. Chem. Soc.*, 2010, **132**, 15359.
3. C. Unverzagt and Y. Kajihara, *Chem. Soc. Rev.*, 2013, **42**, 4408.
4. J. Brailsford, S. J. Danishefsky, A. Fernández-Tejada, C. D. Hall, R. A. Jones, A. R. Katritzky, T. Kawakami, C.-F. Liu, F. Liu, L. R. Malins, J. P. Mayer, M. A. S. Moore, S. S. Panda, R. J. Payne, J.-H. Shieh, Y.-W. Wu, A. Yang, R. Yang, Q. Zhang and L. Zhao, in *Protein Ligation and Total Synthesis I*, ed. L. Liu, Springer, New York, 2nd edn, 2015.

5. R. B. Merrifield, *J. Am. Chem. Soc.*, 1963, **85**, 2149.
6. N. J. Davis and S. L. Flitsch, *Tetrahedron Lett.*, 1991, **46**, 6793.
7. S. Elliott, M. A. Foote and G. Molineux, *Erythropoietins, Erythropoietic Factors, and Erythropoiesis: Molecular, Cellular, Preclinical, and Clinical Biology*, Birkhäuser, Basel, Switzerland, 2nd edn, 2009, p. 3.
8. A. J. Sytkowski, *Erythropoietin*, Wiley-VCH, Weinheim, Germany, 1st edn, 2004.
9. J. Szenajch, G. Wcislo, J. Y. Jeong, C. Szczylik and L. Feldman, *Biochim. Biophys. Acta, Rev. Cancer*, 2010, **1806**, 82.
10. (a) P. H. Lai, R. Everett, F. F. Wang, T. Arakawa and E. Goldwasser, *J. Biol. Chem.*, 1986, **261**, 3116; (b) H. Sasaki, B. Bothner, A. Dell and M. Fukud, *J. Biol. Chem.*, 1987, **262**, 12059; (c) M. Takeuchi, S. Takasaki, H. Miyazaki, T. Kato, S. Hoshi, N. Kochibe and A. Kobata, *J. Biol. Chem.*, 1988, **263**, 3657.
11. P. E. Dawson, T. W. Muir, I. Clark-Lewis and S. B. H. Kent, *Science*, 1994, **266**, 776.
12. Q. Wan and S. J. Danishefsky, *Angew. Chem., Int. Ed.*, 2007, **46**, 9248.
13. (a) V. Y. Dudkin, J. S. Millera and S. J. Danishefsky, *Tetrahedron Lett.*, 2003, **44**, 1791; (b) C. Unverzagt, *Chem.–Eur. J.*, 2003, **9**, 1369; (c) B. Wu, Z. H. Hua, J. D. Warren, K. Ranganathan, Q. Wan, G. Chen, Z. P. Tan, J. H. Chen, A. Endoa and S. J. Danishefsky, *Tetrahedron Lett.*, 2006, **47**, 5577; (d) I. Matsuo, K. Totani, A. Tatamia and Y. Ito, *Tetrahedron*, 2006, **62**, 8262; (e) S. Eller, S. Mezzato, R. Schuberth and C. Unverzagt, *Chem.–Eur. J.*, 2008, **14**, 1304.
14. X. Guinchard, S. Picard and D. Crich, Synthetic Approaches to Bioactive Carbohydrates, in *Modern Tools for the Synthesis of Complex Bioactive Molecules*, ed. J. Cossy and S. Arseniyadis, John Wiley & Sons, Inc., Hoboken, New Jersey, 2012, pp. 395–432.
15. (a) S. Sabcsan and J. C. Paulson, *J. Am. Chem. Soc.*, 1986, **108**, 2068; (b) Y. Ichikawa, Y. C. Lin, D. P. Dumas, G. J. Shen, J. E. Garcia, M. A. Williams, R. Bayer, C. Ketcham, L. E. Walker, J. C. Paulson and C. H. Wong, *J. Am. Chem. Soc.*, 1992, **114**, 9283; (c) K. Witte, P. Sears, R. Martin and C. H. Wong, *J. Am. Chem. Soc.*, 1997, **119**, 2114; (d) K. Akihiko, I. Matsuo, M. Takatani, A. Seko, M. Hachisu, Y. Takeda and Y. Ito, *Angew. Chem., Int. Ed.*, 2013, **52**, 7426.
16. A. Seko, M. Nishizono, Y. Enoki, H. R. Ibrahim, L. R. Juneja, M. Kim, T. Yamamoto and M. Koketsu, *Biochim. Biophys. Acta, Gen. Subj.*, 1997, **1335**, 23.
17. (a) N. Yamamoto, Y. Ohmori, T. Sakakibara, K. Sasaki, L. R. Juneja and Y. Kajihara, *Angew. Chem., Int. Ed.*, 2003, **42**, 2537; (b) Y. Suzuki, N. Yamamoto, K. Sasaki, T. Sakakibara, L. R. Juneja and Y. Kajihara, *Chem.–Eur. J.*, 2004, **10**, 971; (c) N. Yamamoto, T. Sakakibara and Y. Kajihara, *Tetrahedron Lett.*, 2004, **45**, 3287.
18. I. L. Deras, N. Kawasaki and Y. C. Lee, *Carbohydr. Res.*, 1998, **306**, 469.
19. L. Franco-Fraguas, A. Plá, F. Ferreira, H. Massaldi, N. Suárez and F. Batista-Viera, *J. Chromatogr. B.*, 2003, **790**, 365.

20. Y. Makimura, T. Kiuchi, M. Izumi, S. Dedola, Y. Ito and Y. Kajihara, *Carbohydr. Res.*, 2012, **364**, 41.
21. K. Hirano, D. Macmillan, K. Tezuka, T. Tsuji and Y. Kajihara, *Angew. Chem., Int. Ed.*, 2009, **48**, 9557.
22. G. Walsh and R. Jefferis, *Nat. Biotechnol.*, 2006, **24**, 1241.
23. K. Hirano, M. Izumi, D. Macmillan, K. Tezuka, T. Tsuji and Y. Kajihara, *J. Carbohydr. Chem.*, 2011, **30**, 306.
24. (a) J. B. Schwarz, S. D. Kuduk, X. T. Chen, D. Sames, P. W. Glunz and S. J. Danishefsky, *J. Am. Chem. Soc.*, 1999, **121**, 2662; (b) Y. Shin, K. A. Winans, B. J. Backes, S. B. H. Kent, J. A. Ellman and C. R. Bertozzi, *J. Am. Chem. Soc.*, 1999, **121**, 11684.
25. N. Yamamoto, Y. Tanabe, R. Okamoto, P. E. Dawson and Y. Kajihara, *J. Am. Chem. Soc.*, 2008, **130**, 501.
26. N. Yamamoto, A. Takayanagi, A. Yoshino, T. Sakakibara and Y. Kajihara, *Chem.–Eur. J.*, 2007, **13**, 613.
27. (a) F. Effenberger and G. Epple, *Angew. Chem., Int. Ed.*, 1972, **11**, 299; (b) J. P. Tam, W. F. Heath and R. B. Merrifield, *J. Am. Chem. Soc.*, 1986, **108**, 5242.
28. (a) C. Piontek, P. Ring, O. Harjes, C. Heinlein, S. Mezzato, N. Lombana, C. Pöhner, M. Püttner, D. Varún Silva, A. Martin, F. X. Schmid and C. Unverzagt, *Angew. Chem., Int. Ed.*, 2009, **48**, 1936; (b) C. Piontek, D. Varún Silva, C. Heinlein, C. Pöhner, S. Mezzato, P. Ring, A. Martin, F. X. Schmid and C. Unverzagt, *Angew. Chem., Int. Ed.*, 2009, **48**, 1941.
29. S. Mezzato, M. Schaffrath and C. Unverzagt, *Angew. Chem., Int. Ed.*, 2005, **44**, 1650.
30. Y. Kajihara, Y. Suzuki, K. Sasaki and L. R. Juneja, *Methods Enzymol.*, 2003, **362**, 44.
31. C. Heinlein, D. Varón Silva, A. Tröster, J. Schmidt, A. Gross and C. Unverzagt, *Angew. Chem., Int. Ed.*, 2011, **50**, 6406.
32. M. Murakami, R. Okamoto, M. Izumi and Y. Kajihara, *Angew. Chem., Int. Ed.*, 2012, **51**, 3567.
33. (a) D. Bang, B. L. Pentelute and S. B. H. Kent, *Angew. Chem., Int. Ed.*, 2006, **45**, 3985; (b) J. Lee, Y. Kwon, B. L. Pentelute and D. Bang, *Bioconjugate Chem.*, 2011, **22**, 1645.
34. R. Okamoto, M. Kimura, T. Ishimizu, M. Izumi and Y. Kajihara, *Chem.–Eur. J.*, 2014, **20**, 10425.
35. (a) S. R. Chong, F. B. Mersha, D. G. Comb, M. E. Scott, D. Landry, L. M. Vence, F. B. Perler, J. Benner, R. B. Kucera, C. A. Hirvonen, J. J. Pelletier, H. Paulus and M. Q. Xu, *Gene*, 1997, **192**, 271; (b) J. Kang, N. L. Reynolds, C. Tyrrell, J. R. Dorin and D. Macmillan, *Org. Biomol. Chem.*, 2009, **7**, 4918.
36. R. Okamoto, K. Morooka and Y. Kajihara, *Angew. Chem., Int. Ed.*, 2012, **51**, 191.
37. A. Reif, S. Siebenhaar, A. Tröster, M. Schmälzlein, C. Lechner, P. Velisetty, K. Gottwald, C. Pöhner, I. Boos, V. Schubert, S. RoseJohn and C. Unverzagt, *Angew. Chem., Int. Ed.*, 2014, **53**, 12125.

38. G. M. Fang, Y. M. Li, F. Shen, Y. C. Huang, J. B. Li, Y. Lin, H. K. Cui and L. Liu, *Angew. Chem., Int. Ed.*, 2011, **50**, 7645.
39. M. Q. Xu and T. C. Evans Jr, *Methods*, 2001, **24**, 257.
40. V. Ullmann, M. Rdisch, I. Boos, J. Freund, C. Pçhner, S. Schwarzinger and C. Unverzagt, *Angew. Chem., Int. Ed.*, 2012, **51**, 11566.
41. P. Wang, B. Aussedat, Y. Vohra and S. J. Danishefsky, *Angew. Chem., Int. Ed.*, 2012, **51**, 11571.
42. S. T. Cohen-Anisfeld and P. T. Lansbury Jr, *J. Am. Chem. Soc.*, 1993, **115**, 10531.
43. T. Tanaka, H. Nagai, M. Noguchi, A. Kobayashi and S. Shoda, *Chem. Commun.*, 2009, 3378.
44. A. Fernandez-Tejada, J. Brailsford, Q. Zhang, J. Shieh, M. A. S. Moore and S. J. Danishefsky, Total Synthesis of Glycosylated Proteins, in *Protein Ligation and Total Synthesis I*, ed. L. Liu, Springer, New York, 2nd edn, 2015, pp. 1–26.
45. P. Wang, S. Dong, J. A. Brailsford, K. Iyer, S. D. Townsend, Q. Zhang, R. C. Hendrickson, J. Shieh, M. A. S. Moore and S. J. Danishefsky, *Angew. Chem., Int. Ed.*, 2012, **51**, 11576.
46. P. Wang, B. Aussedat, Y. Vohra and S. J. Danishefsky, *Angew. Chem., Int. Ed.*, 2012, **51**, 11571.
47. Q. Zhang, E. V. Johnston, J. H. Shieh, M. A. S. Moore and S. J. Danishefsky, *Proc. Natl. Acad. Sci.*, 2014, **111**(8), 2885.
48. G. Rosner, M. Ross, J. Vredenburgh, B. Meisenberg, C. Gilbert, J. Kurtzberg and W. P. Peters, *Blood*, 1993, **81**, 1709.
49. E. Vacchelli, A. Eggermont, W. H. Fridman, J. Galon, L. Zitvogel, G. Kroemer and L. Galluzzi, *OncoImmunology*, 2013, **2**, e24850.
50. K. S. Clive, J. A. Tyler, G. T. Clifton, J. P. Holmes, E. A. Mittendorf, S. Ponniah and G. E Peoples, *Expert Rev. Vaccines*, 2010, **9**, 519.
51. P. Nagorny, B. Fasching, X. Li, G. Chen, B. Aussedat and S. J. Danishefsky, *J. Am. Chem. Soc.*, 2009, **131**, 5792.
52. D. J. Becker and J. B. Lowe, *Glycobiology*, 2003, **13**, 41R.
53. P. Wang, S. Dong, J. H. Shieh, E. Peguero, R. Hendrickson, M. A. S. Moore and S. J. Danishefsky, *Science*, 2013, **342**, 1357.
54. J. Pierce and T. F. Parsons, *Annu. Rev. Biochem.*, 1981, **50**, 465.
55. B. Aussedat, B. Fasching, E. Johnston, N. Sane, P. Nagorny and S. J. Danishefsky, *J. Am. Chem. Soc.*, 2012, **134**, 3532.
56. A. Fernandez-Tejada, P. A. Vadola and S. J. Danishefsky, *J. Am. Chem. Soc.*, 2014, **136**, 8450.
57. P. Nagorny, N. Sane, B. Fasching, B. Aussedat and S. J. Danishefsky, *Angew. Chem., Int. Ed.*, 2012, **51**, 975.
58. I. Sakamoto, K. Tezuka, K. Fukae, K. Ishii, K. Taduru, M. Maeda, M. Ouchi, K. Yoshida, Y. Nambu, J. Igarashi, N. Hayashi, T. Tsuji and Y. Kajihara, *J. Am. Chem. Soc.*, 2012, **134**, 5428.
59. M. Murakami, T. Kiuchi, M. Nishihara, K. Tezuka, R. Okamoto, M. Izumi and Y. Kajihara, *Sci. Adv.*, 2016, **2**, e1500678.
60. J. Zheng, S. Tang, Y. Qi, Z. Wang and L. Liu, *Nat. Protoc.*, 2013, **8**, 2483.

61. R. Okamoto, K. Mandal, M. Ling, A. D. Luster, Y. Kajihara and S. B. H. Kent, *Angew. Chem., Int. Ed.*, 2014, **53**, 5188.
62. R. Okamoto, K. Mandal, M. R. Sawaya, Y. Kajihara, T. O. Yeates and S. B. H. Kent, *Angew. Chem., Int. Ed.*, 2014, **53**, 5194.
63. L. Ellgaard and A. Helenius, *Nat. Rev. Mol. Cell Biol.*, 2003, **4**, 181.
64. (a) K. Totani, Y. Ihara, I. Matsuo, H. Koshino and Y. Ito, *Angew. Chem., Int. Ed.*, 2005, **44**, 7950; (b) K. Totani, Y. Ihara, T. Tsujimoto, I. Matsuo and Y. Ito, *Biochemistry*, 2009, **48**, 2933.
65. M. Izumi, Y. Makimura, S. Dedola, A. Seko, A. Kanamori, M. Sakono, Y. Ito and Y. Kajihara, *J. Am. Chem. Soc.*, 2012, **134**, 7238.
66. S. Dedola, M. Izumi, Y. Makimura, A. Seko, A. Kanamori, M. Sakono, Y. Ito and Y. Kajihara, *Angew. Chem., Int. Ed.*, 2014, **53**, 2883.
67. Y. Tanabe, S. Sasaoka, R. Okamoto and Y. Kajihara, *Chem.–Eur. J.*, 2012, **18**, 5944.

CHAPTER 6

Chemoenzymatic Synthesis of N-*Glycans*

XUEFENG CAO[a], JUN ZHOU[a], YUXI GUO[a], YUNPENG LIU[a], LEI LI[a] AND PENG GEORGE WANG*[a]

[a]Department of Chemistry and Center of Diagnostics & Therapeutics, Georgia State University, 50 Decatur St SE, Atlanta, GA 30303, USA
*E-mail: pwang11@gsu.edu

6.1 Introduction

The concise definition of glycobiology was firstly proposed three decades ago.[1] Significant changes have happened since then and have gradually convinced us of the great importance of carbohydrates in myriad biological processes. The term "glycans" was defined by IUPAC as synonyms meaning "compounds consisting of a large number of monosaccharides linked glycosidically". In practice, the term glycans is always used to refer to medium-sized oligosaccharides or the carbohydrate moiety of a glycoconjugate, including glycoprotein, glycolipid, and proteoglycan.[2] Glycans can be found on the exterior surface of cells in most living systems, attached to proteins as in glycoproteins and proteoglycans. Results of recent studies indicated that glycans are ubiquitous and play diverse roles in a wide range of biological processes, such as cellular recognition and information transfer.[2–6]

O- and *N*-glycans are the major two classes of glycans, depending on the linkage at the reducing terminal. Structurally, *N*-glycans are covalently

Chemical Biology No. 2
Chemical Biology of Glycoproteins
Edited by Zhongping Tan and Lai-Xi Wang

Published by the Royal Society of Chemistry, www.rsc.org

attached to proteins at asparagine (Asn) residues by an *N*-glycosidic bond. Five different *N*-glycan linkages have been reported. However, the most common core structure is an asparagine-connected pentasaccharide sequence, Manα1–6(Manα1–3)Manβ1–4GlcNAcβ1–4GlcNAcβ1-Asn.[7] On the basis of their antennary components, *N*-glycans are usually classified into three types (Figure 6.1): (1) oligomannose type, containing only "pure" oligomannan antennae; (2) complex type, where antennae are initiated by *N*-acetyl glucosamine (GlcNAc) residue; and (3) hybrid type, in which mannose residues are only linked to the Manα1–6 arm of the core with a GlcNAc-initiated antennae on the Manα1–3 arm.[7] From the diversity of antennary components, we can easily deduce that there are large amounts of *N*-glycans with different structures in nature.

It is well known that the structure–activity relationship (SAR) is essential to understanding of the detailed structural functions of *N*-glycans in biological processes.[8] One major precondition for SAR study is obtaining structurally defined *N*-glycans which serve as standard compounds to determine the exact structures of *N*-glycans in complex mixtures and to study the specificities of *N*-glycan protein binding.[9] However, we have not developed powerful methods to obtain all kinds of *N*-glycan libraries until now. Therefore, development of efficient synthesis of *N*-glycan libraries is of considerable interest for chemical glycobiology.

Three major approaches to prepare *N*-glycans have been developed: extraction from natural sources, chemical synthesis, and chemoenzymatic synthesis. The most traditional one is isolation of *N*-glycans from natural sources.[10] By isolation from egg yolk, a disialylated undecasaccharide with fluorenylmethyloxycarbonate (Fmoc)–Asn was obtained through chromatographies, proteinase digestion, treatment with Fmoc-OSulfate, and HPLC

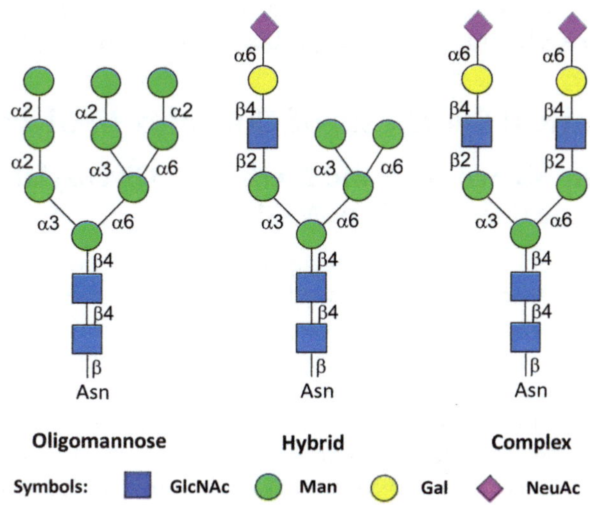

Figure 6.1 Examples of three *N*-glycan types.

purification.[11–13] The disialylated undecasaccharide was also degraded to several valuable compounds at the milligram scale.[14] Moreover, the *N*-glycans isolated from egg yolk can also be further enzymatically extended to *N*-glycan libraries.[11] However, naturally occurring glycans are typically isolated in small quantities as mixtures of closely related structures that are difficult to separate and, therefore, are not a practical source of well-defined *N*-glycans.

Chemical synthesis of *N*-glycans is well developed by many carbohydrate chemists and has become the efficient and commonly used approach to prepare structurally well-defined *N*-glycans with natural and unnatural structures in many cases. However, when it comes to complex target molecules, for example, asymmetrically branched *N*-glycans, chemical synthesis remains technically challenging, tedious, and time-consuming.[15] One of the major challenges encountered in the synthesis of complex *N*-glycans is stereoselective control during the formation of glycosidic linkages. Moreover, the structural diversity of mammalian and microbial glycomes requires a large number of sugar building blocks and makes the chemical synthesis of an *N*-glycan library extremely difficult.

Glycochemists are turning more attention to enzymatic methods because glycosyltransferases (GTs) are natural catalysts to effectively form glycosidic bonds with much stricter regio- and stereo-selectivity than chemical glycosylation. Compared with chemical assembly, enzymatic synthesis of glycans possesses the advantages of requiring mild conditions, being time-saving, and being more cost efficient. The most recently established chemoenzymatic strategy, a combination of chemical and enzymatic approaches, has been successfully applied to the efficient synthesis of *N*-glycan libraries.[8,9,16–18] By using this strategy, *N*-glycan core precursors are first chemically synthesized and further glycosylated in enzymatic extension by glycosyltransferases (GTs) to rapidly produce a large number of complex *N*-glycans.

In this chapter, we will describe recent advances in the two major approaches to *N*-glycans: chemical synthesis and chemoenzymatic methods.

6.2 Chemical Synthesis of Complex *N*-Glycans

6.2.1 Assembly Strategy and Method of Glycosylation in Chemical Synthesis of *N*-Glycans

In the efforts to chemically synthesize complex *N*-glycans, several synthetic strategies and methods have been well established in the past years. Among these studies, Danishefsky and colleagues[19–24] reported the construction of a $Man_3GlcNAc_2$ core pentasaccharide by attaching mannose monomers onto a $ManGlcNAc_2$ trisaccharide and further introduction of antennae to this pentasaccharide. Boons and colleagues[9] developed a general strategy based on an orthogonally protected $Man_3GlcNAc_2$ core pentasaccharide and site-defined individual deprotection–glycosylation transformations for antenna elongation. In the synthetic strategies described above, one branch moiety is constructed in a step-wise manner by starting from $ManGlcNAc_2$ trisaccharide, however, excellent stereoselectivities are conferred in the formation of Man-α-(1-3)/(1-6)-Man glycosidic linkages. In contrast, a highly convergent strategy has been employed in efficient synthesis of complex *N*-glycans, in which the entire α-(1-3) or (1-6) branch moiety is directly coupled onto ManGlcNAc$_2$ trisaccharide. For example, Unverzagt and colleagues[25] accomplished the synthesis of multi-antennary complex type *N*-glycans by glycosylating ManGlcNAc$_2$ trisaccharide with oligosaccharyl Schmidt donors. In the synthesis of bi-, tri-, and tetra-antennary complex-type *N*-glycans, Wong[18] utilized oligosaccharyl fluorides as donors to couple with ManGlcNAc$_2$ trisaccharide. To prepare various glycosyl donors in these approaches, a temporary anomeric protecting group was introduced first and subsequently converted into glycosyl fluoride or imidate, depending on the choice of glycosylation method. The Wang group[8] developed an efficient convergent strategy that directly utilizes oligosaccharyl thioglycoside donors for the glycosylation of a ManGlcNAc$_2$ core trisaccharide, providing *N*-glycan precursors at high yields with good stereoselectivities.

6.2.2 Total Synthesis of Complex *N*-Glycans by Global Glycosylation of a Core Pentasaccharide

Danishefsky and coworkers accomplished the total synthesis of symmetric bi- and tri-antennary fully sialylated *N*-glycans of complex type using a global glycosylation strategy, where multiple antennary blocks were attached to Man$_3$GlcNAc$_2$ core pentasaccharide efficiently in only one procedure.[19,20] To prepare the bi-antennary *N*-glycan (Scheme 6.1), the synthesis started with glycosylation of the 3,6-diol core trisaccharide **1** using a mannose thioglycoside donor **2**, which generated pentasaccharide **3** with an excellent yield of 89%. Subsequent deprotection of 2,5-difluorobenzoyl ester groups under basic conditions allowed for a rapid construction of diol **4** in quantitative yield. The resulting hydroxyl functionalities served as acceptor sites to the *N*-acetylneuraminic acid (Neu5Ac)-terminated trisaccharyl fluoride donor **5**, and the double glycosylation delivered the fully protected undecasaccharide **6** in 74% yield. The total synthesis of free bi-antennary *N*-glycan was completed by a final deprotection sequence. A tri-antennary *N*-glycan was also produced using a similar strategy of global glycosylation.

6.2.3 Convergent Synthesis of Complex *N*-Glycans by Glycosylation of Core Trisaccharide

Wong's group reported a unified convergent strategy, where various oligosaccharyl fluorides are sequentially coupled with ManGlcNAc$_2$ core trisaccharide, allowing for the efficient synthesis of symmetric bi-, tri-, and tetra-antennary complex-type *N*-glycans.[18] For example, in the synthesis of bi-antennary *N*-glycan (Scheme 6.2), the α-(1-3) branch was first installed onto the key trisaccharide **7** by using the trisaccharyl fluoride donor **8**.

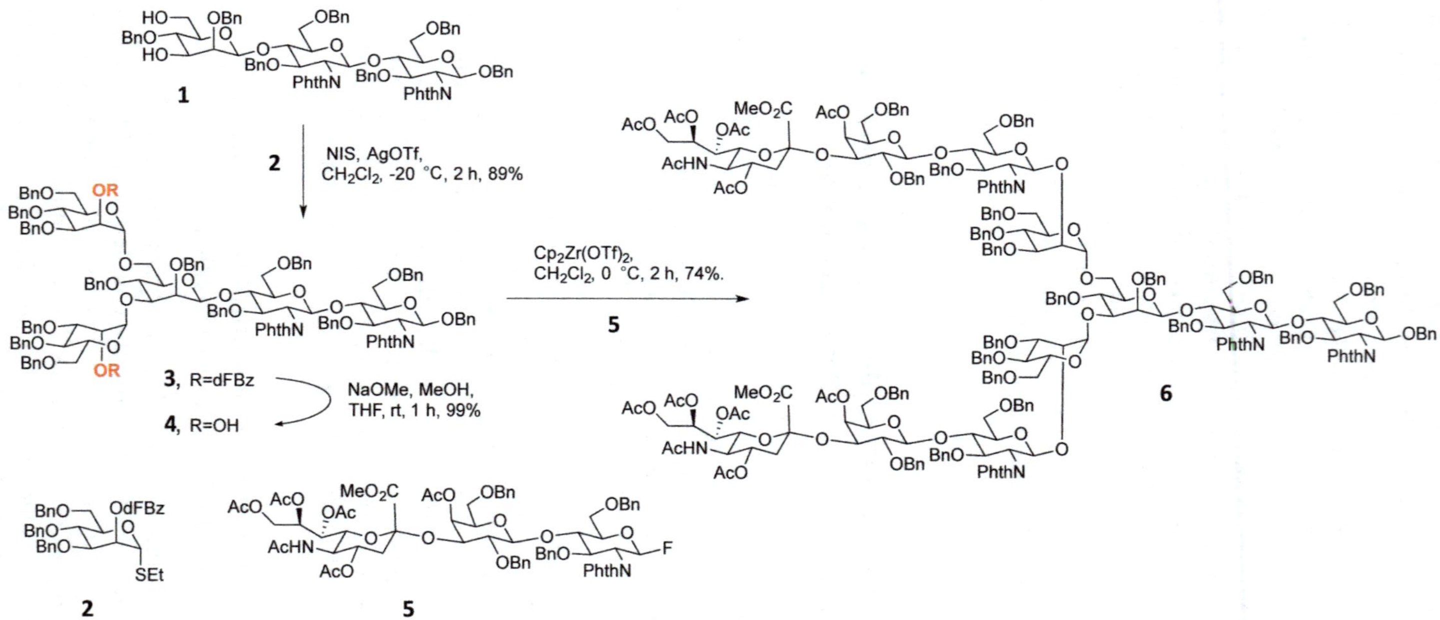

Scheme 6.1 Assembly of a fully protected bi-antennary undecasaccharide by global glycosylation strategy. AgOTf, silver trifluoromethanesulfonate; NaOMe, sodium methoxide; NIS, *N*-iodosuccinimide; rt, room temperature; THF, tetrahydrofuran; dFBz, 2,5-difluorobenzoyl; Bn, benzyl; NPhth, phthalimide; Cp2Zr(OTf)2, bis(cyclopentadienyl)titanium(IV) trifluoromethanesulfonate.

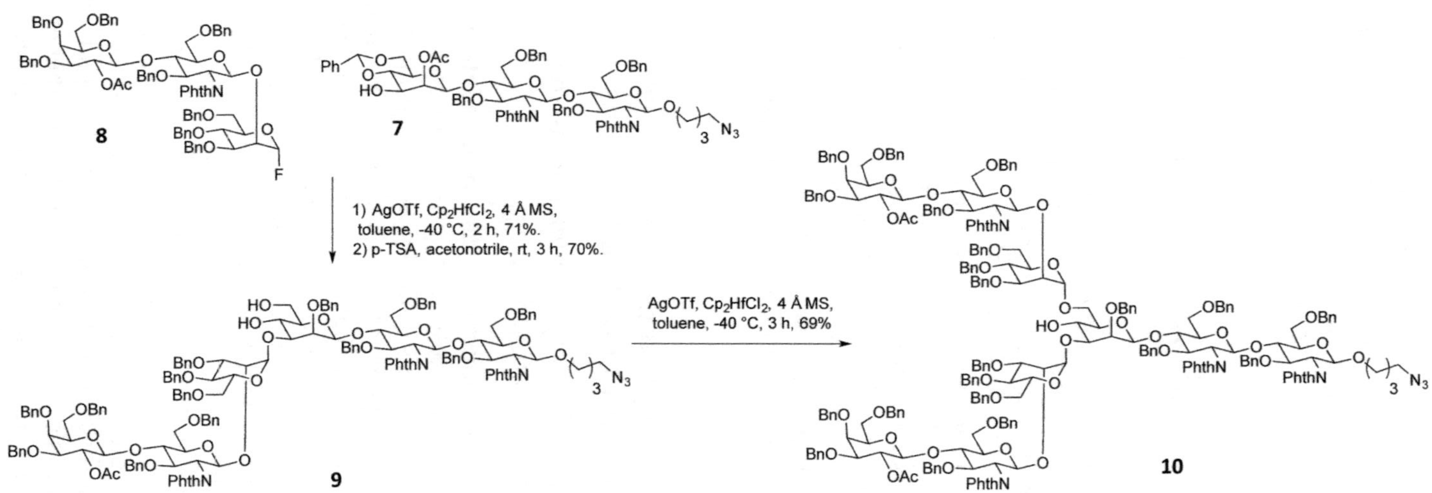

Scheme 6.2 Assembly of the fully protected bi-antennary *N*-glycan by convergent assembly using oligosaccharyl fluoride donors. *p*-TSA, *p*-toluenesulfonic acid.

The resulting hexasaccharide was subjected to cleavage of 4,6-*O*-benzylidene protection to provide the diol acceptor **9**. The subsequent region-selective *O*-6 glycosylation with the use of trisaccharyl fluoride donor **8** delivered the fully protected bi-antennary *N*-glycan **10** with 69% yield. Using a similar convergent strategy, a series of tri- and tetra-antennary complex type *N*-glycans were produced.

6.3 Chemoenzymatic Synthesis of Complex *N*-Glycans

Chemoenzymatic approaches to synthesis of *N*-glycans generally include chemical synthesis of core structures and further enzymatic extension to *N*-glycan libraries, therefore incorporating their advantages. In this respect, chemical method can provide structurally well-defined medium-sized *N*-glycan cores. Glycosyltransferases (GTs) are then used to efficiently add more sugar residues to the cores, generating libraries with a larger number of complex *N*-glycans. GTs are natural catalysts to effectively construct glycan chains with much stricter regio- and stereo-selectivity than chemical glycosylation. Thus, in enzymatic extension, the yield and fidelity are assured, and many tedious chemical steps can be avoided. In the following paragraphs, we will describe two representative studies on chemoenzymatic synthesis of *N*-glycan libraries.

6.3.1 A General Strategy for the Chemoenzymatic Synthesis of Asymmetrically Branched *N*-Glycans

6.3.1.1 Chemical Synthesis of Asymmetric N-*Glycans by Orthogonal Protection Strategy*

The researchers performing studies on glycan–protein interactions are eager for systematic and efficient access to a diverse library of asymmetrically branched *N*-glycans. To solve this difficulty, Boons's group developed a general strategy based on orthogonal protections.[9] A key feature of this strategy is a unique core pentasaccharide that is decorated by a set of orthogonal protecting groups at potential branching points, allowing individual removal of each protecting group and subsequent installation of specific glycan moieties by chemical glycosylation using a diverse set of glycosyl donors. For example, the key pentasaccharide **11** resembles the core structure common to all eukaryotic *N*-glycans (Scheme 6.3). The protecting groups include levulinoyl (Lev), Fmoc, allyloxycarbonate (Alloc), and 2-naphthylmethyl (Nap), which can be selectively deprotected, and subsequent individual chemical glycosylations using trifluoroacetimidate donors **12–15** would provide several *N*-glycan cores (for example, decasaccharide **16**) for further enzymatic extension. Therefore, it would be a general approach to produce a library of bi-, tri-, and tetra-antennary complex *N*-glycan backbones.

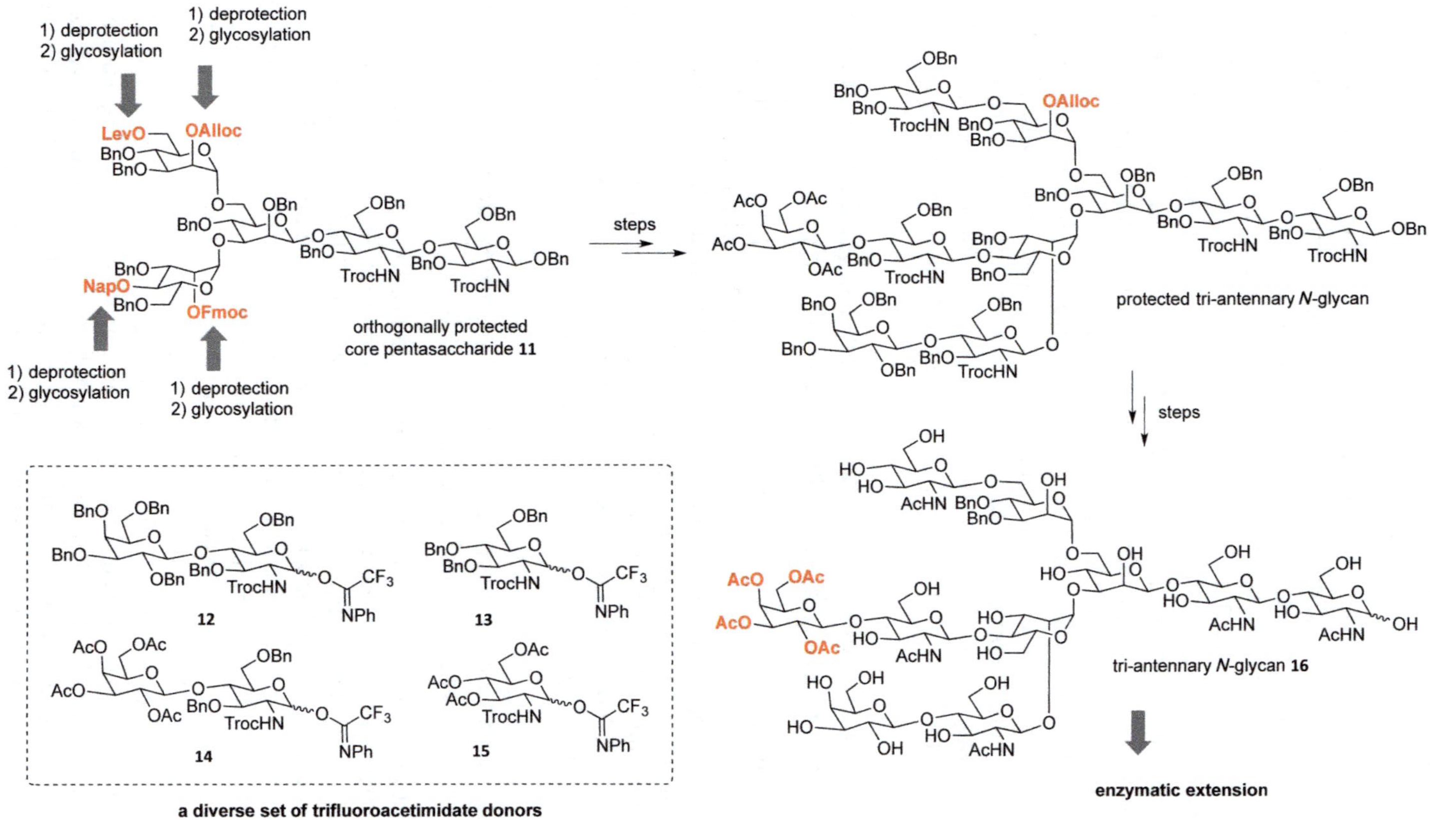

Scheme 6.3 Chemical synthesis of asymmetric *N*-glycans by orthogonal protection strategy.

6.3.1.2 Enzymatic Extension to Synthesize Asymmetrically Branched N-Glycans

The *N*-glycan cores **11** could be selectively and efficiently extended by glycosyltransferases to give a large number of highly complex asymmetrical multi-antennary glycans. For example (Figure 6.2), the free *N*-acetyl-D-lactosamine (LacNAc) non-reducing terminal of decasaccharide **16** was selectively sialylated by α-2,3 sialyltransferase (ST3GalIV) owing to acetyl ester protections, which were then removed by treatment with aqueous ammonia. Compound **18** which had an unmasked LacNAc moiety at the C-4 of the Man-α3 arm was expected to be available for enzymatic glycosylation by α-1,3-fucosyltransferase (α-3FucT). Then β-1,4 galactosyltransferase (GalT-1) attached a Gal residue to GlcNAc from C-6 antenna to give compound **20**. Sequential treatment of **20** with β-1,3-*N*-acetylglucosaminyltransferase (β-1,3GlcNAcT) and GalT-1 yielded compound **22**. The β1,6-branch was further extended by α-2,6 sialyltransferase (ST6Gal-1) to produce target compound **23**, which has distinctive oligosaccharide appendages on each of the three antennae.

6.3.2 Core Synthesis/Enzymatic Extension (CSEE) Strategy for Efficient Synthesis of *N*-Glycan Libraries

More recently, the Wang group developed a facile and practical strategy called Core synthesis/enzymatic extension (CSEE), which has also been successfully applied in efficient synthesis of an *N*-glycan library with structural diversities.[8] By using this strategy, the Wang lab recently prepared 73 *N*-glycans (most of which are asymmetric) using eight chemically synthesized *N*-glycan cores and five GTs. Each of those *N*-glycans was in milligram quantities with high purity (98%) and characterized.

6.3.2.1 Chemical Synthesis of N-Glycan Core Structures Through Convergent Block Coupling

The synthesis commenced with the preparation of the core structures. The core structures include the first few sugar sequences connected to Asn in *N*-glycans. They are modest in size, complex in natural biosynthetic pathway, but their chemical synthesis is well investigated and documented.

To prepare *N*-glycan core structures for enzymatic extension, Wang and his co-workers developed an efficient convergent strategy that directly utilizes oligosaccharyl thioglycoside donors for glycosylations at branching points (Figure 6.3), facilitating the installation of antennae in minimum glycosylation events with excellent yields and good stereoselectivities. The synthetic strategy indicated that core trisaccharide **24** containing a β-mannosidic bond would be a key precursor, where the C4- and C6-hydroxyl groups on β-Man residue are blocked with benzylidene and the C3-OH is unprotected to allow further elongation. Stereoselective construction

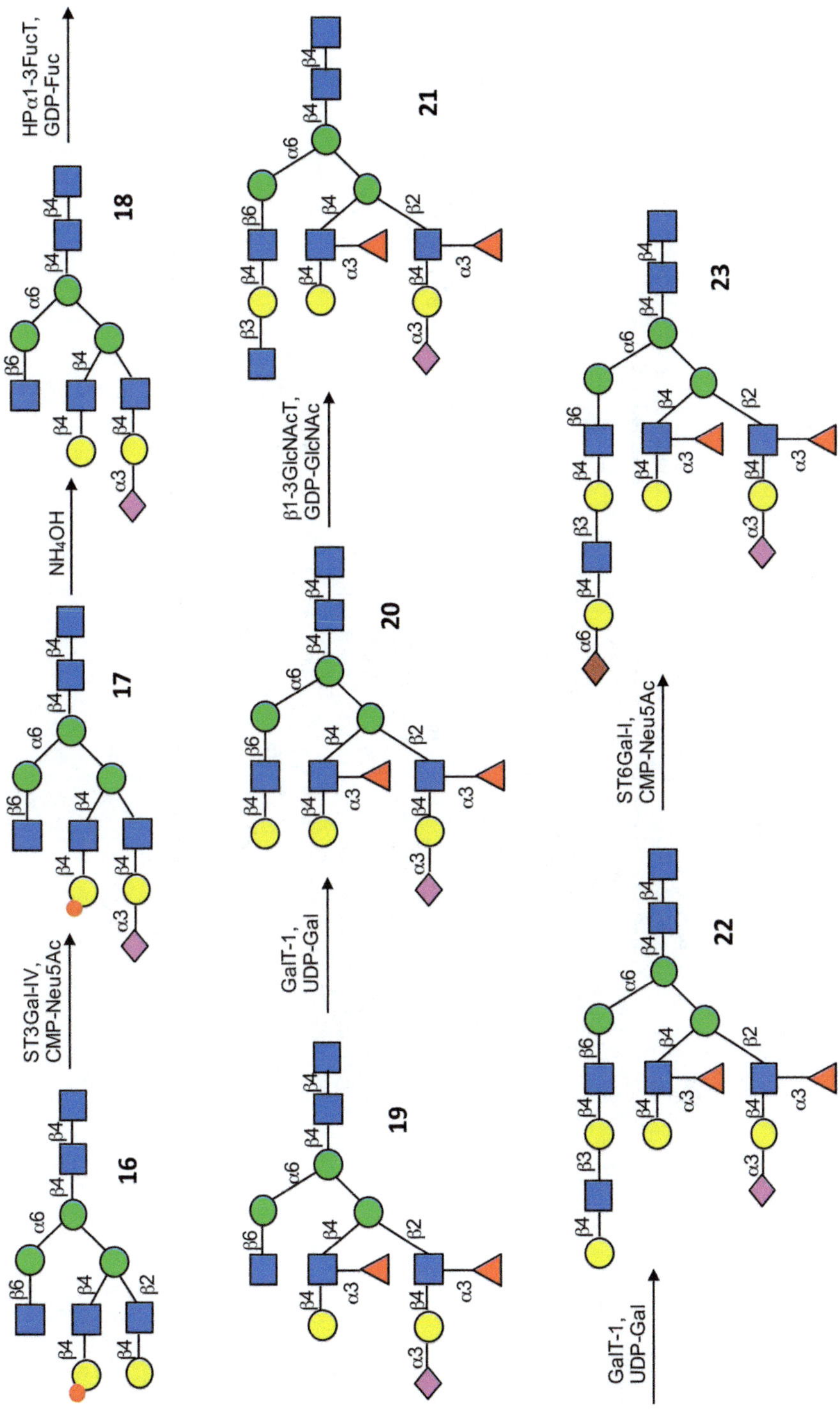

Figure 6.2 Chemoenzymatic synthesis of complex oligosaccharides from **16**.

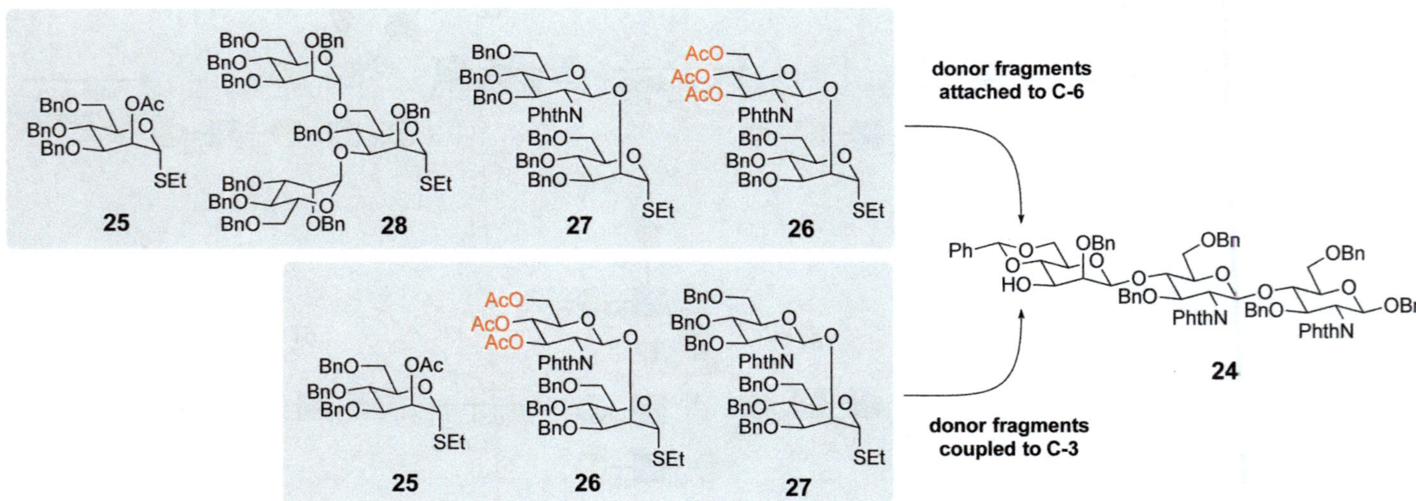

Figure 6.3 Key trisaccharide precursor **24** and four donor fragments (**25–28**) for the convergent block assembly of eight core structures.

of β-mannoside, the most challenging glycosylation in *N*-glycan synthesis, was achieved by using Crich–Kahne conditions with satisfactory yield and good β-selectivity.[26,27] Regioselective ring opening of the benzylidene ring can provide either C6-OH[9] or C4-OH[20] as the branching point for further glycosylation. Thus eight GlcNAc terminated cores (Figure 6.4) were designed and prepared as the starting point for enzymatic extension. Strategically, **N110** and **N210** were partially protected by acetyl groups at the non-reducing end on either the upper or lower branch, allowing the synthesis of asymmetric bi-antennary *N*-glycans. Once the unmasked branch has been extended by glycosyltransferases, the acetyl groups can be readily removed for further transformations.

The key precursor trisaccharide **24**[20] and donor fragments **25**,[28] **26**,[29] **27**, and **28**[30] were prepared according to reported protocols. Using these building blocks, the synthesis of eight core structures were accomplished in a convergent manner (Schemes 6.4–6.7). For example, to synthesize core structure **N010**, the C3-hydroxyl of trisaccharide **11** was first glycosylated with thioglycoside donor **27** in the presence of *N*-iodosuccinimide (NIS)/silver trifluoromethanesulfonate, (AgOTf), providing pentasaccharide **29** at 93% yield with good stereoselectivity. Regioselective ring opening of benzylidene using triethylsilane (Et_3SiH) /phenylboron dichloride ($PhBCl_2$) combination generated pentasaccharide **30** with an excellent yield of 96%. By subsequent installation of fragment **28** onto C6-hydroxyl of acceptor **30**, the fully protected octasaccharide **31** was obtained at 85% yield. Finally, the phthalimide protections in **31** were transformed into acetamide groups, and the benzyl ether groups (Bn) were cleaved by hydrogenolysis under $Pd(OH)_2/H_2$ in methanol/H_2O (10:1), generating the core structure **N010** at a total yield of 63% over three steps (Scheme 6.4). Using a similar approach, bi-antennary cores **N110**, **N030**, **N050**, **N210**, and **N000** (Schemes 6.4–6.6) were all synthesized at high yields with good stereoselectivity. The configurations of all newly generated glycosidic linkages were confirmed by a combination of NMR techniques.

Starting from pentasaccharide intermediate **30**, Core **N020** was conveniently synthesized by amino protecting group transformation and global deprotection without installation of the upper branch (Scheme 6.4). For the synthesis of **N040**, trisaccharide **24** was subjected to simple 3-*O*-benzylation and regioselective cleavage of benzylidene acetal to produce acceptor **41**, which was then glycosylated with **27** to provide pentasaccharide **42** at 91% yield with a satisfactory stereoselectivity ($\alpha/\beta = 3.5:1$). Further protecting group manipulations yielded the core **N040** as previously described (Scheme 6.7).

6.3.2.2 *Enzymatic Extension to Obtain a* N-*Glycan Library*

With eight *N*-glycan cores on hand, an *N*-glycan library was efficiently obtained through enzymatic glycosylations catalyzed by four robust glycosyltransferases. For example, starting with core **N010**, Wang and co-workers prepared five *N*-glycans (**N011** to **N015**) at milligram scales (Figure 6.5A). Reactions were monitored by high-performance liquid chromatography–evaporative laser scattering detection (HPLC–ELSD), and the products were

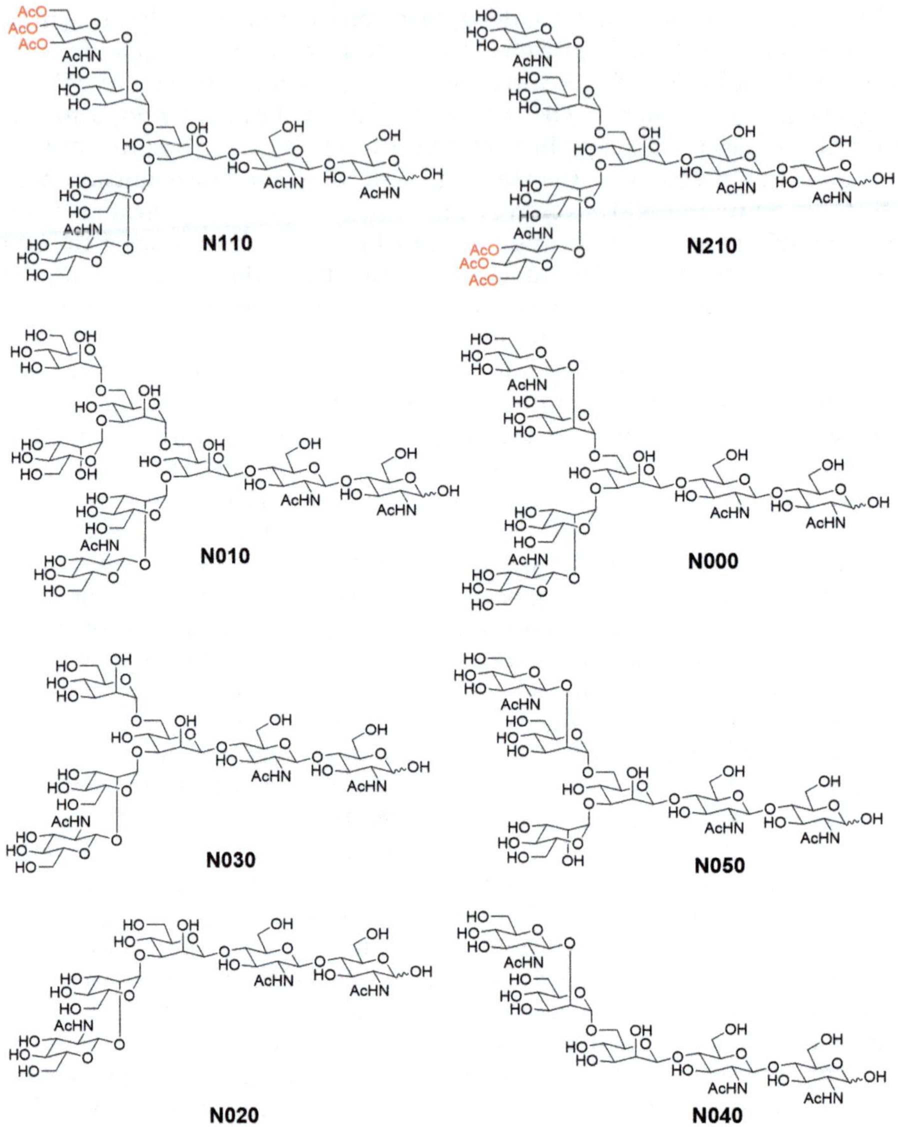

Figure 6.4 Eight GlcNAc-terminated core structures.

purified using a semi-preparative amide column (10 × 250 mm). Up to 5 mg of *N*-glycans can be separated with at least 98% purity by a single run within 30 min. It should be noted that two adjacent peaks were always observed on the HPLC profile, which were later identified as α and β configurations of the reducing terminal GlcNAc. Such a CSEE strategy enables us to prepare 73 structurally defined *N*-glycans (with 5–13 sugars) (Figure 6.5B). A similar strategy was also applied in the preparation of amine-tagged glycans for glycan-microarray studies, yielding over 80 *N*-glycans (2–10 mg).

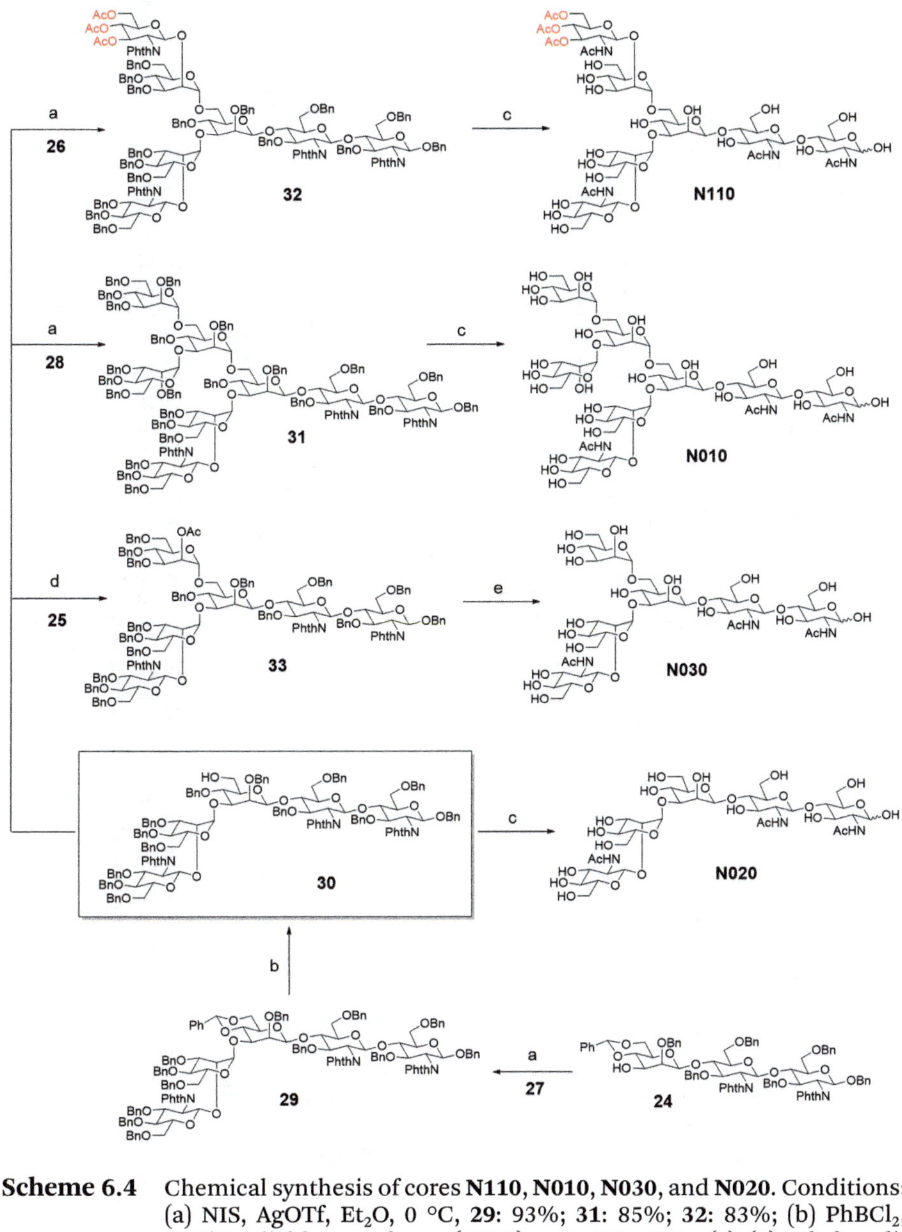

Scheme 6.4 Chemical synthesis of cores **N110**, **N010**, **N030**, and **N020**. Conditions: (a) NIS, AgOTf, Et_2O, 0 °C, **29**: 93%; **31**: 85%; **32**: 83%; (b) $PhBCl_2$, Et_3SiH, dichloromethane (DCM), −78 °C, 96%; (c) (1) ethylenediamine, *n*-butanol, 90 °C; (2) Ac_2O, pyridine, room temperature (rt); (3) $Pd(OH)_2$, H_2, methanol/H_2O (10 : 1), over three steps, **N010**: 63%; **N020**: 67%; **N110**: 69%; (d) NIS, AgOTf, DCM, 0 °C, 91%; (e): (1) ethylenediamine, *n*-butanol, 90 °C; (2) acetic anhydride (Ac_2O), pyridine, rt; (3) sodium methoxide (NaOMe), methanol; (4) $Pd(OH)_2$, H_2, methanol/H_2O (10 : 1), over four steps, 53%.

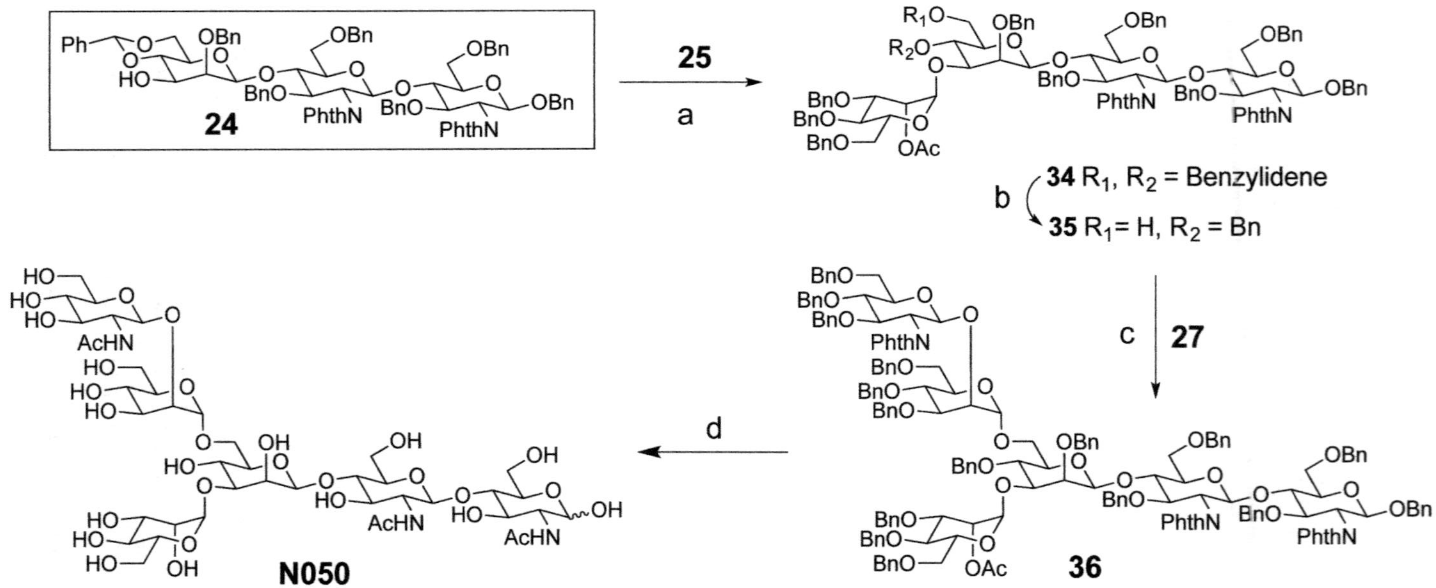

Scheme 6.5 Chemical synthesis of cores **N050**. Conditions: (a) NIS, AgOTf, DCM, 0 °C, 95%; (b) $PhBCl_2$, Et_3SiH, DCM, −78 °C, 92%; (c) NIS, AgOTf, Et_2O, 0 °C, 90% ($\alpha:\beta = 5:1$); (d): (1) ethylenediamine, *n*-butanol, 90 °C; (2) Ac_2O, pyridine, rt; (3) NaOMe, methanol; (4) $Pd(OH)_2$, H_2, methanol/H_2O (10 : 1), over four steps, 55%.

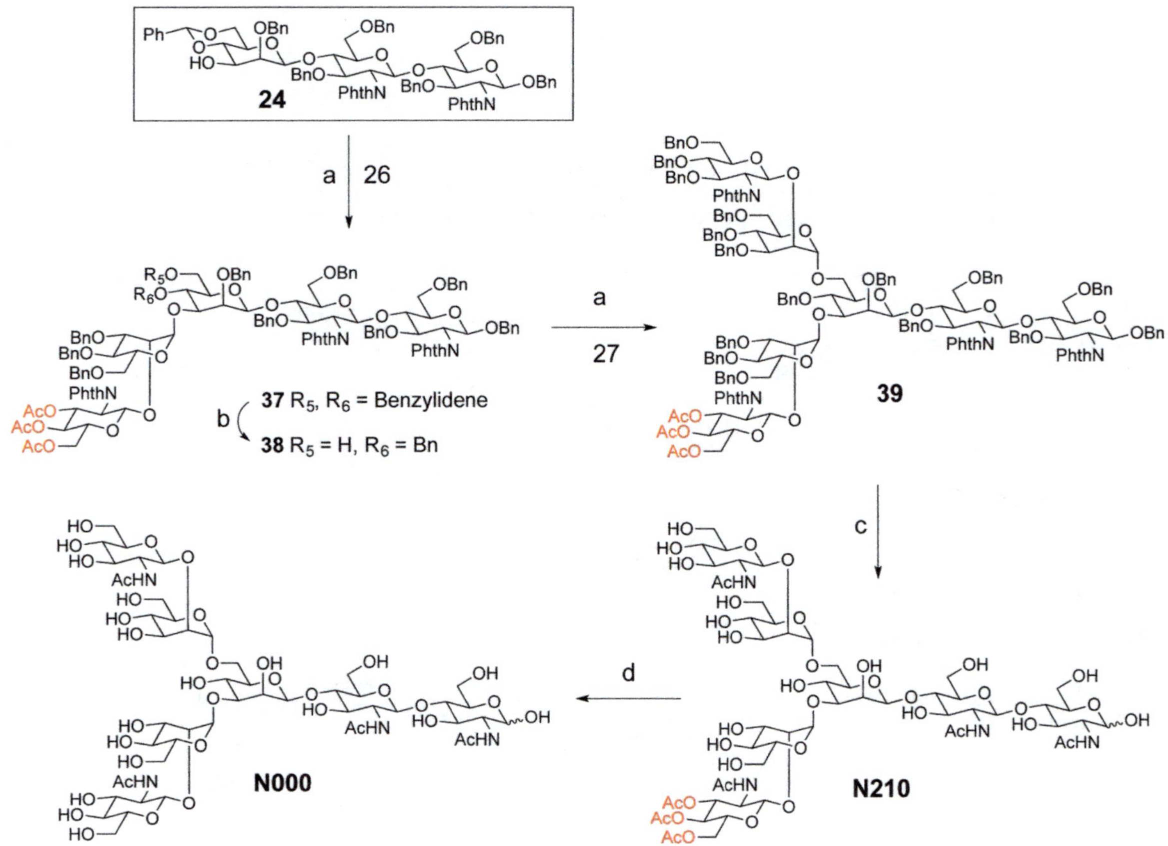

Scheme 6.6 Chemical synthesis of cores **N210** and **N000**. Conditions: (a) NIS, AgOTf, Et_2O, 0 °C, **24**: 94%; **26**: 86% ($\alpha/\beta = 4:1$); (b) $PhBCl_2$, Et_3SiH, DCM, −78 °C, 93%; (c) (1) ethylenediamine, *n*-butanol, 90 °C; (2) Ac_2O, pyridine, rt; (3) $Pd(OH)_2$, H_2, methanol/H_2O (10 : 1), over three steps, 65%; (d) 30% NH_4OH : H_2O (1 : 10), quant.

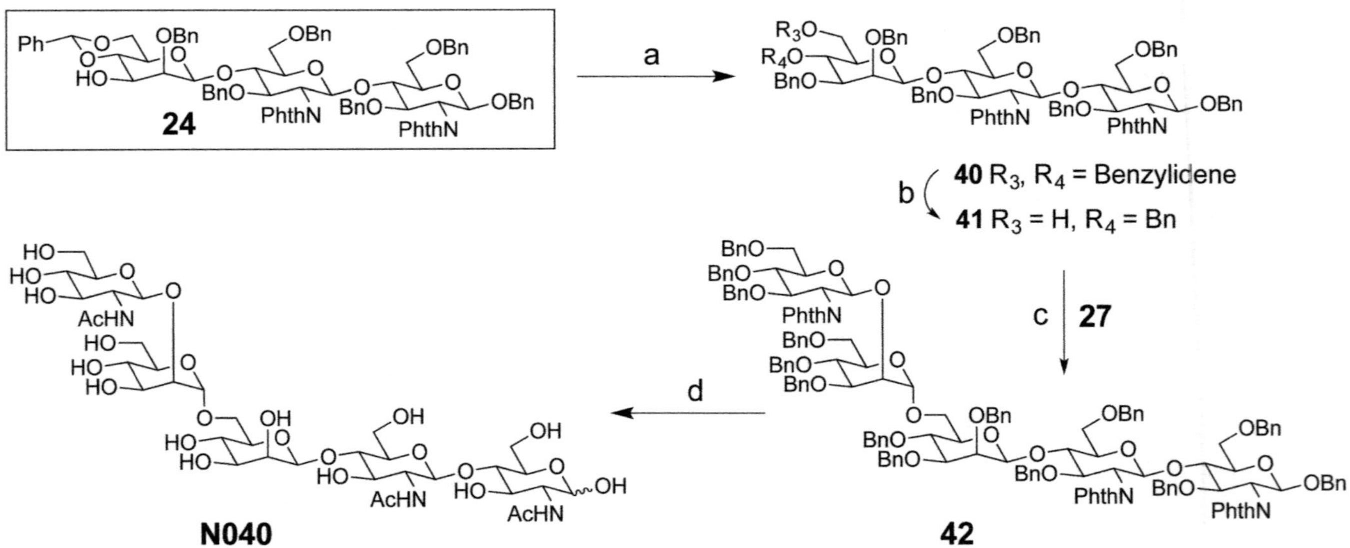

Scheme 6.7 Chemical synthesis of cores **N040**. Conditions: (a): benzyl bromide (BnBr), NaH, dimethylformamide (DMF), 90%; (b) $PhBCl_2$, Et_3SiH, DCM, −78 °C, 96%; (c) NIS, AgOTf, diethyl ether (Et_2O), 0 °C, 91% ($\alpha/\beta = 3.5:1$); (d) (1) ethylenediamine, *n*-butanol, 90 °C; (2) Ac_2O, pyridine, rt; (3) $Pd(OH)_2$, H_2, methanol/H_2O (10 : 1), over three steps, 61%.

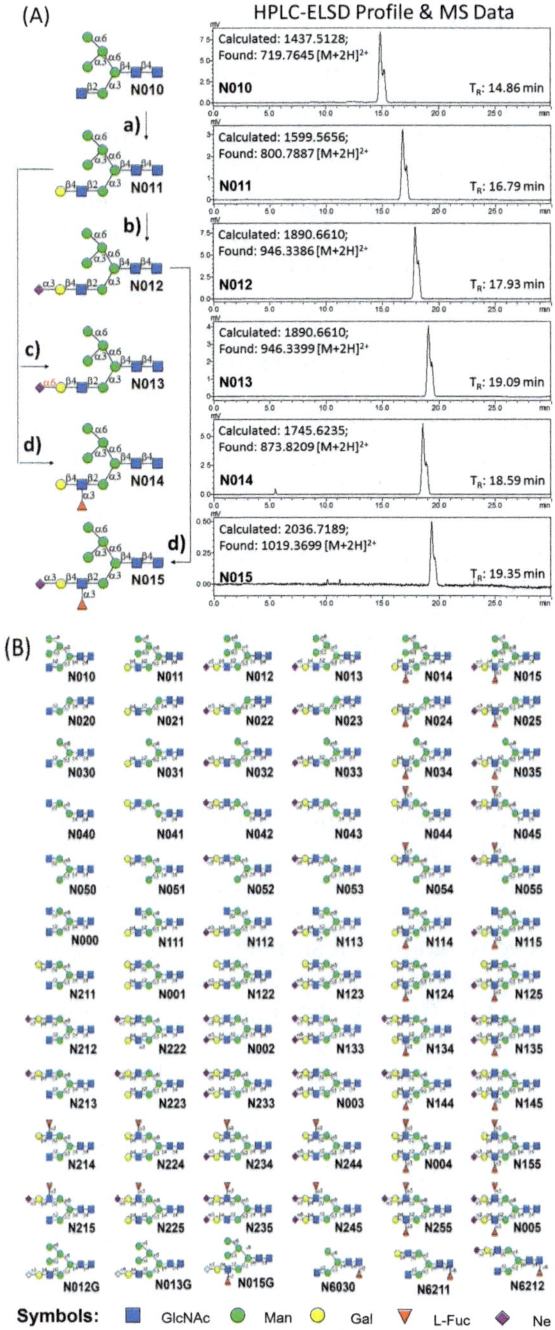

Figure 6.5 *N*-Glycan synthesis *via* enzymatic extension. (A) A typical procedure and product characterization. Conditions: (a) commercially available β1,4GalT (Sigma); (b) 2,3sialyltransferase (PmST1), (c) α2,6sialyltransferase (Pd26ST); (d) and (e) Hpα1,3FucT. (B) The synthetic *N*-glycans.

6.4 Conclusion

SAR investigation of *N*-glycans has been impeded by the lack of adequate, pure, and structurally defined *N*-glycan libraries. Naturally occurring glycans are typically isolated in small quantities as mixtures of closely related structures. Thus the discovery of efficient synthesis of *N*-glycans is of considerable interest for chemical glycobiology. Chemical synthesis of *N*-glycans has been well developed during recent years and has become the most used approach to producing well-defined *N*-glycans. Nevertheless, chemical synthesis remained challenging, tedious, and time-consuming for complex *N*-glycans and especially for *N*-glycan library construction. Compared with chemical assembly, enzymatic glycosylation has the advantages of using mild conditions, being time-saving, and being cost efficient. The CSEE strategy was most recently developed, wherein *N*-glycan cores are first chemically synthesized and further glycosylated by GTs to rapidly produce a complex *N*-glycans library and multi-antennary asymmetrically branched *N*-glycans. The recent successful chemoenzymatic synthesis of *N*-glycans will serve as a foundation for glycoscience. The libraries of *N*-glycans which mimic the natural glycan structure on mammalian cells can be used to establish focused and diverse microarrays for screening lectins, antibodies, or viruses. The synthetic glycoconjugates could also be potential antigens and synthetic vaccine candidates.

References

1. T. Rademacher, R. Parekh and R. Dwek, Glycobiology, *Annu. Rev. Biochem.*, 1988, **57**(1), 785–838.
2. R. A. Dwek, Glycobiology: toward understanding the function of sugars, *Chem. Rev.*, 1996, **96**(2), 683–720.
3. R. Jefferis, Glycosylation as a strategy to improve antibody-based therapeutics, *Nat. Rev. Drug Discovery*, 2009, **8**(3), 226–234.
4. J. D. Marth and P. K. Grewal, Mammalian glycosylation in immunity, *Nat. Rev. Immunol.*, 2008, **8**(11), 874–887.
5. G. W. Hart and R. J. Copeland, Glycomics hits the big time, *Cell*, 2010, **143**(5), 672–676.
6. A. Helenius and M. Aebi, Intracellular functions of N-linked glycans, *Science*, 2001, **291**(5512), 2364–2369.
7. A. Varki, R. D. Cummings, J. D. Esko, H. H. Freeze, P. Stanley, C. R. Bertozzi, G. W. Hart and M. E. Etzler, *Essentials of Glycobiology*, Cold Spring Harbor Laboratory Press, 2nd edn, 2009.
8. L. Li, *et al.*, Efficient chemoenzymatic synthesis of an N-glycan isomer library, *Chem. Sci.*, 2015, **6**(10), 5652–5661.
9. Z. Wang, *et al.*, A general strategy for the chemoenzymatic synthesis of asymmetrically branched N-glycans, *Science*, 2013, **341**(6144), 379–383.
10. H. H. Freeze and C. Kranz, Endoglycosidase and Glycoamidase Release of N-Linked Glycans, *Curr. Protoc. Immunol.*, 2008, 8.15. 1–8.15. 26.

11. A. Seko, *et al.*, Occurrence of a sialylglycopeptide and free sialylglycans in hen's egg yolk, *Biochim. Biophys. Acta*, 1997, **1335**(1), 23–32.
12. Y. Kajihara, *et al.*, Chemoenzymatic Synthesis of Diverse Asparagine-Linked Oligosaccharides, *Methods Enzymol*, Academic Press, 2003, p. 44–64.
13. C. Unverzagt and Y. Kajihara, Chemical assembly of N-glycoproteins: a refined toolbox to address a ubiquitous posttranslational modification, *Chem. Soc. Rev.*, 2013, **42**(10), 4408–4420.
14. Y. Kajihara, *et al.*, Prompt Chemoenzymatic Synthesis of Diverse Complex-Type Oligosaccharides and Its Application to the Solid-Phase Synthesis of a Glycopeptide with Asn-Linked Sialyl-undeca-and Asialo-nonasaccharides, *Chem.–Eur. J.*, 2004, **10**(4), 971–985.
15. T. J. Boltje, T. Buskas and G.-J. Boons, Opportunities and challenges in synthetic oligosaccharide and glycoconjugate research, *Nat. Chem.*, 2009, **1**(8), 611–622.
16. M. R. Pratt and C. R. Bertozzi, Chemoselective ligation applied to the synthesis of a biantennary N-linked glycoform of CD52, *J. Am. Chem. Soc.*, 2003, **125**(20), 6149–6159.
17. W. Huang, *et al.*, Glycosynthases enable a highly efficient chemoenzymatic synthesis of N-glycoproteins carrying intact natural N-glycans, *J. Am. Chem. Soc.*, 2009, **131**(6), 2214–2223.
18. S. S. Shivatare, *et al.*, Efficient convergent synthesis of bi-, tri-, and tetra-antennary complex type N-glycans and their HIV-1 antigenicity, *J. Am. Chem. Soc.*, 2013, **135**(41), 15382–15391.
19. M. A. Walczak, J. Hayashida and S. J. Danishefsky, Building biologics by chemical synthesis: practical preparation of di-and triantennary N-linked glycoconjugates, *J. Am. Chem. Soc.*, 2013, **135**(12), 4700–4703.
20. M. A. Walczak and S. J. Danishefsky, Solving the convergence problem in the synthesis of triantennary N-glycan relevant to prostate-specific membrane antigen (PSMA), *J. Am. Chem. Soc.*, 2012, **134**(39), 16430–16433.
21. P. Wang, *et al.*, Total Synthesis of the 2, 6-Sialylated Immunoglobulin G Glycopeptide Fragment in Homogeneous Form, *J. Am. Chem. Soc.*, 2009, **131**(46), 16669–16671.
22. V. Y. Dudkin, *et al.*, Toward a prostate specific antigen-based prostate cancer diagnostic assay: preparation of keyhole limpet hemocyanin-conjugated normal and transformed prostate specific antigen fragments, *J. Am. Chem. Soc.*, 2008, **130**(41), 13598–13607.
23. V. Y. Dudkin, J. S. Miller and S. J. Danishefsky, Chemical synthesis of normal and transformed PSA glycopeptides, *J. Am. Chem. Soc.*, 2004, **126**(3), 736–738.
24. G. Ragupathi, *et al.*, On the power of chemical synthesis: immunological evaluation of models for multiantigenic carbohydrate-based cancer vaccines, *Proc. Natl. Acad. Sci. U. S. A.*, 2002, **99**(21), 13699–13704.
25. C. Unverzagt, *et al.*, Synthesis of Multiantennary Complex Type N-Glycans by Use of Modular Building Blocks, *Chem.–Eur. J.*, 2009, **15**(45), 12292–12302.

26. Y. Liu, *et al.*, Chemical Synthesis of a Bisphosphorylated Mannose-6-Phosphate N-Glycan and its Facile Monoconjugation with Human Carbonic Anhydrase II for in vivo Fluorescence Imaging, *ChemBioChem*, 2011, **12**(5), 685–690.
27. L. Yan and D. Kahne, Generalizing glycosylation: synthesis of the blood group antigens Lea, Leb, and Lex using a standard set of reaction conditions, *J. Am. Chem. Soc.*, 1996, **118**(39), 9239–9248.
28. T. Bamhaoud, S. Sanchez and J. Prandi, 1,2,5-ortho esters of D-arabinose as versatile arabinofuranosidic building blocks. Concise synthesis of the tetrasaccharidic cap of the lipoarabinomannan of Mycobacterium tuberculosis, *Chem. Commun.*, 2000, (8), 659–660.
29. H. Paulsen, M. Heume and H. Nürnberger, Synthese der verzweigten nonasaccharid-sequenz der "bisected" struktur von N-glycoproteinen, *Carbohydr. Res.*, 1990, **200**, 127–166.
30. Y. Liu, *et al.*, Synthesis of novel bivalent mimetic ligands for mannose-6-phosphate receptors, *Bioorg. Med. Chem. Lett.*, 2013, **23**(8), 2328–2331.

CHAPTER 7

Towards Synthesis of Heparan Sulfate Glycopeptides and Proteoglycans

SHERIF RAMADAN[a,b], WEIZHUN YANG[a] AND XUEFEI HUANG*[a]

[a]Department of Chemistry, Michigan State University, 578 S. Shaw Lane, East Lansing, Michigan, 48824-1322, USA; [b]Chemistry Department, Faculty of Science, Benha University, Benha, Qaliobiya, Egypt
*E-mail: xuefei@chemistry.msu.edu

7.1 Structures, Biological Functions and Biosynthesis of Proteoglycans

Proteoglycans (PGs) are a family of macromolecules consisting of a core protein bearing one or more glycosaminoglycan (GAG) chain(s). PGs are ubiquitously present on the surfaces of mammalian cells, and are also major components of the extracellular matrix.[1] PGs are known to play important roles in many biological processes, such as tissue remodeling during embryonic development, cell proliferation, differentiation, adhesion, and wound repair.[2–5] Understanding of the multi-faceted functions of PGs is of high current interest.

In PGs, the GAG molecules are mainly attached to serine (Ser) residues in the core protein through a tetrasaccharide linker. The linker is composed of a

Chemical Biology No. 2
Chemical Biology of Glycoproteins
Edited by Zhongping Tan and Lai-Xi Wang

Published by the Royal Society of Chemistry, www.rsc.org

glucuronic acid (GlcA) residue, two galactose (Gal) residues, and a xylose (Xyl) residue with the structure of GlcA-*β*-1,3-Gal-*β*-1,3-Gal-*β*-1,4-Xyl–Ser–protein.[6,7] The GAG chains are linear non-branched anionic polysaccharides with repeating disaccharide units of alternating hexosamine and hexuronic acid with variable *N* and/or *O* sulfation.[8] PGs can be classified according to the composition of the GAG sidechains, as summarized in the following (Figure 7.1):

Heparin and heparan sulfates (HS) are composed of disaccharide repeating units of D-GlcA β-1,4-linked with glucosamine (GlcN) (α-1,4-GlcA–β-1,4-GlcN). The GlcA can be epimerized to L-iduronic acid (IdoA) during biosynthesis. The 3-OH and 6-OH positions of GlcN and the 2-OH of the uronic acid can be sulfated. In addition, the amine of the GlcN can be *N*-acetylated, *N*-sulfated or free. Heparins are more highly sulfated than HS with higher percentages of *N*-sulfated GlcN and IdoA.

Chondroitin sulfates (CS) are composed of disaccharide repeating units of β-1,4-GlcA–β-1,3-*N*-acetyl-galactosamine (GalNAc).[9] The 4- and 6-OH positions of GalNAc and the 2-OH of the GlcA are the main positions that can be sulfated.

Dermatan sulfates (DS) have similar structures to CS except that the GlcA in the CS backbone is epimerized to IdoA. The DS disaccharide repeating units have the structure of β-1,4-IdoA–α-1,3-GalNAc.[10]

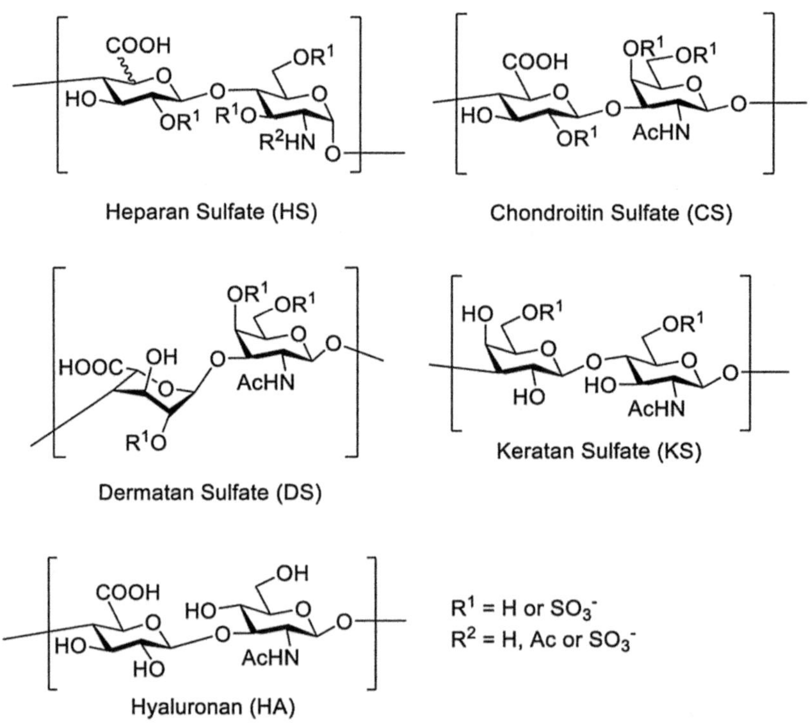

Figure 7.1 Structures of the GAGs.

Keratan sulfates (KS) are composed of disaccharide units of β-1,3-Gal–β-1,4-GlcNAc, which can be sulfated at 6-OH of either or both of the monosaccharides in the backbone. The KS polysaccharide chain can be linked to the core protein through *N*-acetyl glucosamine (GlcNAc), mannose, or GalNAc.[11,12]

Hyaluronic acid (HA) is composed of repeating units of β-1,4-GlcA–β-1,3-GlcNAc. It is a unique type of GAG as it does not have any sulfate groups and nor is it covalently attached to a protein. Its structure is included here for the sake of completeness.

A major type of PG is the HS proteoglycans (HSPG), which have three main subfamilies according to the sequences of their core proteins.[13] The first and the most abundant type of HSPG is the transmembrane protein syndecans, of which there are four members (syndecans 1–4).[14] The second class are the core proteins that carry HS is the glypicans, which are anchored in the membrane through a glycophosphatidylinositol (GPI) anchor.[15] The third class of HSPG is present in extracellular matrix rather than on plasma membranes. These include perlecan, agrin, and collagen XVIII.[3]

The PGs are biosynthesized through successive additions of monosaccharide units to the core protein, where specific glycosyl transferases catalyze the glycan addition reactions using UDP–glycan as the glycosyl donor.[16] The biosynthesis of HSPG is initiated by the addition of a Xyl moiety to the protein backbone, which is followed by sequential transfers of Gal, Gal and GlcA.[17] The HS backbone is then formed by alternating addition of GlcNAc and GlcA to the glycan chain. After elongation, further modifications of the HS backbone take place. GlcNAc *N*-deacetylase/*N*-sulfotransferase converts the *N*-acetyl group of GlcNAc to *N*-sulfates. The C5 epimerase catalyzes the transformation of GlcA residues to IdoA and additional *O*-sulfate groups can be introduced by several *O*-sulfotransferases. These reactions are often incomplete, thus resulting in significant structural diversities of HS in HSPGs (Figure 7.2).

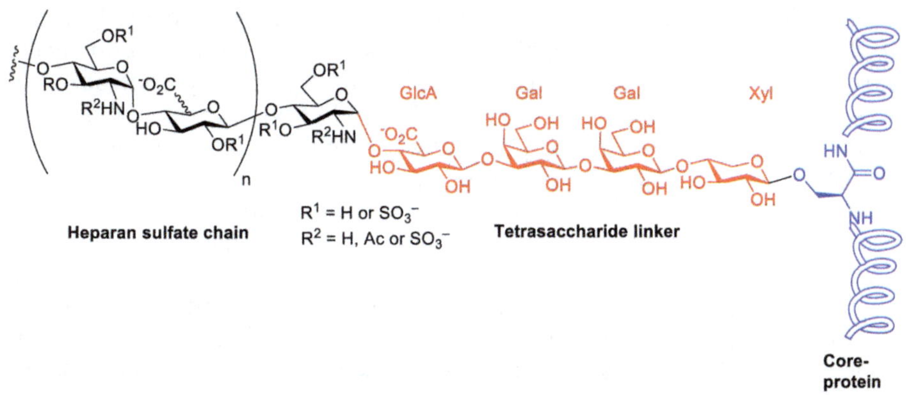

Figure 7.2 Heparan sulfate proteoglycan.

With their complex structures, HS can interact with many biological targets such as fibroblast growth factor,[3] integrin,[18] Wnt proteins,[19] and herpes simplex virus glycoprotein D.[20,21] The core proteins of HSPGs were previously thought to mainly serve as carriers of HS. However, there is increasing evidence indicating that the core protein itself can be biologically active.[18,22,23] For example, the core protein of a HSPG syndecan-1 can regulate invasive properties of cancer cells, with the mutation of a five-amino-acid sequence in the core resulting in a loss of invasive migration abilities of these cells.[24] The effects of HS chain and the core protein can also be synergistic for their biological functions,[23] which adds another dimension to the functional complexity of HSPG. Therefore, the availability of HSPG glycopeptides and glycoproteins bearing structurally defined homogeneous HS will greatly aid in the study of the biological functions of HSPGs. Since separation of sufficient quantities of HSPGs bearing homogenous HS chains from nature is extremely challenging due to the glycan heterogeneity, synthesis of well-defined HSPGs becomes crucial for providing sufficient quantities of pure compounds to advance glycobiology.

To date, HSPG synthesis has been primarily focused on the preparation of HS oligosaccharides, which has been well covered in several reviews.[25–30] In this chapter, we summarize the progress made in synthesis of the tetrasaccharide linker and the linker-bearing glycopeptide as well as efforts towards the preparation of homogenous HS glycopeptides.

7.2 Chemical Synthesis of the PG Linker–Peptide Conjugates

With the identification of carbohydrate–protein linkages in PGs in 1964,[31,32] synthesis towards the tetrasaccharide linker of PGs began shortly after that. Early chemistry relied on the glycosyl bromides as the glycosyl donor using the Koenigs–Knorr reaction to construct the glycosyl linkages.[33] Several fragments from the linkage region of PGs were synthesized, including disaccharides and trisaccharides, such as galactosylxylose, galactosylgalactose, galactosylgalactosylxylose,[34,35] and glucuronosylgalactose,[36] as well as the glycosyl amino acids xylosylserine,[37–40] galactosylxylosylserine,[40–42] and galactosylgalactosylxylosylserine.[42,43] For example, di-Gal glycosyl bromide **1** was activated with silver triflate to glycosylate acceptor **2**, producing trisaccharide **3** at 44% yield along with 10% of the corresponding α anomer (Scheme 7.1).

As harsh reaction conditions are generally required for the preparation of glycosyl bromides, subsequent synthesis of the linker has taken advantage of the more readily accessible glycosyl donors such as trichloroacetimidates and thioglycosides. The Jacquinet group reported the synthesis of glycosylated oligopeptides using trichloacetimidate donors.[44] For the preparation of the galactosylxylose disaccharide, the benzoylated trichloroacetimidate[45] donor **4** was selected (Scheme 7.2a), while the usage of galactosyl bromide donor or peracetylated galactosyl trichloroacetimidate led to the formation

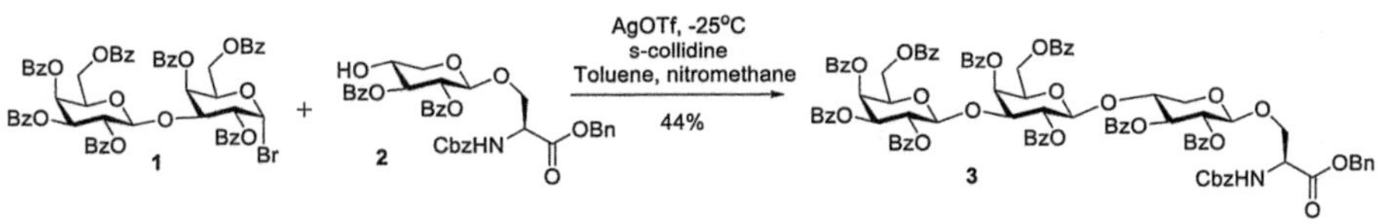

Scheme 7.1 Cbz: Carboxybenzyl.

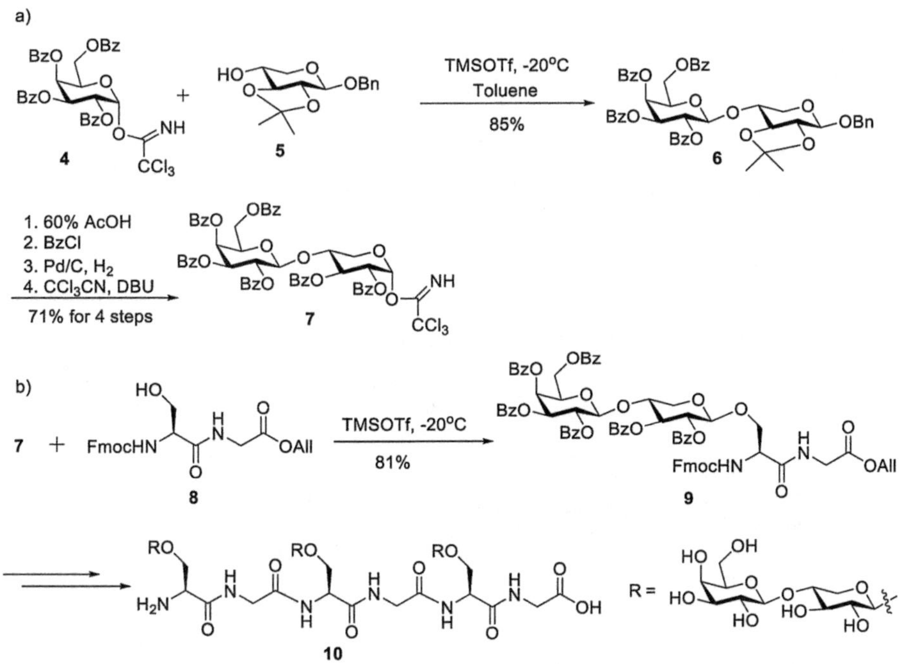

Scheme 7.2 AcOH: acetic acid, BzCl: benzoyl chloride, and DBU: 1,8-diazabicyclo [5.4.0]undec-7-ene.

of orthoesters. Treatment of **5** with the *O*-benzoylated trichloroacetimidate **4** under the catalysis of trimethylsilyl triflate (TMSOTf) in toluene produced the β-linked disaccharide **6** with 85% yield. In order to elongate the chain, after protective group adjustment on disaccharide **6**, the benzyl ether at the reducing end of the disaccharide was selectively removed by catalytic hydrogenolysis. The resulting hemiacetal was subsequently converted to disaccharide trichloroacetimidate donor **7**. Reaction of donor **7** with serine containing dipeptide acceptor **8** led to the 9-fluorenylmethyloxycarbonyl (Fmoc) protected glycosyl peptide **9** in 81% yield (Scheme 7.2b), which was then transformed to glycopeptide **10** through iterative peptide coupling.

With experience in the synthesis of glycosylated oligopeptides, the Jacquinet group went on to assemble glycopeptides bearing the full tetrasaccharide glycan linker from PGs (Scheme 7.3).[46] When the benzoylated galactoside acceptor **11** was used to react with the uroyl imidate donor **12**,

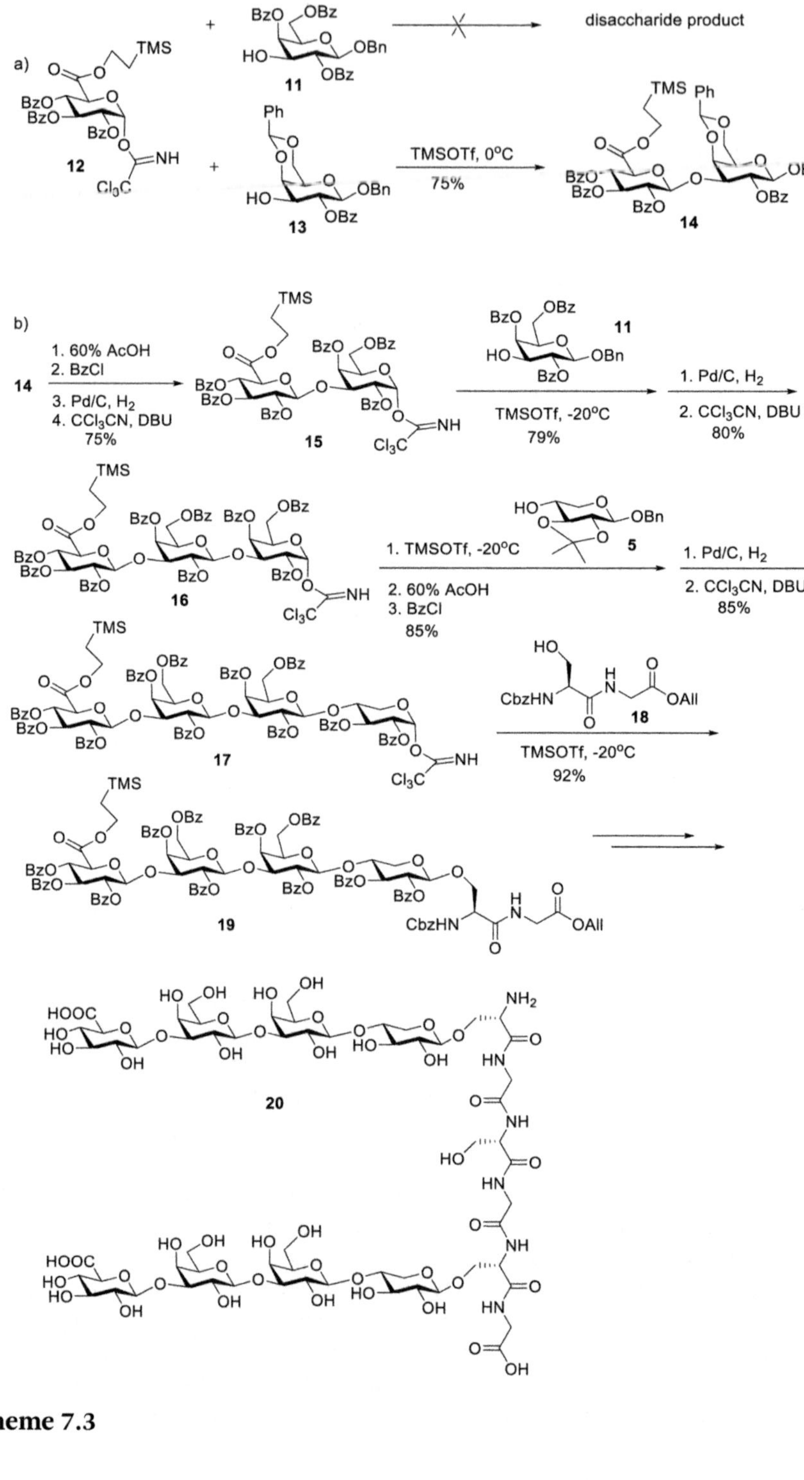
a)
disaccharide product
TMS
BzO
OBz
OBn
HO
11
12
NH
Cl_3C
Ph
13
TMSOTf, 0°C
75%
14
b)
1. 60% AcOH
2. BzCl
3. Pd/C, H_2
4. CCl_3CN, DBU
75%
15
TMSOTf, -20°C
79%
1. Pd/C, H_2
2. CCl_3CN, DBU
80%
16
5
1. TMSOTf, -20°C
2. 60% AcOH
3. BzCl
85%
1. Pd/C, H_2
2. CCl_3CN, DBU
85%
17
18
CbzHN
OAll
TMSOTf, -20°C
92%
19
20
HOOC
NH_2

Scheme 7.3

low yields of the desired disaccharide were obtained, presumably due to the low nucleophilicity of the acceptor (Scheme 7.3a). Switching the acceptor **11** to the benzylidene-bearing galactoside **13** led to the successful formation of disaccharide **14** with 75% yield (Scheme 7.3a). An iterative sequence of protective group manipulation, selective removal of the anomeric benzyl ether, trichloroacetimidate donor formation, and glycosylation led to the glycopeptide **19** with good overall yields and stereoselectivities (Scheme 7.3b).[46] The glycopeptide **19** was then utilized to produce glycopeptide **20** bearing two tetrasaccharide chains.

In nature, selected positions of the tetrasaccharide linkers of PGs can be sulfated or phosphorylated. The Jacquinet group reported a systematic preparation of all possible sulfoforms of the trisaccharides conjugated to biotin to serve as potential probes for the hexosaminyl transferase activities in PG biosynthesis.[47] Donors **21** and **22** and acceptor **23** were used in a divergent manner (Scheme 7.4) and all glycosylations were stereocontrolled by 2-*O*-benzoyl (Bz) through neighboring group participation. Finally, a set of sulfoforms of trisaccharides (**26–30**) were prepared at high yields. In addition, both sulfated and phosphorylated tetrasaccharide dipeptides from the linkage region were synthesized.[48–51]

Instead of using a benzyl ether to mask the anomeric position of the acceptor, the Kong group explored the methoxyphenyl (MP) moiety, which can be

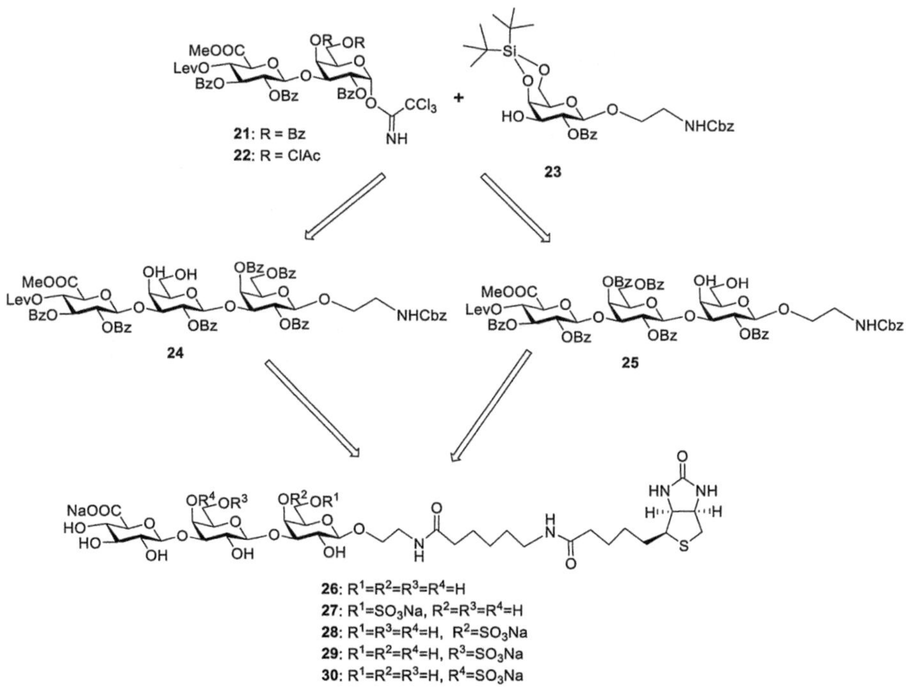

Scheme 7.4

selectively removed under oxidative conditions using cerium ammonium nitrate (CAN).[52] The MP galactoside **31** was converted to the trichloroacetimidate donor **32** with 76% overall yield using a two-step sequence (Scheme 7.5a).[52] At the same time, the allyl group in **31** was removed to generate a galactoside acceptor **33**, which was glycosylated by donor **32** producing disaccharide **34** with 54% yield (Scheme 7.5b). Interestingly, despite the presence of the 2-OBz in donor **32**, 31% of the *α*-anomer was also formed. The reason for the erosion of 1,2-*trans* selectivity was not clear. Subsequently, disaccharide **34** was deprotected and condensed with glucuronic methyl ester imidate donor **35** generating trisaccharide **36** with 55% yield, which underwent oxidative cleavage of 1-OMP (OMP: O-methoxyphenyl) with CAN[53] followed by trichloroacetimidation[54] to give donor **37** (Scheme 7.5b). Trisaccharide **37** was coupled with the xylose acceptor **38** to give tetrasaccharide **39**, which was deprotected to generate the target tetrasaccharide **40**.

Instead of sequential addition of monosaccharides, Tamura and coworkers[55] showed that the tetrasaccharide linker could be prepared through a more convergent 2+2 strategy. Coupling of disaccharide trichloroacetimidate

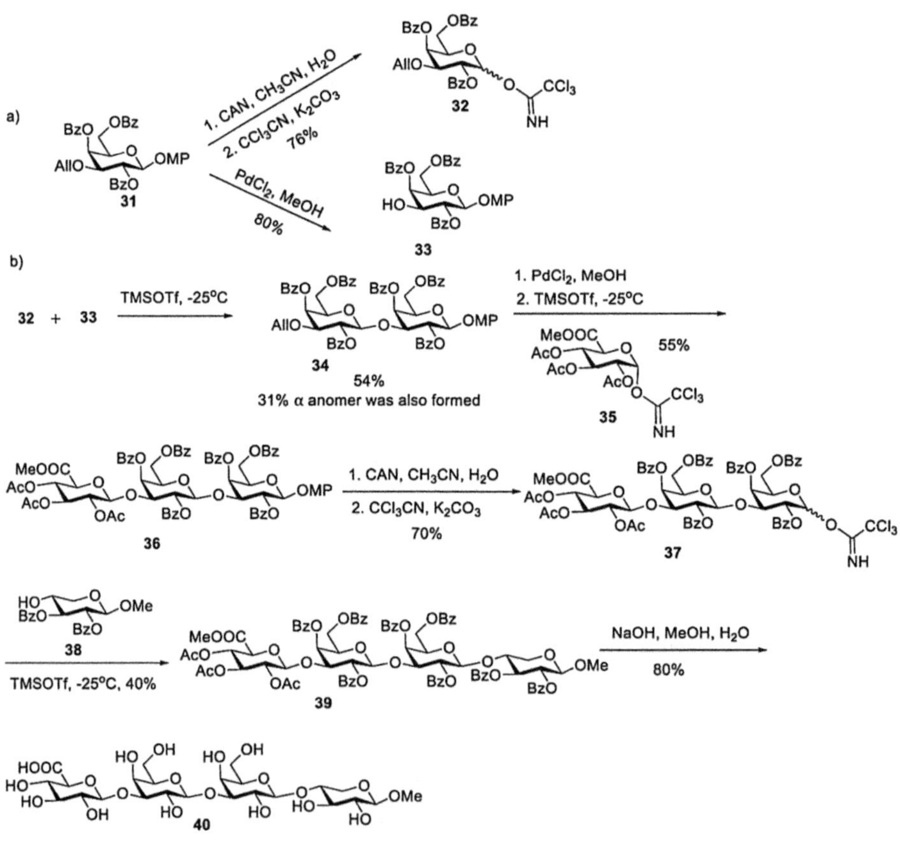

Scheme 7.5

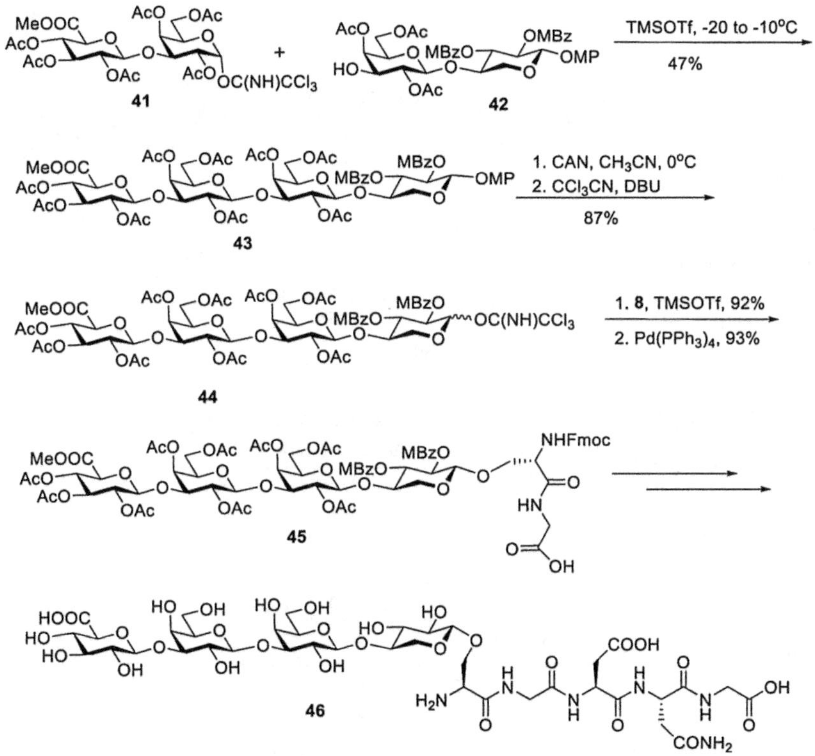

Scheme 7.6 CAN: cerium (IV) ammonium nitrate.

donor **41** with disaccharide **42** acceptor bearing anomeric OMP produced tetrasaccharide β-isomer with 47% yield (Scheme 7.6). Similar to Kong's synthesis,[52] even with the presence of an acetyl group at the 2-position, the α-anomer was isolated in 24% yield. The tetrasaccharide β-anomer was converted to trichloroacetimidate donor **44** through oxidative removal of OMP and subsequent trichloroacetimidation. Tetrasaccharide **44** glycosylated the serylglycine allyl ester **8**, which was then subjected to peptide synthesis and deprotection, leading to glycopeptide **46**. This type of glycopeptide was utilized to probe peptide substrate specificity of glycosyl transferases involved in PG synthesis.[55]

Besides trichloroacetimidates, thioglycosides are another popular type of building blocks for PG synthesis. The Ogawa group reported the preparation of glycotetrasoyl tetrapeptide in a stereocontrolled manner using thioglycosides.[56–58] The synthesis was performed from the reducing end to the non-reducing end.[59] *N*-Iodosuccinimide (NIS) and trifluoromethanesulfonic acid (TfOH)-promoted glycosylation of methyl thiogalactoside donor **47** with xyloside **48** gave 92% yield of **49** (Scheme 7.7). Protective group manipulations of **49** produced the disaccharide acceptor **50**. Subsequent iterations of glycosylation, protective group manipulations and

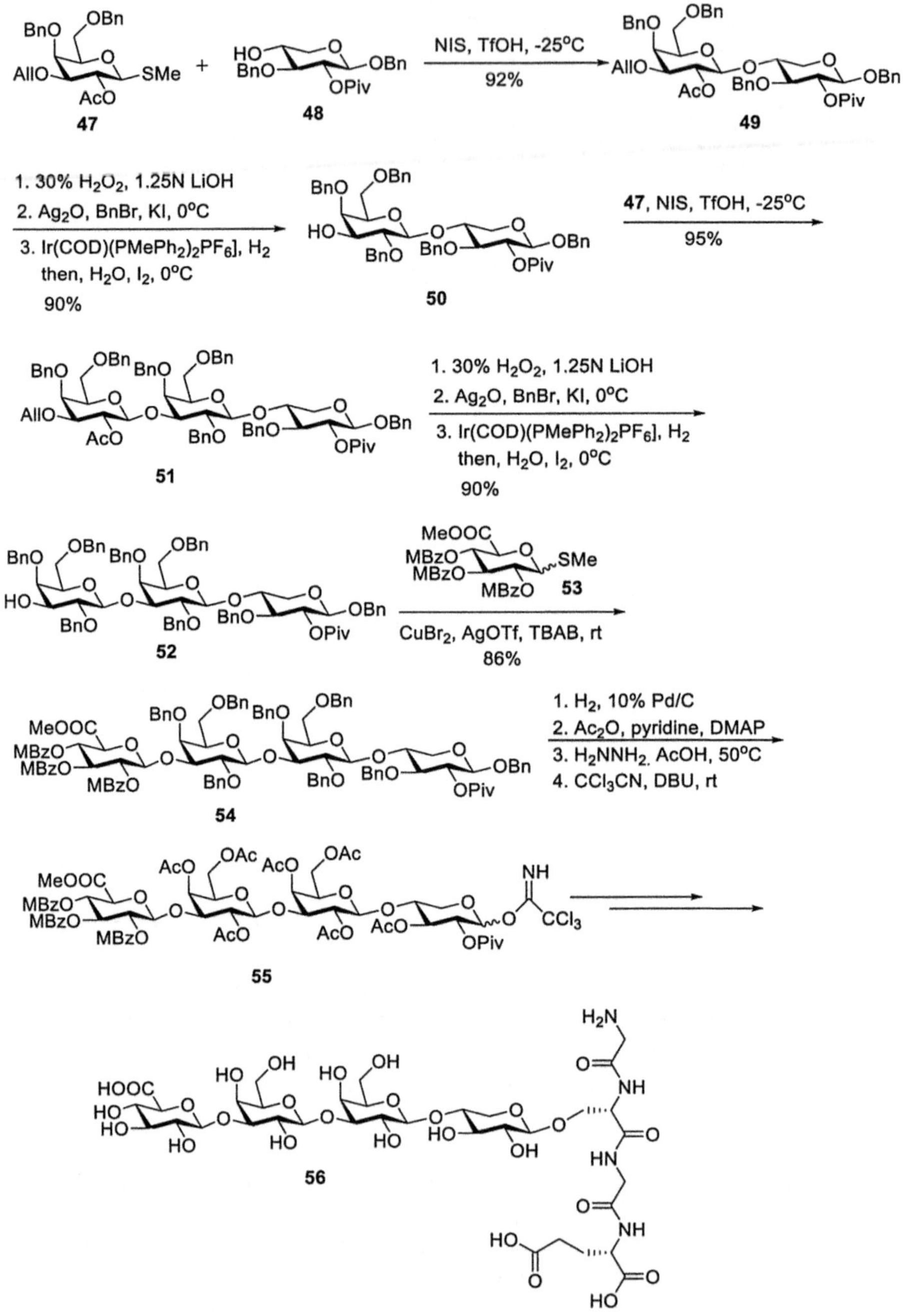

Scheme 7.7 Ir(COD)(PMePh2)2PF6: (1,5-cyclooctadiene)bis(methyldiphenylphosphine)iridium(I) hexafluorophosphate.

glycosylation led to tetrasaccharide **54**. The anomeric benzyl ether of **54** was deprotected, which, followed by peracetylation, selective removal of anomeric acetate and trichloroacetimidation, generated a tetrasaccharide donor **55**. The tetrasaccharide **55** was subsequently transformed to glycopeptide **56**.[60,61]

To improve the synthetic efficiency, Fraser-Reid and co-workers developed an active–latent glycosylation approach utilizing the *N*-pentenyl glycosides (NPGs) pioneered by their group.[62] Selective activation of trichloroacetimidate donor **57** was performed in the presence of galactosyl NPG acceptor **58** with triethylsilyl triflate (TESOTf) as the promoter (Scheme 7.8a). The resulting disaccharide product **59** could be directly utilized as a glycosyl donor without further

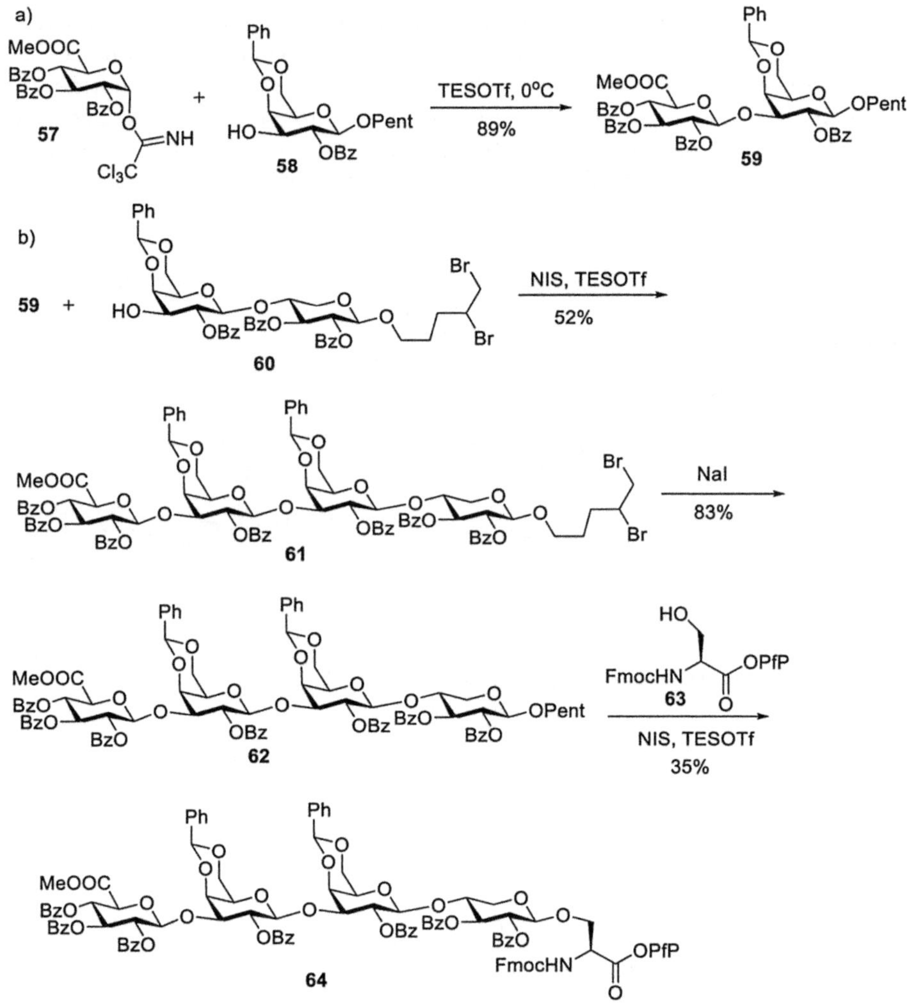

Scheme 7.8

manipulation. Glycosylation of disaccharide acceptor **60** by donor **59** was promoted with NIS and TESOTf producing tetrasaccharide **61** (Scheme 7.8b). Compound **61** was a latent donor, the 1,2-dibromopentane aglycon of which could be transformed into a *N*-pentenyl moiety, resulting in the active NPG donor **62** with 83% yield. The coupling of tetrasaccharide donor **62** with serine acceptor **63** was carried out to give the glycosyl amino ester **64** with 35% yield.

Rather than starting from monosaccharide building blocks, Nishimura and coworkers[63] explored the alternative of using disaccharide lactose for synthesis of Galβ(1→4)Xyl that can be extended to linkage region of the proteoglycans. 1,6-Anhydro-β-lactoside **65** was obtained from lactose,[64] which then underwent deacetylation, introduction of allyl group at the 3′ position and perbenzylation to yield compound **66** (Scheme 7.9). Treating **66** with phenylthio(trimethyl) silane (TMSSPh) and trimethylsilyltrifluoromethanesulfonate (TMSOTf) gave thiophenyl lactoside **67**. The reducing end glucosyl unit of **67** was transformed to Xyl through an 11-step procedure, generating the Galβ(1→4)Xyl disaccharide building block **69**, which was then used to construct glycopeptide **70**. While this synthesis utilized the more readily available disaccharide lactoside **65** as the starting material, the lengthy transformation needed to convert it to the required Galβ(1→4)Xyl disaccharide was a drawback.

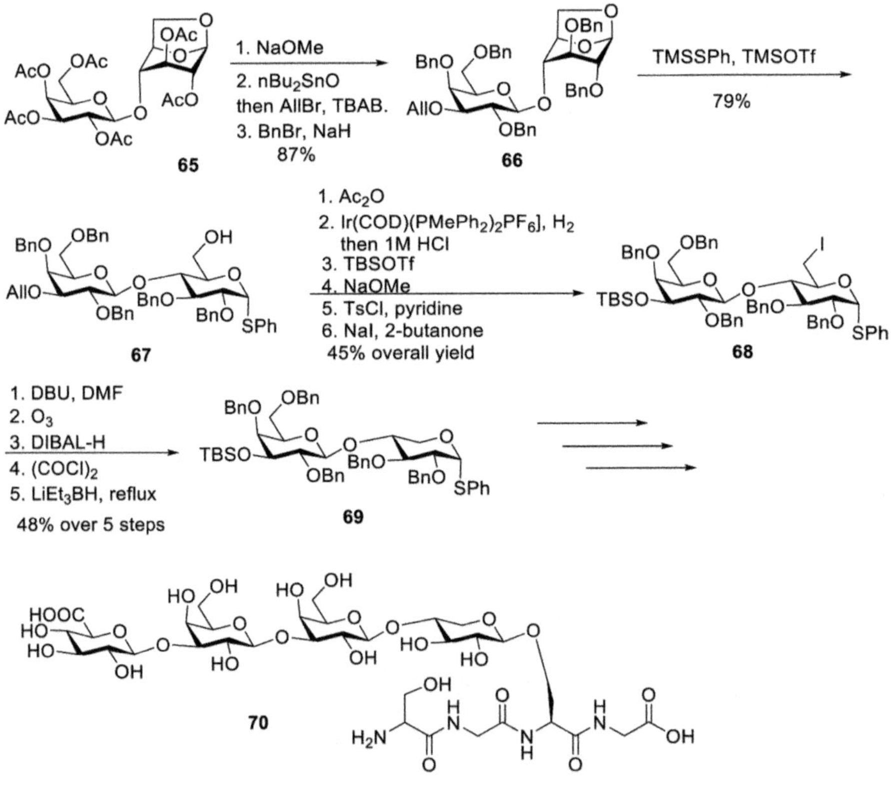

Scheme 7.9 nBu2SnO: dibutyltin oxide, AllBr: allyl bromide, TBAB: tetrabutylammonium bromide, DIBAL-H: diisobutylaluminium hydride, and LiEt3BH: lithium triethylborohydride.

To improve the overall synthetic efficiency and reduce the number of synthetic manipulations needed, Hung and coworkers developed regioselective one-pot preparations of the requisite monosaccharide building blocks **72**, **75**, and **77**.[65] As shown in Scheme 7.10a, treatment of trimethylsilyl (TMS)-protected thiotolyl xyloside **71** with triethylsilane, TMSOTf, and benzaldehyde followed by benzoyl anhydride and ytterbium triflate in the same reaction flask led to the bifunctional xyloside **72** with 45% yield. Similarly, starting from silyl ether-protected thiogalactoside **73** and thioglucoside **76**, the requisite galactoside and glucoside building blocks **75** and **77** were generated (Scheme 7.10b and c). In addition, the assembly of the target tetrasaccharide was achieved *via* one-pot procedures (Scheme 7.10d). The preparation of disaccharide **80** began *via* chemoselective glycosylation of xyloside **72** by galactoside **78**. The resulting disaccharide directly glycosylated linker acceptor **79** in the same reaction flask, which was followed by selective removal of TBS (TBS: tert-butyl(dimethyl)silyl) protecting group in one-pot affording acceptor **80** with 32% overall yield. Then coupling between disaccharide **80** and thiogalactoside **75** was performed. It

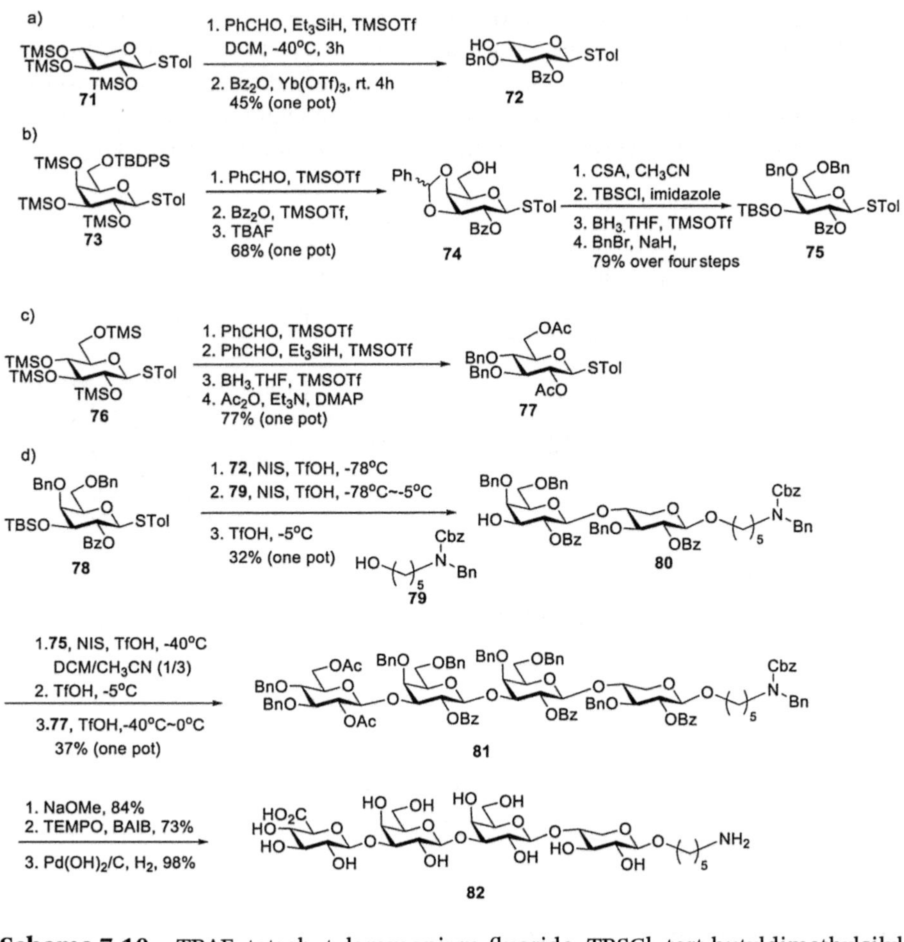

Scheme 7.10 TBAF: tetrabutylammonium fluoride, TBSCl: tert-butyldimethylsilyl chloride, Et3SiH: triethylsilane, and DCM: dichloromethane.

was found that the stereoselectivity was improved from a β:α ratio of 3:1 with CH_2Cl_2 as solvent to 14:1 with the CH_2Cl_2:CH_3CN (1:3) solvent mixture. The trisaccharide formed was treated with TfOH to cleave the TBS group followed by glycosylation with glucose donor **77** generating tetrasaccharide **81** in one pot with a 37% overall yield. Tetrasaccharide **81** underwent deacylation and debenzoylation followed by oxidation and global debenzylation, leading to the desired tetrasaccharide **82** (Scheme 7.10d).

Enzymatic synthesis can be a powerful approach to glycan assembly due to the potentially high regio- and stereo-selectivities of enzymes. However, glycosyl transferases involved in biosynthesis of the PG linkage region are not yet available for synthetic purposes. As an alternative, glycosidases have been utilized by the Kusumoto group,[66] although these enzymes are not as regio-specific as the glycosyl transferases (Scheme 7.11). Using 4-methylumbelliferyl-β-D-xyloside (Xyl-MU) **83** as the acceptor and *p*-nitrophenyl galactoside **84** as the glycosyl donor, the transfer of the galactosyl unit to the xyloside **83** was catalyzed with β-galactosidase from *Escherichia coli*. The MU moiety facilitated the product purification through HPLC due to its strong absorbance at 320 nm. The desired disaccharide product **85** was obtained at 17% yield along with 5% of a regioisomer disaccharide with the Gal linked to the 3-*O* position of the xyloside. The primary hydroxyl group of **85** was converted to an acetoxy group by a lipase-catalyzed acetate transfer reaction. The blocking of the more reactive primary hydroxyl group was necessary to improve the regioselectivity of the

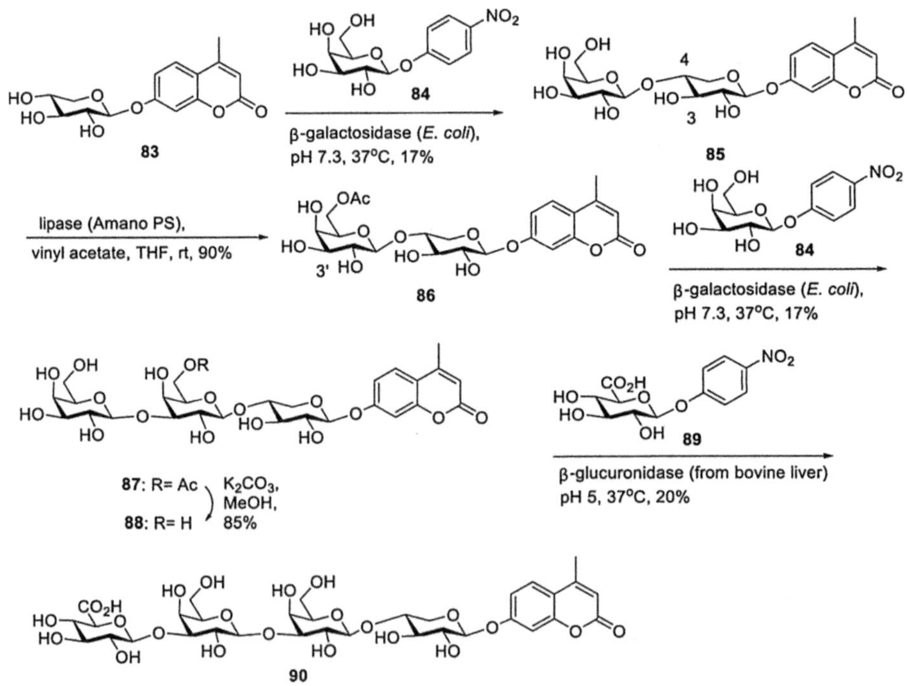

Scheme 7.11 Amano PS: lipase from *Pseudomonas cepacia*.

subsequent enzymatic reaction. Addition of a galactosyl unit to the 3′-OH position of disaccharide **86** was catalyzed using the β-galactosidase from *Escherichia coli* leading to trisaccharide **87** in 17% yield, which was deprotected. Subsequent transfer of a glucuronic acid to trisaccharide **88** promoted by a β-glucuronidase from bovine liver produced tetrasaccharide **90** with a 20% yield.

The low yields of glycosidase-catalyzed reactions were partly attributed to the competitive hydrolysis of the glycosyl donor or glycoside product. The reactions should be carried out with high concentrations of donor to improve the yield of the desired glycoside product. The aqueous solubilities of *p*-nitrophenyl donor such as **84** are not high due to the hydrophobicity of the *p*-nitrophenyl group. Kusumoto and his team[67] explored *β*-galactosyl fluoride **92** as the donor for β-galactosidase-catalyzed transglycosylation (Scheme 7.12). Disaccharide **93** was obtained under optimized conditions when the concentration of **92** was increased to 600 mM by warming the solution to 50 °C, where a higher buffer concentration (1.0 M) was needed to neutralize HF liberated during the reaction and maintain the solution at pH 7.3. The desired disaccharide **93** was prepared with a yield of 35% together with a minor regioisomeric

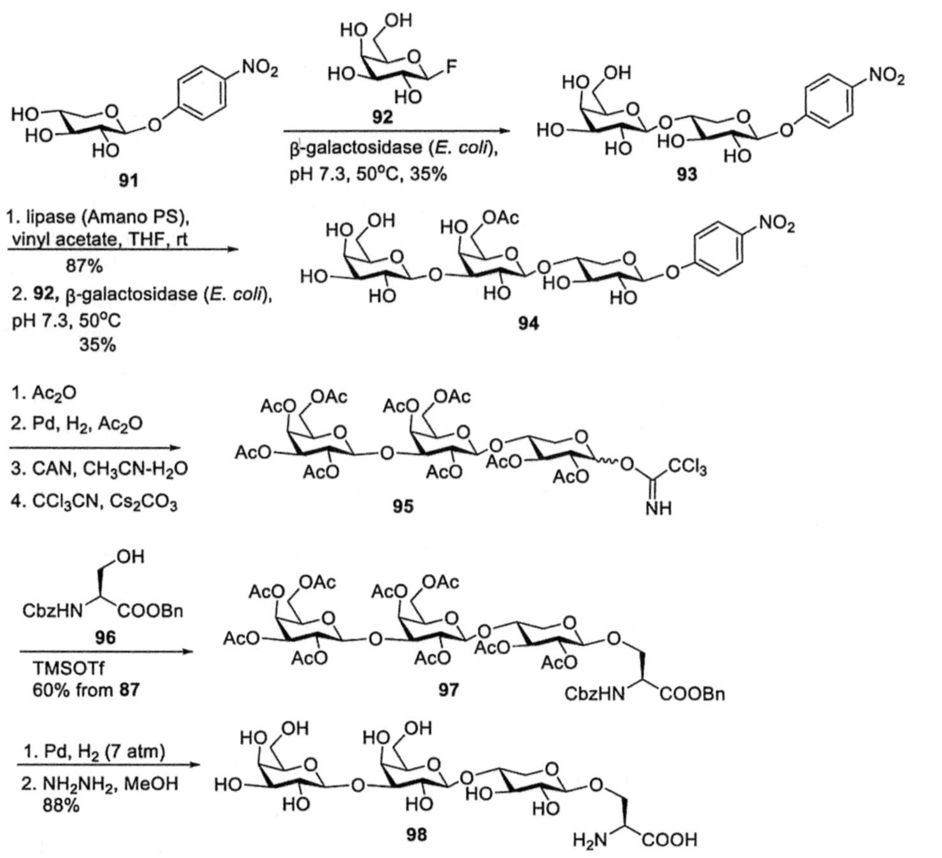

Scheme 7.12

product with glycosylation at O-3 (4%). Enzymatic galactosylation of the 3′-OH of the disaccharide **93** after protection of its 6′-OH with an acetyl group was carried out to generate trisaccharide **94** with a 35% yield, which was much improved compared with that obtained using the *p*-nitrophenyl donor.[68] Trisaccharide **94** was then converted to a trichloroacetimitate donor **95**, which glycosylated serine **96** and produced glycopeptide **98** after deprotection (Scheme 7.12). While the results of these studies indicate the potential of enzymatic reactions, the relatively low yields and the lack of regiospecificity in transglycosylation reactions indicate that for large-scale applications, enzymes with higher selectivity and glycosylation activity need to be developed.

7.3 Synthesis of HS Glycopeptides

Naturally existing HSPGs contain one or more HS chains attached to the core protein through the tetrasaccharide linker. It has becoming increasingly recognized that both HS chains and the core proteins can significantly affect biological functions of HSPGs.[23,69,70] Synthesis of HSPGs bearing full-length HS chains is still beyond the capabilities of current synthetic methodologies. As a step towards HSPGs, we have begun to investigate strategies to prepare HS glycopeptides.

The synthesis of HS glycopeptides has many challenges. These include the difficult glycosylation reactions, stereochemical control, regiospecific sulfation, and integration of the complex oligosaccharide with the peptide chain. Traditional peptide synthesis relies heavily on side-chain protective groups such as trityl and *t*-butyl that are removable under acidic conditions. However, *O*- and *N*-sulfates in HS are highly labile under acidic conditions, which indicates that the usage of strong acids for deprotections of glycopeptides must be avoided. In addition, the linkage between glycan and peptide backbone is prone to base-promoted β-elimination. Thus, suitable protective groups and deprotection sequences must be carefully evaluated and established.

The first target molecule was the HS glycopeptide: syndecan-1 **99**,[71] which bears a sulfated octasaccharide sequence on a tridecapeptide backbone (Figure 7.3). To prepare this glycopeptide, a cassette approach was adopted where the octasaccharide serine conjugate **100** was prepared first, which was followed by deprotection, sulfation and peptide chain elongation.

In order to assemble the octasaccharide module **100**, the 2 + 5+1 glycosylation route was tested first. However, the (2 + 5) coupling between units B and C failed and the (5 + 1) reaction between units G and H was not successful either. This led to the design of a new (3 + 2+3) synthetic route. To achieve this, ABC trisaccharide **101**, DE disaccharide **102**, and FGH trisaccharide **104** were prepared. The 3 + 2 coupling between donor **101** and acceptor **102** was performed using the pre-activation protocol[72] with the promoter system of *para*-toluenesulfenyl chloride–silver triflate (*p*-TolSCl–AgOTf) (Scheme 7.13), generating pentasaccharide **103** with an excellent 93% yield. Subsequently, glycosylation of **104** by **103** gave the desired oligosaccharide **105**. The protective groups and oxidation states of **105** were adjusted producing the free-amine-bearing glycosyl serine **106**.

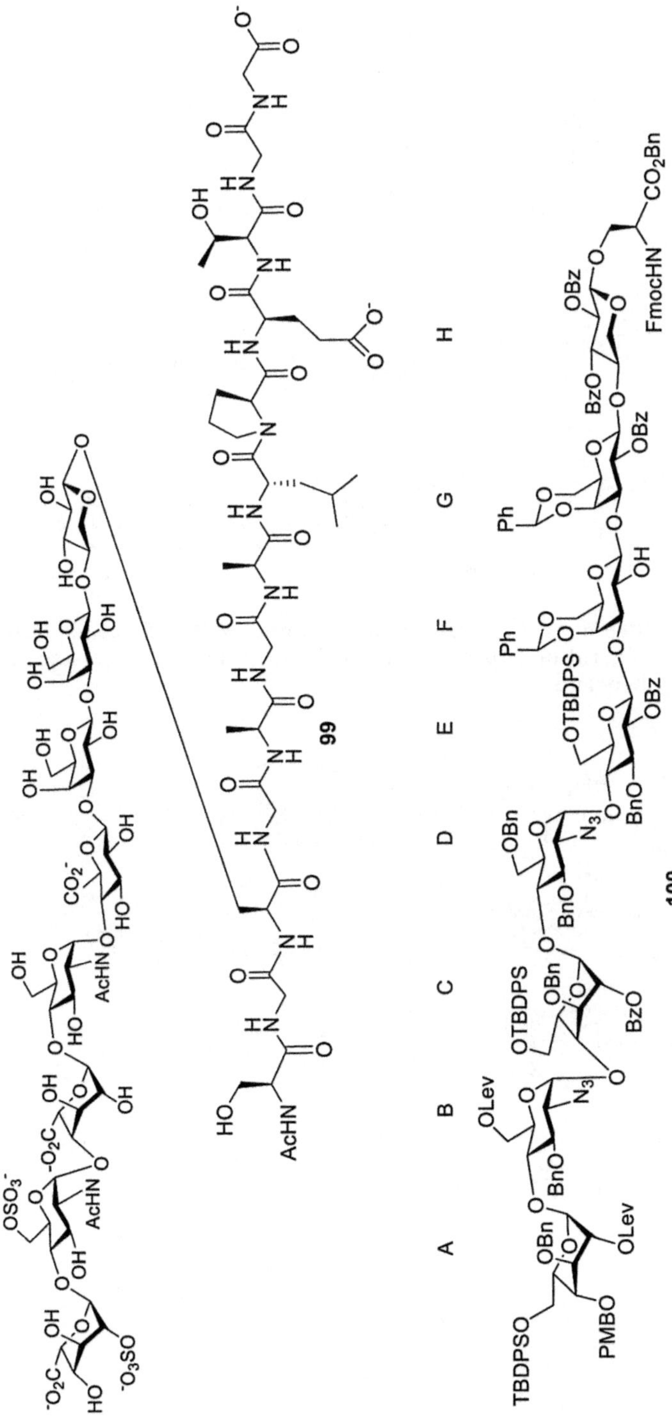

Figure 7.3 HS glycopeptide syndecan-1.

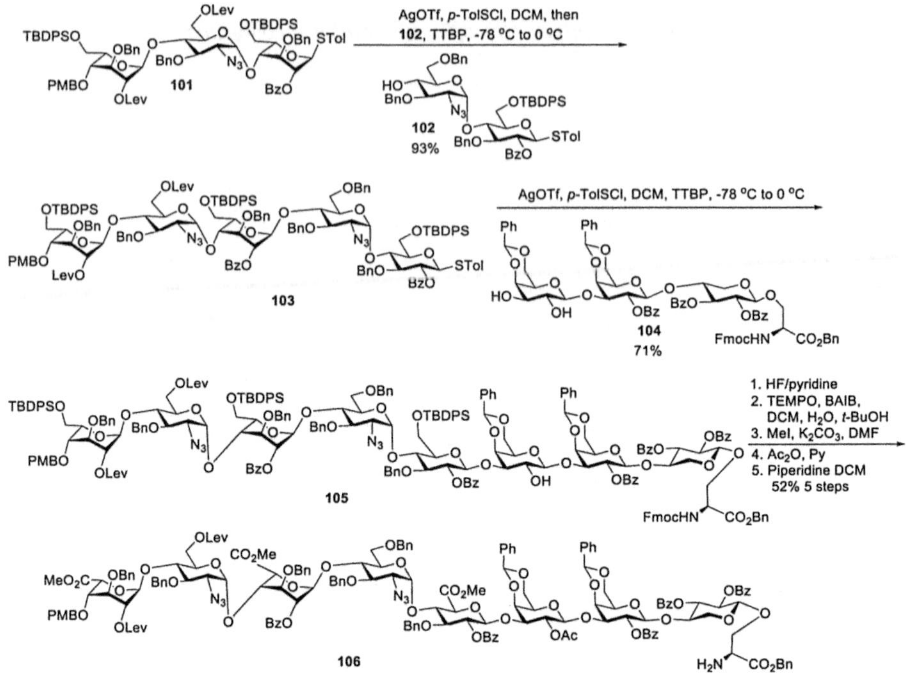

Scheme 7.13 TTBP: 2,4,6-tri-tert-butylpyrimidine, TEMPO: 2,2,6,6-tetramethylpiperidine 1-oxyl, MeI: methyl iodide, and BAIB: [bis(acetoxy)iodo] benzene.

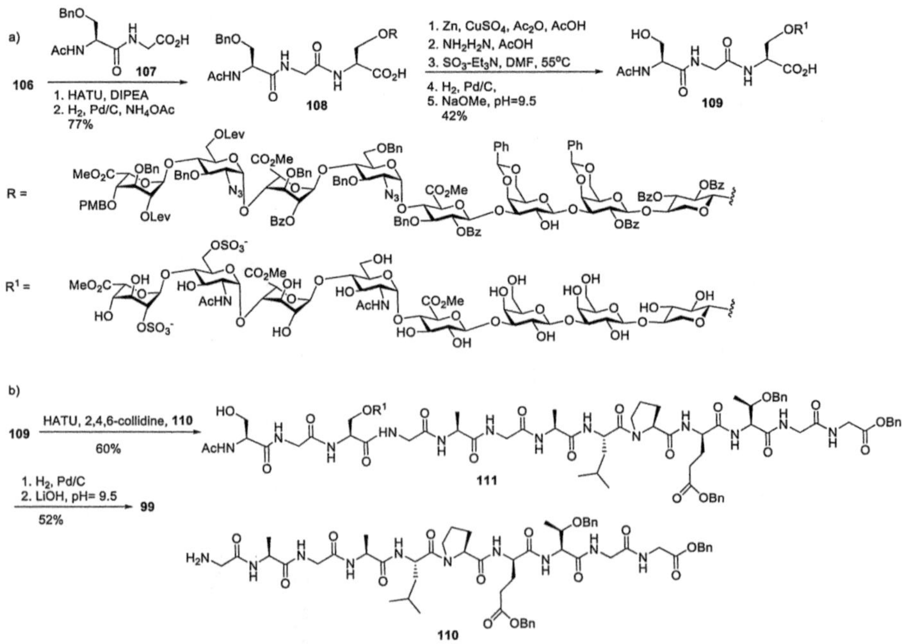

Scheme 7.14 DIPEA: N,N-diisopropylethylamine and HATU: 1-[bis(dimethylamino) methylene]-1H-1,2,3-triazolo[4,5-b]pyridinium 3-oxid hexafluorophosphate.

Coupling between **106** and Ac-Ser(OBn)-Gly dipeptide **107** was carried out and the resulting glycopeptide underwent selective hydrogenolysis[73] to generate glycopeptide **108** with a free carboxylic acid (Scheme 7.14a). As it was difficult to synthetically manipulate the full-length glycopeptide due to functional group compatibilities, the glycopeptide **108** was partially deprotected (azide reduction/acetylation, levulinoyl (Lev) removal, *O* sulfation, hydrogenation and transesterification) leading to **109**. For transesterification, it was important to keep the pH value at 9.5 to prevent the elimination of the glycan chain from the peptide backbone. The partially protected glycopeptide **109** and peptide **110** were successfully coupled with 60% yield (Scheme 7.14b). Subsequent hydrogenation and methyl ester hydrolysis yielded the HS glycopeptide **99**.

With the suitable route to syndecan-1 glycopeptide carrying one heparan sulfate chain established, the synthesis of a more complex proteoglycans that bears two glycan chains, syndecan-3 extracellular domain glycopeptide **112**, was explored (Figure 7.4).[74] To access this molecule, glucuronic acid

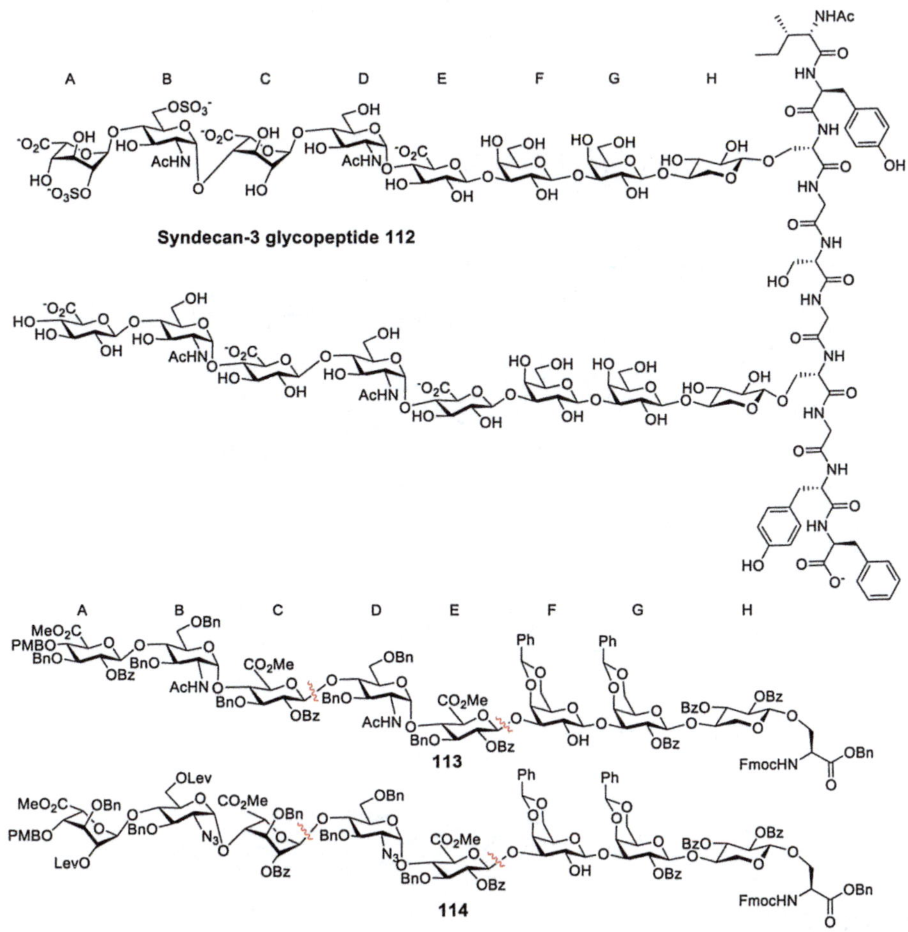

Syndecan-3 glycopeptide 112

113

114

Figure 7.4 HS glycopeptide syndecan-3.

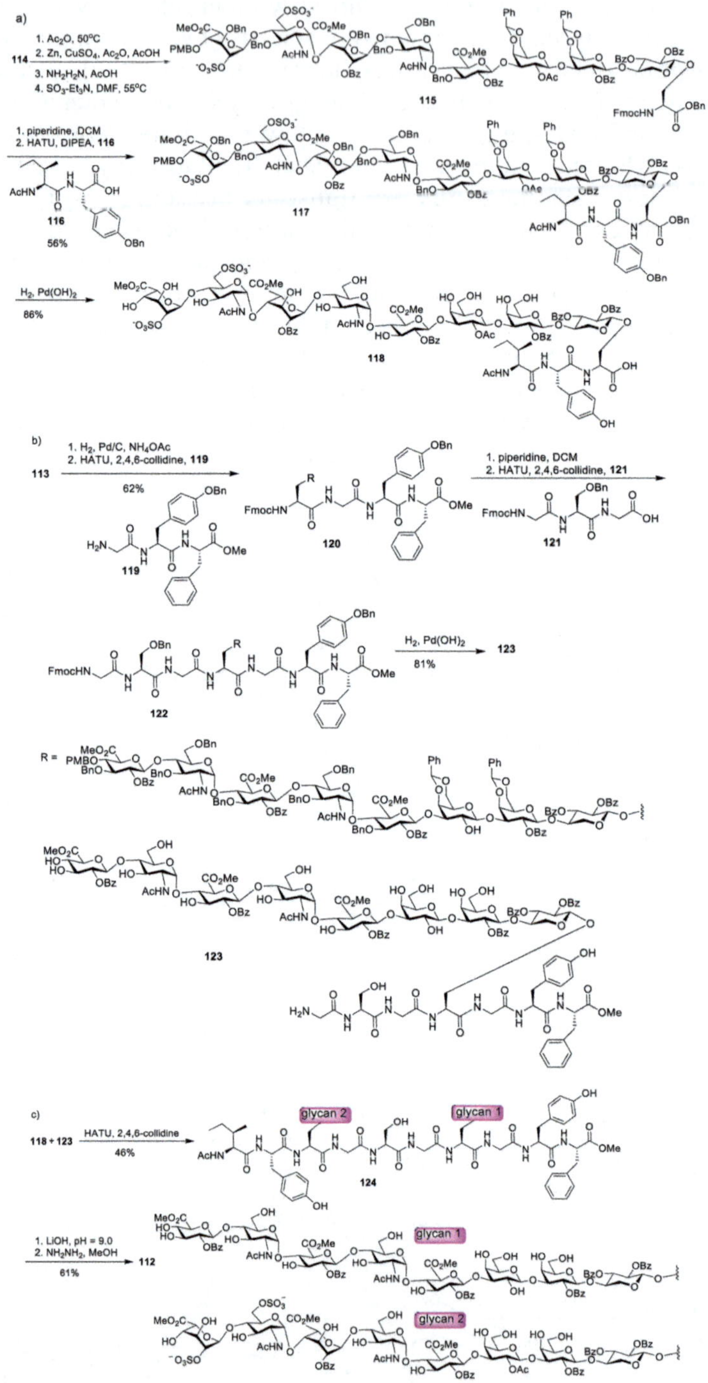
a)
1. Ac2O, 50°C
2. Zn, CuSO4, Ac2O, AcOH
3. NH2H2N, AcOH
4. SO3-Et3N, DMF, 55°C
114
115
1. piperidine, DCM
2. HATU, DIPEA, 116
116
56%
117
H2, Pd(OH)2
86%
118
b)
1. H2, Pd/C, NH4OAc
2. HATU, 2,4,6-collidine, 119
113
62%
119
120
1. piperidine, DCM
2. HATU, 2,4,6-collidine, 121
121
122
H2, Pd(OH)2
81%
123
R =
123
c)
118 + 123
HATU, 2,4,6-collidine
46%
glycan 2
glycan 1
124
1. LiOH, pH = 9.0
2. NH2NH2, MeOH
61%
112
glycan 1
glycan 2

Scheme 7.15

containing octasaccharide **113** and iduronic acid modules **114** were prepared. In a similar manner to the synthesis of syndecan-1 glycopeptide **99**, the 3 + 2+3 strategy for generating **113** and **114** was adapted.

Multiple difficulties were encountered during assembly and subsequent deprotections during synthesis of syndecan-3 **112**, which include base-promoted β-elimination of glycans from the glycopeptide, reduced reactivities of large peptides and glycopeptides for amide bond formation, and decomposition of glycopeptides bearing two glycan chains during the catalytic hydrogenation process. After many trials, a successful route was finally established with the key consideration of minimizing the number of synthetic manipulations on full-length glycopeptides (Scheme 7.15). The two octasaccharide-serine modules **114** and **113** were transformed to glycopeptides **117** and **122**, which were subjected to catalytic hydrogenation, generating partially deprotected glycopeptides **118** and **123** respectively (Scheme 7.15a and b). Coupling of these two fragments gave glycopeptide **124**, which was then treated with LiOH at pH 9 and hydrazine hydrate to yield the desired syndean-3 HS glycopeptide **112** (Scheme 7.15c).

7.4 Conclusion and Future Outlook

Since the establishment of the PG linkage structure and the first synthesis of the linkage-region disaccharide in the 1960s, significant advances have been made in synthesis of glycopeptides from PGs. HS glycopeptides bearing multiple HS chains with well-defined sulfation patterns can now be accessed. However, the sizes of glycopeptides that can be produced *via* total synthesis are still relatively small compared with naturally existing PGs. The HS chains installed in glycopeptides to date lack *N*-sulfation, a critical structural element of naturally existing HSPGs. The peptide backbone of synthetic HS glycopeptides does not yet contain all possible naturally existing amino acid residues, such as aspartic acid and lysine, which will require additional considerations in protective group design. Innovative methodologies including glycosylation technology, protective group chemistry, sulfation, peptide chemistry and enzymology are urgently needed to produce synthetic PGs with increased structural complexities to facilitate the glycobiological studies of these interesting molecules.

Acknowledgements

We are grateful for financial support by the National Institute of General Medical Sciences, NIH (R01GM072667, U01GM116262 and U01GM102137) and the National Science Foundation (CHE 1507226).

References

1. M. Ly, T. N. Laremore and R. J. Linhardt, *OMICS*, 2010, **14**, 389–399.
2. M. D. Bass, M. R. Morgan and M. J. Humphries, *Sci. Signaling*, 2009, **2**, pe18.
3. X. Lin, *Development*, 2004, **131**, 6009–6021.

4. M. Bernfield, M. Gotte, P. W. Park, O. Reizes, M. L. Fitzgerald, J. Lincecum and M. Zako, *Annu. Rev. Biochem.*, 1999, **68**, 729–777.
5. C. Kirn-Safran, M. C. Farach-Carson and D. D. Carson, *Cell. Mol. Life Sci.*, 2009, **66**, 3421–3434.
6. J. D. Esko and L. Zhang, *Curr. Opin. Struct. Biol.*, 1996, **6**, 663–670.
7. L. Fransson, I. Silverberg and I. Carlstedt, *J. Biol. Chem.*, 1985, **260**, 14722–14726.
8. J. D. Esko and S. B. Selleck, *Annu. Rev. Biochem.*, 2002, **71**, 435–471.
9. T. Mikami and H. Kitagawa, *Biochim. Biophys. Acta*, 2013, **1830**, 4719–4733.
10. K. Meyer and E. Chaffee, *J. Biol. Chem.*, 1941, **138**, 491–499.
11. T. Krusius, J. Finne, R. K. Margolis and R. U. Margolis, *J. Biol. Chem.*, 1986, **261**, 8237–8242.
12. J. L. Funderburgh, *Glycobiology*, 2000, **10**, 951–958.
13. U. Hacker, K. Nybakken and N. Perrimon, *Nat. Rev. Mol. Cell Biol.*, 2005, **6**, 530–541.
14. J. R. Bishop, M. Schuksz and J. D. Esko, *Nature*, 2007, **446**, 1030–1037.
15. J. Filmus, M. Capurro and J. Rast, *Genome Biol.*, 2008, **9**, 224, DOI: 210.1186/gb-2008-1189-1185-1224.
16. J. Kuhn, C. Gotting, M. Schnolzer, T. Kempf, T. Brinkmann and K. Kleesiek, *J. Biol. Chem.*, 2001, **276**, 4940–4947.
17. J. D. Esko, K. Kimata and U. Lindahl, in *Essentials of Glycobiology*, ed. A. Varki, R. D. Cummings, J. D. Esko, H. H. Freeze, P. Stanley, C. R. Bertozzi, G. W. Hart and M. E. Etzler, Cold Spring Harbor, NY, 2nd edn, 2009.
18. M. R. Morgan, M. J. Humphries and M. D. Bass, *Nat. Rev. Mol. Cell Biol.*, 2007, **8**, 957–969.
19. B. De Cat, S. Y. Muyldermans, C. Coomans, G. Degeest, B. Vanderschueren, J. Creemers, F. Biemar, B. Peers and G. David, *J. Cell Biol.*, 2003, **163**, 625–635.
20. S. Bacsa, G. Karasneh, S. Dosa, J. Liu, T. Valyi-Nagy and D. Shukla, *J. Gen. Virol.*, 2011, **92**, 733–743.
21. Y.-P. Hu, S.-Y. Lin, C.-Y. Huang, M. M. L. Zulueta, J.-Y. Liu, W. Chang and S.-C. Hung, *Nat. Chem.*, 2011, **3**, 557–563.
22. S. Choi, H. Lee, J. R. Choi and E. S. Oh, *BMB Rep.*, 2010, **4**, 305–310.
23. R. V. Iozzo, *J. Clin. Invest.*, 2001, **108**, 165–167.
24. J. K. Langford, Y. Yang, T. Kieber-Emmons and R. D. Sanderson, *J. Biol. Chem.*, 2005, **280**, 3467–3473.
25. S. B. Dulaney and X. Huang, *Adv. Carbohydr. Chem. Biochem.*, 2012, **67**, 95–136 and references cited therein.
26. L. Poletti and L. Lay, *Eur. J. Org. Chem.*, 2003, **2003**, 2999–3024.
27. N. A. Karst and R. J. Linhardt, *Curr. Med. Chem.*, 2003, **10**, 1993–2031.
28. M. Petitou and C. A. A. van Boeckel, *Angew. Chem., Int. Ed.*, 2004, **43**, 3118–3133 and references cited therein.
29. J. D. C. Codée, H. S. Overkleeft, G. A. van der Marel and C. A. A. van Boeckel, *Drug Discovery Today: Technol.*, 2004, **1**, 317–326.

30. C. Noti, J. L. de Paz, L. Polito and P. H. Seeberger, *Chem.–Eur. J.*, 2006, **12**, 8664–8686.
31. J. D. Gregory, T. C. Laurent and L. Roden, *J. Biol. Chem.*, 1964, **239**, 3312–3320.
32. U. Lindahl and L. Roden, *Biochem. Biophys. Res. Commun.*, 1964, **17**, 254–259.
33. B. Lindberg and B. G. Silvander, *Acta Chem. Scand.*, 1965, **19**, 530–531.
34. B. Lindberg, L. Rodén and B.-G. Silvander, *Carbohydr. Res.*, 1966, **2**, 413–417.
35. L. Benzingnguyen and L. Roden, *Carbohydr. Res.*, 1977, **53**, 123–128.
36. H. M. Flowers, *Carbohydr. Res.*, 1967, **4**, 312–317.
37. K. Kum and S. Roseman, *Biochemistry*, 1966, **5**, 3061–3065.
38. K. Brendel and E. A. Davidson, *Carbohydr. Res.*, 1966, **2**, 42–48.
39. J. M. Lacombe, A. A. Pavia and J. M. Rocheville, *Can. J. Chem.*, 1981, **59**, 473–481.
40. G. Ekborg, M. Klinger, L. Roden, J. W. Jensen, J. S. Schutzbach, D. H. Huang, N. R. Krishna and G. M. Anantharamaiah, *Glycoconjugate J.*, 1987, **4**, 255–266.
41. B. Erbing, B. Lindberg and T. Norberg, *Acta Chem. Scand., Ser. B*, 1978, **32**, 308–310.
42. P. J. Garegg, B. Lindberg and T. Norberg, *Acta Chem. Scand., Ser. B*, 1979, **33**, 449–452.
43. G. Ekborg, T. Curenton, N. R. Krishna and L. Roden, *J. Carbohydr. Chem.*, 1990, **9**, 15–37.
44. S. Rio, J. M. Beau and J.-C. Jacquinet, *Carbohydr. Res.*, 1991, **219**, 71–90.
45. R. R. Schmidt, *Angew. Chem., Int. Ed.*, 1986, **25**, 212–235.
46. S. Rio, J. M. Beau and J.-C. Jacquinet, *Carbohydr. Res.*, 1993, **244**, 295–313.
47. K. Aït-Mohand, C. Lopin-Bon and J.-C. Jacquinet, *Carbohydr. Res.*, 2012, **353**, 33–48.
48. S. Rio, J. M. Beau and J.-C. Jacquinet, *Carbohydr. Res.*, 1994, **255**, 103–124.
49. J.-C. Jacquinet, *Carbohydr. Res.*, 2004, **339**, 349–359.
50. B. Thollas and J.-C. Jacquinet, *Org. Biomol. Chem.*, 2004, **2**, 434–442.
51. J.-C. Jacquinet, *Carbohydr. Res.*, 2006, **341**, 1630–1644.
52. L. Chen and F. Kong, *Carbohydr. Res.*, 2002, **337**, 1373–1380.
53. T. Fukuyama, A. A. Laird and L. M. Hotchkiss, *Tetrahedron Lett.*, 1985, **26**, 6291–6292.
54. R. R. Schmidt and W. Kinzy, *Adv. Carbohydr. Chem. Biochem.*, 1994, **50**, 21–123.
55. J.-i. Tamura, T. Nakamura-Yamamoto, Y. Nishimura, S. Mizumoto, J. Takahashi and K. Sugahara, *Carbohydr. Res.*, 2010, **345**, 2115–2123.
56. R. J. Ferrier, R. W. Hay and N. Vethaviyasar, *Carbohydr. Res.*, 1973, **27**, 55–61.
57. J. D. C. Codée, R. E. Litjens, L. J. van den Bos, H. S. Overkleeft and G. A. van der Marel, *Chem. Soc. Rev.*, 2005, **34**, 769–782.
58. G. Lian, X. Zhang and B. Yu, *Carbohydr. Res.*, 2015, **403**, 13–22.

59. K. W. Neumann, J. Tamura and T. Ogawa, *Bioorg. Med. Chem.*, 1995, **3**, 1637–1650.
60. F. Goto and T. Ogawa, *Tetrahedron Lett.*, 1992, **33**, 5099–5102.
61. K. W. Neumann, J. Tamura and T. Ogawa, *Glycoconjugate J.*, 1996, **13**, 933–936.
62. J. G. Allen and B. Fraser-Reid, *J. Am. Chem. Soc.*, 1999, **121**, 468–469.
63. K. Shimawaki, Y. Fujisawa, F. Sato, N. Fujitani, M. Kurogochi, H. Hoshi, H. Hinou and S.-I. Nishimura, *Angew. Chem., Int. Ed.*, 2007, **46**, 3074–3079.
64. A. Morikawa and H. Kuzuhara, *J. Carbohydr. Chem.*, 1990, **9**, 167–179.
65. T.-Y. Huang, M. M. L. Zulueta and S.-C. Hung, *Org. Lett.*, 2011, **13**, 1506–1509.
66. T. Yasukochi, K. Fukase, Y. Suda, K. Takagaki, M. Endo and S. Kusumoto, *Bull. Chem. Soc. Jpn.*, 1997, **70**, 2719–2725.
67. K. Fukase, T. Yasukochi and S. Kusumoto, *Bull. Chem. Soc. Jpn.*, 2001, **74**, 1123–1128.
68. K. Fukase, T. Yasukochi, Y. Suda, M. Yoshida and S. Kusumoto, *Tetrahedron Lett.*, 1996, **37**, 6763–6766.
69. Y. Choi, H. Chung, H. Jung, J. R. Couchman and E. S. Oh, *Matrix Biol.*, 2011, **30**, 93–99.
70. B. De Cat, S. Y. Muyldermans, C. Coomans, G. Degeest, B. Vanderschueren, J. Creemers, F. Biemar, B. Peers and G. David, *J. Cell Biol.*, 2003, **163**, 625–635.
71. B. Yang, K. Yoshida, Z. Yin, H. Dai, H. Kavunja, M. H. El-Dakdouki, S. Sungsuwan, S. B. Dulaney and X. Huang, *Angew. Chem., Int. Ed.*, 2012, **51**, 10185–10189.
72. X. Huang, L. Huang, H. Wang and X.-S. Ye, *Angew. Chem., Int. Ed.*, 2004, **43**, 5221–5224.
73. K. S. Kim, J. H. Kim, Y. J. Lee, Y. J. Lee and J. Park, *J. Am. Chem. Soc.*, 2001, **123**, 8477–8481.
74. K. Yoshida, B. Yang, W. Yang, Z. Zhang, J. Zhang and X. Huang, *Angew. Chem., Int. Ed.*, 2014, **53**, 9051–9058.

CHAPTER 8

Chemoenzymatic Synthesis of Low-Molecular-Weight Heparin and Heparan Sulfate

XING ZHANG[a] AND ROBERT J. LINHARDT*[a]

[a]Department of Chemistry and Chemical Biology, Department of Chemical and Biological Engineering, Department of Biology, and Department of Biomedical Engineering, Center for Biotechnology and Interdisciplinary Studies, Rensselaer Polytechnic Institute, Troy, New York 12180, USA
*E-mail: linhar@rpi.edu

8.1 Introduction

8.1.1 What are Heparan Sulfate, Heparin and Heparin-Derivatives?

Heparan sulfate (HS) belongs to the family of linear, highly sulfated anionic polysaccharides known as glycosaminoglycans (GAGs). HS consists of repeating, 1 → 4-linked disaccharide units of alternating α-D-glucosamine with a β-D-glucuronic acid (GlcA) or α-L-iduronic acid (IdoA) (Figure 8.1, structures **1** and **2**).[1] HS is widely found on the surfaces of animal cells and in the extracellular matrix and exhibits a wide range of biological functions controlling angiogenesis, inflammatory responses, lipid metabolism, tumor metastasis, and blood coagulation.[2]

Chemical Biology No. 2
Chemical Biology of Glycoproteins
Edited by Zhongping Tan and Lai-Xi Wang

Published by the Royal Society of Chemistry, www.rsc.org

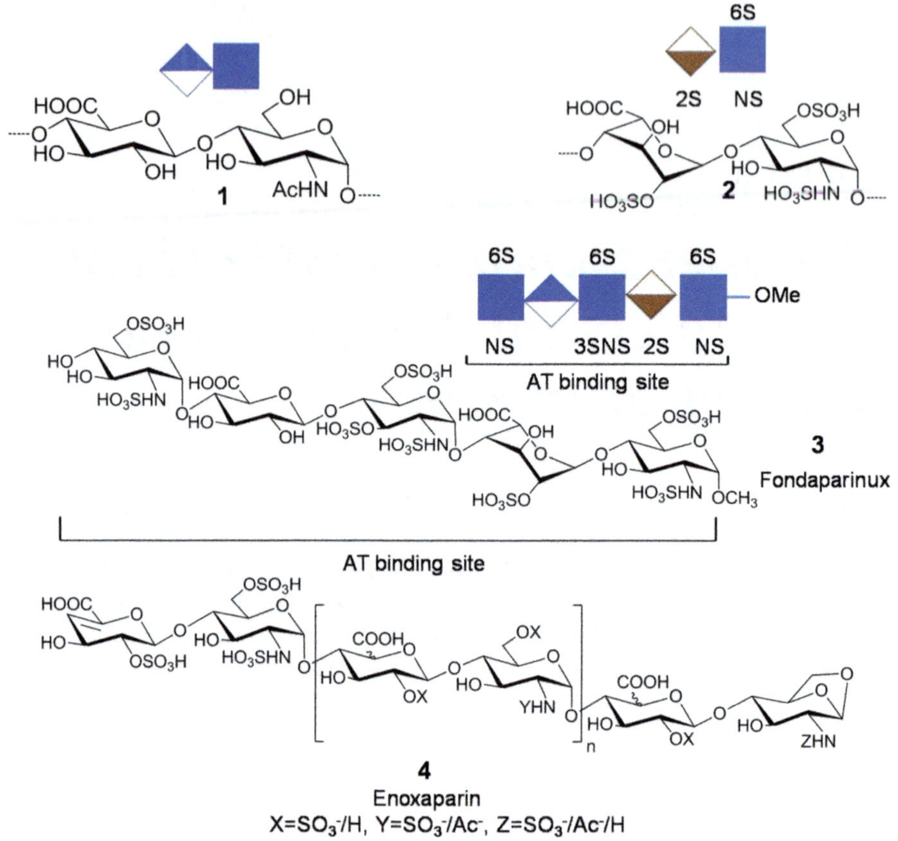

Figure 8.1 Structures of HS, heparin and heparin derivatives. The disaccharide repeating units of HS (primarily **1**) and heparin (primarily **2**), the structure of Arixtra® (fondaparinux) **3**, a ULMWH containing the AT-binding site, and a representative chain of the LMWH, Lovenox® (enoxaparin) **4**, are shown. The chemical structures of the saccharide residues as well as their symbolic representations are shown ref. 1,12.

Heparin is a special form of HS, found within specialized granulated cells, which has a higher level of sulfation, a higher IdoA content, and displays strong anticoagulant activity.[1] There are an average of approximately 2.6 sulfo groups per disaccharide unit in heparin, compared with only approximately 0.6 sulfo groups per disaccharide unit in a typical HS.[1] Approximately 80–90% of the disaccharide units in heparin contain IdoA compared with approximately 20% IdoA in HS.[1] The locations of both the sulfo groups and the IdoA and GlcA residues within heparin and HS play profound role in their activities. A pentasaccharide sequence present within these polysaccharides having the structure, →GlcNS6S → GlcA → GlcNS3S6S → IdoA2S → GlcNS6S → (NS is N-sulfo; Figure. 8.1 **3**, or several other structural variants[3]), is responsible for its specific binding to the serine protease inhibitor (Serpin) antithrombin III (AT).

A diverse collection of HS and heparin oligosaccharides and polysaccharides provide structural leads for developing new therapeutic agents and represent a frontier in medicinal chemistry and in understanding the biology of blood coagulation and other important physiological processes.

Heparin was discovered a century ago by McLean and Howell at Johns Hopkins University, and was soon after introduced as a clinical anticoagulant.[3,4] After 90 years of use, heparin remains the first choice for controlling blood coagulation and treating arterial thrombotic disorders.[5] The Serpin AT binds heparin at a pentasaccharide sequence, containing a central 3-*O*-sulfoglucosamine residue (Figure 8.1, **3**), to form an AT-heparin complex, which inhibits the blood-coagulation proteases to prevent clotting.[6,7] The clinical application of heparin greatly promoted the development of many medical and surgical procedures and in the absence of heparin modern medicine is not possible.

Heparins are classified into three forms, each of which have been approved by the US Food and Drug Administration (FDA): (1) unfractionated heparin (UFH, average molecular weight (MW_{avg}) 17 000 Da); (2) low-molecular-weight heparin (LMWH, MW_{avg} 3500–6000 Da); and (3) ultralow-molecular-weight heparin (ULMWH, $MW_{avg} < 2000$ Da).[8] UFH is primarily isolated from porcine intestine and represents approximately 40% of all the heparin used in the USA.[9] UFH is an intravenous drug used in surgery and kidney dialysis but prone to result in the development of a life-threatening adverse event known as heparin-induced thrombocytopenia (HIT).[10] LMWH is prepared from UFH by various depolymerization methods and is the most widely prescribed heparin in the USA (approximately 55%) due to its predictable subcutaneous bioavailability, longer half-life and slightly reduced incidence of HIT and bleeding (see Figure 8.1 **4** for an example of a LMWH).[8] ULMW heparins are synthetic, subcutaneously bioavailable products, corresponding to five to ten saccharide units, and because of expense and limited applications still represent a small percentage (less than 5%) of clinically used heparin.[9] ULMWH shows greatly reduced risks of HIT. The only clinically approved ULMWH, Arixtra® (fondaparinux, introduced in 2003), is a pentasaccharide containing the AT-binding region of heparin (Figure 8.1, **3**).[11]

8.1.2 Why are Synthetic Heparins and Heparan Sulfates Needed?

UFH and LMWH are polypharmacological agents, complex mixtures of molecules prepared from animal tissues.[13] The supply chain of pharmaceutical heparin has been poorly regulated and has presented safety concerns. Due to the similar nature of glycosaminoglycans (GAGs), their polydispersity, molecular heterogeneity, limitations in purification methods and the lack of simple and reliable analytical methods, heparin often contains HS, dermatan sulfate (DS) and chondroitin sulfate (CS). Worldwide distribution of contaminated heparin in 2007, caused by its adulteration with a semi-synthetic oversulfated chondroitin sulfate (OSCS) contaminant, adversely affected the purity and safety of animal-sourced UFH and LMWH, and was associated with over 200 deaths in the USA.[14,15] Another potential threat is that other bioactive entities, such as viruses, prions or growth-modulating factors, may remain associated with the HS chains in animal extracts. Thus, the cost-effective preparation of a more structurally defined heparin from non-animal sources is highly desirable as are reliable structural characterization techniques.[14]

Until now, the preparation of homogeneous ULMW heparin using a purely synthetic chemical method, based on repetitive steps of protection, activation, coupling, and de-protection, is very challenging. For instance, Arixtra® (fondaparinux, Figure 8.1 **3**), a relatively simple pentasaccharide, requires as many as 60 chemical steps, and the overall yield is only 0.1%.[16,17] In addition, during preparation, there are many separation steps needed to remove undesirable isomers (especially the incorrect anomers) as well as the unavoidable failure products that add cost and decrease overall yields. Moreover, complex LMWHs and UFHs are usually not amenable to chemical synthesis, as it is virtually impossible to commercially prepare heparin oligosaccharides longer than five sugar units using currently available chemical approaches.[18]

Chemoenzymatic synthesis, relying on combined enzymatic and chemical methods, mimics the biosynthetic pathway of heparin, and represents a promising strategy to solve these synthetic challenges.[3] In this strategy, enzymatic synthesis is usually performed under mild temperatures (20–60 °C) using "green" aqueous conditions[19] catalyzing the reaction with exquisite regioselectivity and stereoselectivity (α- or β- glycosidic linkages) without the need for repetitive protection or deprotection steps. Chemical reactions are required to prepare unnatural enzymatic reaction substrates (unnatural uridine diphosphate (UDP)–sugars or acceptors), modify the backbone of enzyme-catalyzed heparin products or to control the regioselectivity of the enzymatic modifications taking place along the polysaccharide backbone.[1,18]

8.1.3 Types of Chemoenzymatic Synthesis

The total-synthesis of heparin and HS includes both backbone elongation and saccharide modification and starts from available monosaccharides or disaccharides. Glycosyltransferases (KfiA and pmHS2) utilize UDP-activated sugars (**6**, **7**, **8**) as donors to transfer saccharides onto an acceptor (**5**) to build the HS backbone structure (Figure 8.2A).[5] The nascent HS chain subsequently undergoes a series of modifications through the action of *N*-deacetylase (ND)/*N*-sulfotransferase (NST) (or the bifunctional enzymes NDSTs), C5-epimerase (C5-epi), 2-*O*-sulftransferase (2-OST), 6-*O*-sulftransferases (6-OSTs) and 3-*O*-sulftransferases (3-OSTs) result in the target HS or heparin structure.[8] A number of structurally well-defined HS and heparin oligosaccharides have been prepared and these synthetic heparins can exhibit potent bioactivity.

Semi-synthesis of heparin and HS employs naturally occurring polydisperse polysaccharide, obtained from bacterial fermentation, as starting material.[20] After its partial *N*-deacetylation and *N*-sulfation, this polysaccharide is then transformed into heparin or HS using recombinant biosynthetic enzymes,

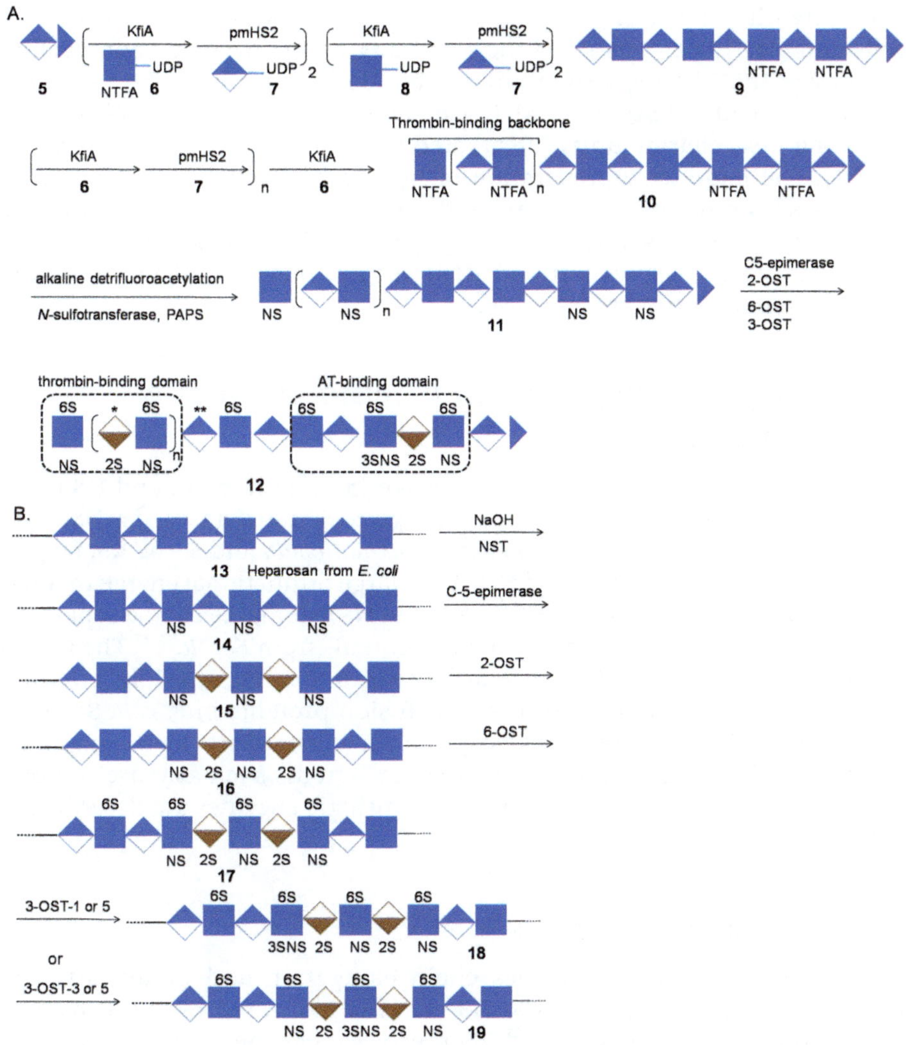

Figure 8.2 Chemoenzymatic synthesis of heparins and heparan sulfates and their oligosaccharides. (A) Total synthesis of heparin and heparan sulfate oligosaccharides (**12**) through the addition of UDP-activated donor sugars to a disaccharide acceptor prepared through the nitrous-acid-catalyzed depolymerization of *N*-deacteylated heparosan. The residue at the position indicated with * can also be an IdoA or a GlcA residue and the residue at the position indicated with ** can be an IdoA residue. (B) The semi-synthesis, starting from heparosan (**13**) of heparin and heparan sulfate (**19**) using a chemoenzymatic approach. NTFA: *N*-trifluoroacetyl.

mimicking the *in vivo* biosynthesis of these GAGs (Figure 8.2B).[21] Rosenberg and colleagues[22] first reported the small-scale semi-synthesis of AT-binding-site-containing HS using cloned enzymes mimicking those present in cells and clearly demonstrated the feasibility of an enzymatic approach to synthesize HS. The heparosan polysaccharide is comprised of a [→4) β-D-glucuronic acid (GlcA) (1 → 4) *N*-acetyl-α-D-glucosamine (GlcNAc) (1→]$_n$ repeating disaccharide units (unsulfated heparan sulfate, Figure 8.1 structure **1**). The large-scale preparation of heparosan from *Escherichia coli* K5 strain has been described.[22] Compared with total synthesis of heparins, the semi-synthetic approach is more easily scaled and provides a cost saving by not requiring the use of purified UDP–sugars, but the polymer size and compositional control are sacrificed.

8.2 Enzymes Required for Chemoenzymatic Synthesis

The enzymes used for chemoenzymatic synthesis of heparin and HS come from a variety of different sources. The glycosyltransferases, pmHS2 and KfiA, are bacterial enzymes from *Pasteurella multocida* and *E. coli* K5, respectively.[23,24] The mammalian Golgi NDST is a large bifunctional enyme that has not been actively expressed, but the ND and NST domains can be separately expressed in their active forms as fusion proteins from *E. coli*.[21,25] The C5-epi and the specialized *O*-sulfotransferases are also mammalian Golgi enzymes and have been cloned and expressed as fusion proteins in *E. coli*. Since all these mammalian enzymes are glycoproteins, recent efforts have focused on their expression in glycosylation-competent systems such as yeast and baculovirus-infected insect cells.[26] These biosynthetic enzymes are described in Table 8.1.

8.2.1 Glycosyltransferases

The biosynthesis of HS includes chain elongation and modification of the sugar backbone. *In vivo*, the HS chain is elongated by the Exostosin gene products, Ext1 and Ext2.[27] This process can be mimicked *in vitro* due to the availability of two recombinant expressed bacterial glycosyltransferases, KfiA and pmHS2. KfiA, prepared from *E. coli* K5 strain, is an *N*-acetylglucosaminyltransferase polymerizing UDP–GlcNAc (or the unnatural UDP–glucosamintrifluoroacetate (UDP–GlcTFA)) donor to a monosaccharide or disaccharide acceptor in a controlled, stepwise manner until the desired oligosaccharide length is reached. The UDP–GlcNAc can be produced by uridyltransferases such as GlmU, which allows the synthesis of some unnatural UDP–sugars for introduction into heparin oligosaccharides.[28] PmHS2 is a homologous heparosan synthase readily obtained from *P. multocida*, which can transfer UDP–GlcA donor to build the HS backbone. KifiA and pmHS2 are usually used in an alternating fashion during chemoenzymatic synthesis.

Table 8.1 Enzymes utilized in heparin and HS synthesis.

Enzyme	Abbreviation	Expression system	Enzymatic function
N-acetyl-D-glucosaminyltransferase	KfiA	*E. coli* K5	Transfers a GlcNAc (or a GlcNTFA) residue to the backbone
Heparosan synthase 2	PmHS2	*P. multocida*	Transfers a GlcA and a GlcNAc (or a GlcNTFA) residue to the backbone
N-acetyl-glucosamine-1-phosphate uridyltransferase	GlmU	*E. coli* K5	Converts a GlcNAc-1-phosphate to an UDP–GlcNAc
C5-epimerase	C5-epi	*Cricetulus griseus* (CHO cell)	Converts a GlcA to an IdoA residue
2-*O*-sulfotransferase	2-OST-1	*C. griseus*	Introduces a sulfo group to the 2-OH position of an IdoA or a GlcA residue
6-*O*-sulfotransferase 1	6-OST-1	*Mus musculus*	Introduces a sulfo group to the 6-OH position of a GlcNS(GlcNAc) residue
6-*O*-sulfotransferase 3	6-OST-3	*M. musculus*	The same as 6OST-1
3-*O*-sulfotransferase 1	3-OST-1	*M. musculus*	Introduces a sulfo group to the 3-OH position of a GlcNS ± 6S residue that is linked to a GlcA (or an IdoA) on the non-reducing end
3-*O*-sulfotransferase 5	3-OST-5	*M. musculus*	Has both 3-OST-1 and 3-OST-3A substrate specificities
3-*O*-sulfotransferase 3	3-OST-3	*M. musculus*	Introduces a sulfo group to the 3-OH position of a GlcNS ± 6S residue that is linked to an IdoA2S on the non-reducing end
N-deacetylase/*N*-sulfotransferase-1	NDST-1	*Rattus norvegicus*	Converts a GlcNAc to a GlcNS residue

8.2.2 Sulfotransferases and C5-Epimerase

Several sulfotransferases involved in heparin biosynthesis have been expressed and characterized. These sulfotransferases transferred sulfo groups from 3′-phosphoadenosine-5′-phosphosulfate (PAPS), the natural sulfate donor, to heparin and HS oligosaccharides with excellent regiospecificity.[29] The introduction of *N*-sulfo groups is catalyzed by *N*-deacetylase/*N*-sulfotransferase (NDST), a bifunctional enzyme having two active sites. The *N*-deacetylase removes the *N*-acetyl group from a GlcNAc residue and the *N*-sulfotransferase activity transfers a sulfo group to the resulting GlcN residue. The expression of an active bacterial NDST has been problematic. This has resulted in the development of an alternative method to position GlcNS residues using the unnatural UDP–GlcNTFA donor to introduce a GlcNTFA residue that can be easily chemically de-*N*-trifluoroacetylated to GlcN for subsequent

N-sulfation. The N-deacetylase/N-sulfotransferase (NDST) is believed to be the first enzyme to modify an intact heparosan chain during biosynthesis, and its action is crucial to all subsequent sulfation reactions.[21]

The 2-OST is only present in one isoform, transferring a sulfo group to the 2-position of both IdoA and GlcA residues, with a preference for IdoA.[30] Thus, IdoA2S is commonly found in heparin and HS but GlcA2S is rarely observed.[31] Three 6-OST isoforms (6-OST-1, 6-OST-2, 6-OST-3) and seven 3-OST isoforms (3-OST-1-7) have been identified and transfer sulfo groups to the 6- and 3- positions of the GlcN residues, respectively.[32,33] The different isoforms of enzymes exhibit slightly different specificities. For example, 6-OST-1 and 6-OST-2 prefer to transfer 6-*O*-sulfo groups to a GlcNS that is next to a GlcA residue and an IdoA2S residue, respectively.[13] The 3-OST-1 isoform transfers a sulfo group to a GlcNS that is linked to a GlcA or IdoA at the non-reducing end, while the 3-OST-5 isoform prefers a GlcNS that is linked to an IdoA2S, GlcA, or IdoA.[34]

C5-epimerase is responsible for converting GlcA into its C5-epimer, IdoA. Although both the natural UDP–GlcA and the unnatural UDP–IdoA have been tested as donors, only the GlcA residue is incorporated.[35] Thus, C5-epi is indispensable in preparing heparin and HS containing IdoA residues. C5-epi catalyzes both the forward and reverse reaction, generating an equilibrium mixture of GlcA and IdoA residues. Since the GlcA residue is thermodynamically favored, the IdoA residue is only generated in large amounts in the presence of 2-OST resulting in the formation of IdoA2S, which cannot be converted by C5-epi to GlcA2S.[36]

8.3 Building Blocks Prepared for Chemoenzymatic Synthesis

8.3.1 Acceptors

In the total synthesis of HS and heparin oligosaccharides, a number of monosaccharide and disaccharide acceptors have been developed that have greatly facilitated the chemoenzymatic synthesis (Figure 8.2A **5** and Figure 8.3A **20**, **21**, **5**). The addition of an ultraviolet-detectable, hydrophobic tag to the acceptor synthesis has improved both detection and purification. The *p*-nitrophenyl (*p*-NP), *p*-nitrobenzyl (*p*-NPM), *p*-methoxybenzyl (*p*-MPM) and *p*-pivaloylaminobenzyl (*p*-PAB) β-glucuronides **20**, represent a series of important commercially available acceptors.[37,38] These aromatic groups are UV-visible and hydrophobic and they can be reversibly bound to reversed-phase chromatography resins, enabling the easy purification of the resulting oligosaccharides using a C18 resin. In addition, these tags can be easily removed or modified using standard chemical methods. For example, by reducing the *p*-NP group to a free amine followed by amidation with fluorescent tag, Alexa Fluor 488 dye, the resulting complex could be used as a sensitive probe in live or fixed cells.[39] However, the potential toxicity of *p*-NP is a concern, especially in clinical applications.

Compared with the *p*-NP tagged acceptor, a monosaccharide or disaccharide acceptor with a reducing end α-*O*-methyl glycoside (Figure 8.3 **21**), is

8.3.2.2.2 Other Unnatural UDP–Sugars. The application of unnatural donors can also play an important role in synthesizing unnatural heparins, as temporary protecting groups, as chemical tags, or as a precursor for further modifications before their selective removal under mild basic or acidic conditions. Developing various unnatural UDP–sugars is of great interests, and many novel donors have been proposed. Chemical synthesis is still very important in preparing unnatural UDP–sugars and can still require a many steps. Linhardt and colleagues[35] have prepared UDP–IdoA and UDP–hexuronic acid (HexA) as potential glucuronosyltransferse substrates. Chen and coworkers[50] successfully used a 6-azide substituted UDP–GlcNAc derivative to synthesize a heparin-like oligosaccharide. The resulting 6-azide-GlcNAc residue can be either converted to a special moiety 6 *N*-sulfo or form an oligosaccharide cluster using click chemistry. We anticipate that many additional unnatural UDP–sugars, such as thiol-, azide- or amine-substituted donors, will one day be designed and synthesized. These resulting unnatural HS oligosaccharide analogs might be resistant to catabolism by blocking the action of heparanase and lysosomal degradative enzymes,[51,52] increasing the biological half-life of the oligosaccharides prepared using these donors.

8.3.2.2.3 Challenges for Unnatural Donors. Unnatural UDP–sugars have been successfully applied in synthesis of heparin analogs, establishing a platform for developing new heparin-based therapeutics and unraveling mechanisms underlying sugar-related biological processes. However, a pure chemical approach requires a long and tedious synthesis and the yields of UDP–sugar are generally low.[18] Enzymatic or cell-based syntheses of unnatural donors need to be developed for their inexpensive and large-scale preparation.[53] The ability of natural synthases and glycosyltransferases to accept novel unnatural UDP–sugars may be limited, thus, protein engineering strategies may be required to make these enzymes more tolerant of unnatural UDP–sugar donors.[18]

8.3.3 Polysaccharide and Oligosaccharide Backbone

Polysaccharides serve as the building blocks in semi-synthesis of HS and heparin. In particular, heparosan, prepared from fermentation of *Escherichia coli* K5, and its derivatives are promising substrates for the preparation of heparin-based therapeutics from non-animal sources. Heparosan is typically modified sequentially through chemical *N*-deacetylation and *N*-sulfonation steps (due to difficulties in the bacterial expression of NDST). The subsequent actions of C5-epi, and 2-OST, 6-OST-1, 6-OST-3, and 3-OST-1 result in bioengineered heparin.[8] A lower level of enzymatic modification or the use of alternative sulfotransferase isoforms can lead to a wide variety of bioengineered HS polysaccharides with differing biological activities.

Chemically synthesized oligosaccharide building blocks can also provide flexibility in the preparation of structurally defined HS oligosaccharides with desired biological activities. Huang, Liu, and colleagues[54] used the pre-activation-based one-pot approach to chemically prepare a single hexasaccharide precursor, which was further enzymatically extended to produce a series of oligosaccharides with different sulfation patterns and sizes, indicating that interfacing chemical and enzymatic synthesis could be a powerful technique to elucidate the structure and activity relationship of HS.

8.4 Control of Product Through Sequential Enzymatic Modification

The chemoenzymatic synthesis of bioengineered heparin mimics the HS biosynthetic pathway, involving of a series of enzymatic reaction steps, and requiring careful reaction scheme design, especially an appropriate modification sequence. The most effective schemes are those closely related to natural heparin synthesis. The ideal reaction order can greatly affect the enzyme activity, while an improper sequence may result in poor yield and purity. For example, the introduction of 2-*O*-sulfo group into the IdoA residue can only occur following the introduction of *N*-sulfo group in adjacent GlcN residues and prior to the introduction of 6-*O*-sulfo groups into GlcN residues.[1] The critical order of these steps has been demonstrated from ternary crystal structural data.[1,55] C5-epi works best when used in tandem with 2-OST, as the introduction of a 2-*O*-sulfo group can block the reversible activity of C5-epimerase, possibly due to the steric-hindrance. Moreover, C5-epi must be used after the introduction of *N*-sulfo group, because C5-epi only acts on a GlcA residue that is between two GlcNS residues.

ULMWHs **30** and **35** are chemoenzymatically synthesized oligosaccharides that contain the AT-binding domains of porcine and bovine heparin (Figure 8.4A).[41] Disaccharide **5**, prepared from a heparosan disaccharide acceptor, was elongated by two bacterial glycosyltransferases (PmHS2 and KfiA) in alternating fashion to produce the heptasaccharide **27**. In this synthesis, selective C5-epimerization/2-*O*-sulfation takes place on residue **D** but not residue **B**, because residue **D** is flanked by two GlcNS residues. Similarly, as residue **C** is flanked by a GlcA residue at its non-reducing end, 3-OST-1 selectively adds a 3-*O*-sulfo group to residue **C** rather than residue **E** to generate ULMW heparin **30**. The structure of oligosaccharide **35** is a little different from that of **30**, which should introduce a GlcNS6S residue to the non-reducing end. The GlcNTFA residue is critical because it not only can prevent the action of C5-epi and 2-OST on the GlcA but also act a precursor of GlcNS. Finally, the synthesis of oligosaccharide **35** was achieved by rearranging the sequence of the modification and elongation steps without employing additional enzymes, thus demonstrating that both structural control and target diversification are possible in chemoenzymatic synthesis.

It is known that C5-epi is a two-way catalyst, transferring a GlcA to an IdoA residue and also converting an IdoA to a GlcA residue, which results in a final product containing a mixture of GlcA and IdoA.[36] This unique property of C5-epimerase makes the synthesis of a structurally defined LMWH

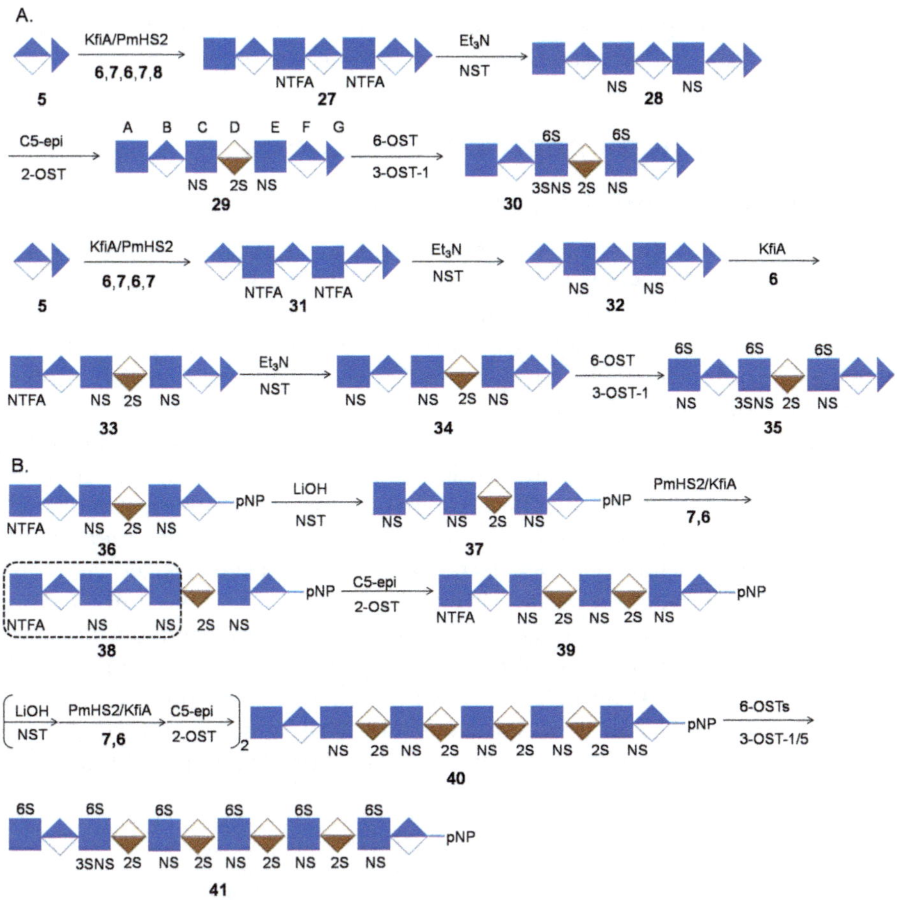

Figure 8.4 Controlled chemoenzymatic synthesis of synthetic heparin HS oligosaccharides. (A) ULMWH **30** and **35** are chemoenzymatically synthesized on acceptor **5**. (B) LMWH **41** is chemoenzymatically synthesized from *p*-NP–hexasaccharide acceptor **36**..

or ULMWH challenging, particularly one containing repeating -IdoA2S–GlcNS6S- units. However, a unique distribution of GlcNAc and GlcNS residues of the substrate can force C5-epi to act as a one-way catalyst. Linhardt, Liu and colleagues[49] have designed an alternative strategy, placement of a pentasaccharide domain GlcNTFA–GlcA–GlcNS–GlcA–GlcNS, for synthesizing repeating -IdoA2s-GlcNS6S- motifs without incomplete conversion (Figure 8.4B). The desired pentasaccharide domain of compound **38** was synthesized from known hexasaccharide **36** through *N*-detrifluoroacetylation, NST and two glycosyl transferase steps. Then C5-epimerization of GlcA was performed and followed by the introduction of a 2-*O*-sulfo group, generating IdoA2S, removing further C5-epi modification. Repeating these steps several times it is possible to obtain repeating -IdoA2S–GlcNS6S- units.

The resulting compound **40** was then converted to the target LMWH **41** after the introduction of *N*-sulfo, 6-*O*-sulfo and 3-*O*-sulfo groups.

8.5 Novel Chemoenzymatic Synthesis

8.5.1 One-Pot Multienzyme System

Glycosyltransferases are widely used in the initial steps of reaction sequences followed by other enzymes such as C5-epi and OSTs (Figures 8.2A). In some instances, it might be more efficient to transform complicated multiple-enzyme-catalyzed reaction sequences into one-pot reactions, where oligosaccharides or polysaccharides are prepared without the required isolation or purification of intermediates.

Chen and colleagues[50] have developed an efficient OPME system utilizing a series of unnatural sugar nucleotides to synthesize heparin-like oligosaccharides (Figure 8.5A). Using GlcNTFA as the donor precursor, glucuronide GlcAβ2AA with a 2-aminoacridone (2AA) fluorescent label as the acceptor substrate **43**, disaccharide GlcNTFAα1-4GlcAβ2AA **44** was synthesized *via* a one-pot four-enzyme reaction containing NahK_ATCC55813, PmGlmU, PmPpA and PmHS2. The resulting disaccharide **44** was used in a one-pot three-enzyme GlcA activation and transfer system to produce the trisaccharide **46**. In this scheme, Glc-1-P **45** and UTP were coupled using *E. coli* glucose-1-P uridylyltransferase EcGalU to generate UDP–Glc. UDP–glucose dehydrogenase, PmUgd, in the presence of the coenzyme nicotinamide adenine dinucleotide

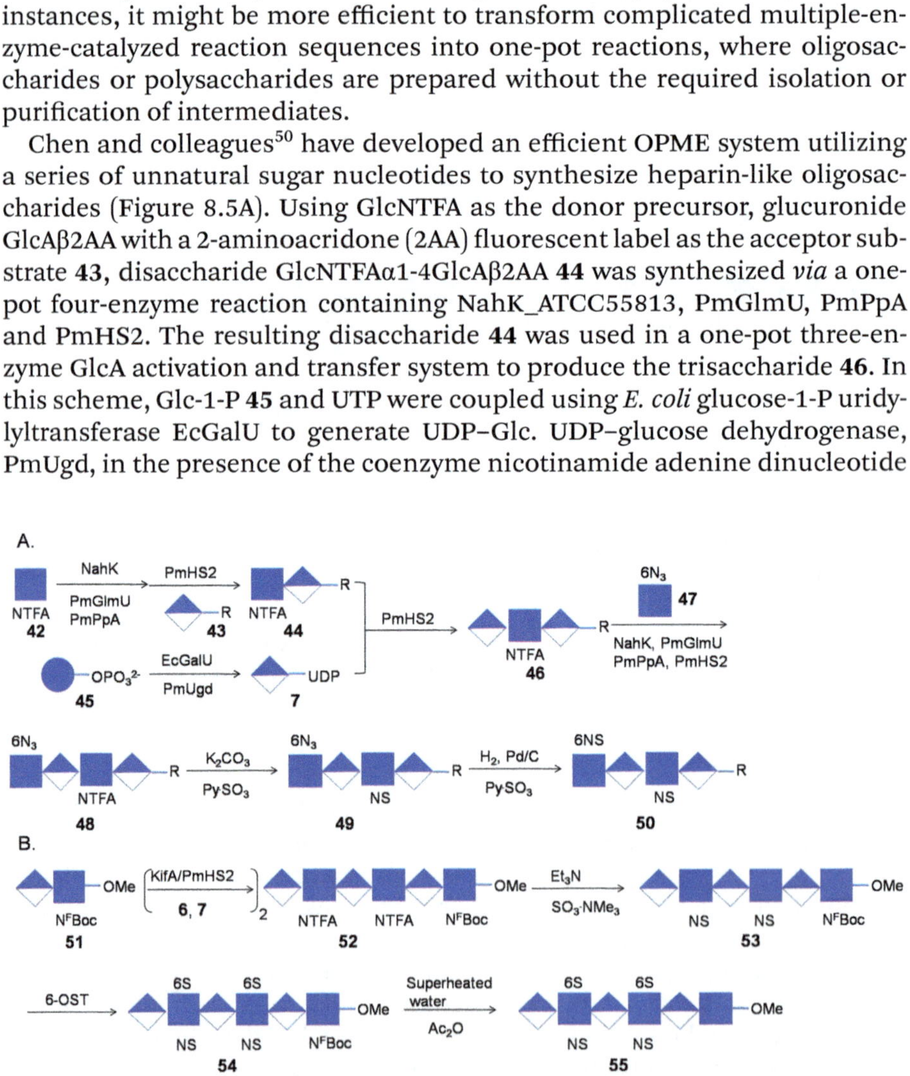

Figure 8.5 Novel approaches to chemoenzymatic synthesis of heparan sulfate oligosaccharides. (A) One-pot multi-enzyme approach where R is 2-aminoacridone. (B) Fluorous-assisted chemoenzymatic synthesis where ^{F}Boc is a fluorinated Boc protecting group.

(NAD^+) oxidized the C6 position of Glc in UDP–Glc forming UDP–GlcA **7** and pmHS2 catalyzed the synthesis of trisaccharide **46**. The trisaccharide **46** obtained was used as a substrate in the one-pot four-enzyme GlcNAc6N_3 activation and transfer system to form tetrasaccharide **48**. The *N*-TFA group and 6-N_3 group were converted into free amino groups and sequential *N*-sulfonation generated the HS tetrasaccharide analog **50**. This HS tetrasaccharide analog may have improved therapeutic potential as it should be more resistant to catabolism by heparanase.

8.5.2 Fluorous-Tagging Techniques

Fluorous chemistry is a new tool for solution-phase high-throughput organic synthesis, developed in the late 1990s.[56] Based on the high affinity of perfluoroalkyl chains toward fluorous surfaces and solvent, fluorous separation techniques can significantly facilitate the isolation and purification of intermediates and products. Tagging organic compounds with a light fluorous group such as a perfluorooctyl (C_8F_{17}) or a perfluorohexyl (C_6F_{13}) group followed by product purification using fluorous solid-phase extraction (FSPE) has found increasing synthetic uses.[57] These advances in fluorous chemistry have been successfully applied by Pohl and coworkers[58,59] to the chemical synthesis of oligosaccharides. The combination of fluorous methodology with chemoenzymatic approaches can further enhance synthetic and purification processes.

Linhardt and colleagues[40] have successfully applied a temporary fluorous *tert*-butyl dicarbonate ^{F}Boc linker and FSPE technique in the chemoenzymatic synthesis of HS oligosaccharides (Figure 8.5B). The synthetic fluorous-tagged disaccharide acceptor **51** was repetitively elongated with glycosyltransferase KfiA and UDP–GlcNTFA, PmHS2 and UDP–GlcA, to form the hexasaccharide **52**. With the HS backbone constructed, base-catalyzed deprotection of the TFA group was followed by chemical *N*-sulfation with $SO_3 \cdot MeN_3$ to generate the *N*-sulfate hexasaccharide **53**. Then incubation of **53** with 6-OSTs and PAPS was used to obtain the 6-*O*-sulfo hexasaccharide **54**. Finally, superheated water, a green solvent,[19] was employed to remove the ^{F}Boc tag and was followed by *N*-acetylation to obtain the HS hexasaccharide **55**. This effort reveals that the fluorous tag did not interfere with the action of these glycosyltransferases and the FSPE technique was suitable for the purification of HS oligosaccharides.

Chen and colleagues[60] have combined the OPME system with fluorous solid-phase extraction cartridges purification to highly efficiently synthesize a series of sialosides and galactosides. This method may be applied to more complex HS oligosaccharides synthesis in the future.

8.5.3 Solid-Phase Synthesis

Since Frechet and colleagues[61] initiated the evolution of solid-phase synthesis of oligosaccharides on a polymer supports in 1971, a number of advances, particularly by the Seeberger group[62,63] have led to the automated chemical synthesis of more complex targets such as glycosaminoglycan oligosaccharides. This approach, could greatly simplify purification while at the same time affording exquisite stereoselectivity and regioselectivity. Since chemoenzymatic synthesis is performed in aqueous solvents, it is critical that the solid support be hydrophilic to allow good swelling and that the appropriate length of linker, between the acceptor saccharides and the solid support, is used to make the acceptor sites enzyme-accessible. Although promising, to date there have been few examples of solid-phase chemoenzymatic synthesis of carbohydrate reported. Linhardt and colleagues[64] have immobilized HS substrate onto the surface of the dextran-modified gold channel microfluidic surface plasmon resonance sensor chip and on streptavidin-modified magnetic nanoparticles in a digital microfluidic device, called an artificial Golgi, for subsequent 3-OST-1 modification. While demonstrating proof-of-concept, the multi-step, solid-phase, chemoenzymatic synthesis of more complex targets still need to be demonstrated.

8.5.4 Immobilized Enzymes

While enzymes are conveniently employed as highly efficient biocatalysts under mild reaction conditions they often lack long-term operational stability and often present a technically challenging recovery process.[65,66] Moreover, difficulty in reusing enzyme catalysts has hampered their widespread industrial application. Enzyme immobilization, first reported in 1916 and commercialized in the 1960s, represents a chemical engineering approach to improve enzyme utilization in biotechnological processes.[67] Immobilized enzymes are physically confined catalysts, localized in a defined region of space, which retain catalytic activity, allowing repeated and continuous use. Thus, immobilization provides for the facile separation of enzyme from products, and results in a remarkable reduction of processing costs, and is expected to be of great utility in chemoenzymatic synthesis of heparins. Linhardt and colleagues[68] reported the successful immobilization of the multiple enzymes involved in heparin synthesis, including C5-epi, 2-OST, and arylsulfotransferase-IV. These recombinant heparin biosynthetic enzymes were immobilized on amino-linked agarose gel beads at a loading of 20 mg per ml of gel and showed enhanced thermal stability while still maintaining over 80% of activity after their immobilization. One particular challenge is that these immobilized enzymes need to act in concert to transform a complex substrate into a much more complex product, while also recycling a required cofactor and preventing product inhibition.

8.5.5 Immobilized Enzyme Cofactors

The successful use of immobilized enzyme cofactors, such as NAD^+–NADH, has been demonstrated in oxidation–reduction reactions.[69,70] An immobilized NAD^+ might be applied to the chemoenzymatic synthesis of heparin oligosaccharides described by Chen and colleagues,[50] to oxidize UDP–Glc to UDP–GlcA using PmUgd (Figure 8.5A). It seems fair to speculate that other enzyme cofactors might also prove important in the chemoenzymatic synthesis of heparins.

For example, in the semi-synthesis approach it might be possible to immobilize a tethered PAPS that can be regenerated using aryl sulfotransferase (AST)-IV and an inexpensive sacrificial sulfo group donor, such as *p*-nitrophenylsulfate (PNPS). In the total synthesis approach it might also be possible to tether UDP-sugars and regenerate these *in situ*, greatly reducing synthetic costs.

8.6 Conclusion and Future Perspectives

HS and heparins are important glycosaminoglycans that perform numerous physiological and pharmacological functions. The chemoenzymatic synthesis of LMWH and ULMWH has demonstrated a novel efficient approach to prepare structurally defined heparin oligosaccharides that have been difficult to prepare by traditional chemical synthesis approaches. Over the past five years, there has been a steady improvement in the chemoenzymatic synthesis of such HS and heparin oligosaccharides. These improvements have included higher yields, reduced numbers of steps, greater purities, easier purification, and larger and more structurally complex targets. The advances leading to these improvements have included better enzyme expression, better enzymes resulting from protein engineering, optimization of the order of enzymatic transformations, increases in reaction scale, higher efficiency use of enzyme co-factors, and a better understanding of the substrate specificities of the HS biosynthetic enzymes.

Many challenges remain in this young field. Large-scale chemoenzymatic synthesis for the anticoagulant drug market remains the major challenge, requiring up to a 1 000 000-fold scale-up from micrograms to kilograms. The promises of this technology, however, are great, offering new, safer, and improved drugs for clinicians as well as oligosaccharides for research scientists, leading to a more complete understand of heparin and HS structure–activity relationships.

References

1. J. Liu and R. J. Linhardt, *Nat. Prod. Rep.*, 2014, **31**, 1676–1685.
2. J. R. Bishop, M. Schuksz and J. D. Esko, *Nature*, 2007, **446**, 1030–1037.
3. S. Masuko and R. J. Linhardt, *Future Med. Chem.*, 2012, **4**, 289–296.
4. R. J. Linhardt, *Chem. Ind.*, 1991, **2**, 45–50.
5. Y. Xu, E. H. Pempe and J. Liu, *J. Biol. Chem.*, 2012, **287**, 29054–29061.
6. J. Choay, M. Petitou, J. C. Lormeau, P. Sinaÿ, B. Casu and G. Gatti, *Biochem. Biophys. Res. Commun.*, 1983, **116**, 492–499.
7. L. Thunberg, G. Bäckström and U. Lindahl, *Carbohydr. Res.*, 1982, **100**, 393–410.
8. R. J. Linhardt and J. Liu, *Curr. Opin. Pharmacol.*, 2012, **12**, 217–219.
9. D. M. Dickinson, J. Liu and R. J. Linhardt, in *Glycoscience: Biology and Medicine*, eds. N. Taniguchi, T. Endo, G. Hart, P. Seeberger and C.-H. Wong, Springer, 2015, pp 419–426.
10. G. M. Arepally and T. L. Ortel, *N. Engl. J. Med.*, 2006, **355**, 809–817.

11. J. I. Weitz and L. A. Linkins, *Expert Opin. Investig. Drugs*, 2007, **16**, 271–282.
12. L. Li, F. Zhang, J. Zaia and R. J. Linhardt, *Anal. Chem.*, 2012, **84**, 8822–8829.
13. U. Bhaskar, E. Sterner, A. M. Hickey, A. Onishi, F. Zhang, J. S. Dordick and R. J. Linhardt, *Appl. Microbiol. Biotechnol.*, 2012, **93**, 1–16.
14. H. Liu, Z. Zhang and R. J. Linhardt, *Nat. Prod. Rep.*, 2009, **26**, 313–321.
15. J. E. Turnbull, *Science*, 2011, **334**, 462–463.
16. M. Petitou and C. A. van Boeckel, *Angew. Chem., Int. Ed.*, 2004, **43**, 3118–3133.
17. M. Petitou, J. C. Jacquinet, J. Choay, J. C. Lormeau and M. Nassr, U.S. Patent, 4,818,816, 1989.
18. P. L. DeAngelis, J. Liu and R. J. Linhardt, *Glycobiology*, 2013, **23**, 764–777.
19. A. Farrán, C. Cai, M. Sandoval, Y. Xu, J. Liu, M. J. Hernáiz and R. J. Linhardt, *Chem. Rev.*, 2015, **115**, 6811–6853.
20. J. Chen, F. Y. Avci, E. M. Munoz, L. M. McDowell, M. Chen, L. C. Pedersen, L. Zhang, R. J. Linhardt and J. Liu, *J. Biol. Chem.*, 2005, **280**, 42817–42825.
21. S. Peterson, A. Frick and J. Liu, *Nat. Prod. Rep.*, 2009, **26**, 610–627.
22. B. Kuberan, D. L. Beeler, M. Lech, Z. L. Wu and R. D. Rosenberg, *J. Biol. Chem.*, 2003, **278**, 52613–52621.
23. M. Chen, A. Bridges and J. Liu, *Biochemistry*, 2006, **45**, 12358–12365.
24. A. E. Sismey-Ragatz, D. E. Green, N. J. Otto, M. Rejzek, R. A. Field and P. L. DeAngelis, *J. Biol. Chem.*, 2007, **282**, 28321–28327.
25. R. Liu, Y. Xu, M. Chen, M. Weiwer, X. Zhou, A. S. Bridges, P. L. DeAngelis, Q. Zhang, R. J. Linhardt and J. Liu, *J. Biol. Chem.*, 2010, **285**, 34240–34249.
26. E. P. Chappell and J. Liu, *Bioorg. Med. Chem.*, 2013, **21**, 4786–4792.
27. C. McCormick, G. Duncan, K. T. Goutsos and F. Tufaro, *Proc. Natl. Acad. Sci. U. S. A.*, 2000, **97**, 668–673.
28. S. Masuko, S. Bera, D. E. Green, M. Weïwer, J. Liu, P. L. DeAngelis and R. J. Linhardt, *J. Org. Chem.*, 2012, **77**, 1449–1456.
29. M. Suflita, L. Fu, W. He, M. Koffas and R. Linhardt, *Appl. Microbiol. Biotechnol.*, 2015, **99**, 7465–7479.
30. J. Rong, H. Habuchi, K. Kimata, U. Lindahl and M. Kusche-Gullberg, *Biochemistry*, 2001, **40**, 5548–5555.
31. P. H. Hsieh, Y. Xu, D. A. Keire and J. Liu, *Glycobiology*, 2014, **24**, 681–692.
32. H. Habuchi, M. Kobayashi and K. Kimata, *J. Biol. Chem.*, 1998, **273**, 9208–9213.
33. N. W. Shworak, J. Liu, L. M. Fritze, J. J. Schwartz, L. Zhang, D. Logeart and R. D. Rosenberg, *J. Biol. Chem.*, 1997, **272**, 28008–28019.
34. J. Liu, A. F. Moon, J. Sheng and L. C. Pedersen, *Curr. Opin. Struct. Biol.*, 2012, **22**, 550–557.
35. M. Weïwer, T. Sherwood, D. E. Green, M. Chen, P. L. DeAngelis, J. Liu and R. J. Linhardt, *J. Org. Chem.*, 2008, **73**, 7631–7637.
36. J. P. Li, F. Gong, K. El Darwish, M. Jalkanen and U. Lindahl, *J. Biol. Chem.*, 2001, **276**, 20069–20077.
37. C. Vaxelaire, F. Souquet, M. I. Lannou, J. Ardisson and J. Royer, *Eur. J. Org. Chem.*, 2009, 3138–3140.

38. C. Cai, L. Li, C. Harvey, J. Liu and R. J. Linhardt, *Tetrahedron Lett.*, 2013, **54**, 4471–4474.
39. E. S. Trindade, C. Oliver, M. C. Jamur, H. A. O. Rocha, C. R. C. Franco, R. I. Boucas, T. R. Jarrouge, M. A. S. Pinhal, I. L. S. Tersariol, T. C. Gouvea, C. P. Dietrich and H. B. Nader, *J. Cell. Physiol.*, 2008, **217**, 328–337.
40. C. Cai, D. M. Dickinson, L. Li, S. Masuko, M. Suflita, V. Schultz, S. D. Nelson, U. Bhaskar, J. Liu and R. J. Linhardt, *Org. Lett.*, 2014, **16**, 2240–2243.
41. Y. Xu, S. Masuko, M. Takieddin, H. Xu, R. Liu, J. Jing, S. A. Mousa, R. J. Linhardt and J. Liu, *Science*, 2011, **334**, 498–501.
42. C. Cai, K. Edgar, J. Liu and R. J. Linhardt, *Carbohydr. Res.*, 2013, **372**, 30–34.
43. R. J. Linhardt, J. S. Dordick, P. L. Deangelis and J. Liu, *Semin. Thromb. Hemostasis*, 2007, **33**, 453–465.
44. E. J. Toone, E. S. Simon and G. M. Whitesides, *J. Org. Chem.*, 1991, **56**, 5603–5606.
45. W. Guan, L. Cai, J. Fang, B. Wu and P. George Wang, *Chem. Commun.*, 2009, 6976–6978.
46. Y. Chen, V. Thon, Y. Li, H. Yu, L. Ding, K. Lau, J. Qu, L. Hie and X. Chen, *Chem. Commun.*, 2011, **47**, 10815–10817.
47. M. M. Muthana, J. Qu, M. Xue, T. Klyuchnik, A. Siu, Y. Li, L. Zhang, H. Yu, L. Li, P. G. Wang and X. Chen, *Chem. Commun.*, 2015, **51**, 4595–4598.
48. E. Sterner, S. Masuko, G. Li, L. Li, D. E. Green, N. J. Otto, Y. Xu, P. L. DeAngelis, J. Liu, J. S. Dordick and R. J. Linhardt, *J. Biol. Chem.*, 2014, **289**, 9754–9765.
49. Y. Xu, C. Cai, K. Chandarajoti, P.-H. Hsieh, L. Li, T. Q. Pham, E. M. Sparkenbaugh, J. Sheng, N. S. Key, R. Pawlinski, E. N. Harris, R. J. Linhardt and J. Liu, *Nat. Chem. Biol.*, 2014, **10**, 248–250.
50. Y. Chen, Y. Li, H. Yu, G. Sugiarto, V. Thon, J. Hwang, L. Ding, L. Hie and X. Chen, *Angew. Chem., Int. Ed.*, 2013, **52**, 11852–11856.
51. C. Freeman and J. Hopwood, *Adv. Exp. Med. Biol.*, 1992, **313**, 121–134.
52. I. Vlodavsky, N. Ilan, A. Naggi and B. Casu, *Curr. Pharm. Des.*, 2007, **13**, 2057–2073.
53. N. A. Karst and R. J. Linhardt, *Curr. Med. Chem.*, 2003, **10**, 1993–2031.
54. Y. Xu, Z. Wang, R. Liu, A. S. Bridges, X. Huang and J. Liu, *Glycobiology*, 2012, **22**, 96–106.
55. C. Liu, J. Sheng, J. M. Krahn, L. Perera, Y. Xu, P. H. Hsieh, W. Dou, J. Liu and L. C. Pedersen, *J. Biol. Chem.*, 2014, **289**, 13407–13418.
56. D. P. Curran, *Angew. Chem., Int. Ed.*, 1998, **37**, 1174–1196.
57. W. Zhang and D. P. Curran, *Tetrahedron*, 2006, **62**, 11837–11865.
58. F. A. Jaipuri, B. Y. Collet and N. L. Pohl, *Angew. Chem., Int. Ed.*, 2008, **47**, 1707–1710.
59. F. A. Jaipuri and N. L. Pohl, *Org. Biomol. Chem.*, 2008, **6**, 2686–2691.
60. J. Hwang, H. Yu, H. Malekan, G. Sugiarto, Y. Li, J. Qu, V. Nguyen, D. Wu and X. Chen, *Chem. Commun.*, 2014, **50**, 3159–3162.
61. C. Schuerch and J. M. Frechet, *J. Am. Chem. Soc.*, 1971, **93**, 492–496.

62. C. Noti and P. H. Seeberger, *Chem. Biol.*, 2005, **12**, 731–756.
63. J. L. de Paz, C. Noti and P. H. Seeberger, *J. Am. Chem. Soc.*, 2006, **128**, 2766–2767.
64. J. G. Martin, M. Gupta, Y. Xu, S. Akella, J. Liu, J. S. Dordick and R. J. Linhardt, *J. Am. Chem. Soc.*, 2009, **131**, 11041–11048.
65. R. A. Sheldon, *Biochem. Soc. Trans.*, 2007, **35**, 1583–1587.
66. N. R. Mohamad, N. H. C. Marzuki, N. A. Buang, F. Huyop and R. A. Wahab, *Biotechnol. Biotechnol. Equip.*, 2015, **29**, 205–220.
67. S. Hari Krishna, *Biotechnol. Adv.*, 2002, **20**, 239–267.
68. J. Xiong, U. Bhaskar, G. Li, L. Fu, L. Li, F. Zhang, J. S. Dordick and R. J. Linhardt, *J. Biotechnol.*, 2013, **167**, 241–247.
69. Y. Yamazaki and H. Maeda, *Agric. Biol. Chem.*, 1982, **46**, 1571–1581.
70. W. Liu, S. Zhang and P. Wang, *J. Biotechnol.*, 2009, **139**, 102–107.

CHAPTER 9

Synthetic Studies of GPI-Anchored Peptides, Glycopeptides, and Proteins

JIAN GAO[a] AND ZHONGWU GUO*[b]

[a]National Glycoengineering Research Center, Shandong Provincial Key Laboratory of Carbohydrate Chemistry and Glycobiology, Shandong University, 27 Shanda Nan Lu, Jinan 250100, China; [b]Department of Chemistry, University of Florida, 214 Leigh Hall, P.O. Box 117200, Gainesville, FL 32611, USA
*E-mail: zguo@chem.ufl.edu

9.1 Introduction

Glycosylphosphatidylinositols (GPIs) are a group of natural glycolipids, which anchor surface proteins and glycoproteins onto the cell membrane. GPI attachment to proteins is one of the most common posttranslational modifications in eukaryotes.[1–3] Results of genomic analyses have indicated that as many as 0.5–1.0% of all eukaryotic proteins may be attached to the cell surface *via* GPIs.[4,5] GPIs and GPI-anchored proteins or glycoproteins play a pivotal role in various biological processes, such as cell recognition and interaction,[6] signal transduction,[7–9] infection,[10] and cell surface enzymatic reaction,[11] or acting as cell surface markers.[1]

In 1988, Ferguson and coworkers reported the first fully characterized GPI anchor obtained from *Trypanosoma brucei*.[12,13] Since then, more than 50 GPI anchors and over 250 GPI-anchored proteins and glycoproteins have

Chemical Biology No. 2
Chemical Biology of Glycoproteins
Edited by Zhongping Tan and Lai-Xi Wang

Published by the Royal Society of Chemistry, www.rsc.org

been isolated and characterized from various eukaryotic species ranging from yeasts and parasites to humans.[3,4,14] All of these GPI anchors share a highly conserved core structure, *i.e.*, $H_2NEt-(PO_4)$-6-Manα(1→2)Manα(1→6) Manα(1→4)GlcNα(1→6)-*myo*-inositol-1-(PO_4)-glycerolipid (**1**, Figure 9.1; H_2NEt-: ethanolamine; Man: mannosyl; GlcN: glucosaminyl). Moreover, all proteins have their *C*-termini covalently attached to the conserved phosphoethanolamine moiety at the non-reducing end of the GPI core glycan. GPI-anchored molecules are attached to the cell membrane through the insertion of GPI lipid chains into the membrane bilayer. The structural diversity amongst known GPIs primarily results from variations in the lipid chains, as well as additional phosphoethanolamine residues and carbohydrate branches linked to Man I, II, and III.[1]

To study GPIs and GPI-anchored proteins and glycoproteins, it is essential to have access to these biomolecules in homogeneous forms and sufficient quantities, which is difficult to achieve *via* isolation of natural products due to their structural microheterogeneity. In this context, chemical and chemo-enzymatic synthesis of these biomolecules has become a hot topic.[15–17] In 1991, the Ogawa group[18–20] finished the first synthesis of the intact GPI anchor of *T. brucei*. Thereafter, numerous total syntheses of various GPIs and their functionalized derivatives have been reported by several research groups, including Ley,[21,22] Fraser-Reid,[23–26] Martin-Lomas,[27,28] Schmidt,[29–32] Seeberger and Varon Silva,[33–38] Vishwakarma,[39,40] Nikolaev,[41,42] and Guo,[43–45] *etc.* The recent advances in the field of GPI synthesis has been exhaustively reviewed in several articles.[15–17] Therefore, this chapter focuses on the recent

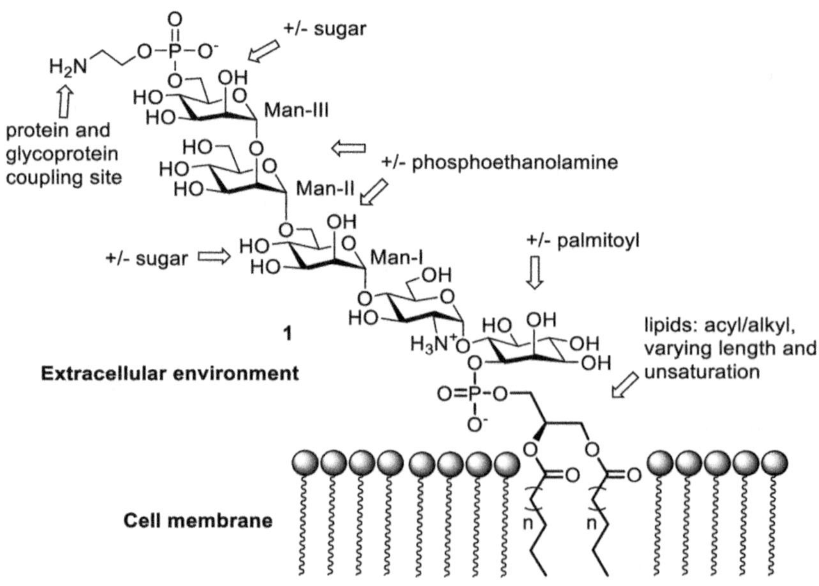

Figure 9.1 The conserved core structure (**1**) of GPI anchors and their anchorage of surface proteins and glycoproteins onto the cell membrane.

developments working towards the synthesis of GPI-anchored peptides, glycopeptides, proteins, and glycoproteins, as well as other related topics, such as the biosynthesis of GPI-anchored proteins and the study of GPI-anchored proteomics.

9.2 Biosynthesis of GPIs and GPI-Anchored Proteins

The biosynthetic pathways of GPIs and GPI-anchored proteins have been largely elucidated,[46,47] although several issues remain to be clarified. Overall, GPI anchors are preassembled in the endoplasmic reticulum (ER) and then linked to target proteins by GPI transamidase (GPI-T) as a type of posttranslational modification. Subsequently, the inositol is deacylated and the fatty acids are then remodeled in the Golgi apparatus, before the resultant GPI-anchored proteins are delivered onto the plasma membrane through the Golgi network.[3,4,6]

9.2.1 Biosynthesis of GPI Anchors

As shown in Figure 9.2, an early intermediate of GPI biosynthesis is glucosaminyl–phosphatidylinositol (GlcN–PI), which is generated in two steps on the cytoplasmic side of the ER membrane. The first step is transfer of *N*-acetyl-α-D-glucosamine (GlcNAc) from UDP–GlcNAc to the endogenous phosphatidylinositol, forming GlcNAc–PI, catalyzed by an enzyme complex consisting of at least seven components, including PIG-A, PIG-C, PIG-H, PIG-P, PIG-Q, PIG-Y, and DPM2 (step 1).[48–50] The newly formed intermediate, GlcNAc–PI, is deacetylated by PIG-L to yield GlcN–PI (step 2).[51,52] Next, GlcN–PI is flipped across the ER membrane to the luminal side (step 3) to finish the remaining steps of the biosynthesis. This translocation process is believed to be mediated by a "flippase", which has not been identified yet.[53] Thereafter, the inositol 2-*O*-position in GlcN–PI is palmitoylated by an acyltransferase, PIG-W, in the presence of acyl-CoA, to produce GlcN–PI(acyl) (step 4).[54,55] The inositol 2-*O*-palmitoylation is required for GPI biosynthesis, but in most cases this acyl group is removed after GPI biosynthesis is finished. Lipid remodeling of GlcN–PI(acyl) then converts the diacyl PI to other diacyl PI or 1-alkyl-2-acyl PI or a mixture of the two (step 5).[56] However, the details about the remodeling reaction and enzymes involved are yet to be clarified.

Subsequently, Man I and II residues are consecutively installed to the remodeled GlcN–PI(acyl). Man I is transferred to GlcN–PI(acyl) from dolichol-phosphate–Man (Dol-P–Man), catalyzed by GPI mannosyltransferase 1 consisting of the catalytic unit PIG-M and the stabilizing unit PIG-X, to give Man–GlcN–PI(acyl) (step 6).[57,58] Man II is also transferred from Dol-P–Man, but the involved enzyme is GPI mannosyltransferase 2, PIG-V, resulting in Man–Man–GlcN–PI(acyl) (step 7).[59] In mammalian cells, the Man-I 2-*O*-position is then modified with a phosphoethanolamine (EtNP) functionality, catalyzed by GPI–EtNP transferase 1, PIG-N, using phosphatidylethanolamine

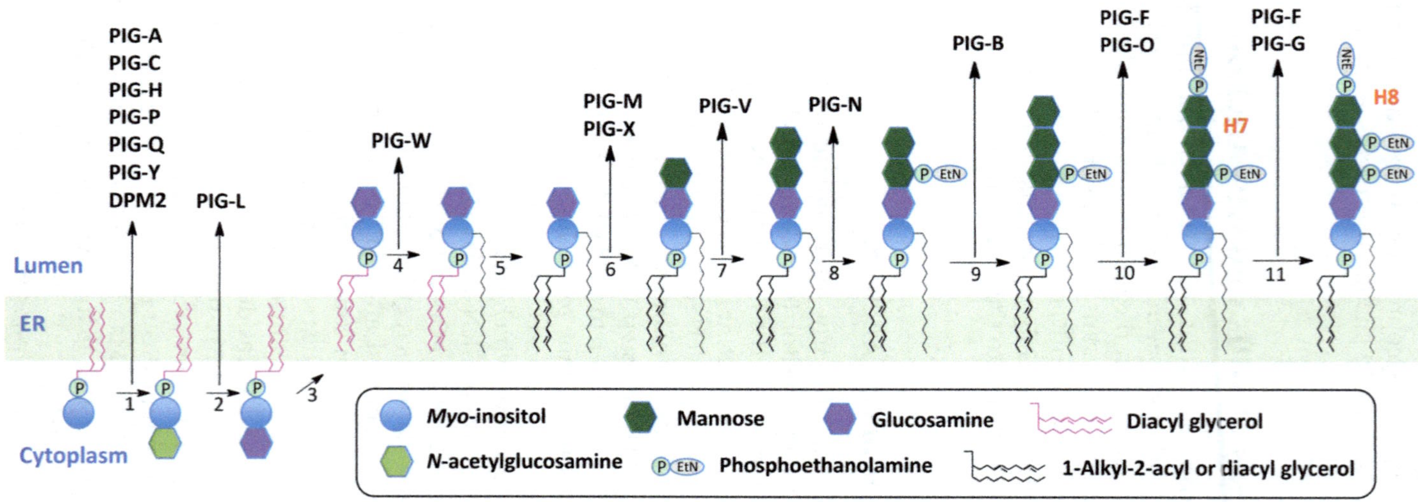

Figure 9.2 Biosynthesis of GPI anchor is achieved in the ER. More than 20 genes are involved in this complex pathway, which generates the mature GPI anchors, H7 and H8.

as the substrate donor, producing Man–(EtNP)Man–GlcN–PI(acyl) (step 8).[60] Man III attachment is mediated by PIG-B, that is, GPI mannosyltransferase 3, generating Man–Man–Man(EtNP)–GlcN-PI(acyl) (step 9).[61] It is noteworthy that the addition of the three Man units requires three distinct mannosyltransferases, PIG-M, PIG-V, and PIG-B, and PIG-M also needs PIG-X as an essential counterpart. These three mannosyltransferases use the same substrate donor Dol-P–Man, which is biosynthesized on the cytoplasmic side of the ER membrane by an enzyme complex consisting of DPM1, DPM2, and DPM3 and then translocated into the luminal side of the ER by the enzyme SL15.[62,63]

The attachment of the "bridging" EtNP moiety, the linker between GPI anchor and protein, to the 6-*O*-position of Man III is achieved by GPI-EtNP transferase 2 consisting of PIG-O, the catalytic subunit, and PIG-F, the stabilizing subunit, using phosphatidylethanolamine as the substrate (step 10).[64–66] The resultant EtNP–Man–Man–Man(EtNP)–GlcN–PI(acyl), *i.e.*, H7, is one of the mature GPI anchor precursors for protein attachment. However, H7 is often modified with an EtNP unit at the Man II 6-*O*-position, resulting in EtNP–Man–Man(EtNP)–Man(EtNP)–GlcN–PI(acyl), namely H8 (step 11), which is another mature GPI anchor precursor for protein attachment. The latter transformation is occasionally observed in mammalian cells, catalyzed by GPI-EtNP transferase 3 that is composed of PIG-G, the catalytic subunit, and PIG-F, the stabilizing subunit.[67,68] H8 is sometimes further modified with Man IV. In yeast, Man IV transfer is catalyzed by GPI mannosyltransferase 4, PIG-Z, and this reaction occurs before the attachment of the "bridging" EtNP.[69,70] However, as for the mammalian system, the exact step at which Man IV is attached has not been clarified yet.[46]

It is worth noting that the Man I 4-*O*-position can be modified by a β-1,4-*N*-acetylgalactosamine (β-1,4-GalNAc), which was found in rat brain Thy-1, human erythrocyte CD59, and porcine and human renal membrane dipeptidase. This GalNAc side chain may be further elongated with β-1,3-Gal and/or sialic acid (Neu5Ac) residues.[13,71,72] Interestingly, the modifications have never been found in GPI precursors, indicating that they may happen after attachment of GPI to proteins. However, the genes involved in generating the Neu5Ac–Gal-β3–GalNAcβ4 side chain and its biological functions have not been identified.[73]

9.2.2 Posttranslational Attachment of GPIs to Proteins

After mature GPI anchors are assembled, they are attached to target proteins on the luminal side of the ER membrane as well (Figure 9.3). The nascent proteins destined for GPI anchorage contain an *N*-terminal signal sequence to facilitate ER translocation and a *C*-terminal signal sequence for GPI attachment.[74] Therefore, after these proteins are synthesized, they are translocated and attached to the luminal side of the ER membrane, where the *N*-terminal signal is removed and GPI is attached. For GPI attachment, the *C*-terminal signal sequence is recognized by an enzyme known as

GPI-T residing in the ER membrane, cleaved at the ω amino acid by GPI-T to form a reactive protein–GPI-T complex, and eventually link the protein *C*-terminus to the amine group of the "bridging" EtNP in the pre-assembled GPI anchor.[2,11] Although the nascent proteins are structurally diverse, their GPI attachment signals share some common characteristics. For example, (i) GPI anchor is attached to the ω-amino acid of the signal peptides, and the ω-amino acids are usually those with small side chains, such as Asn, Asp, Gly, Ala, and Cys; (ii) $\omega + 2$ amino acids, such as Gly, Ala, and Ser, also contain small side chains; (iii) five to ten hydrophilic amino acids starting from the $\omega + 3$ site are present in the signal peptide; (iv) the hydrophilic sequence is followed by a *C*-terminal hydrophobic sequence containing 15–20 amino acids, long enough to span the ER membrane.[75,76] After proteins are attached to GPI anchors, the resultant GPI-anchored proteins depart from the ER and are transported to the plasma membrane *via* the Golgi apparatus, where inositol deacylation and fatty acid remodeling are accomplished with the proteins post-GPI attachment to proteins 3 (PGAP3) and PGAP2, respectively.[77–79]

GPI-T, which catalyzes GPI attachment to proteins, is a complex membrane-bound multi-subunit enzyme consisting of PIG-K (also known as GPI8), glycosylphosphatidylinositol anchor attachment 1 (GPAA1), PIG-S, PIG-T and PIG-U.[62,80–83] PIG-K is the catalytic subunit, which has sequence homology to members of the cysteine protease family. It cleaves the peptide bond between the ω and $\omega + 1$ amino acids to generate a substrate–enzyme complex, in which the ω-site amino acid is attached to the catalytic cysteine unit *via* a thioester bond (Figure 9.3).[81,84] Next, GPAA1 presents the pre-assembled GPI anchor with its conserved phosphoethanolamine group as a nucleophile to attack at the substrate–enzyme complex and complete the transamidation step.[81,85] The PIG-T unit, which is coupled with PIG-K *via* a disulfide bond, probably stabilizes the enzyme complex.[86] The functions of PIG-S and PIG-U are not clarified yet, but they are also essential for GPI-T.[47]

9.3 Synthesis of GPI-Anchored Proteins and Glycoproteins

All naturally occurring GPI-anchored proteins and glycoproteins identified so far have their *C*-terminus linked to the conserved phosphoethanolamine moiety at the glycan non-reducing end of the GPI core structure. Therefore, a potentially generally applicable synthetic strategy for GPI-anchored proteins and glycoproteins is to prepare GPIs and proteins or glycoproteins separately and then stitch them together by a site-specific conjugation reaction. Along this line, several synthetic methods have been developed to access structurally well-defined GPI-anchored peptides, glycopeptides, and proteins by means of regioselective chemical condensation, native chemical ligation (NCL), and enzymatic ligation.

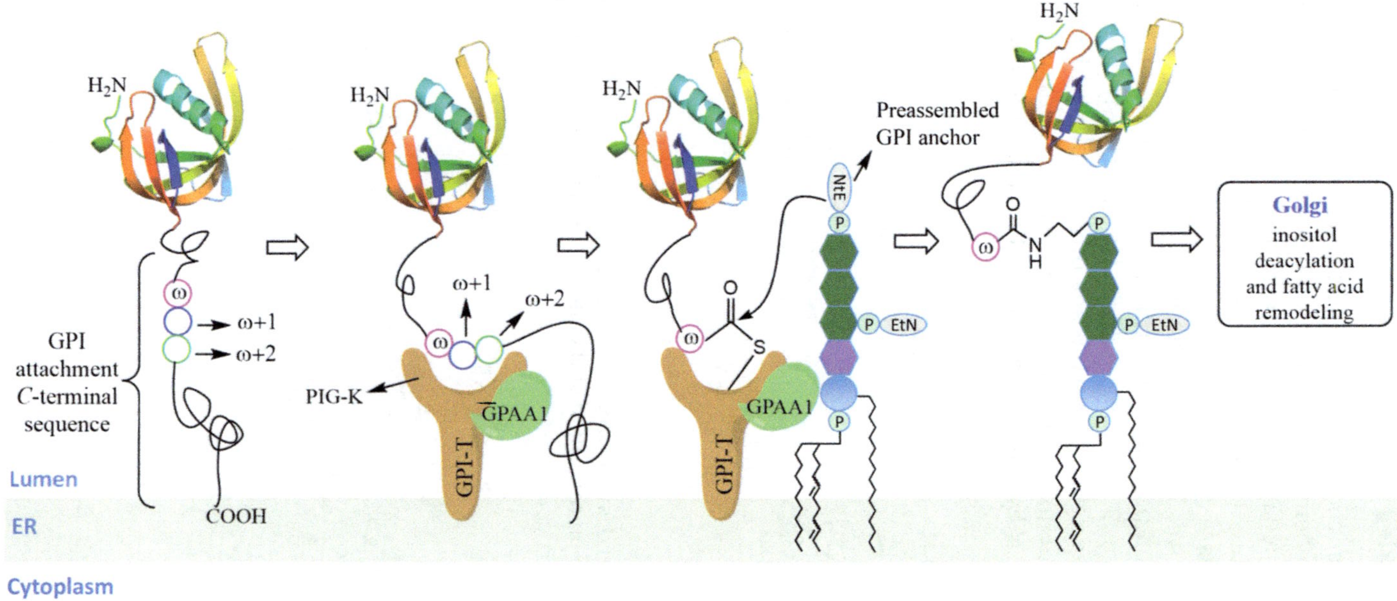

Figure 9.3 GPI attachment to proteins occurs in the ER lumen and is catalyzed by GPI-T. The resultant GPI-anchored proteins are then transported to the plasma membrane *via* the Golgi apparatus, where they undergo inositol deacylation and fatty acid remodeling.

9.3.1 Chemical Total Synthesis of GPI-Anchored Peptides and Glycopeptides

The synthesis of GPI-anchored peptides and glycopeptides based on conventional, regioselective chemical condensation between GPIs and peptides or glycopeptides is outlined in Scheme 9.1. In this regard, the GPI and peptide/glycopeptide substrates should be properly protected to ensure regioselective coupling reactions.

The above strategy was first demonstrated by Guo and coworkers in their synthesis of GPI-linked peptide **2**, a partial structure of the human sperm CD52 antigen (Scheme 9.2).[87] Separately prepared properly-protected GPI analog **3** carrying a free amino group was smoothly coupled with dipeptide **4** having a free *C*-terminus using 1-ethyl-3-(3-dimethylaminopropyl)carbodiimide (EDC·HCl) and 1-*N*-hydroxybenzotriazole (HOBt) as the condensation

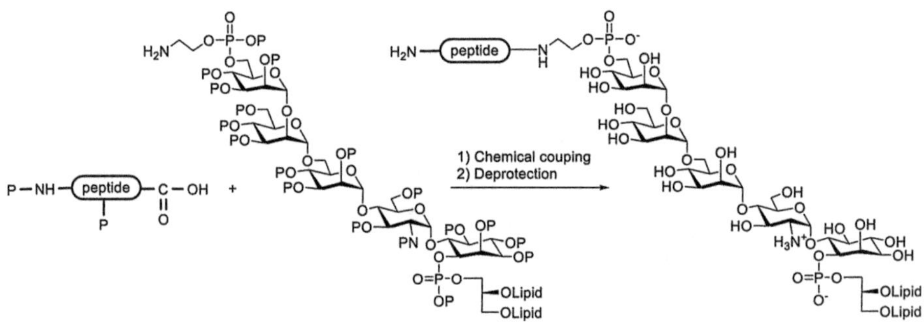

Scheme 9.1 The general strategy for GPI-anchored peptide or glycopeptide synthesis based on conventional chemical condensation reactions.

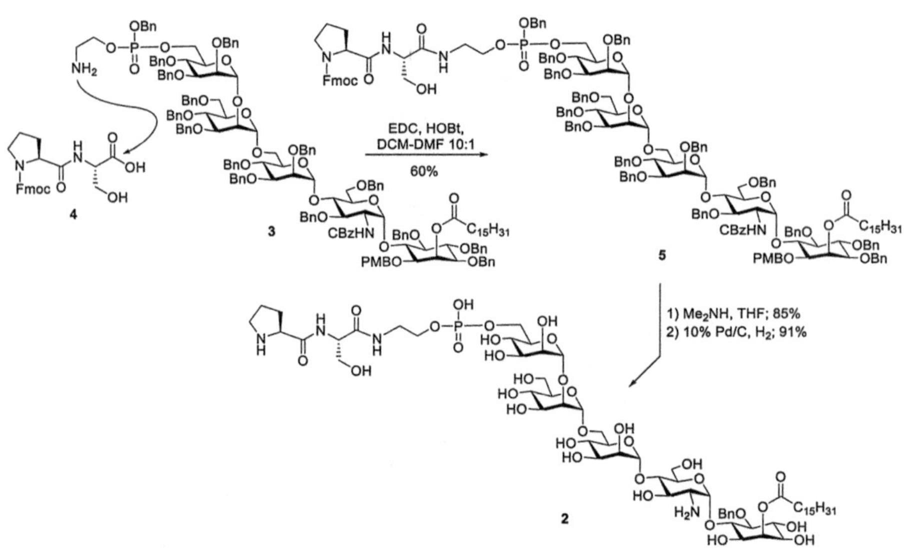

Scheme 9.2 Synthesis of GPI-anchored peptide **2**.

reagents. This reaction was proved to be racemization-free to give fully protected **5** with a 60% yield, which was globally deprotected in two steps, including removal of the *N*-fluorenylmethyloxycarbonyl (*N*-Fmoc) group with dimethylamine and the carboxybenzyl and benzyl groups *via* Pd-catalyzed hydrogenolysis, to produce the target molecule **2**.

The strategy was also used to successfully synthesize a skeleton structure of sperm CD52 antigen **6** (Scheme 9.3), carrying the entire peptide sequence

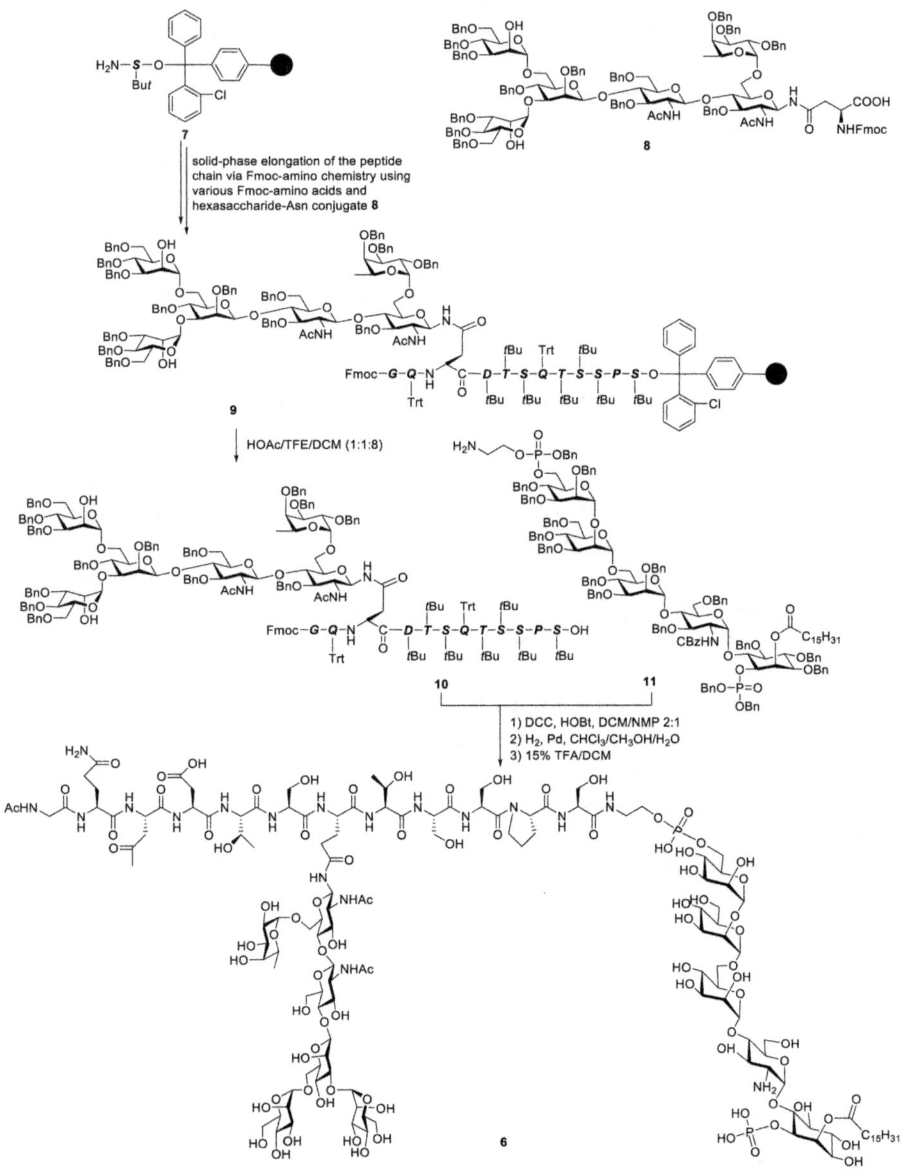

Scheme 9.3 Synthesis of GPI-anchored glycopeptide **6**.

of the antigen and an *N*-linked fucosylated hexasaccharide.[88] This work represented the first chemical synthesis of a natively linked GPI-anchored glycopeptide. One of the key building blocks used in this synthesis was protected glycopeptide **10**, which was assembled from commercial Fmoc-amino acids and the synthetic hexasaccharide–Asn conjugate **8** by automatic solid-phase peptide synthesis (SPPS) employing acid-sensitive 2-chlorotrityl resin as the solid-phase support. The desired glycopeptide **10** was retrieved from the resin by 10% acetic acid in trifluoroethanol and dichloromethane (DCM) (1 : 8), which did not affect the acid-sensitive fucosyl linkage and protections on the peptide and glycan. The coupling reaction between **10** and the properly protected GPI **11** was accomplished in DCM and *N*-methylpyrrolidinone (NMP) (2 : 1) using EDC and HOBt as the condensation reagents, which was smooth and did not result in observable racemization. Finally, all of the protecting groups were removed in two steps, including Pd-catalyzed hydrogenolysis of benzyl ethers and acidic cleavage of amino acid side chain protections with 15% trifluoroacetic acid (TFA) in DCM, to give the target molecule **6**. It is noteworthy that the deprotection sequence was important because the glycosidic linkage of free fucose was stable to diluted TFA but not that of the benzylated fucose.[88]

9.3.2 Synthesis of GPI-Anchored Peptides and Proteins *via* NCL

For the synthesis of large GPI-anchored peptides and proteins, NCL, a powerful strategy widely utilized in protein synthesis,[89–91] can be very useful, because NCL can regiospecifically ligate a thioester with any molecule carrying a terminal cysteine residue under mild reaction conditions. As depicted in Scheme 9.4, the key building blocks for NCL are a peptide-α-thioester **12** and a peptide **13** having an *N*-terminal cysteine residue. The side chain thiol group in **13** can attack the thioester **12** reversibly in a chemoselective and regioselective manner *via* transthioesterification to generate the thioester intermediate **14**. Subsequently, **14** can rearrange spontaneously *via* a favorable five-membered ring intramolecular nucleophilic attack by the cysteine α-amino group to produce thermodynamically stable **15**, thus having the two peptide fragments coupled together *via* a native peptide bond.

Nakahara and coworkers[92] explored NCL for GPI-anchored peptide synthesis first (Scheme 9.5). A cysteine-containing GPI analog **19** was obtained by coupling **17** with the pentafluorophenyl ester of protected cysteine **18** in the presence of HOBt, followed by removal of the *N*-Fmoc group. Thereafter, **19** was successfully coupled with a tridecapeptide thioester in the presence of thiophenol at room temperature in a phosphate buffer (pH 8.0) to produce the target molecule **16** with a 62% yield.

The Bertozzi group[93] employed NCL to ligate a full-size protein, green fluorescent protein (GFP), with GPI derivatives bearing modified glycan units to prepare a series of GPI–GFP conjugates, as exampled by the preparation of **20** (Scheme 9.6). The expressed protein ligation (EPL) technology[94,95] was used to prepare GFP as an intein fusion, which was then converted into a

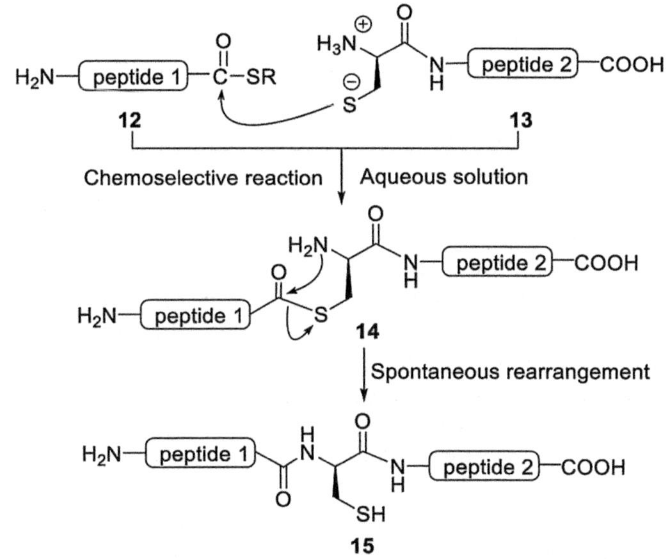

Scheme 9.4 The principle of NCL for protein synthesis.

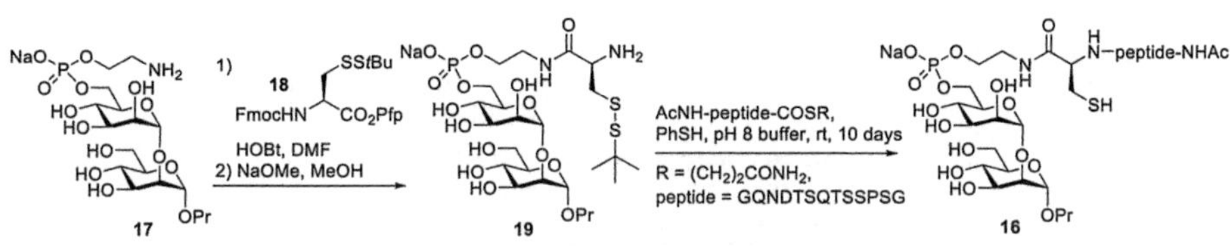

Scheme 9.5 Synthesis of GPI-anchored peptide **16** *via* NCL.

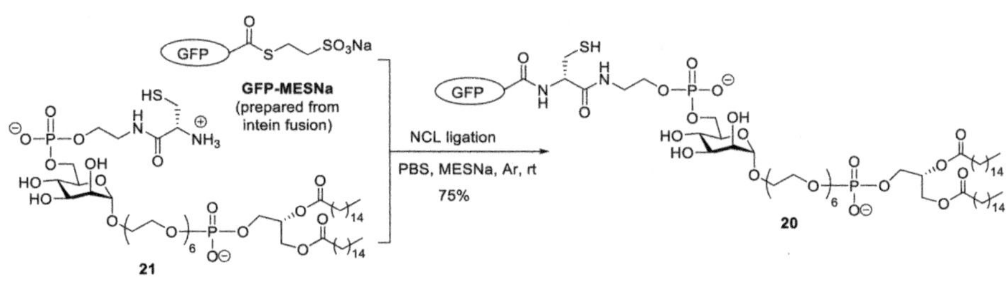

Scheme 9.6 Synthesis of GPI–GFP conjugate **20** through NCL.

C-terminal thioester (GFP–MESNa; MESNa = 2-mercaptoethane sulfonate) through a transthioesterification reaction. The attachment of GFP–MESNa to cysteine-bearing GPI analog **21** by NCL was smooth and produced the GPI–GFP conjugate **20** with a 75% yield. Next, they employed supported lipid bilayers as models of cell membrane together with the synthesized GPI–GFP conjugates to analyze the influence of various glycan residues on the behavior of GPI-proteins on membranes. The results indicated that the GPI–GFP conjugate containing two mannose units diffused more rapidly in the supported lipid bilayers than the GPI–GFP conjugates containing one or no mannose residues. It was concluded that the GPI anchor may be more than a simple membrane-anchoring device and may serve to prevent transient interactions between the attached protein and the lipid bilayer, thus permitting rapid diffusion in the bilayer.

NCL was also explored by Seeberger and coworkers[96] in the synthesis of GPI-anchored protein **22**, which had the recombinant prion protein (rPrP) attached to a GPI anchor (Scheme 9.7). The cysteine-tagged GPI anchor **24** was synthesized by hydrogenolysis of **23** with Pd/C and formic acid in methanol to deprotect benzyl ethers and removal of the acid-labile *tert*-butyloxycarbonyl (Boc) and *tert*-butyl (*t*Bu) groups with TFA and $Hg(OTFA)_2$. The ligation between **24** and the *C*-terminal MESNa ester of rPrP, which was also prepared using EPL technology, proceeded smoothly in 6 M guanidine hydrochloride at pH 7.8 in the presence of 1% (v/v) thiophenol to give GPI-anchored rPrP **22** with a 50% yield.

Next, Seeberger and coworkers[96] investigated the influence of a GPI anchor on rPrP association with the membrane using the synthetic GPI-anchored rPrP **22**. Accordingly, **22** was incorporated into small unilamellar

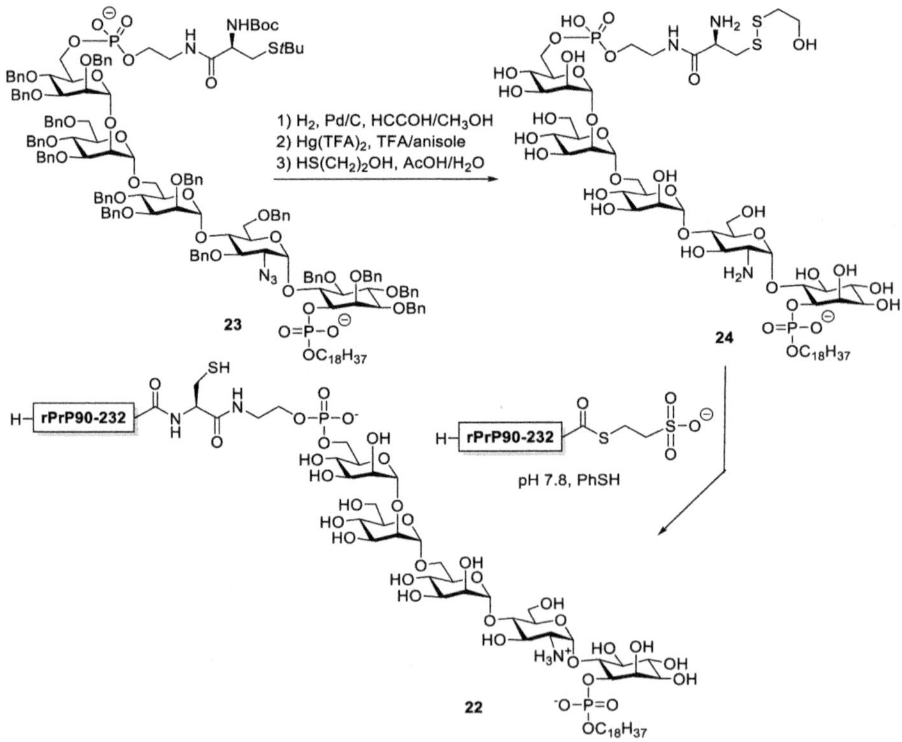

Scheme 9.7 Synthesis of GPI-anchored rPrP **22** bearing a cysteine unit *via* NCL.

1,2-dioleoyl-*sn*-glycero-3-phosphocholine (DOPC) vesicles (SUVs) for a vesicle-spin-down study. The experiment revealed that **22** could be attached to the vesicles quantitatively, even though the GPI anchor in **22** contained only one C_{18} lipid chain, while rPrP without a GPI anchor could not be affixed. This study should be very useful for elucidating the influence of GPI-mediated membrane association on the conversion of PrP into its pathogenic isoform PrP scrapie (PrP^{Sc}).

Additionally, Morris and coworkers[97] found that the GPI anchor could significantly affect the conformation of Thy-1 protein, which contains a single variable-type immunoglobulin domain attached to the plasma membrane external surface *via* a GPI anchor. Upon delipidation of the GPI anchor by phospholipase C or D, the reactivity of Thy-1 with various antibodies, including those known to be determined by amino acid residues, is spoiled. Results of molecular dynamics studies implied that the conformational change of the GPI glycan chain was induced by delipidation and subsequently affected the protein in the region of the antigenic epitopes.

9.3.3 Synthesis of GPI-Anchored Peptides, Glycopeptides, and Proteins *via* Enzymatic Ligation

Guo and coworkers[98–102] have recently developed a biomimetic strategy for the synthesis of GPI-anchored peptides, glycopeptides, and proteins based on coupling peptides, glycopeptides, and proteins with GPIs *via* enzyme-catalyzed ligation. However, instead of GPI-T, which should be the ideal enzyme for this application but is too complex (refer to the above discussion), sortase A (SrtA), a simple bacterial transpeptidase derived from *Staphylococcus aureus*, was utilized in the strategy. Sortases are enzymes used by bacteria to anchor surface proteins to the bacterial cell walls. Similar to GPI-T, SrtA also recognizes a protein *C*-terminal signal, a pentapeptide LP*X*TG (*X* is variable) known as the sorting signal, reacts with this signal peptide to generate a thioester with the carboxyl group of T, and then links the protein *C*-terminus to the pentaglycine *N*-terminus of the cell wall peptidoglycan.[103–105]

To probe the application of SrtA to GPI-anchored protein synthesis, its water-soluble extracellular segment was expressed and tested for the catalysis of the coupling between the short peptide **25** and the simple GPI analog **26** (Scheme 9.8).[98] This reaction gave a hydrolyzed peptide **27**, instead of the coupling product, suggesting that GPI itself may not be a good SrtA substrate. However, after one or two glycine residues were introduced into GPI, the resulting GPI analogs **28** and **29** were readily coupled with peptide **25** by SrtA to give conjugates **30** and **31** with excellent yields (>95%). Clearly, glycine-modified GPI analogs can be accepted by SrtA.

Structure–reactivity relationship studies of substrate peptides revealed that both **32** and **33** were accepted by SrtA for coupling with **28** to form conjugates **35** and **36** with good yields, but not peptide **34** (Scheme 9.9).[99] The results indicated that the sorting signal in substrate peptides or proteins should not be directly exposed at the peptide *C*-terminus. It was further proved that SrtA could accept more complex GPI analogs such as **37** for coupling with peptides (Scheme 9.9).

SrtA was then utilized to conjugate polypeptides **39** and **40** with GPI derivatives **41** and **42** to obtain several analogs **43–46** of the GPI-anchored human antigens CD52 and CD24 that carried small proteins at 45%, 48%, 69%, and 51% yields, respectively (Scheme 9.10).[100] It is noteworthy that **43–46** were still unnatural analogs of the human CD52 and CD24 antigens since they contain the sorting signal at their polypeptide *C*-terminus.

SrtA was also shown to accept sterically hindered glycopeptides, such as **47**, as substrates for ligation with GPI derivatives to produce GPI-anchored glycopeptides **48** and **49** (Scheme 9.11).[101] The glycopeptide substrate **47** contained a partial sequence of MUC1. Moreover, after *N*-terminal deprotection to expose the glycine residue in **48**, its glycopeptide chain was further elongated by SrtA-catalyzed ligation with **47** to produce the GPI-anchored mucin product **50** with a 59% yield, indicating that SrtA can deal with sterically hindered peptide substrates.

Eventually, SrtA was employed to achieve the attachment of a large protein, GFP **51**, to a GPI **52** to produce GPI-anchored GFP **53** (Scheme 9.12).

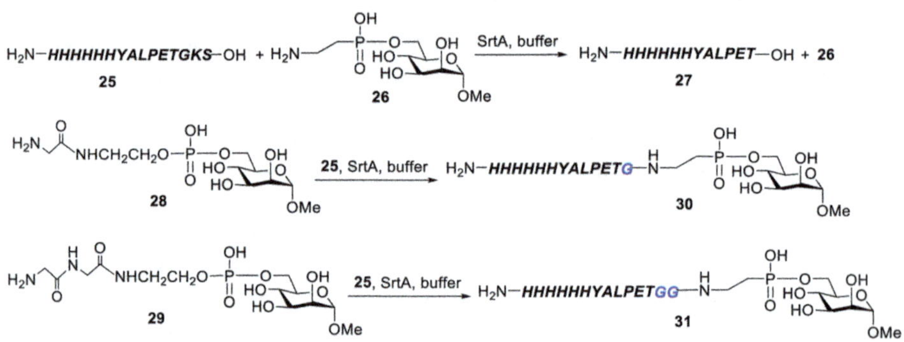

Scheme 9.8 SrtA-catalyzed coupling of simple peptides and GPI analogs.

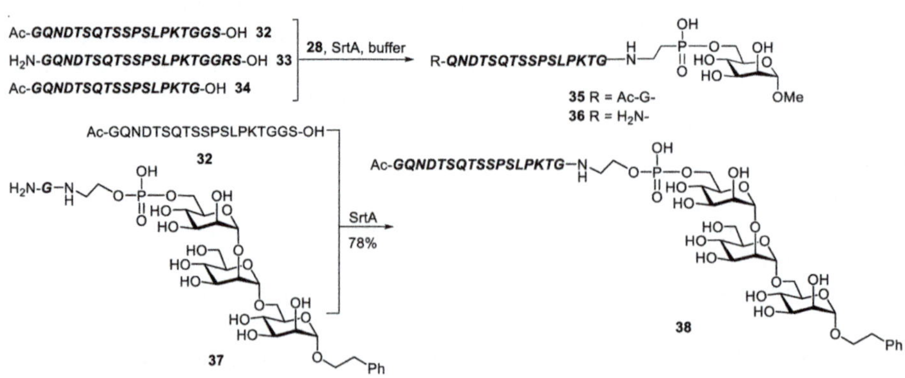

Scheme 9.9 SrtA-catalyzed coupling of more complex peptides and GPI analogs.

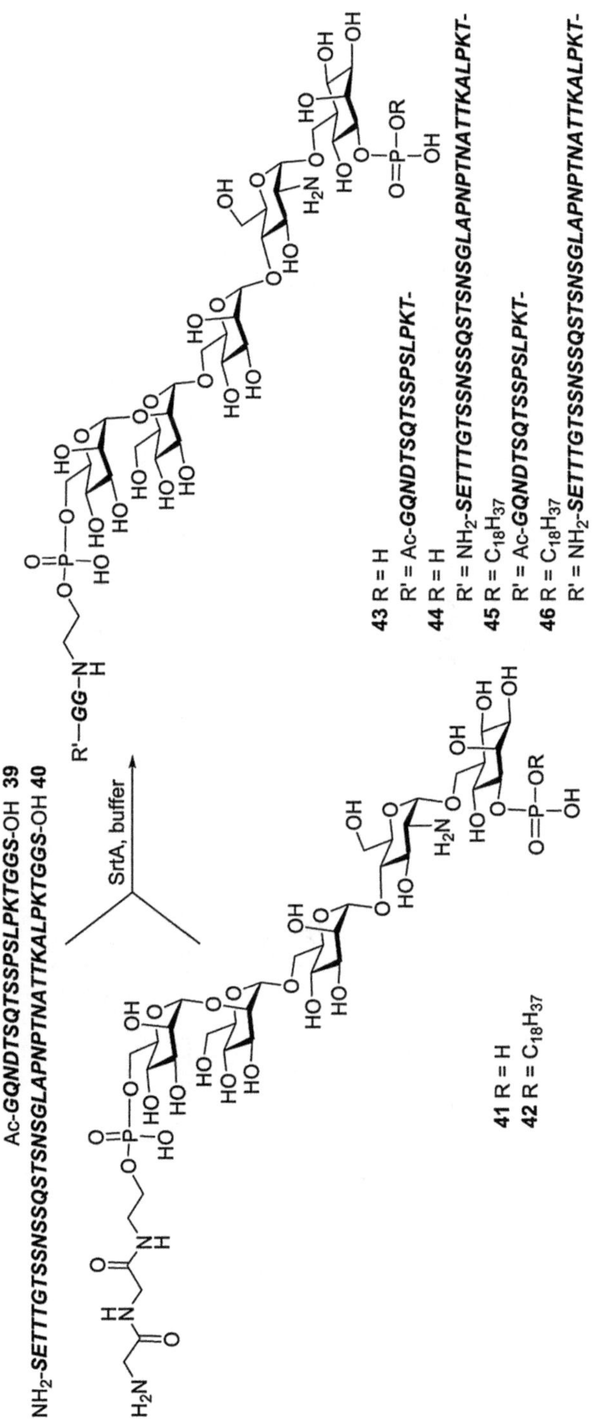

Scheme 9.10 SrtA-mediated synthesis of human CD52 and CD24 antigen analogs.

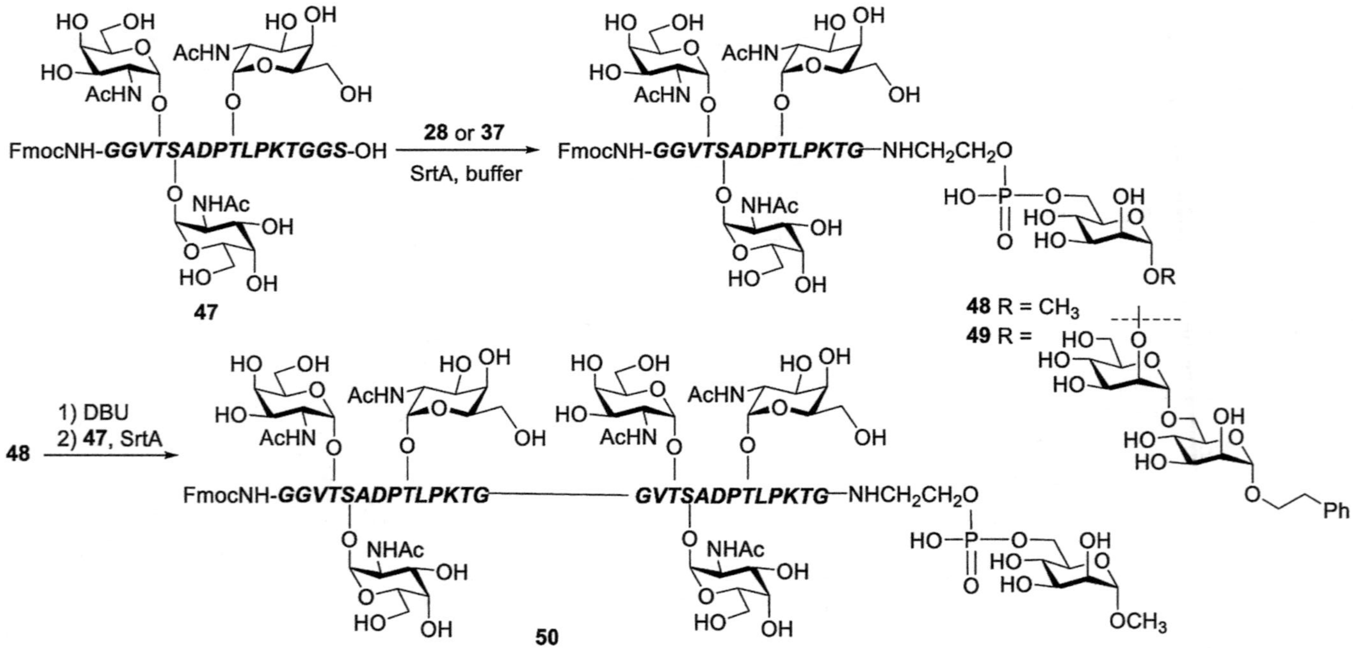

Scheme 9.11 SrtA-catalyzed synthesis of GPI-anchored glycopeptides **48–50**.

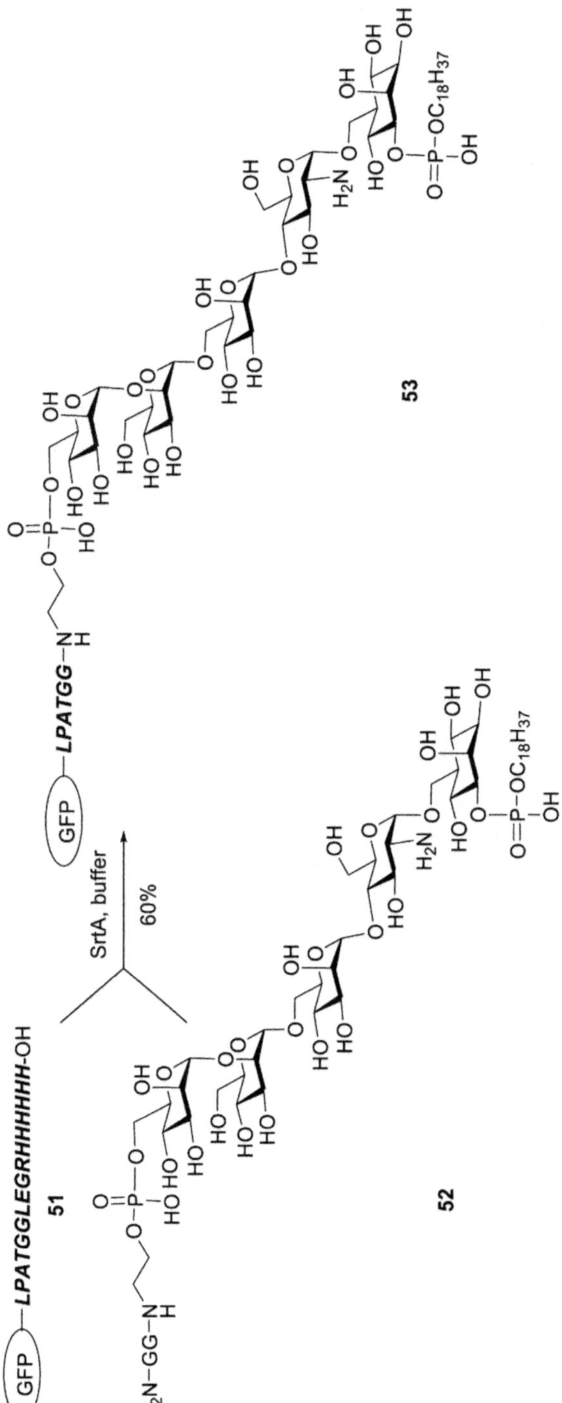

Scheme 9.12 SrtA-mediated synthesis of GPI-anchored GFP 53.

This ligation was performed under conventional reaction conditions to give a 60% yield.[102] The resultant product **53** with a full-sized fluorescent protein, instead of small fluorescent tags, could be a very useful tool for various biological studies of GPIs and GPI-anchored proteins.

9.4 GPI-Anchored Proteomics Studies

GPI-anchored proteomics is an emerging field. Its analysis can help identify novel GPI-anchored proteins, study their biological functions, and discover unique GPI-anchored molecular markers related to diseases. For example, it has been noticed that breast carcinoma has increased levels of some GPI-T subunits, such as GPAA1 and PIG-T, which are related to tumorigenicity *in vitro* and *in vivo*.[106] To identify the elevated GPI-anchored proteins in breast carcinoma and understand their functions, Abbott and coworkers[107] have developed a strategy to isolate GPI-anchored protein for proteomics analysis. Biotinylated alpha toxin derived from *Clostridium septicum*, which is a member of aerolysin-like pore-forming toxins that can selectively bind GPI anchors, was utilized to capture GPI-anchored proteins in the total membrane proteins extracted from cells or tissues. The alpha-toxin-bound proteins were then purified with Streptavidin magnetic beads, followed by nano-electrospray ionization-tandem mass spectrometry (ESI-MS/MS) analysis to analyze GPI-anchored proteins quantitatively. Comparative membrane proteomics studies have led to the discovery of several new molecular targets, such as fermitin family member 3 (FERMT3) and filamin A (FLNA), potentially useful for breast cancer diagnosis and therapy.[107] In addition, this strategy was also employed to find elevated GPI-anchored proteins in the plasma induced by other human tumors such as the ovarian, kidney, liver, lung, colon, and brain cancer.[108]

Guo and coworkers[109] have recently developed another versatile strategy that can be utilized to study GPI-anchored proteomics, and this strategy is based on metabolic engineering of the cell surface GPI anchor. As outlined in Figure 9.4, upon incubation with an inositol derivative bearing a chemically reactive bioorthogonal azido group, which should be still acceptable by enzymes involved in GPI biosynthesis, the unnatural inositol is taken up by the cell and incorporated into GPI anchors and GPI-anchored proteins. The labeled GPIs and GPI-anchored proteins on the cell surface can be further elaborated through bioorthogonal click reaction at the azido group for the attachment of various molecular tags, such as biotin as an affinity tag to facilitate the pull-down of GPI-anchored proteins for proteomics analysis or fluorescent molecules to make cell-surface GPIs and GPI-anchored proteins visible for various biological studies.

After studying a series of azide-modified inositol derivatives, Guo and coworkers[109] found that compounds **54–56** (Figure 9.4), especially **54**, could be effectively incorporated by *Saccharomyces cerevisiae* yeast cells and various human cancer cell lines, such as lung cancer cell A549, breast adenocarcinoma cell MCF-7, cervix epitheloid carcinoma cell Hela, melanoma cell

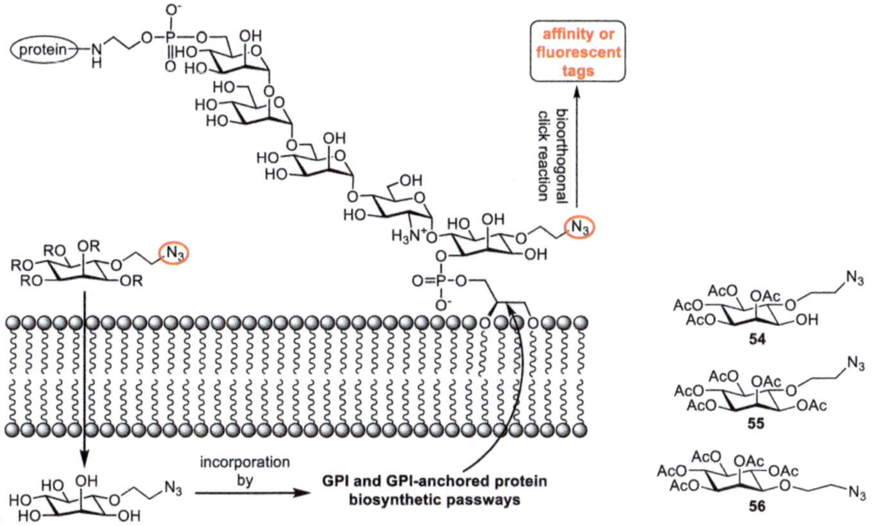

Figure 9.4 Labeling of cell-surface GPIs and GPI-anchored proteins *via* metabolic engineering of GPIs.

SKM28, and leukemia cell K562, to express azide-labeled GPI anchors and GPI-anchored proteins on the cell surface. The labeled GPIs on cells were further elaborated *via* a CuI-catalyzed click reaction to attach a biotin molecule, which enabled cell staining with streptavidin–allophycocyanin (APC), a fluorescent conjugate, for flow cytometry studies. Moreover, images of these cells had revealed that the fluorophores were concentrated on the cell surface. In the meantime, the azide-labeled GPI-anchored proteins expressed by metabolically engineered cancer cell A549 were released from the cell surface on treatment with phosphatidylinositol-specific phospholipase C (PI-PLC) and then subjected to click reaction to attach a biotin molecule. These biotinylated GPI-anchored proteins were probed by western blot using streptavidin-conjugated peroxidase to discover a series of unique surface proteins. These results indicated that this technique can be used to pull down GPI-anchored proteins for proteomics analysis.

Mootz and coworkers[110] also achieved the modification of GPI-anchored proteins on living cells by means of protein *trans*-splicing, which is a post-translational reaction and an excellent method for labeling proteins site-specifically. In this process, the internal polypeptide sequences, known as the split intein fragments IntN and IntC, excise themselves out through joining their flanking sequences by a peptide bond. The naturally occurring *Npu* DnaE intein, a split intein, has by far the highest rate constant in the protein *trans*-splicing reaction. The IntC fragment of *Npu* DnaE intein was fused with the GPI signal sequence Gas1p from yeast and subsequently expressed on the cell surface of mouse neuro-2a cells. Using the *E. coli* expression system, the GFP was fused as the *N*-terminal extein to the IntN fragment.

The resultant fusion protein was purified and added to the media of the mouse neuro-2a cells. Protein *trans*-splicing on the cells would enable the transfer of GFP that was fused as the *N*-extein to the *Npu* DnaE on the GPI anchor that deposited in the plasma membrane.

9.5 Concluding Remarks

In the past two and half decades, studies on the biosynthetic pathways of GPI anchors and GPI-anchored proteins have made great progresses. However, detailed investigation of GPI-anchored proteins has been hindered by the difficulty of accessing these biomolecules in homogeneous and structurally defined forms and sufficient quantity. To overcome the problem, a number of groups have been working on the development of methods for GPI-anchored-protein synthesis.

Total synthesis of GPI conjugates through traditional regioselective coupling reactions between properly protected GPIs and peptides or glycopeptides has been proved feasible.[87,88] One of the main advantages of this technique is the flexibility to control final product structures and obtain truly natural products. However, this strategy is probably only applicable to short peptides and glycopeptides but not to large proteins or glycoproteins, as it is almost impossible to selectively protect large proteins or glycoproteins to only leave the *C*-terminus free. To address this issue, NCL have been explored, which can couple unprotected GPIs with free peptides and proteins in aqueous solution. Whilst this can be developed into a very powerful strategy for the synthesis of GPI-anchored peptides, proteins, and glycoproteins, it also has some drawbacks. For instance, the preparation of required protein *C*-terminal thioesters and cysteine-bearing GPIs is not trivial, and more importantly, all final products contain a cysteine reside between the GPI and the polypeptide chain, which may affect their properties.[92,93,96]

Another powerful and distinctive strategy for GPI-anchored conjugate synthesis is based on the SrtA-catalyzed ligation reactions between GPI derivatives and peptides, glycopeptides, and proteins.[98–102] This strategy turns out to be highly flexible and widely applicable because any GPI bearing one or two Gly residues on its conserved phosphoethanolamine moiety and any peptide, glycopeptide or protein bearing the sorting signal at its *C*-terminus can be accepted by SrtA as substrates to undertake regioselective ligation. However, an issue for this strategy is that it produces unnatural products because, similar to the NCL strategy, its final products contain an artificial peptide, the sorting signal, between the GPI and the polypeptide chain.

In conclusion, in the past few years, several strategies have been developed for the synthesis of structurally defined GPI-anchored peptides, glycopeptides, and proteins, which has significantly promoted the study and understanding of GPI anchorage to the cell membrane. However, all of these methods, except for chemical total synthesis, produce unnatural GPI-anchored conjugates. Even though these conjugates are useful as analogs of natural GPI-anchored proteins for various biological studies, the effects of the artificial amino acids or peptides in these products on their biological functions are unclear. Therefore, currently a very important and urgent issue in the field is to develop practical methods for the synthesis of truly natural GPI-anchored proteins, which is critical for the understanding of the functions of these biomolecules and gaining more insights into their structure–activity relationships.

Another important topic related to GPI research is GPI-anchored proteomics analysis,[107–110] which can help not only identify the identities and quantities of GPI-anchored proteins expressed by various cells but also discover unique or overexpressed GPI-anchored molecular markers associated with human diseases, such as cancer. These molecular markers can be very useful for the development of novel diagnostic and therapeutic methods. In this context, strategies employing GPI-binding bacterial toxins or using GPI metabolic engineering techniques to tag GPI-anchored proteins on the cell surface have been exploited for the isolation, accumulation or pull-down of natural GPI-anchored proteins to facilitate GPI-anchored proteomics investigation. We believe that with the broader applications of these techniques and development of new methods this area will witness further significant progress in the near future.

Acknowledgements

Our research work has been supported by the National Science Foundation (NSF CHE-1053848) and the National Institutes of Health (NIH, R01 GM090270) of the USA.

References

1. M. G. Paulick and C. R. Bertozzi, The glycosylphosphatidylinositol anchor: a complex membrane-anchoring structure for proteins, *Biochemistry*, 2008, **47**, 6991–7000.
2. H. Ikezawa, Glycosylphosphatidylinositol (GPI)-anchored proteins, *Biol. Pharm. Bull.*, 2002, **25**, 409–417.
3. P. T. Englund, The structure and biosynthesis of glycosyl phosphatidylinositol protein anchors, *Annu. Rev. Biochem.*, 1993, **62**, 121–138.
4. M. A. Ferguson, The structure, biosynthesis and functions of glycosylphosphatidylinositol anchors, and the contributions of trypanosome research, *J. Cell Sci.*, 1999, **112**, 2799–2809.
5. M. A. Ferguson and A. F. Williams, Cell-surface anchoring of proteins via glycosyl-phosphatidylinositol structures, *Annu. Rev. Biochem.*, 1988, **57**, 285–320.
6. H. T. He, J. Finne and C. Goridis, Biosynthesis, membrane association, and release of N-CAM-120, a phosphatidylinositol-linked form of the neural cell adhesion molecule, *J. Cell Biol.*, 1987, **105**, 2489–2500.
7. P. J. Robinson, Signal transduction via GPI-anchored membrane proteins, *Adv. Exp. Med. Biol.*, 1997, **419**, 365–370.

8. D. D. Eardley and M. E. Koshland, Glycosylphosphatidylinositol: a candidate system for interleukin-2 signal transduction, *Science*, 1991, **251**, 78–81.
9. S. Ilangumaran, H. T. He and D. C. Hoessli, Microdomains in lymphocyte signalling: beyond GPI-anchored proteins, *Immunol. Today*, 2000, **21**, 2–7.
10. G. A. Cross, Cellular and genetic aspects of antigenic variation in trypanosomes, *Annu. Rev. Immunol.*, 1990, **8**, 83–110.
11. K. Y. Hwa, Glycosyl phosphatidylinositol-linked glycoconjugates: structure, biosynthesis and function, *Adv. Exp. Med. Biol.*, 2001, **491**, 207–214.
12. M. A. Ferguson, S. W. Homans, R. A. Dwek and T. W. Rademacher, Glycosylphosphatidylinositol moiety that anchors Trypanosoma brucei variant surface glycoprotein to the membrane, *Science*, 1988, **239**, 753–759.
13. S. W. Homans, M. A. Ferguson, R. A. Dwek, T. W. Rademacher, R. Anand and A. F. Williams, Complete structure of the glycosyl phosphatidylinositol membrane anchor of rat brain Thy-1 glycoprotein, *Nature*, 1988, **333**, 269–272.
14. Y. H. Tsai, X. Liu and P. H. Seeberger, Chemical biology of glycosylphosphatidylinositol anchors, *Angew. Chem., Int. Ed. Engl.*, 2012, **51**, 11438–11456.
15. Z. W. Guo and L. Bishop, Chemical synthesis of GPIs and GPI-anchored glycopeptides, *Eur. J. Org. Chem.*, 2004, 3585–3596.
16. A. V. Nikolaev and N. Al-Maharik, Synthetic glycosylphosphatidylinositol (GPI) anchors: how these complex molecules have been made, *Nat. Prod. Rep.*, 2011, **28**, 970–1020.
17. Z. Guo, Synthetic Studies of Glycosylphosphatidylinositol (GPI) Anchors and GPI-Anchored Peptides, Glycopeptides, and Proteins, *Curr. Org. Synth.*, 2013, **10**, 366–383.
18. C. Murakata and T. Ogawa, Synthetic Studies on Cell-Surface Glycans .78. A Total Synthesis of Gpi Anchor of Trypanosoma Brucei, *Tetrahedron Lett.*, 1991, **32**, 671–674.
19. C. Murakata and T. Ogawa, Stereoselective total synthesis of the glycosyl phosphatidylinositol (GPI) anchor of Trypanosoma brucei, *Carbohydr. Res.*, 1992, **235**, 95–114.
20. C. Murakata and T. Ogawa, Stereoselective synthesis of glycobiosyl phosphatidylinositol, a part structure of the glycosyl-phosphatidylinositol (GPI) anchor of Trypanosoma brucei, *Carbohydr. Res.*, 1992, **234**, 75–91.
21. D. K. Baeschlin, A. R. Chaperon, V. Charbonneau, L. G. Green, S. V. Ley, U. Lucking and E. Walther, Rapid assembly of oligosaccharides: Total synthesis of a glycosylphosphatidylinositol anchor of Trypanosoma brucei, *Angew. Chem., Int. Ed.*, 1998, **37**, 3423–3428.
22. D. K. Baeschlin, A. R. Chaperon, L. G. Green, M. G. Hahn, S. J. Ince and S. V. Ley, 1,2-diacetals in synthesis: Total synthesis of a glycosylphosphatidylinositol anchor of Trypanosoma brucei, *Chem.–Eur. J.*, 2000, **6**, 172–186.

23. J. Lu, K. N. Jayaprakash and B. Fraser-Reid, First synthesis of a malarial prototype: a fully lipidated and phosphorylated GPI membrane anchor, *Tetrahedron Lett.*, 2004, **45**, 879–882.
24. J. Lu, K. N. Jayaprakash, U. Schlueter and B. Fraser-Reid, Synthesis of a malaria candidate glycosylphosphatidylinositol (GPI) structure: A strategy for fully inositol acylated and phosphorylated GPIs, *J. Am. Chem. Soc.*, 2004, **126**, 7540–7547.
25. U. E. Udodong, R. Madsen, C. Roberts and B. Fraserreid, A Ready, Convergent Synthesis of the Heptasaccharide Gpi Membrane Anchor of Rat-Brain Thy-1 Glycoprotein, *J. Am. Chem. Soc.*, 1993, **115**, 7886–7887.
26. A. S. Campbell and B. Fraserreid, First Synthesis of a Fully Phosphorylated Gpi Membrane Anchor – Rat-Brain Thy-1, *J. Am. Chem. Soc.*, 1995, **117**, 10387–10388.
27. N. C. Reichardt and M. Martin-Lomas, A practical solid-phase synthesis of glycosylphosphatidylinositol precursors, *Angew. Chem., Int. Ed.*, 2003, **42**, 4674–4677.
28. J. Lopez-Prados and M. Martin-Lomas, Inositolphosphoglycan mediators: An effective synthesis of the conserved linear GPI anchor structure, *J. Carbohydr. Chem.*, 2005, **24**, 393–414.
29. K. Pekari and R. R. Schmidt, A variable concept for the preparation of branched glycosyl phosphatidyl inositol anchors, *J. Org. Chem.*, 2003, **68**, 1295–1308.
30. K. Pekari, D. Tailler, R. Weingart and R. R. Schmidt, Synthesis of the fully phosphorylated GPI anchor pseudohexasaccharide of Toxoplasma gondii, *J. Org. Chem.*, 2001, **66**, 7432–7442.
31. T. G. Mayer, B. Kratzer and R. R. Schmidt, Synthesis of a Gpi Anchor of Yeast (Saccaromyces-Cerevisiae), *Angew. Chem., Int. Ed.*, 1994, **33**, 2177–2181.
32. T. G. Mayer and R. R. Schmidt, Glycosyl phosphatidylinositol (GPI) anchor synthesis based on versatile building blocks - Total synthesis of a GPI anchor of yeast, *Eur. J. Org. Chem.*, 1999, 1153–1165.
33. M. C. Hewitt, D. A. Snyder and P. H. Seeberger, Rapid synthesis of a glycosylphosphatidylinositol-based malaria vaccine using automated solid-phase oligosaccharide synthesis, *J. Am. Chem. Soc.*, 2002, **124**, 13434–13436.
34. L. Schofield, M. C. Hewitt, K. Evans, M. A. Siomos and P. H. Seeberger, Synthetic GPI as a candidate anti-toxic vaccine in a model of malaria, *Nature*, 2002, **418**, 785–789.
35. Y. H. Tsai, S. Gotze, N. Azzouz, H. S. Hahm, P. H. Seeberger and D. V. Silva, A general method for synthesis of GPI anchors illustrated by the total synthesis of the low-molecular-weight antigen from Toxoplasma gondii, *Angew. Chem., Int. Ed. Engl.*, 2011, **50**, 9961–9964.
36. X. Liu, Y. U. Kwon and P. H. Seeberger, Convergent synthesis of a fully lipidated glycosylphosphatidylinositol anchor of Plasmodium falciparum, *J. Am. Chem. Soc.*, 2005, **127**, 5004–5005.

37. Y. U. Kwon, X. Liu and P. H. Seeberger, Total syntheses of fully lipidated glycosylphosphatidylinositol anchors of Toxoplasma gondii, *Chem. Commun. (Cambridge, U. K.)*, 2005, 2280–2282.
38. Y.-H. Tsai, S. Gotze, I. Vilotijevic, M. Grube, D. V. Silva and P. H. Seeberger, A general and convergent synthesis of diverse glycosylphosphatidylinositol glycolipids, *Chem. Sci.*, 2013, **4**, 468–481.
39. A. Ali, D. C. Gowda and R. A. Vishwakarma, A new approach to construct full-length glycosylphosphatidylinositols of parasitic protozoa and [4-deoxy-Man-III]-GPI analogues, *Chem. Commun.*, 2005, 519–521.
40. A. Ali and R. A. Vishwakarma, Total synthesis of the fully lipidated glycosylphosphatidylinositol (GPI) anchor of malarial parasite Plasmodium falciparum, *Tetrahedron*, 2010, **66**, 4357–4369.
41. D. V. Yashunsky, V. S. Borodkin, M. A. Ferguson and A. V. Nikolaev, The chemical synthesis of bioactive glycosylphosphatidylinositols from Trypanosoma cruzi containing an unsaturated fatty acid in the lipid, *Angew. Chem., Int. Ed. Engl.*, 2006, **45**, 468–474.
42. D. V. Yashunsky, V. S. Borodkin, P. G. McGivern, M. A. J. Ferguson and A. V. Nikolaev, The Chemical Synthesis of Glycosylphosphatidylinositol Anchors from Trypanosoma cruzi Trypomastigote Mucins, *Frontiers in Modern Carbohydrate Chemistry*, American Chemical Society, 2007, pp. 285–306.
43. B. M. Swarts and Z. Guo, Synthesis of a glycosylphosphatidylinositol anchor bearing unsaturated lipid chains, *J. Am. Chem. Soc.*, 2010, **132**, 6648–6650.
44. S. Burgula, B. M. Swarts and Z. Guo, Total synthesis of a glycosylphosphatidylinositol anchor of the human lymphocyte CD52 antigen, *Chem.–Eur. J.*, 2012, **18**, 1194–1201.
45. B. M. Swarts and Z. Guo, Chemical synthesis and functionalization of clickable glycosylphosphatidylinositol anchors, *Chem. Sci.*, 2011, **2**, 2342–2352.
46. T. Kinoshita and M. Fujita, Biosynthesis of GPI-anchored proteins: special emphasis on GPI lipid remodeling, *J. Lipid Res.*, 2016, **57**, 6–24.
47. T. Kinoshita, M. Fujita and Y. Maeda, Biosynthesis, remodelling and functions of mammalian GPI-anchored proteins: recent progress, *J. Biochem.*, 2008, **144**, 287–294.
48. R. Watanabe, N. Inoue, B. Westfall, C. H. Taron, P. Orlean, J. Takeda and T. Kinoshita, The first step of glycosylphosphatidylinositol biosynthesis is mediated by a complex of PIG-A, PIG-H, PIG-C and GPI1, *EMBO J.*, 1998, **17**, 877–885.
49. Y. Murakami, U. Siripanyaphinyo, Y. Hong, Y. Tashima, Y. Maeda and T. Kinoshita, The initial enzyme for glycosylphosphatidylinositol biosynthesis requires PIG-Y, a seventh component, *Mol. Biol. Cell*, 2005, **16**, 5236–5246.
50. R. Watanabe, Y. Murakami, M. D. Marmor, N. Inoue, Y. Maeda, J. Hino, K. Kangawa, M. Julius and T. Kinoshita, Initial enzyme for glycosylphosphatidylinositol biosynthesis requires PIG-P and is regulated by DPM2, *EMBO J.*, 2000, **19**, 4402–4411.

51. N. Nakamura, N. Inoue, R. Watanabe, M. Takahashi, J. Takeda, V. L. Stevens and T. Kinoshita, Expression cloning of PIG-L, a candidate N-acetylglucosaminyl-phosphatidylinositol deacetylase, *J. Biol. Chem.*, 1997, **272**, 15834–15840.
52. R. Watanabe, K. Ohishi, Y. Maeda, N. Nakamura and T. Kinoshita, Mammalian PIG-L and its yeast homologue Gpi12p are N-acetylglucosaminylphosphatidylinositol de-N-acetylases essential in glycosylphosphatidylinositol biosynthesis, *Biochem. J.*, 1999, **339**, 185–192.
53. R. A. Vishwakarma and A. K. Menon, Flip-flop of glycosylphosphatidylinositols (GPI's) across the ER, *Chem. Commun. (Cambridge, U. K.)*, 2005, 453–455.
54. W. T. Doerrler, J. Ye, J. R. Falck and M. A. Lehrman, Acylation of glucosaminyl phosphatidylinositol revisited. Palmitoyl-CoA dependent palmitoylation of the inositol residue of a synthetic dioctanoyl glucosaminyl phosphatidylinositol by hamster membranes permits efficient mannosylation of the glucosamine residue, *J. Biol. Chem.*, 1996, **271**, 27031–27038.
55. Y. Murakami, U. Siripanyapinyo, Y. J. Hong, J. Y. Kang, S. Ishihara, H. Nakakuma, Y. Maeda and T. Kinoshita, PIG-W is critical for inositol acylation but not for flipping of glycosylphosphatidylinositol-anchor, *Mol. Biol. Cell*, 2003, **14**, 4285–4295.
56. N. Kanzawa, Y. Maeda, H. Ogiso, Y. Murakami, R. Taguchi and T. Kinoshita, Peroxisome dependency of alkyl-containing GPI-anchor biosynthesis in the endoplasmic reticulum, *Proc. Natl. Acad. Sci. U. S. A.*, 2009, **106**, 17711–17716.
57. Y. Maeda, R. Watanabe, C. L. Harris, Y. Hong, K. Ohishi, K. Kinoshita and T. Kinoshita, PIG-M transfers the first mannose to glycosylphosphatidylinositol on the lumenal side of the ER, *EMBO J.*, 2001, **20**, 250–261.
58. H. Ashida, Y. Hong, Y. Murakami, N. Shishioh, N. Sugimoto, Y. U. Kim, Y. Maeda and T. Kinoshita, Mammalian PIG-X and yeast Pbn1p are the essential components of glycosylphosphatidylinositol-mannosyltransferase I, *Mol. Biol. Cell*, 2005, **16**, 1439–1448.
59. J. Y. Kang, Y. Hong, H. Ashida, N. Shishioh, Y. Murakami, Y. S. Morita, Y. Maeda and T. Kinoshita, PIG-V involved in transferring the second mannose in glycosylphosphatidylinositol, *J. Biol. Chem.*, 2005, **280**, 9489–9497.
60. Y. Hong, Y. Maeda, R. Watanabe, K. Ohishi, M. Mishkind, H. Riezman and T. Kinoshita, Pig-n, a mammalian homologue of yeast Mcd4p, is involved in transferring phosphoethanolamine to the first mannose of the glycosylphosphatidylinositol,*J. Biol. Chem.*, 1999, **274**, 35099–35106.
61. M. Takahashi, N. Inoue, K. Ohishi, Y. Maeda, N. Nakamura, Y. Endo, T. Fujita, J. Takeda and T. Kinoshita, PIG-B, a membrane protein of the endoplasmic reticulum with a large lumenal domain, is involved in transferring the third mannose of the GPI anchor, *EMBO J.*, 1996, **15**, 4254–4261.
62. P. Orlean and A. K. Menon, Thematic review series: lipid posttranslational modifications. GPI anchoring of protein in yeast and mammalian

cells, or: how we learned to stop worrying and love glycophospholipids, *J. Lipid Res.*, 2007, **48**, 993–1011.
63. P. A. Colussi, C. H. Taron, J. C. Mack and P. Orlean, Human and Saccharomyces cerevisiae dolichol phosphate mannose synthases represent two classes of the enzyme, but both function in Schizosaccharomyces pombe, *Proc. Natl. Acad. Sci. U. S. A.*, 1997, **94**, 7873–7878.
64. Y. Hong, Y. Maeda, R. Watanabe, N. Inoue, K. Ohishi and T. Kinoshita, Requirement of PIG-F and PIG-O for transferring phosphoethanolamine to the third mannose in glycosylphosphatidylinositol, *J. Biol. Chem.*, 2000, **275**, 20911–20919.
65. N. Inoue, T. Kinoshita, T. Orii and J. Takeda, Cloning of a human gene, PIG-F, a component of glycosylphosphatidylinositol anchor biosynthesis, by a novel expression cloning strategy, *J. Biol. Chem.*, 1993, **268**, 6882–6885.
66. C. H. Taron, J. M. Wiedman, S. J. Grimme and P. Orlean, Glycosylphosphatidylinositol biosynthesis defects in Gpi11p-and Gpi13p-deficient yeast suggest a branched pathway and implicate Gpi13p in phosphoethanolamine transfer to the third mannose, *Mol. Biol. Cell*, 2000, **11**, 1611–1630.
67. N. Shishioh, Y. J. Hong, K. Ohishi, H. Ashida, Y. Maeda and T. Kinoshita, GPI7 is the second partner of PIG-F and involved in modification of glycosylphosphatidylinositol, *J. Biol. Chem.*, 2005, **280**, 9728–9734.
68. S. Hirose, G. M. Prince, D. Sevlever, L. Ravi, T. L. Rosenberry, E. Ueda and M. E. Medof, Characterization of putative glycoinositol phospholipid anchor precursors in mammalian cells. Localization of phosphoethanolamine, *J. Biol. Chem.*, 1992, **267**, 16968–16974.
69. B. Taron, P. Colussi, J. Wiedman, P. Orlean and C. Taron, A human homologue of yeast Smp3p adds a fourth mannose to yeast and mammalian glycosylphosphatidylinositol precursors in vivo, *Glycobiology*, 2003, **13**, 875.
70. B. W. Taron, P. A. Colussi, J. M. Wiedman, P. Orlean and C. H. Taron, Human Smp3p adds a fourth mannose to yeast and human glycosylphosphatidylinositol precursors in vivo, *J. Biol. Chem.*, 2004, **279**, 36083–36092.
71. P. M. Rudd, B. P. Morgan, M. R. Wormald, D. J. Harvey, C. W. van den Berg, S. J. Davis, M. A. Ferguson and R. A. Dwek, The glycosylation of the complement regulatory protein, human erythrocyte CD59, *J. Biol. Chem.*, 1997, **272**, 7229–7244.
72. I. A. Brewis, M. A. Ferguson, A. Mehlert, A. J. Turner and N. M. Hooper, Structures of the glycosyl-phosphatidylinositol anchors of porcine and human renal membrane dipeptidase. Comprehensive structural studies on the porcine anchor and interspecies comparison of the glycan core structures, *J. Biol. Chem.*, 1995, **270**, 22946–22956.
73. N. Singh, L. N. Liang, M. L. Tykocinski and A. M. Tartakoff, A novel class of cell surface glycolipids of mammalian cells. Free glycosyl phosphatidylinositols, *J. Biol. Chem.*, 1996, **271**, 12879–12884.

74. S. E. Maxwell, S. Ramalingam, L. D. Gerber, L. Brink and S. Udenfriend, An active carbonyl formed during glycosylphosphatidylinositol addition to a protein is evidence of catalysis by a transamidase, *J. Biol. Chem.*, 1995, **270**, 19576–19582.
75. B. Eisenhaber, P. Bork and F. Eisenhaber, Post-translational GPI lipid anchor modification of proteins in kingdoms of life: analysis of protein sequence data from complete genomes, *Protein Eng.*, 2001, **14**, 17–25.
76. B. Eisenhaber, P. Bork and F. Eisenhaber, Prediction of potential GPI-modification sites in proprotein sequences, *J. Mol. Biol.*, 1999, **292**, 741–758.
77. F. Reggiori, E. Canivenc-Gansel and A. Conzelmann, Lipid remodeling leads to the introduction and exchange of defined ceramides on GPI proteins in the ER and Golgi of Saccharomyces cerevisiae, *EMBO J.*, 1997, **16**, 3506–3518.
78. H. Murakami, Y. Wang, H. Hasuwa, Y. Maeda, T. Kinoshita and Y. Murakami, Enhanced response of T lymphocytes from Pgap3 knockout mouse: Insight into roles of fatty acid remodeling of GPI anchored proteins, *Biochem. Biophys. Res. Commun.*, 2012, **417**, 1235–1241.
79. Y. Maeda, Y. Tashima, T. Houjou, M. Fujita, T. Yoko-o, Y. Jigami, R. Taguchi and T. Kinoshita, Fatty acid remodeling of GPI-anchored proteins is required for their raft association, *Mol. Biol. Cell*, 2007, **18**, 1497–1506.
80. D. Hamburger, M. Egerton and H. Riezman, Yeast Gaa1p Is Required for Attachment of a Completed Gpi Anchor onto Proteins, *J. Cell Biol.*, 1995, **129**, 629–639.
81. K. Ohishi, N. Inoue, Y. Maeda, J. Takeda, H. Riezman and T. Kinoshita, Gaa1p and Gpi8p are components of a glycosylphosphatidylinositol (GPI) transamidase that mediates attachment of GPI to proteins, *Mol. Biol. Cell*, 2000, **11**, 1523–1533.
82. K. Ohishi, N. Inoue and T. Kinoshita, PIG-S and PIG-T, essential for GPI anchor attachment to proteins, form a complex with GAA1 and GPI8, *EMBO J.*, 2001, **20**, 4088–4098.
83. Y. J. Hong, K. Ohishi, J. Y. Kang, S. Tanaka, N. Inoue, J. Nishimura, Y. Maeda and T. Kinoshita, Human PIG-U and yeast Cdc91p are the fifth subunit of GPI transamidase that attaches GPI-anchors to proteins, *Mol. Biol. Cell*, 2003, **14**, 1780–1789.
84. M. Benghezal, A. Benachour, S. Rusconi, M. Aebi and A. Conzelmann, Yeast gpi8p is essential for GPI anchor attachment onto proteins, *EMBO J.*, 1996, **15**, 6575–6583.
85. B. Eisenhaber, S. Eisenhaber, T. Y. Kwang, G. Gruber and F. Eisenhaber, Transamidase subunit GAA1/GPAA1 is a M28 family metallo-peptide-synthetase that catalyzes the peptide bond formation between the substrate protein's omega-site and the GPI lipid anchor's phosphoethanolamine, *Cell Cycle*, 2014, **13**, 1912–1917.
86. K. Ohishi, K. Nagamune, Y. Maeda and T. Kinoshita, Two subunits of glycosylphosphatidylinositol transamidase, GPI8 and PIG-T, form a

functionally important intermolecular disulfide bridge, *J. Biol. Chem.*, 2003, **278**, 13959–13967.
87. N. Shao, J. Xue and Z. W. Guo, Chemical synthesis of CD52 glycopeptides containing the acid-labile fucosyl linkage, *J. Org. Chem.*, 2003, **68**, 9003–9011.
88. N. Shao, B. Xue and Z. W. Guo, Chemical synthesis of a skeleton structure of sperm CD52-A GPI-anchored glycopeptide, *Angew. Chem., Int. Ed.*, 2004, **43**, 1569–1573.
89. E. C. B. Johnson and S. B. H. Kent, Insights into the mechanism and catalysis of the native chemical ligation reaction, *J. Am. Chem. Soc.*, 2006, **128**, 6640–6646.
90. P. E. Dawson, T. W. Muir, I. Clark-Lewis and S. B. Kent, Synthesis of proteins by native chemical ligation, *Science*, 1994, **266**, 776–779.
91. D. Bang and S. B. Kent, A one-pot total synthesis of crambin, *Angew. Chem., Int. Ed.*, 2004, **43**, 2534–2538.
92. Y. Tanaka, Y. Nakahara, H. Hojo and Y. Nakahara, Studies directed toward the synthesis of protein-bound GPI anchor, *Tetrahedron*, 2003, **59**, 4059–4067.
93. M. G. Paulick, A. R. Wise, M. B. Forstner, J. T. Groves and C. R. Bertozzi, Synthetic analogues of glycosylphosphatidylinositol-anchored proteins and their behavior in supported lipid bilayers, *J. Am. Chem. Soc.*, 2007, **129**, 11543–11550.
94. B. Ziaco, S. Pensato, L. D. D'Andrea, E. Benedetti and A. Romanelli, Semisynthesis of dimeric proteins by expressed protein ligation, *Org. Lett.*, 2008, **10**, 1955–1958.
95. T. W. Muir, Semisynthesis of proteins by expressed protein ligation, *Annu. Rev. Biochem.*, 2003, **72**, 249–289.
96. C. F. Becker, X. Liu, D. Olschewski, R. Castelli, R. Seidel and P. H. Seeberger, Semisynthesis of a glycosylphosphatidylinositol-anchored prion protein, *Angew. Chem., Int. Ed.*, 2008, **47**, 8215–8219.
97. E. Barboni, B. P. Rivero, A. J. George, S. R. Martin, D. V. Renoup, E. F. Hounsell, P. C. Barber and R. J. Morris, The glycophosphatidylinositol anchor affects the conformation of Thy-1 protein, *J. Cell Sci.*, 1995, **108**, 487–497.
98. X. Guo, Q. Wang, B. M. Swarts and Z. Guo, Sortase-catalyzed peptide-glycosylphosphatidylinositol analogue ligation, *J. Am. Chem. Soc.*, 2009, **131**, 9878–9879.
99. Z. Wu, X. Guo, Q. Wang, B. M. Swarts and Z. Guo, Sortase A-catalyzed transpeptidation of glycosylphosphatidylinositol derivatives for chemoenzymatic synthesis of GPI-anchored proteins, *J. Am. Chem. Soc.*, 2010, **132**, 1567–1571.
100. Z. M. Wu, X. Q. Guo, G. F. Gu and Z. W. Guo, Chemoenzymatic Synthesis of the Human CD52 and CD24 Antigen Analogues, *Org. Lett.*, 2013, **15**, 5906–5908.
101. Z. M. Wu, X. Q. Guo and Z. W. Guo, Chemoenzymatic synthesis of glycosylphosphatidylinositol-anchored glycopeptides, *Chem. Commun.*, 2010, **46**, 5773–5774.

102. Z. M. Wu, X. Q. Guo, J. Gao and Z. W. Guo, Sortase A-mediated chemoenzymatic synthesis of complex glycosylphosphatidylinositol-anchored protein, *Chem. Commun.*, 2013, **49**, 11689–11691.
103. B. A. Frankel, R. G. Kruger, D. E. Robinson, N. L. Kelleher and D. G. McCafferty, Staphylococcus aureus sortase transpeptidase SrtA: insight into the kinetic mechanism and evidence for a reverse protonation catalytic mechanism, *Biochemistry*, 2005, **44**, 11188–11200.
104. X. Huang, A. Aulabaugh, W. Ding, B. Kapoor, L. Alksne, K. Tabei and G. Ellestad, Kinetic mechanism of Staphylococcus aureus sortase SrtA, *Biochemistry*, 2003, **42**, 11307–11315.
105. L. A. Marraffini, A. C. Dedent and O. Schneewind, Sortases and the art of anchoring proteins to the envelopes of gram-positive bacteria, *Microbiol. Mol. Biol. Rev.*, 2006, **70**, 192–221.
106. G. Wu, Z. Guo, A. Chatterjee, X. Huang, E. Rubin, F. Wu, E. Mambo, X. Chang, M. Osada, M. Sook Kim, C. Moon, J. A. Califano, E. A. Ratovitski, S. M. Gollin, S. Sukumar, D. Sidransky and B. Trink, Overexpression of glycosylphosphatidylinositol (GPI) transamidase subunits phosphatidylinositol glycan class T and/or GPI anchor attachment 1 induces tumorigenesis and contributes to invasion in human breast cancer, *Cancer Res.*, 2006, **66**, 9829–9836.
107. P. Zhao, A. V. Nairn, S. Hester, K. W. Moremen, R. M. O'Regan, G. Oprea, L. Wells, M. Pierce and K. L. Abbott, Proteomic identification of glycosylphosphatidylinositol anchor-dependent membrane proteins elevated in breast carcinoma, *J. Biol. Chem.*, 2012, **287**, 25230–25240.
108. S. Dolezal, S. Hester, P. S. Kirby, A. Nairn, M. Pierce and K. L. Abbott, Elevated levels of glycosylphosphatidylinositol (GPI) anchored proteins in plasma from human cancers detected by C. septicum alpha toxin, *Cancer Biomarkers*, 2014, **14**, 55–62.
109. L. L. Lu, J. Gao and Z. W. Guo, Labeling Cell Surface GPIs and GPI-Anchored Proteins through Metabolic Engineering with Artificial Inositol Derivatives, *Angew. Chem., Int. Ed.*, 2015, **54**, 9679–9682.
110. T. Dhar and H. D. Mootz, Modification of transmembrane and GPI-anchored proteins on living cells by efficient protein trans-splicing using the Npu DnaE intein, *Chem. Commun.*, 2011, **47**, 3063–3065.

CHAPTER 10

Chemical Approaches to Image Protein Glycosylation

WEI LIN[a], LING GAO[a], KE QIN[a] AND XING CHEN*[a]

[a]College of Chemistry and Molecular Engineering, Peking-Tsinghua Center for Life Sciences, Peking University, Beijing 100871, China
*E-mail: xingchen@pku.edu.cn

10.1 Background

Protein glycosylation plays essential roles in regulating protein folding, translocation, and activity.[1,2] Visualization of protein glycans in live cells and in living organisms is invaluable for elucidating the biological functions of protein glycosylation. However, it has proven challenging to develop tools for molecular imaging of glycans.[3] First of all, the chemical structures of glycans are highly complex and heterogonous within cells, and glycosylation occurs not only on proteins but also on other biomolecules, such as lipids. Secondly, effective methods for labeling glycans with imaging probes are relatively limited, in comparison to the available tools for protein labeling. Finally, glycans with the same or similar structures are often attached to numerous proteins and/or lipids, which makes it extremely difficult to analyze the glycosylation state of a specific protein in the physiological environment.[4] In this chapter, we summarize recent advances in developing chemical strategies for imaging protein glycosylation.

Chemical Biology No. 2
Chemical Biology of Glycoproteins
Edited by Zhongping Tan and Lai-Xi Wang

Published by the Royal Society of Chemistry, www.rsc.org

10.2 Structure, Biosynthesis, and Function of Protein Glycans

There are various types of glycans that differ in structure and linkage. In vertebrates, glycans are synthesized from nine monosaccharide building blocks: D-glucose (Glc), *N*-acetyl-D-glucosamine (GlcNAc), D-galactose (Gal), *N*-acetyl-D-galactosamine (GalNAc), D-mannose (Man), D-xylose (Xyl), L-fucose (Fuc), D-glucuronic acid (GlcA), and sialic acid (Sia) (L-Iduronic acid (IdoA) is epimerized from GlcA in dermatan sulfate and heparin sulfate by GlcA C5 epimerase.). Membrane-associated and secreted proteins are modified with *N*-linked or/and mucin-type *O*-linked glycans (Figure 10.1).[5,6] Intracellular proteins are ubiquitously modified with *O*-GlcNAc.[7] These are three most common types of protein glycans and are the major focus of this chapter.

There are other kinds of protein glycosylation that occur only in specific protein families. Some of the typical examples include *O*-mannose modification of α-dystroglycan,[8] and *O*-fucosylation and *O*-glucosylation on Notch receptors.[9,10] Moreover, glycosylphosphatidylinositol (GPI)-anchored

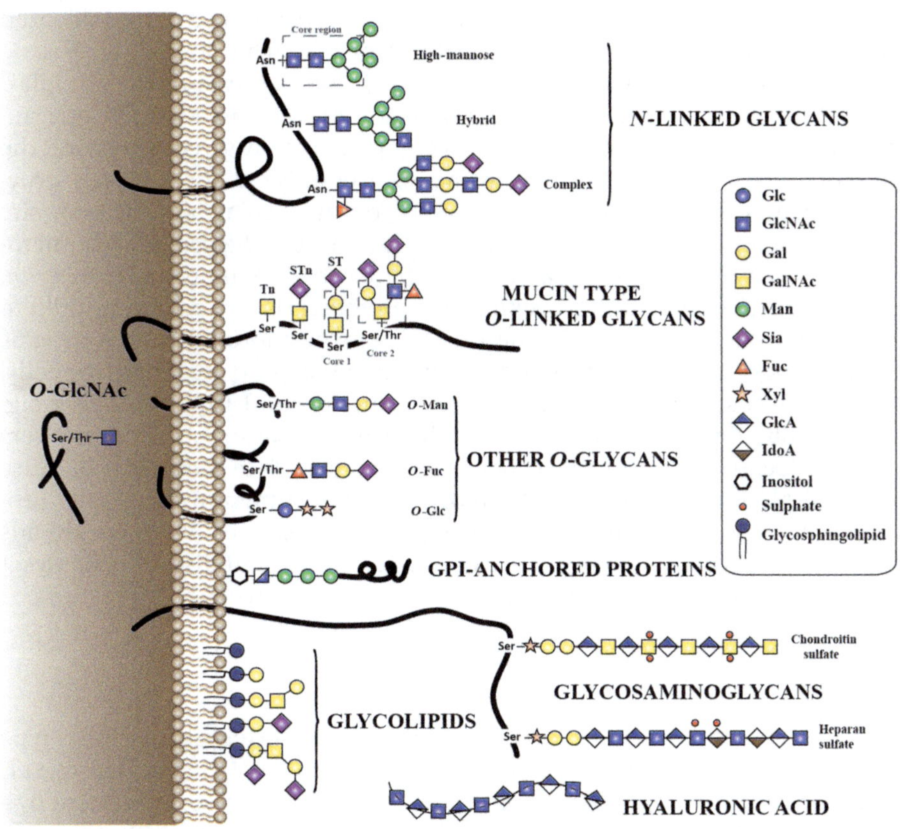

Figure 10.1 Common classes of glycans in vertebrate cells.

proteins are a group of proteins containing a glycan within GPI, which is attached to the protein C-terminus and responsible for attaching the proteins to the cell membrane.[11] In addition, proteoglycans are made of a protein core attached with one or more glycosaminoglycans, which are linear polysaccharides containing repeated units of acidic disaccharides.[12] Methods for imaging of those types of protein glycosylation are underdeveloped and are not discussed herein.

10.2.1 *N*-Linked Glycans

When trafficking through the secretory pathway, most of the secreted and membrane proteins are subjected to *N*-glycosylation within the endoplasmic reticulum (ER) and the Golgi apparatus.[5,13] In mammals, oligosaccharyltransferase (OST) residing on the ER membrane catalyzes the transfer of a 14-sugar oligosaccharide $Glc_3Man_9GlcNAc_2$ onto asparagine (Asn) within the consensus sequence Asn–!Pro–Ser/Thr (where !Pro is any amino acid other than proline) of the translating proteins in the lumen. A panel of glycosidases and glycosyltransferases lined up in the ER or Golgi then remodel the *N*-glycans by trimming and adding sugars while the glycoproteins are translocated to the plasma membrane or extracellular matrix. This assembly process results in three subtypes of *N*-linked glycans: the high-mannose, complex, and hybrid classes, each of which contains a common core Man_3Glc_2 glycan. Intriguingly, each subtype of *N*-glycans comprises many glycoforms, and the protein *N*-glycans are highly heterogeneous. Not only can the *N*-glycan structure vary among different proteins, but also a particular modification site within a protein is dynamically occupied with variable *N*-glycans. This microheterogeneity of *N*-linked glycans makes them very difficult to analyze *in situ*.

N-Linked glycosylation is essential for proteostasis by mediating protein folding and quality control in the ER.[5,14,15] Furthermore, protein *N*-glycans on cell surfaces function as ligands for binding receptors, neighboring cells, and pathogens, which regulates protein function and the underlying signal transduction events. In addition, the *N*-glycosylation state often alters the translocation and turnover dynamics of cell-surface receptors. Because *N*-linked glycosylation affects the biogenesis and function of various proteins, aberrant metabolism (biosynthesis/catabolism) of *N*-linked glycans has been implicated in many diseases such as congenital disorders of glycosylation (CDG),[16] lysosomal storage diseases,[17] and cancers.[18]

10.2.2 Mucin-Type *O*-Linked Glycans

Being the most abundant protein *O*-glycan in vertebrates, mucin-type *O*-linked glycans are characterized by an α-linked GalNAc attached to Ser/Thr.[6] The biosynthesis of *O*-glycans is initiated in the Golgi compartment by a family of enzymes known as polypeptide *N*-acetylgalactosaminyltransferases (ppGalNAcTs), which catalyze the transfer of GalNAc from the nucleotide sugar UDP–GalNAc to the proteins.[19] The simplest mucin-type *O*-glycan is a single GalNAc residue linked to Ser/Thr (GalNAcα1–S/T), which is named Tn antigen. Longer and more complex oligosaccharide chains are built upon the Tn antigen, such as the T antigen (Galβ1-3GalNAcα1–S/T), which is also termed core 1 *O*-GalNAc glycan. Both Tn and T antigens can be modified by sialic acids to form sialyl Tn or sialyl T antigens, respectively.[20] There are additional core structures including core 2, core 3, and core 4, each of which is controlled by specific subsets of transferases.[21]

In mucosal tissues, especially intestinal tract, the mucus layer forms a protective homeostatic barrier between the resident microbiota and the underlying immune cells. *O*-Glycans of mucins, which are the main components of mucus, play important roles in mucus barrier function and promote mutualism with resident microbiota.[22] *O*-Glycans are also found on mucin-like glycoproteins, such as Glycosylation-dependent cell adhesion molecule-1 (GlyCAM-1) and CD34, and are important for mediating cell adhesion and trafficking.[23,24] Elevated expression of Tn and STn antigens is often detected in a number of diseases, such as thrombocytopenia, IgA nephropathy, and cancers. Visualization of specific glycoforms of mucin-type *O*-linked proteins is useful for disease diagnosis.

10.2.3 Sialic Acids

Sialic acids are a family of nine-carbon monosaccharides, including *N*-acetylneuraminic acid (NeuAc), *N*-glycolylneuraminic acid (NeuGc), and 2-keto-3-deoxynonic acid (KDN).[25] Sialic acids are negatively charged under physiological conditions and often serve as the terminal sugar for *N*-linked and mucin-type *O*-linked glycans. Moreover, sialic acids are attached to glycolipids such as gangliosides. In addition, sialic acids form polysialic acid (PSA), a homopolymer of sialic acids. In the *de novo* biosynthesis of sialic acids, UDP–GlcNAc is enzymatically converted to *N*-acetyl-D-mannosamine (ManNAc), which serves as a committed precursor of sialic acid. Through several steps of enzymatic reactions, ManNAc is converted to CMP–Sia, which is the sugar donor for sialyltransferases in the Golgi. Cell-surface sialic acids play important functional roles in various biological processes including molecular recognition, leukocyte trafficking, and cell–cell interactions. In particular, changes in the sialylation level of a cell often correlate with its physiological and pathological states, which raises significant interest in visualizing sialoglycans in living system.[26,27]

10.2.4 *O*-GlcNAc

Discovered about three decades ago, *O*-linked-*N*-acetylglucosamine (*O*-GlcNAc) modification is one of the most ubiquitous posttranslational modifications, with thousands of cytoplasmic, nuclear, and mitochondrial proteins that have been identified to be *O*-GlcNAcylated.[7,28] Two enzymes, β-*N*-acetylglucosaminyltransferase (*O*-GlcNAc transferase, OGT) and β-*N*-acetylglucosaminidase (*O*-GlcNAcase, OGA), are responsible for catalyzing the addition of GlcNAc onto the Ser/Thr residues of substrate proteins and cleavage of *O*-GlcNAc from the modified proteins, respectively. In comparison with *N*-linked glycosylation and mucin-type *O*-linked glycosylation, *O*-GlcNAc has several unique features. It modifies intracellular proteins with a GlcNAc monosaccharide using UDP–GlcNAc as the sugar donor. The *O*-GlcNAc modification is reversible and dynamic. *O*-GlcNAc has a complex interplay with other PTMs, such as phosphorylation and ubiquitination.[29,30]

Protein *O*-GlcNAcylation plays essential roles in mediating many biological processes, including transcription regulation, stress response, and cell differentiation.[31] Dysregulation of *O*-GlcNAcylation has been implicated in several diseases, such as neurodegeneration, diabetes, and cancers.[32] Many of the biological functions of *O*-GlcNAc are exerted by affecting protein stability and translocation. A variety of proteins, including p53, β-catenin, Snail1, and CLOCK, are stabilized by *O*-GlcNAcylation through blocking ubiquitin-dependent protein degradation.[30] It is likely that this is a general mechanism for regulation the stability of many more proteins. Since phosphorylation is well known to regulate protein translocation, *O*-GlcNAc-phosphate crosstalk can play an important role in this process.[33] Notably, *O*-GlcNAc also participates in many other aspects of protein function, such as protein interactions and enzymatic activity.[34] *O*-GlcNAc-recognizing antibodies, such as RL2[35] and CTD110.6,[36] have greatly facilitated the biochemical characterizations of protein *O*-GlcNAcylation. To elucidate the cellular functions of *O*-GlcNAc, methods to visualize *O*-GlcNAcylation at the cellular level and correlate the modification state to protein functions are currently an active research area.

10.3 Methods for Glycan Labeling and Imaging

Genetic manipulation of glycosyltransferases and biochemical analysis of isolated glycoprotein samples have been traditionally employed to study glycosylation, which has shed light on the biological functions of glycans. Although molecular imaging of proteins *in vivo* has been greatly advanced over recent decades, the development of imaging methods suitable for glycans remains lagging behind. One of the main reasons is the non-genetically encoded nature of glycosylation. As a result, the genetically encoded labeling methods that are widely used for tagging proteins with fluorophores cannot be directly applied to glycan imaging. To overcome this limitation, several chemical approaches have been developed.[3] Together with the classical lectin and antibody staining methods, these chemical tools have greatly expanded the toolbox for glycan labeling and imaging. Notably, many of the methods introduced in this section are not specifically targeting protein glycans; rather, glycans have often been labeled and imaged as a separate class of biomolecules without information on the attaching aglycones (*e.g.*, proteins or lipids).

10.3.1 Lectins and Antibodies

Lectins, which are in nature glycan-binding proteins, have been widely used for glycan detection and analysis.[37,38] A variety of lectins with distinct specificities have been isolated and are commercially available (Table 10.1). Lectins can be used to identify the glycan patterns through high-throughput platforms, such as lectin microarrays.[39] Lectin-based methods have been proven useful in detecting glycosylation changes related to human diseases such as diabetes and cancers.[40,41] However, the binding affinity of lectins with glycans, especially monosaccharides, is usually weak.[42] Moreover, lectins are tissue-impermeable and often toxic to the cells. Therefore, visualization of glycans *via* lectins is mostly limited to *ex vivo* experiments.

Similar to the lectins, antibodies are another common biochemical tool for glycan analysis. Generally, the specificity of antibodies is better than that of lectins. Monoclonal antibodies have been raised for glycan epitopes such as *O*-GlcNAc, sialyl Lewis^X and heparan sulfate.[35,36,43–45] Glycan antibodies serve as an essential tool for biochemical studies of protein glycosylation. Moreover, although *in vivo* imaging using antibodies suffers from the same issues as

Table 10.1 Representative lectins with distinct specificities.[a]

Lectin	Lectin name	Combining motif
N-Acetylglucosamine-binding lectins		
WGA	Wheat germ agglutinin	GlcNAcβ1-4GlcNAcβ1-4GlcNAc
GSL II	*Griffonia simplicifolia* lectin II	Terminal-α,β-GlcNAc; Glycogen
Sialic acid-binding lectins		
WGA	Wheat germ agglutinin	Neu5Ac
SNA	Elderberry lectin	Neu5Acα2-6Gal(NAc)-R
MAL	*Maackia amurensis* lectin	Neu5Ac/Gcα2,3Galβ1,4Glc(NAc)
Mannose-binding lectins		
Con A	Concanavalin A	Branched α-mannosidic structures; high-mannose type, hybrid type, and bi-antennary complex type *N*-Glycans
LCH	Lentil lectin	Fucosylated core region of bi- and tri-antennary complex type *N*-Glycans
GNA	Snowdrop lectin	α1-3- and α1-6-linked high-mannose structures
Galactose or *N*-acetylgalactosamine-binding lectins		
PNA	Peanut agglutinin	Galβ1-3GalNAcα1-Ser/Thr (T-antigen)
VVL	Hairy vetch lectin	GalNAcα-Ser/Thr (Tn-antigen)
BPL	*Bauhinia purpurea alba*	Galβ1-3GalNAc
RCA120	*Ricinus communis* agglutinin	Galβ1-4GalNAcβ1-R
AIL	Jacalin	(Sia)Galβ1-3GalNAcα1-Ser/Thr (T-antigen)
Fucose-binding lectins		
UEA	*Ulex europaeus* agglutinin	Fucα1-2Gal-R
AAL	*Aleuria aurantia* lectin	Fucα1-2Galβ1-4(Fucα1-3/4)Galβ1-4GlcNAc; R2-GlcNAcβ1-4(Fucα1-6) GlcNAc-R1

[a]Note: for a more comprehensive database of lectins, refer to the Lectin Frontier Database (http://jcggdb.jp/rcmg/glycodb/LectinSearch).

lectins, immunofluorescence microscopy using glycan antibodies can be readily performed on fixed cells. However, it remains difficult and cumbersome to raise glycan-specific antibodies and the glycan antibodies usually have affinities much lower than those of antibodies recognizing protein epitopes.

10.3.2 Metabolic Glycan Labeling

Metabolic glycan labeling exploits the underlying biosynthetic pathways to introduce unnatural monosaccharides containing subtle chemical modifications into cellular glycans.[3] When the unnatural monosaccharides are modified with a bioorthogonal chemical reporter, imaging probes can be chemoselectively conjugated onto glycans *via* bioorthogonal chemistry, which enables imaging of glycans in living systems (Figure 10.2). A handful of chemical reporters have been used for metabolic glycan labeling, with the azides and alkynes being the two most commonly used ones.[27] Once installed on the glycans, azides and alkynes can be derivatized *via* click chemistry with alkyne- and azide-containing probes, respectively. Copper-catalyzed click chemistry, *i.e.*, Cu(I)-catalyzed azide–alkyne cycloaddition (CuAAC), between azides and terminal alkynes, has emerged as one of the most popular bioorthogonal reactions.[46,47] Alleviating the copper cytotoxicity by using Cu(I) ligands facilitates its applications in biology.[48] A copper-free variant of click chemistry, strain-promoted azide–alkyne cycloaddition between azides and cyclooctynes, has been developed and become widely used, especially for *in vivo* labeling.[49]

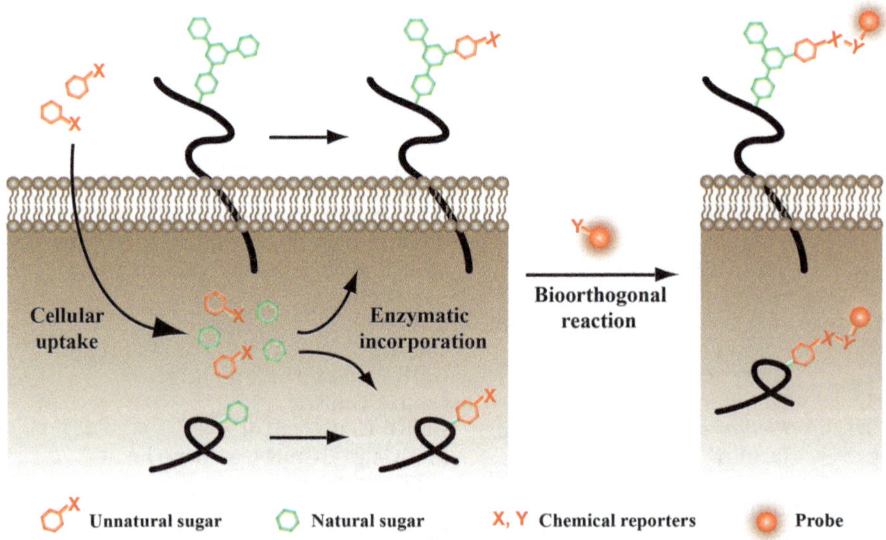

Figure 10.2 Glycans on the cell surface or inside the cell can be metabolically labeled with unnatural sugars bearing a chemical reporter group, which can be detected in a second step by covalent bioorthogonal reaction with a probe.

Various unnatural monosaccharides have been developed for being incorporated into several kinds of glycans (Figure 10.3). The specificity of the unnatural sugars hinges on the pathways they enter and how they are tolerated by the enzymes involved. Sialoglycans can be labeled with azido or alkynyl analogs of sialic acids or its metabolic precursor ManNAc by hijacking the sialic acid biosynthetic pathway.[27] These analogs are enzymatically converted to the corresponding analogs of CMP–sialic acid and are believed not to be metabolically converted to other sugars.[50] Fucosylated glycans have been metabolically labeled by using peracetylated 6-azidofucose ($Ac_4$6AzFuc) or peracetylated 6-alkynylfucose (Ac_4alkynyl fucose), both of which exploit the fucose salvage pathway, and no crosstalk to other pathways has been reported.[51,52] The azido or alkynyl analogs of nucleotide sugars may also be used through microinjection.[53,54] In mammalian cells, metabolic crosstalk occurs between the salvage pathways of GalNAc and GlcNAc. As a result, the analogs of GalNAc and GlcNAc often label multiple glycans.[55–59] For example, tetraacylated *N*-azidoacetylgalactosamine (Ac_4GalNAz) is processed by enzymes in the GalNAc pathway to generate UDP–GalNAz, which is interconvertible with UDP–*N*-azidoacetylgalactosamine (UDP–GlcNAz). In principle, glycans containing either GalNAc or GlcNAc can therefore be labeled. Glycoproteomic analysis has confirmed that Ac_4GalNAz results in azide-incorporation into mucin-type *O*-linked glycans, *N*-glycans, and *O*-GlcNAc.[60] Varying the chemical structure of the unnatural sugars can be used to tune the labeling specificity.[59] Notably, the incorporation efficiency of unnatural sugars is probably dependent on cell types and culture conditions.

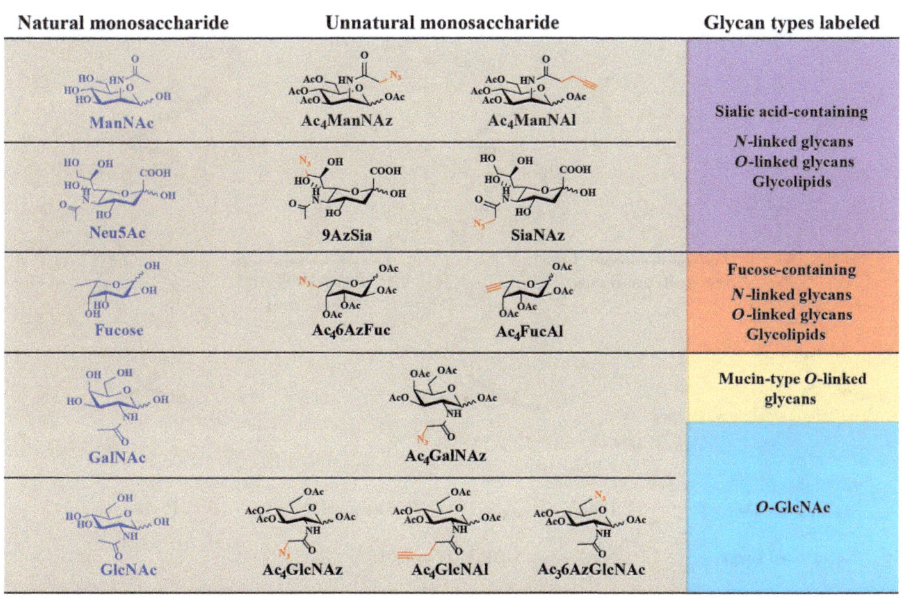

Figure 10.3 Representative unnatural sugars bearing an azide or alkyne and their labeling specificities.

10.3.3 Chemoenzymatic Labeling

The bioorthogonal sugar reporters can also be installed onto glycans using a chemoenzymatic method. In this approach, recombinant or engineered glycosyltransferases are exploited to enzymatically transfer the unnatural sugars from the supplied azido or alkynyl analogs of nucleotide sugars to the substrate glycans (Figure 10.4). The labeling specificity is determined by the glycosyl acceptor glycans that the enzyme recognizes. Although many glycosyltransferases can tolerate chemical modifications on sugar donors, they usually have a very stringent specificity for the glycan substrates, which enables specific labeling of glycans particularly oligosaccharides. Metabolic glycan labeling mostly targets monosaccharides. Recombinant ST6Gal I sialyltransferase has been used to transfer unnatural sialic acids onto cell-surface *N*-linked glycans.[61,62] An engineered mutant of β-1,4 galactosyltransferase (Y289L GalT), which can selectively transfer the GalNAz moiety from the UDP–GalNAz substrate onto *O*-GlcNAc, has been used to detect and image the *O*-GlcNAc-glycosylated proteins.[63] The chemoenzymatic approach has been used to selectively label several disaccharide epitopes. For example, *N*-acetyllactosamine (LacNAc) is labeled by using α-1,3-fucosyltransferase (FucT) and GDP–FucAl;[64] terminal Fuc-α(1-2)-Gal is labeled by using *N*-acetylgalactosaminetransferase BgtA and UDP–GalNAz;[65] Thomsen–Freidenreich Antigen is labeled by two tandem enzymatic reactions using fucosyltransferase Wbwk with GDP–fucose and BgtA with UDP–GalNAz.[66]

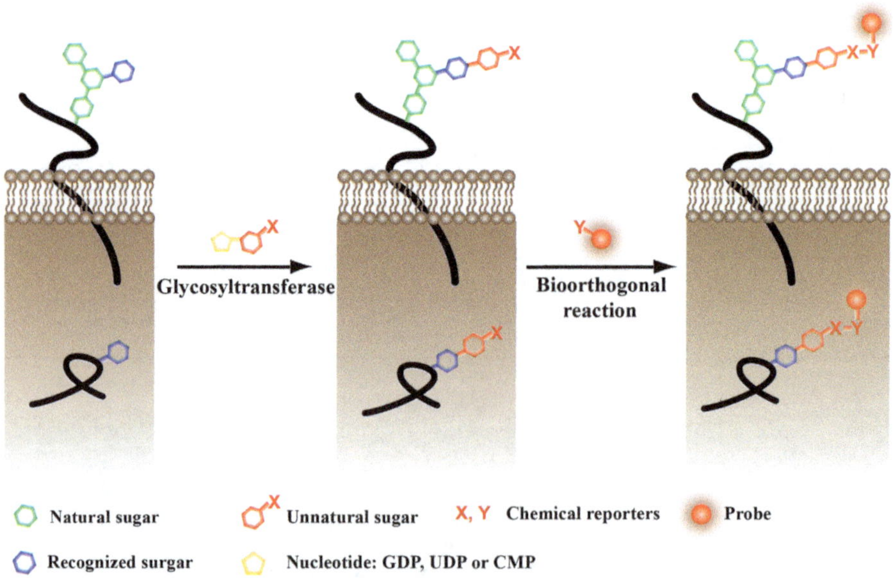

Figure 10.4 Glycans can be specifically recognized and extended with an unnatural sugar by chemoenzymatic reactions. The unnatural sugar can be detected in a second step by bioorthogonal reaction with a probe.

10.4 Imaging Protein Glycosylation

Glycosylation is one of the most ubiquitous protein translational modifications. A particular class of glycosylation often occurs on hundreds or even thousands of proteins in a cell, and the function of proteins can be individually regulated by glycosylation. It is therefore of interest to develop tools for visualizing specific glycoforms in live cells, which is invaluable for dissecting the functional significance of protein glycosylation at the molecular level. However, most of the glycan labeling methods aforementioned, such as metabolic glycan labeling, report on the global changes in glycosylation and do not distinguish the identity of the proteins to which the glycans are attached. On the other hand, a specific protein can be conveniently labeled with imaging probes by using genetically encoded protein tags, such as fluorescent proteins and enzyme-recognizing peptide tags (*e.g.*, LplA acceptor peptide (LAP) and sortase tags).[67,68] Integrating methods for labeling proteins and glycans, several chemical approaches have recently been developed for protein-specific imaging of protein glycosylation in living cells.

10.4.1 General Principles of the Dual-Labeling-Based Methods

To distinguish the glycans on a particular protein within the complex cellular context, one strategy that is under extensive exploration is to integrate global glycan labeling with specific protein labeling (Figure 10.5). By dually labeling both the protein of interest and the glycans, the target glycoprotein is exclusively labeled with both the protein tag and the glycan tag, while other glycoproteins are only labeled with the glycan tag. The spatial proximity between the two tags labeled on the target glycoprotein is exploited to generate a proximity-dependent signal. Several technical variants based on Förster resonance energy transfer (FRET), proximity ligation assay (PLA), proximity-enhanced bioorthogonal ligation (PEBL), and surface enhanced Raman scattering (SERS) have been developed for imaging glycosylation in a protein-specific manner.[69–77] All of these techniques hinge on stringent nanometer proximity for imaging, thus ensuring the signal specificity.

10.4.2 FRET-Based Protein-Specific Imaging of Glycosylation

FRET is a distance-sensitive technique which has been widely used for resolving protein interactions and conformations.[78] In a FRET process, the donor fluorophore is photo-activated to the electronic excited state and transfers energy non-radiatively to the acceptor fluorophore through a long-range dipole–dipole coupling mechanism. The efficiency of energy transfer (E) is dependent on the inverse of the sixth power of the spatial distance between the donor and receptor (d), as described by eqn (10.1):

$$E = R_0{}^6/(R_0{}^6 + d^6) \qquad (10.1)$$

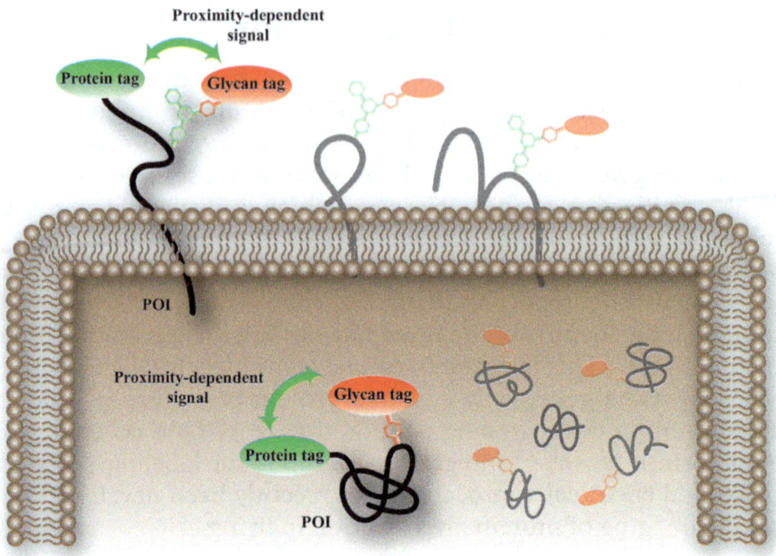

Figure 10.5 Proximity-enabled protein-specific imaging of PTMs. In the context of the whole-cell proteome, a glycan tag is installed onto various glycoproteins in a pan-specific manner, while the protein of interest (POI) is specifically labeled with a protein tag on the peptide scaffold. The protein and glycan tags on the protein of interest are in close proximity, which is exploited to *in situ* generate signal through FRET, PLA, PEBL, and SERS.

where R_0 is the Förster distance at which E equals 50%. Typical FRET pairs have R_0 values within the range of 3–7 nm, so effective FRET can only occur when d is within the range of 1–10 nm. In addition to the distance requirement, the donor and acceptor fluorophores need to have sufficient spectral overlap and a favorable dipole–dipole orientation. Through FRET, donor fluorescence is quenched and the acceptor fluorescence is increased, based on which intensity-based FRET microscopy is performed, typically measuring the acceptor fluorescence signal. Alternatively, the occurrence of FRET results in a shortening of fluorescence lifetime (τ) of the donor, which can be measured by fluorescence lifetime imaging microcopy (FLIM).[79]

The FRET-based methods have been developed for protein-specific imaging of cell-surface sialoglycans. Most of the cell-surface proteins are modified with *N*-glycans or/and mucin-type *O*-linked glycans, which are often terminated with sialic acids. The sialic acids can be metabolically labeled with azides or alkynes by using Ac_4ManNAz or Ac_4ManNAl, respectively. Subsequent click reaction conjugates a fluorophore, which serves as the FRET acceptor, onto the various cell-surface sialoglycans. For labeling the protein with a donor, it can be genetically fused with an enhanced green fluorescent protein (EGFP) or its variants.[71] For membrane-associated proteins, fluorescent proteins usually have to be fused to the cytosolic termini because fusion to the other end may compromise the translocation of the target proteins into the membrane. Using such a *trans*-membrane configuration, the FRET donor is distant from the acceptor residing on the other side of the membrane, which results in low FRET efficiency. Nevertheless, the *trans*-membrane FRET-based method can be used for imaging sialylation on proteins with a short cytoplasmic tail, such as GLUT4.[71] The limitations on FRET efficiency can be overcome by conjugating the donor onto the extracellular termini of proteins by using site-specific labeling methods, such as lipoic acid ligase-based probe incorporation mediated by enzymes (PRIME) labeling,[73] or using a Fab fragment of antibodies recognizing the extracellular domain of the protein.[72] By using the Fab to label proteins, it enables imaging of the endogenous proteins, an advantage to be taken into consideration when choosing the labeling methods. The *cis*-membrane FRET-based method is applicable to visualizing glycosylation of proteins with a big cytosolic domain such as sialylation of integrins.[73] Using *cis*-membrane FRET imaging on live cells, sialylation was found to be important for activation of $\alpha_X\beta_2$ integrin (Figure 10.6).

Because the acceptor is in large excess in the dual-labeling methods, the major pitfall that needs to be carefully considered in using intensity-based FRET microscopy for protein-specific imaging of glycosylation is acceptor spectral bleed-through, *i.e.*, direct acceptor excitation by the donor excitation light due to its overlap with part of the acceptor absorption spectrum. In this regard, FLIM–FRET microscopy is advantageous in avoiding the acceptor spectral bleed-through, because fluorescence lifetime measures the rate of decay of the emission and is unaffected by the bleed-through. Although it has become readily accessible, the instrumentation of FLIM–FRET is more sophisticated than that used for intensity-based FRET. FLIM–FRET microscopy was successfully used for imaging the sialylation state of integrin $\alpha_V\beta_3$.[72] *O*-GlcNAc modification can also be imaged using this strategy.[76,77] The *O*-GlcNAc is metabolically labeled with unnatural sugar reporters to install the FRET acceptor, a process similar to the metabolic labeling of sialic acids. EGFP can be fused to either the N-termini or the C-termini of *O*-GlcNAcylated proteins, since they are intracellular proteins. By measuring the lifetime of EGFP, imaging the *O*-GlcNAc state using FLIM–FRET microscopy has been demonstrated for a panel of proteins including tau, β-catenin, OGT, foxo1, p53, and Akt1.[76,77]

10.4.3 PLA-Based Protein-Specific Imaging of Glycosylation

PLA utilizes two distinct antibodies, each of which is conjugated with a DNA strand, as a pair of proximity probes.[80,81] When the two DNA molecules are brought into close proximity, they hybridize with two connector DNA molecules that are ligated to form a circular DNA. The circular DNA is then amplified by rolling circle amplification (RCA). The RCA products are detected by *in situ* hybridizing with a fluorescently labeled DNA probe (*i.e.*, *in situ* PLA) (Figure 10.7). The two antibodies can recognize two sites of the same protein,

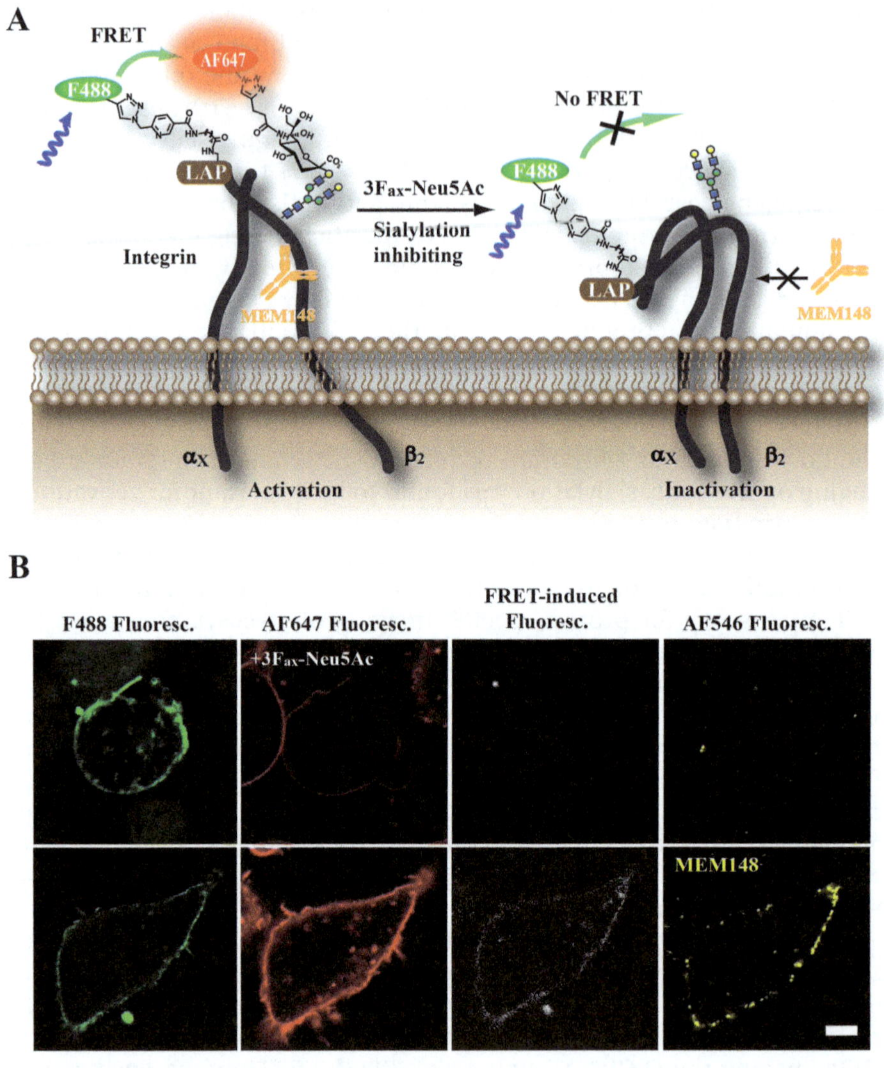

Figure 10.6 *Cis*-Membrane FRET imaging of the changes in sialylation of $\alpha_X\beta_2$ on live cells. (A) Schematic of correlating the sialylation states to the integrin activation states. (B) FRET imaging of the sialylation state of integrin $\alpha_X\beta_2$ on live cells while the activation state of $\alpha_X\beta_2$ is probed by the activation-dependent antibody MEM148. Scale bar: 5 μm. The images are adapted with permission from ref. 67, Figure 8 (http://pubs.acs.org/doi/full/10.1021/ja410086d).

enabling detection of proteins with improved specificity and sensitivity. Detection of protein–protein interactions is achieved by directing two antibodies to the respective interacting partners. *In situ* PLA can also be exploited for imaging PTMs by using a protein-recognizing antibody and a PTM-recognizing antibody.

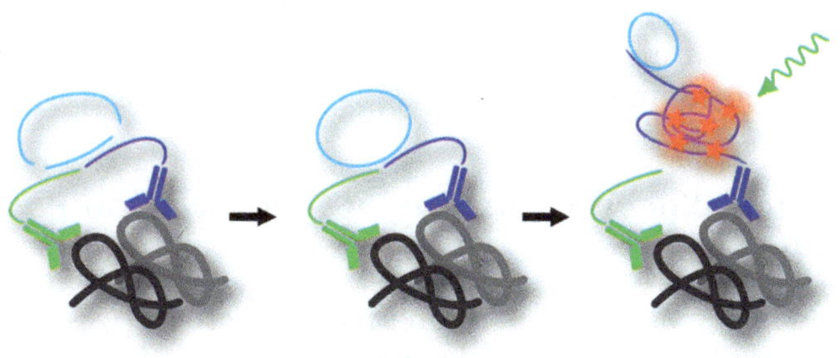

Figure 10.7 Schematic of proximity ligation assay (PLA). Two proximity probes bind close to each other. Then subsequently added linear connector oligonucleotides are guided to form a circular structure joined by enzymatic DNA ligation. After ligation, one of the proximity probes is used as primer to initiate rolling-circle amplification (RCA). Finally, the fluorescence-labeled oligonucleotides hybridize with a tag sequence in the RCA product which is detected using a fluorescence microscope.

In addition to metabolic labeling, glycans can be labeled using lectins or antibodies, which can be readily implemented in the PLA-based method. By using two antibodies which recognize MUC2 and the sialyl-Tn glycan, respectively, PLA-based protein-specific imaging of glycans was performed, which identified MUC2 as the major carrier for sialyl-Tn antigen in intestinal metaplasia and gastric cancer.[69] Various glycan-specific antibodies can be explored. For example, antibodies specific for Tn, Sialyl-Tn, T, Sialyl-Le^a, and Sialyl-Le^X were used in combinations with antibodies against MUC1, MUC2, MUC5AC, and MUC6 to image distinct mucin glycoforms in mucinous adenocarcinomas for the purpose of identifying new cancer biomarkers.[70]

10.4.4 Protein-Specific Imaging of Glycosylation Based on PEBL

The proximity-dependent signal can also be generated by taking advantage of the enhanced local concentrations of the two tags to accelerate a bioorthogonal chemical reaction.[74] To realize such a design, the sialylated glycans are metabolically labeled with azides by using $Ac_4ManNAz$. An aptamer recognizing the protein of interest is functionalized with a fluorophore and a cyclooctyne. Upon binding of the aptamer to the protein, the proximity between the cyclooctyne and the azide on the same protein drives the ligation to each other preferentially due to the increased local concentration. Although this method currently suffers from low signal-to-background ratio and applying it to fluorescence microscopy has been unsuccessful, it represents an appealing methodological development. This method is compatible with detecting protein glycoforms using mass spectrometry, western blotting, and flow cytometry. Using this strategy, a novel glycoform of PTK7 was discovered.[74] Future improvements and optimizations are anticipated to make this method compatible with imaging applications.

10.4.5 Protein-Specific Imaging of Glycosylation Based on SERS

Imaging modalities other than fluorescence microscopy such as Raman microscopy may also be explored for glycan imaging.[82] SERS amplifies the intrinsically weak Raman signal by electromagnetic enhancement when the Raman probes are close to the surface of gold or silver nanoparticles.[83] The enhancement effect is precipitously dependent on the distance between the probes and the nanoparticle surface, a phenomenon that can be exploited for protein-specific glycan imaging.[75] The cell-surface sialylated glycans metabolically incorporated with azides are reacted with cyclooctyne-functionalized gold nanoparticles (AuNPs) load with a Raman probe, 5,5′-dithiobis(2-nitrobenzoic acid) (DTBN). The DTBN–AuNPs has a diameter of 10 nm and is incapable of producing SERS, because the optimal size of AuNPs for effective SERS is in the range of 30–100 nm. The protein is dually labeled with a second AuNP with a diameter of 30 or 40 nm. The zone-controllable SERS configuration enables selective enhancement of the Raman signal of DTBN residing on the glycans of the protein of interest. Although quantification of SERS signals is notoriously complicated, this method opens an avenue for using different kinds of imaging modalities for protein-specific imaging of glycosylation.

10.5 Conclusion and Perspective

The importance of glycosylation to the biological functions of proteins has been well appreciated. A long-standing question in the field has been how to decipher the glycan code. The ubiquity of glycosylation in biological systems imposes a challenge to track specific glycans and discover critical function effects. The chemical approaches to visualizing specific protein glycans in living systems, such as those discussed in this chapter, offer a powerful tool to correlate the glycosylation state with protein functions. From the technical perspective, developing chemical approaches for imaging protein glycosylation is still in its infancy. Improving the current methods and inventing new tools are directions to pursue.

References

1. R. G. Spiro, *Glycobiology*, 2002, **12**, 43r.
2. K. W. Moremen, M. Tiemeyer and A. V. Nairn, *Nat. Rev. Mol. Cell Biol.*, 2012, **13**, 448.
3. S. T. Laughlin and C. R. Bertozzi, *Proc. Natl. Acad. Sci. U. S. A.*, 2009, **106**, 12.

4. W. Lin, L. Gao and X. Chen, *Curr. Opin. Chem. Biol.*, 2015, **28**, 156.
5. A. Helenius and M. Aebi, *Science*, 2001, **291**, 2364.
6. H. C. Hang and C. R. Bertozzi, *Bioorg. Med. Chem.*, 2005, **13**, 5021.
7. L. Wells, *Science*, 2001, **291**, 2376.
8. L. Wells, *J. Biol. Chem.*, 2013, **288**, 6930.
9. N. Haines and K. D. Irvine, *Nat. Rev. Mol. Cell Biol.*, 2003, **4**, 786.
10. P. Stanley and T. Okajima, *Curr. Top. Dev. Biol.*, 2010, **92**, 131.
11. M. G. Paulick and C. R. Bertozzi, *Biochemistry*, 2008, **47**, 6991.
12. J. R. Couchman, *Annu. Rev. Cell Dev. Biol.*, 2010, **26**, 89.
13. D. F. Zielinska, F. Gnad, J. R. Wisniewski and M. Mann, *Cell*, 2010, **141**, 897.
14. D. N. Hebert, L. Lamriben, E. T. Powers and J. W. Kelly, *Nat. Chem. Biol.*, 2014, **10**, 902.
15. G. W. Hart and R. J. Copeland, *Cell*, 2010, **143**, 672.
16. J. Jaeken and G. Matthijs, *Annu. Rev. Genomics Hum. Genet.*, 2007, **8**, 261.
17. A. H. Futerman and G. van Meer, *Nat. Rev. Mol. Cell Biol.*, 2004, **5**, 554.
18. S. S. Pinho and C. A. Reis, *Nat. Rev. Cancer*, 2015, **15**, 540.
19. E. P. Bennett, U. Mandel, H. Clausen, T. A. Gerken, T. A. Fritz and L. A. Tabak, *Glycobiology*, 2012, **22**, 736.
20. T. Ju, V. I. Otto and R. D. Cummings, *Angew. Chem., Int. Ed.*, 2011, **50**, 1770.
21. K. S. B. Bergstrom and L. J. Xia, *Glycobiology*, 2013, **23**, 1026.
22. M. A. McGuckin, S. K. Linden, P. Sutton and T. H. Florin, *Nat. Rev. Microbiol.*, 2011, **9**, 265.
23. S. Hemmerich, H. Leffler and S. D. Rosen, *J. Biol. Chem.*, 1995, **270**, 12035.
24. T. Satomaa, O. Renkonen, J. Helin, J. Kirveskari, A. Makitie and R. Renkonen, *Blood*, 2002, **99**, 2609.
25. X. Chen and A. Varki, *ACS Chem. Biol.*, 2010, **5**, 163.
26. A. Varki, *Trends Mol. Med.*, 2008, **14**, 351.
27. B. Cheng, R. Xie, L. Dong and X. Chen, *ChemBioChem*, 2016, **17**, 11.
28. J. Ma and G. W. Hart, *Clin. Proteomics*, 2014, **11**, 8.
29. Q. Zeidan and G. W. Hart, *J. Cell Sci.*, 2010, **123**, 13.
30. H. B. Ruan, Y. Nie and X. Yang, *Mol. Cell. Proteomics*, 2013, **12**, 3489.
31. D. J. Vocadlo, *Curr. Opin. Chem. Biol.*, 2012, **16**, 488.
32. M. R. Bond and J. A. Hanover, *Annu. Rev. Nutr.*, 2013, **33**, 205.
33. Q. Zeidan, Z. Wang, A. De Maio and G. W. Hart, *Mol. Biol. Cell*, 2010, **21**, 1922.
34. M. R. Bond and J. A. Hanover, *J. Cell Biol.*, 2015, **208**, 869.
35. C. M. Snow, A. Senior and L. Gerace, *J. Cell Biol.*, 1987, **104**, 1143.
36. F. I. Comer, K. Vosseller, L. Wells, M. A. Accavitti and G. W. Hart, *Anal. Biochem.*, 2001, **293**, 169.
37. R. D. Cummings, *Methods Enzymol.*, 1994, **230**, 66.
38. N. Sharon and H. Lis, *Glycobiology*, 2004, **14**, 53r.
39. J. Hirabayashi, M. Yamada, A. Kuno and H. Tateno, *Chem. Soc. Rev.*, 2013, **42**, 4443.
40. J. Carlsson, C. Gullstrand, J. Ludvigsson, I. Lundstrom and F. Winquist, *Talanta*, 2008, **76**, 333.

41. N. Okuyama, Y. Ide, M. Nakano, T. Nakagawa, K. Yamanaka, K. Moriwaki, K. Murata, H. Ohigashi, S. Yokoyama, H. Eguchi, O. Ishikawa, T. Ito, M. Kato, A. Kasahara, S. Kawano, J. Gu, N. Taniguchi and E. Miyoshi, *Int. J. Cancer*, 2006, **118**, 2803.
42. E. J. Toone, *Curr. Opin. Struct. Biol.*, 1994, **4**, 719.
43. K. Fukushima, M. Hirota, P. I. Terasaki, A. Wakisaka, H. Togashi, D. Chia, N. Suyama, Y. Fukushi, E. Nudelman and S. Hakomori, *Cancer Res.*, 1984, **44**, 5279.
44. C. F. Teo, S. Ingale, M. A. Wolfert, G. A. Elsayed, L. G. Not, J. C. Chatham, L. Wells and G. J. Boons, *Nat. Chem. Biol.*, 2010, **6**, 338.
45. G. David, X. M. Bai, B. Van der Schueren, J. J. Cassiman and H. Van den Berghe, *J. Cell Biol.*, 1992, **119**, 961.
46. V. V. Rostovtsev, L. G. Green, V. V. Fokin and K. B. Sharpless, *Angew. Chem., Int. Ed.*, 2002, **41**, 2596.
47. C. W. Tornoe, C. Christensen and M. Meldal, *J. Org. Chem.*, 2002, **67**, 3057.
48. C. Besanceney-Webler, H. Jiang, T. Zheng, L. Feng, D. Soriano del Amo, W. Wang, L. M. Klivansky, F. L. Marlow, Y. Liu and P. Wu, *Angew. Chem., Int. Ed.*, 2011, **50**, 8051.
49. E. M. Sletten and C. R. Bertozzi, *Acc. Chem. Res.*, 2011, **44**, 666.
50. S. J. Luchansky, H. C. Hang, E. Saxon, J. R. Grunwell, C. Yu, D. H. Dube and C. R. Bertozzi, *Methods Enzymol.*, 2003, **362**, 249.
51. D. Rabuka, S. C. Hubbard, S. T. Laughlin, S. P. Argade and C. R. Bertozzi, *J. Am. Chem. Soc.*, 2006, **128**, 12078.
52. M. Sawa, T. L. Hsu, T. Itoh, M. Sugiyama, S. R. Hanson, P. K. Vogt and C. H. Wong, *Proc. Natl. Acad. Sci. U. S. A.*, 2006, **103**, 12371.
53. K. W. Dehnert, B. J. Beahm, T. T. Huynh, J. M. Baskin, S. T. Laughlin, W. Wang, P. Wu, S. L. Amacher and C. R. Bertozzi, *ACS Chem. Biol.*, 2011, **6**, 547.
54. H. Jiang, L. Feng, D. Soriano del Amo, R. D. Seidel Iii, F. Marlow and P. Wu, *J. Visualized Exp.*, 2011, **52**, e2686.
55. H. C. Hang, C. Yu, D. L. Kato and C. R. Bertozzi, *Proc. Natl. Acad. Sci. U. S. A.*, 2003, **100**, 14846.
56. D. J. Vocadlo, H. C. Hang, E. J. Kim, J. A. Hanover and C. R. Bertozzi, *Proc. Natl. Acad. Sci. U. S. A.*, 2003, **100**, 9116.
57. M. Boyce, I. S. Carrico, A. S. Ganguli, S. H. Yu, M. J. Hangauer, S. C. Hubbard, J. J. Kohler and C. R. Bertozzi, *Proc. Natl. Acad. Sci. U. S. A.*, 2011, **108**, 3141.
58. B. W. Zaro, Y. Y. Yang, H. C. Hang and M. R. Pratt, *Proc. Natl. Acad. Sci. U. S. A.*, 2011, **108**, 8146.
59. K. N. Chuh, B. W. Zaro, F. Piller, V. Piller and M. R. Pratt, *J. Am. Chem. Soc.*, 2014, **136**, 12283.
60. K. K. Palaniappan, M. J. Hangauer, T. J. Smith, B. P. Smart, A. A. Pitcher, E. H. Cheng, C. R. Bertozzi and M. Boyce, *Cell Rep.*, 2013, **5**, 546.
61. T. N. Ramya, E. Weerapana, L. Liao, Y. Zeng, H. Tateno, L. Liao, J. R. Yates, 3rd, B. F. Cravatt and J. C. Paulson, *Mol. Cell. Proteomics*, 2010, **9**, 1339.

62. N. E. Mbua, X. Li, H. R. Flanagan-Steet, L. Meng, K. Aoki, K. W. Moremen, M. A. Wolfert, R. Steet and G. J. Boons, *Angew. Chem., Int. Ed.*, 2013, **52**, 13012.
63. P. M. Clark, J. F. Dweck, D. E. Mason, C. R. Hart, S. B. Buck, E. C. Peters, B. J. Agnew and L. C. Hsieh-Wilson, *J. Am. Chem. Soc.*, 2008, **130**, 11576.
64. T. Zheng, H. Jiang, M. Gros, D. S. del Amo, S. Sundaram, G. Lauvau, F. Marlow, Y. Liu, P. Stanley and P. Wu, *Angew. Chem., Int. Ed.*, 2011, **50**, 4113.
65. J. L. Chaubard, C. Krishnamurthy, W. Yi, D. F. Smith and L. C. Hsieh-Wilson, *J. Am. Chem. Soc.*, 2012, **134**, 4489.
66. Q. Li, Z. Li, X. Duan and W. Yi, *J. Am. Chem. Soc.*, 2014, **136**, 12536.
67. R. Y. Tsien, *Angew. Chem., Int. Ed.*, 2009, **48**, 5612.
68. M. Rashidian, J. K. Dozier and M. D. Distefano, *Bioconjugate Chem.*, 2013, **24**, 1277.
69. T. Conze, A. S. Carvalho, U. Landegren, R. Almeida, C. A. Reis, L. David and O. Soderberg, *Glycobiology*, 2010, **20**, 199.
70. R. Pinto, A. S. Carvalho, T. Conze, A. Magalhaes, G. Picco, J. M. Burchell, J. Taylor-Papadimitriou, C. A. Reis, R. Almeida, U. Mandel, H. Clausen, O. Soderberg and L. David, *J. Cell. Mol. Med.*, 2012, **16**, 1474.
71. Y. Haga, K. Ishii, K. Hibino, Y. Sako, Y. Ito, N. Taniguchi and T. Suzuki, *Nat. Commun.*, 2012, **3**, 907.
72. B. Belardi, A. de la Zerda, D. R. Spiciarich, S. L. Maund, D. M. Peehl and C. R. Bertozzi, *Angew. Chem., Int. Ed.*, 2013, **52**, 14045.
73. W. Lin, Y. Du, Y. Zhu and X. Chen, *J. Am. Chem. Soc.*, 2014, **136**, 679.
74. P. V. Robinson, G. de Almeida-Escobedo, A. E. de Groot, J. L. McKechnie and C. R. Bertozzi, *J. Am. Chem. Soc.*, 2015, **137**, 10452.
75. Y. Chen, L. Ding, W. Song, M. Yang and H. Ju, *Chem. Sci.*, 2016, **7**, 569.
76. W. Lin, L. Gao and X. Chen, *ChemBioChem*, 2015, **16**, 2571.
77. F. Doll, A. Buntz, A. K. Spate, V. F. Schart, A. Timper, W. Schrimpf, C. R. Hauck, A. Zumbusch and V. Wittmann, *Angew. Chem., Int. Ed.*, 2016, **55**, 2262.
78. E. A. Jares-Erijman and T. M. Jovin, *Nat. Biotechnol.*, 2003, **21**, 1387.
79. H. Wallrabe and A. Periasamy, *Curr. Opin. Biotechnol.*, 2005, **16**, 19.
80. O. Soderberg, K. J. Leuchowius, M. Gullberg, M. Jarvius, I. Weibrecht, L. G. Larsson and U. Landegren, *Methods*, 2008, **45**, 227.
81. I. Weibrecht, K. J. Leuchowius, C. M. Clausson, T. Conze, M. Jarvius, W. M. Howell, M. Kamali-Moghaddam and O. Soderberg, *Expert Rev. Proteomics*, 2010, **7**, 401.
82. L. Lin, X. Tian, S. Hong, P. Dai, Q. You, R. Wang, L. Feng, C. Xie, Z. Q. Tian and X. Chen, *Angew. Chem., Int. Ed.*, 2013, **52**, 7266.
83. S. Schlucker, *Angew. Chem., Int. Ed.*, 2014, **53**, 4756.

CHAPTER 11

Targeting Glycans of HIV Envelope Glycoproteins for Vaccine Design

ANNA-JANINA BEHRENS[a], GEMMA E. SEABRIGHT[a] AND MAX CRISPIN*[a]

[a]Oxford Glycobiology Institute, Department of Biochemistry, University of Oxford, South Parks Road, Oxford OX1 3QU, UK
*E-mail: max.crispin@bioch.ox.ac.uk

11.1 The Human Immunodeficiency Virus

The human immunodeficiency virus (HIV) causes acquired immunodeficiency syndrome (AIDS) and remains one of the major health challenges worldwide. It is estimated that 37 million people were living with HIV by the end of 2014, most of them in sub-Saharan Africa (average prevalence of 7.4%).[1] Incidences of infection, however, are continuously growing in Eastern Europe and Central Asia. Although there have been huge developments in antiviral therapy, a vaccine is likely to be an essential component in the control of the epidemic.[2]

HIV research began in 1981, when Michael S. Gottlieb observed and linked five clinical cases of homosexual men contracting a rare lung infection (*Pneumocystis carinii* pneumonia) in Los Angeles.[3] In June 1981, the Centers for Disease Control and Prevention (CDC) in Atlanta published the first public

Chemical Biology No. 2
Chemical Biology of Glycoproteins
Edited by Zhongping Tan and Lai-Xi Wang

Published by the Royal Society of Chemistry, www.rsc.org

report of clinical observations of what would later be known as AIDS.[4] Since then, over 30 years of intense HIV research have led to an increasing understanding of the virus and significant advances in the development of an effective treatment. What was once a lethal infection, has been described as a "manageable chronic disease".[5] A recent report by UNAIDS stated that more than 15 million people are currently receiving antiretroviral treatment.[6] Despite this achievement, more than half the people living with the virus still do not have access to medication, thus, there is a need for the development of an effective vaccine in order to bring the HIV/AIDS pandemic under control.

There are two distinct types of human immunodeficiency virus, HIV-1 and HIV-2, which share about 60% amino acid sequence identity.[7] Both types originate from zoonotic (animal to human) transmission of simian immunodeficiency viruses (SIVs), and both are capable of causing AIDS.[8] HIV-2 is generally associated with a longer asymptomatic stage, lower virus titers and a lower transmission rate compared with HIV-1.[8] Being only endemic in West Africa, HIV-2 is currently not the main focus of attention of the HIV research community. HIV-1 is classified into four groups (M, N, O and P), all of which emerged from different zoonotic infections.[9,10] About 90% of all HIV-1 infections are caused by infections of the 'major' group M, which is the only group that has established a pandemic spread. It comprises nine genetically distinctive clades (A, B, C, D, F, G, H, J and K), which vary in their geographical prevalence across the world. Clades A, B and C are the most prevalent genetic forms, with clade C being responsible for about half of all HIV-1 infections.[9] Clade A is common in Eastern and Central Africa as well as in Eastern Europe.[9] Clade B is the dominant form in Western and Central Europe, USA, Thailand, Australia and Japan, whereas clade C is dominant in Southern and Eastern Africa and India.[9]

11.1.1 Structure, Genome and Viral Lifecycle

HIV-1 is a single-stranded, enveloped RNA retrovirus belonging to the genus *Lentivirus* (Figure 11.1). The virus particles are 100–120 nm in diameter and contain two copies of the positive-stranded RNA genome. The viral genome encodes three major gene products (gag, pol and env). All of them are translated as polyproteins, with Gag providing the structural lattice of the viral particle (the viral matrix, the capsid and the nucleocapsid) and Pol the viral enzymes (protease, reverse transcriptase and integrase). The enveloped virus is surrounded by a lipid bilayer gained while budding from the host cell. The surface transmembrane protein, Envelope (Env), is embedded into this bilayer (Figure 11.1B). Env is a trimer of heterodimers of gp120 and gp41, which is generated by furin cleavage of the pro-protein gp160. It is one of the most heavily glycosylated proteins, with about half of the mass of gp120 being composed of *N*-linked glycans.[11] Env initiates host-cell infection and displays the main target for an antibody-mediated immune response. We will discuss the structure and function of Env in detail in Section 11.2.2.

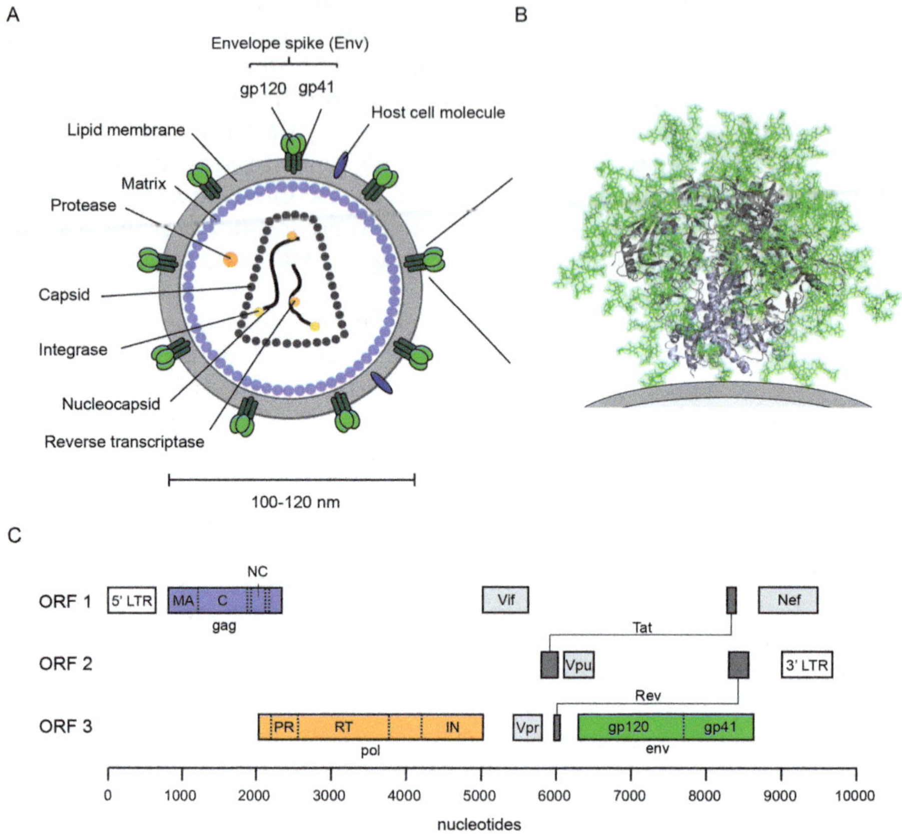

Figure 11.1 Structure and genome of the HIV-1 virion. (A) Virions assemble into spherical particles about 100–120 nm in diameter.[384] The two copies of the positive-sense, single-stranded RNA genome are surrounded by nucleocapsid proteins and the enzymes integrase and reverse transcriptase. The genome is enclosed by a conical capsid, which itself is surrounded by a matrix ensuring structural integrity. The envelope spike (Env), composed of a trimer of heterodimers of gp120 and gp41, is embedded into the host-cell-derived lipid membrane. Around 14 Env molecules are incorporated into each virion.[176] (B) Fully glycosylated model (glycans shown as green sticks) of Env based on PDB ID 5ACO.[232] Env is a trimer of heterodimers of non-covalently associated gp120 and gp41. (C) The RNA HIV-1 genome of approximately 9.8 kb is transcribed in multiple open reading frames (ORFs). The structural matrix (MA), capsid (C) and nucleocapsid (NC) proteins are encoded by gag (blue) and cleaved by the viral protease (PR). PR also cleaves itself and the viral enzymes (encoded by pol, orange) reverse transcriptase (RT) and integrase (IN). Env (green) encodes the envelope glycoproteins gp120 and gp41. Dark grey, Regulatory proteins Tat and Rev; Accessory proteins Vpr, Vfu, Vif and Nef are shown in light grey. The long terminal repeats (LTRs) have a function in transcription regulation. Sequence information was extracted from the Los Alamos HIV sequence database.[235]

The tropism of HIV-1 is defined by the presence of its main receptor, cluster of differentiation 4 (CD4), which is expressed on the surface of macrophages, dendritic cells and T helper cells ($CD4^+$ T lymphocytes), as well as its co-receptors, C–C chemokine receptor 5 (CCR5) or CXC-chemokine receptor 4 (CXCR4).[12–15] The majority of HIV-1 infections are established by CCR5-tropic viruses.[16,17] Viral binding to CD4 receptors is mediated through the gp120 subunit of the envelope spike. Upon binding, conformational changes in Env are initiated, particularly in gp41, which allow subsequent co-receptor binding and fusion of the viral and host-cell membranes.[18] The viral capsid can now enter the host-cell cytoplasm, where the viral reverse transcriptase (RT) produces double-stranded DNA from the viral RNA. The viral DNA is next imported into the nucleus, where the viral integrase (IN) integrates it into the host cell genome.[19] The host cell's translation machinery then takes over protein production, *i.e.* transcription and translation of viral genes, and thus enables the formation of new virions, which bud from the host-cell membrane and undergo maturation.[20] Alternatively, the integrated provirus DNA can latently persist in memory T cells, hence impeding the complete eradication of the virus by antiretroviral treatments.[21–23] It has recently been shown that replication-competent proviral reservoirs mainly derive from quiescent T cells, in contrast to those that undergo clonal expansion.[24]

11.1.2 Transmission and Pathogenesis

The major route of HIV-1 transmission is *via* entry through the vaginal or rectal mucosa during sexual intercourse. Thus, the virus needs to penetrate the mucosal and epithelial barriers covering the genital and gastrointestinal tracts in order to encounter host cells prone to infection. This is thought to be achieved by several means including transcytosis, endocytosis or passing through small gaps within the epithelia.[25,26] It is still not clear if HIV-1 is transmitted in a cell-bound state or only as a free virus. SIV, however, can be transmitted in both ways.[27] Interestingly, an infection with HIV-1 is in most cases established by one single transmitted/founder virus meaning that a successful vaccine needs to induce an immune response capable of eliminating this initial virus.[28–30]

During the first approximately 10 days after infection, termed the eclipse phase, viral RNA in the plasma remains undetectable.[31] Following this initial phase virus levels increase exponentially, peaking at between 21 and 28 days. During this acute phase $CD4^+$ T cell numbers decrease, $CD8^+$ T cells are activated, antibody production starts (seroconversion), and patients may present influenza-like symptoms. The virus spreads into the lymph nodes, in particular into the gut-associated lymphoid tissue (GALT), where numbers of $CCR5^+/CD4^+$ T lymphocytes are high, and where a grave depletion of T lymphocytes is observed during acute infection, independent of the route of transmission.[32–35] The migration and retention of the virus in the gut is thought to be mediated by the gut-homing receptor, integrin $\alpha_4\beta_7$, on the surface of lymphocytes that interacts with a tripeptide in the V2 loop of gp120.[36–38] Engagement of gp120 rapidly activates production of the integrin LFA-1, which facilitates cell-to-cell transmission of HIV-1 through the initiation of virological synapses.[36] It has been proposed that the protective effects observed in the RV144 vaccine trial (described in Section 11.1.5) are partly attributed to a high titer of antibodies that target the V2 loop of gp120 and thus block the gut-homing receptor binding site.[39] Over the following weeks of infection, viral levels decrease to reach the viral set point under the pressure of a rising adaptive immune response, which can be first detected just before peak viremia.[40–43] $CD4^+$ T cell counts then return to almost normal levels, although levels in the GALT remain low, in both HIV-1 and SIV.[31,35,44,45] HIV-1 is highly mutative and manages to repeatedly escape the immune system, *e.g.* by presenting non-neutralizing epitopes or by shifting the position of its surface glycans. The immune system itself is progressively weakened as a result of gradually decreasing T cell counts. Subsequently, the latter stage of the disease is marked by vulnerability to opportunistic infections, indicating progression to AIDS.

11.1.3 The Viral Envelope is the Main Target for the Immune System

Assembled HIV-1 virions obtain the lipid envelope and the Env protein spikes as they bud from the host cell's plasma membrane. Env and Gag are both enriched in lipid raft microdomains in the plasma membrane, which very probably represent the location of virus assembly.[20,46] This is supported by the increased presence of sphingolipids, cholesterol and phosphatidylinositol-4,5-bisphosphate in the viral envelope.[47,48] The carbohydrates of the viral envelope shield the virus from immune recognition. As they originate from the host, they limit the freedom of action for B cell responses due to antigen self-tolerance. Additionally, sialylated viral envelope glycolipids (termed gangliosides) have been demonstrated to interact with the cellular lectin receptor sialic acid-recognizing Ig superfamily lectins-1 (Siglec-1) on the surface of dendritic cells and therefore to potentially enhance HIV-1 infection.[49,50] One notable exception can be observed during the transmission of primary virions, which can carry non-self ABO blood group antigens of the infected human. HIV-1 is able to incorporate host cell ABO carbohydrates into its membrane during budding, hence virus from A- or B-positive donors showed sensitivity to anti-A- or anti-B-antibodies from blood-group-discordant recipients.[51–53] These blood group antigens possibly originate from glycolipids and/or membrane-bound glycoproteins present in the host cell's membrane and thus constitute a host mechanism for reduced viral infectivity. Nevertheless, the discovery and isolation of broadly neutralizing antibodies (bNAbs) from patients has established the general consensus that the glycans on the HIV-1 surface not only offer protection but are also themselves a target for immune surveillance.[2] Many of these Env-targeting bNAbs have glycan-dependent epitopes.[54–68] Results from numerous studies have indicated that bNAbs provide passive protection from viral challenge to non-human primates, highlighting their potential for therapy and vaccine development.[69–74] We will discuss the development of bNAbs and their glycan epitopes in detail in Section 11.3.

11.1.4 Immune Response

Upon infection, invading pathogens are first challenged by the host's innate immune response. HIV-1 host cell entry occurs *via* interaction of Env with the entry receptors described above, but virions can also be captured *via* pattern recognition receptors (PRRs). Soluble and membrane-bound PRRs recognize viral pathogen-associated molecular patterns (PAMPs) and thereupon induce the expression of antiviral genes as well as inflammation *e.g.* by secretion of interferons and proinflammatory cytokines and chemokines.[75] Immune sensing mainly occurs in cells targeted by HIV-1, where exogenous viral particles, viral RNA and intermediates of the reverse transcription process are recognized.[76] However, HIV-1 has evolved to limit detection by PRRs and to inhibit immune signalling *e.g.* by physically shielding viral nucleic acids in the cytoplasm before the dsDNA is translocated into the nucleus.[76] Moreover, the viral accessory protein Vpu inhibits nuclear factor kappa B (NF-κB) signalling and hence weakens antiviral gene expression.[76] The processes by which the innate immune system senses HIV-1 and how the virus evades this response were reviewed recently in great detail.[76–78]

Among the first immune cells that HIV-1 encounters are CD4$^+$ memory T cells, macrophages and dendritic cells (in particular Langerhans cells), which express various groups of PRRs, and vary in their response to HIV-1 invasion. Recognition of HIV-1 glycans through specific carbohydrate-binding PRRs, termed C-type lectin receptors (CLRs), plays a key role in HIV-1 transmission and replication. The CLR family shares a conserved structural motif in one or more carbohydrate recognition domains.[79] CLRs mediate pathogen internalization and endosomal degradation and induce antigen presentation and immune signalling cascades.[80] They display distinct carbohydrate specificities linked to their carbohydrate recognition domains.[81] The particularly high content of oligomannose-type glycans on HIV-1 Env (see Section 11.2.3) allows recognition by mannose-binding CLRs. Among the lectin receptors that recognize mannose residues on the HIV-1 surface are dendritic cell-specific intracellular cell adhesion molecule 3-grabbing non-integrin (DC-SIGN; expressed on dendritic cells and macrophages) and langerin (expressed on Langerhans cells).[82–84] The outcome of glycan receptor recognition depends on the immune cells and the CLR.[85] Binding of HIV-1 to dendritic cells *via* DC-SIGN is thought to promote viral infection and transmission to CD4$^+$ T cells.[86–89] However, the role of Langerhans cells in infection and transmission through langerin expression is controversial. It is known that binding of gp120 to langerin leads to virus internalization into Birbeck granules, but opinions about whether or not this protects or promotes infection and transfer to T cells vary.[82,90–94]

The complement system plays a diverse role in HIV-1 infection.[95] It exhibits a protective role upon HIV-1 infection by complement-dependent viral lysis and neutralization, which are mainly activated through the classical pathway by HIV-1-specific antibodies.[96–98] Additionally, direct binding of complement factor C1q to gp41 as well as the interaction of gp120 glycans with mannose-binding lectin (MBL) is reported to activate the complement system.[95,99,100] Although, as part of the lectin pathway, MBL only mediates low levels of HIV-1 neutralization, its binding can inhibit DC-SIGN uptake of virions and hence protect from viral infection.[101,102] The presence of regulators of complement activation in the viral envelope, such as the membrane-associated CD59 and soluble factor H, limits complement-dependent lysis of virions and infected cells.[95,103,104] In contrast, the complement system also plays a critical role in augmenting viral spread and promoting infection. For example, complement activation products like C3 fragments can be deposited on the viral surface which facilitates binding of HIV-1 to cells expressing complement receptors CR3 and CR4 (*i.e.* dendritic cells and macrophages).[95,105,106]

The adaptive immune response is triggered and regulated by interactions between HIV-1 and dendritic cells. Two distinct types of dendritic-cell-mediated HIV-1 transmission have been described—*cis*- and *trans*-infection. During *cis*-infection, dendritic cells produce and release progeny virions that allow infection of target cells through viral-receptor-mediated entry. During *trans*-infection, transmission of endocytosed or surface-adhered virions occurs through the exosomal secretion pathway and *via* the formation of infectious synapses (with the participation of DC-SIGN) with the target cell, respectively.[107,108]

The first $CD8^+$ T cell responses arise close to the peak of HIV-1 viremia, and peak around one to two weeks later. The earliest targets are very often Env and Nef.[109] The peak in cell count of specific $CD8^+$ T cells triggers the virus sequence to change dramatically and produce $CD8^+$ epitope escape mutants.[31,110] As $CD4^+$ T cells themselves are targeted by HIV-1, related immune responses are hard to verify and cell numbers gradually decline during infection.[31,111] The initial antibody response targeting HIV-1 Env starts around two weeks after transmission and is non-neutralizing, which is likely to be explained by the presence of non-functional Env displaying immunogenic surfaces usually not accessible in functional Env trimers.[112–114] An HIV-1-specific B cell response producing autologous-neutralizing antibodies starts around three months after infection.[115,116] These antibodies also induce escape mutants of the virus. Antibodies that neutralize heterologous viruses to some degree can be identified in about one fourth of all infected people after several years of infection.[117] However, potent broadly neutralizing antibodies targeting conserved regions of Env are very rare. In a recent longitudinal study on an HIV-1-infected cohort in Eastern and South Africa, only about 15% of individuals developed an antibody response with more than 50% neutralization breadth after two to four years of infection.[118] Additionally, acute HIV-1 infection leads to early damage to the mucosal B cell generative microenvironments by inducing lysis and apoptosis.[31,119] In summary, neither the innate nor adaptive immune responses are unfortunately strong or fast enough to clear an HIV-1 infection before manifestation of a latent disease state.

11.1.5 Current Therapies and Steps towards a Vaccine

The first antiretroviral drug against HIV-1 was Azidothymidine, which was launched only a few years after the discovery of the virus.[120] As of 2016, there were 39 FDA-approved drugs available, which can be divided into six distinct classes according to their mechanisms of action: Nucleoside-analog RT inhibitors, non-nucleoside RT inhibitors, integrase inhibitors, protease inhibitors, fusion inhibitors and entry inhibitors. The large number of available drugs can be rationalized by the speed at which HIV-1 has evolved resistance to individual drugs.[121] The viral genome was shown to mutate with an extremely high *in vivo* rate of about 4×10^{-3} mutations per base per cell.[122] High mutation frequencies are the result of an error-prone reverse transcriptase and induction of mutations through cellular cytidine deaminases combined with high viral replication rates.[122–126] This leads to the generation and selection of drug-resistant escape mutants – termed quasi-species.[123,127] The original mono-drug therapy was therefore replaced by highly active antiretroviral therapy (HAART; or cART (combined antiretroviral therapy)), which is a combination therapy of at least three different antiretrovirals targeting two or more distinct molecular targets.[128] HAART has significantly reduced the morbidity and mortality of infected patients; decreasing viral loads and increasing $CD4^+$ T cell counts.[129–133] In some parts the world, HAART has increased life expectancies to almost normal levels. It can also be used as a preventive measure for high-risk groups such as relationships where one partner is infected or mother-to-child transmission during pregnancy.[129,134]

Although the development of HAART has been a great success in fighting the HIV-1/AIDS pandemic, it isn't a cure. The combination of side effects associated with antiretrovirals and very complex treatment regimens often leads to nonadherence, which can quickly result in an increase in viral load as HAART is unable to eliminate latent virus reservoirs.[135,136] Additionally, preventive measures such as condoms and male circumcision can significantly lower the risk of an HIV-1 infection, and trials of some vaginal microbicide creams have also yielded promising results.[137,138] However, all these preventative methods depend highly on personal behavior, and none can fully eradicate the disease. The best hope lies in the development of a successful vaccine. In general, vaccines aim to provide adaptive immunity to a particular pathogen. As described above, most HIV-1 infections originate from one single transmitted/founder virus that needs to be stopped before reservoirs of latent virus are established in cells. This has been shown to happen very early during infection, leaving only a 5–10 day window during which the virus could be eradicated.[31,139]

Four vaccination approaches have been investigated during previous vaccine trials: Protein-subunit vaccines, recombinant adenovirus vectors, canarypox vector priming followed by a protein-subunit boost and DNA priming followed by a recombinant adenovirus vector boost.[140] The first two major trials, Vax003 and Vax004, explored the application of AIDSVAX, which contained recombinant gp120 and the adjuvant alum. Solely non-neutralizing antibodies were developed and no protection was observed.[141,142] The STEP trial and the Phambili trial investigated three recombinant, attenuated adenovirus type 5 vectors expressing Gag, Pol and Nef. A stimulated $CD8^+$ T cell response was observed, however no protection was achieved and the studies were stopped early due to concerns that the risk of an HIV-1 infection was actually increased.[143–146] During the RV144 trial in Thailand a canarypox viral vector (ALVAC) expressing gp120, Gag and Pol was used to prime, followed by two booster injections of the AIDSVAX gp120 vaccine. This trial provided small protective effects for the first time, with approximately 30% fewer infections observed among vaccine recipients.[147] It is not clear how exactly this protection was mediated, but the leading hypothesis is that the reduced risk of infection is associated with V1/V2-targeting antibodies. One possibility is that part of the protective effect can be attributed to non-neutralizing, anti-HIV-1 Fc-mediated antibody effector functions.[148] The efficacy of a DNA prime (consisting of six plasmids encoding Env, Gag, Pol and Nef from different strains) and adenovirus type 5 vector boost (expressing a Gag–Pol fusion protein and Env from different strains) was investigated during the HVTN 505 trial.[149] This trial was also terminated before completion, again due to no protection being observed and concerns about adenovirus vectors increasing the risk of infection.[150] Adenoviral vectors are currently still the most promising $CD8^+$ T cell response-generating vectors, but as clear explanations for the potential increase in susceptibility to HIV-1 infections are lacking, both additional human safety tests and the simultaneous development of alternative vector systems are required.[151]

Currently, follow-up trials building on the promising results of the RV144 study are being developed and are under investigation, testing, for example, redesigned vector platforms.[148] Furthermore, much attention of vaccine research is at present centered on the development of therapeutic Tat vaccines, application of inactivated whole-virus depleted of gp120 and the induction of bNAbs (which we will discuss in detail in Section 11.4).[148]

11.2 The Viral Envelope Spike

The viral spike's crucial role in mediating viral infection and its exposed presence on the virion surface predestine the viral envelope spike as the main target for vaccine development.

11.2.1 Biosynthesis

Mature and functional envelope spikes are trimers of non-covalently associated heterodimers of gp120 and gp41. The precursor pro-protein gp160 contains an N-terminal signal peptide, which targets the nascent peptide to the membrane of the endoplasmic reticulum (ER). Surprisingly, this signal peptide has a down-regulatory and inhibitory effect on Env folding as its co-translational cleavage is very slow and inefficient compared with that of other glycoproteins.[152] This is likely to be explained by the elevated number of charged amino acids present in the typically very hydrophobic signal peptide sequence.[153] Complete release of Env into the ER lumen as well as its membrane-anchored position are determined by the presence of a hydrophobic ribosomal stop-transfer sequence in gp41.[154,155] The envelope is co-translationally modified by the addition of *N*-linked glycans within Asn–X–Ser/Thr sequons, with X being any amino acid except proline.[156] While the potential position of the *N*-linked glycan is determined by the protein sequence, the structure of the glycan is controlled by the host cell's glycosylation machinery. The mammalian *N*-glycosylation pathway in the ER and the Golgi apparatus is shown in Figure 11.2, beginning with the *en bloc* transfer of $Glc_3Man_9GlcNAc_2$ (Glc, glucose; Man, mannose; GlcNAc, *N*-acetylglucosamine) to amide groups of asparagines at potential *N*-glycosylation sites (PNGSs).

The ER-resident α-glucosidases I and II create monoglucosylated *N*-linked glycans, which interact with the folding chaperons and lectins, calnexin and calreticulin.[157] Once α-glucosidase II removes the final glucose, the glycoprotein is released from calnexin and calreticulin. Unless it is recognized as misfolded and re-glucosylated by the UDP–glucose glucosyltransferase, the protein can then leave the ER.[158] Re-glucosylated proteins undergo further chaperone-mediated folding cycles until they are either correctly folded, or disposed of *via* the ER-associated degradation pathway. The α-glucosidase I and II inhibitor *N*-butyldeoxynojirimycin (NB-DNJ) has been investigated as a candidate anti-HIV-1 drug that targets this folding process.[159] Blocking entry of gp160 into the calnexin–calreticulin cycle leads to the formation of viral particles that are less infective, probably due to a conformational defect within the V1/V2 region that then inhibits membrane fusion.[159–161] An *in vitro* study into the use of NB-DNJ as an antiretroviral treatment yielded promising results, in that no resistance mutations were detected after 15-weeks cultivation of HIV-1 in the presence of NB-DNJ.[159] However, the off-target inhibition of gastrointestinal α-glucosidase has prevented clinical development.[162]

Folding of gp160 has been described as extremely slow, with gp160 residing in the ER longer than other glycoproteins.[163] Next to the above described effect of the signal peptide, ER maturation is prolonged by the correct formation and extensive isomerization of the intramolecular disulfide bonds of Env which typically number around ten per protomer (of which nine are located in gp120).[163,164] Properly folded gp160 carrying $Man_9GlcNAc_2$ glycans moves through the ER-Golgi network, during which the *N*-linked glycans can be further trimmed by ER and Golgi α-mannosidases (Figure 11.2). As we will highlight later, the *N*-glycosylation trimming of multiple sites of Env stalls at the oligomannose series $Man_{5-9}GlcNAc_2$, leading to an *N*-linked glycosylation pattern dominated by minimally trimmed glycans that differs from what it normally observed in human glycoproteins.[164–168] Complex-type glycosylation

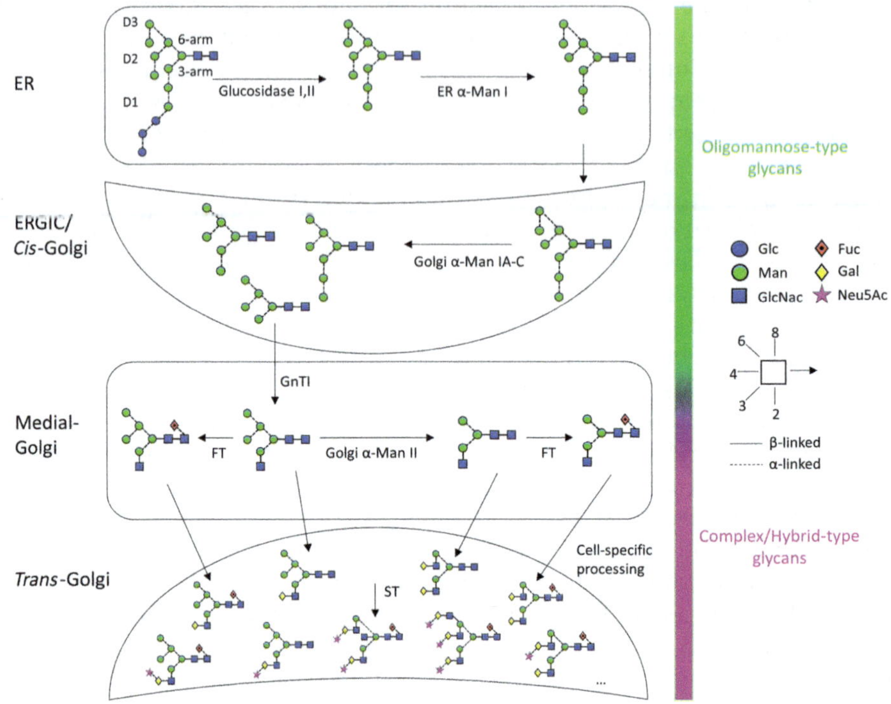

Figure 11.2 Mammalian *N*-glycosylation pathway. In the ER, $Glc_3Man_9GlcNAc_2$ is transferred from a dolichol membrane anchor and asparagine-linked to the glycoprotein. The three terminal glucoses are removed by α-glucosidases I and II, followed by ER α-mannosidase I (ER α-Man I) removing the terminal mannose from the D2 arm. Golgi α-mannosidases IA-C (Golgi α-Man IA-C) are responsible for further mannose trimming leading to $Man_5GlcNAc_2$ in the ER–Golgi intermediate compartment (ERGIC) and *cis*-Golgi. $Man_5GlcNAc_2$ is then further processed by *N*-acetylglucosaminyltransferase I (GnTI) which initiates hybrid and complex glycan processing. $Man_5GlcNAc_2$ is also a substrate for Golgi-resident α1,6-fucosyltransferase (FT). Alternatively, further mannose trimming through the Golgi α-mannosidase II (Golgi α-Man II) can occur leading to $GlcNAcMan_3GlcNAc_2$. In the *trans*-Golgi, glycans are further processed by the action of cell-specific glycosyltransferases, sialyltransferases (ST) and fucosyltransferases. Glycan representation is shown according to the combined nomenclature of Harvey and colleagues and the Center of Functional Glycomics.[385,386]

is initiated by the addition of a β1,2-linked *N*-acetylglucosamine to the 3-arm of $Man_5GlcNAc_2$. This is followed by further processing by Golgi-resident glycosidases and glycosyltransferases. The wide spectrum of hybrid and complex-type glycans is cell type- or tissue-specific and can lead to a range of different glycoforms and microheterogeneity at individual glycan sites.[169]

Gp160 monomers trimerize in the ER, although the presence of non-functional dimers and tetramers has also been described.[155,170,171] While Env traffics through the secretory pathway, it is cleaved into gp120 and gp41 by the Golgi-resident protease furin, at the C-terminus of the consensus sequence Arg-Glu-Lys-Arg.[172] Cleavage of gp160 is crucial for the structural and functional integration of the envelope spike and thus for viral infectivity.[168,173,174] Non-covalently associated gp120 and gp41 trimers then assemble at the plasma membrane, where they are associated with Gag and incorporated into new virions. However, the complete mechanism for the incorporation of Env is not yet fully understood.[20] Once assembled at the plasma membrane, Env is rapidly subjected to clathrin-mediated endocytosis.[175] This, as well as the shedding of gp120 caused by the rather weak gp41–gp120 interactions, leads to an exceptionally low number (about 14) of envelope spikes per virion.[176] Since a higher number of Env molecules has been reported to increase infectivity, as tested in a SIV model, HIV-1 must have other advantages explaining why it has evolved to carry a low number of spikes.[177,178] One explanation might be to avoid neutralization as the binding of bivalent immunoglobulin G (IgG) antibodies is disfavoured.[179] It has also been suggested that the particularly high mean epitope spacing of Env spikes slows down the development of broadly neutralizing antibodies as the distance between two spikes is too high to overcome B cell tolerance mechanisms and hence impedes the development of self-reactive antibody intermediates.[178]

11.2.2 Structure and Function of Env

The type I fusion machine Env is metastable in its pre-fusion state and responsible for CD4 receptor recognition followed by fusion of the viral and host cell membranes. Figure 11.3A shows the overall schematic structure of Env. Gp120 has a highly variable surface consisting of five variable loops (V1–V5) that are interspersed by five constant regions. The sequence of the fusion-mediating gp41 subunit is more conserved, consisting of an ectodomain, a membrane proximal external region (MPER), a transmembrane domain and a cytoplasmic tail.

Structural studies of Env are relatively difficult because of the conformational dynamics of the molecule, the high degree of *N*-glycosylation and the presence of variable loops. Despite these challenges, the publication of the first high-resolution X-ray structure of a monomeric gp120 core in complex with CD4 almost 20 years ago by the laboratory of Peter Kwong, and a number of follow-up studies, have revealed fine structural details about the receptor-bound conformation of gp120.[180–185] Furthermore, the laboratories of Ian Wilson, Andrew Ward and Peter Kwong have recently generated structures of intact trimers by X-ray crystallography and cryo-electron microscopy (cryo-EM), respectively.[186–191]

The structures of monomeric gp120 all show similar folding, consisting of an N-terminal inner domain and a C-terminal outer domain, connected by the bridging sheet—a four-stranded antiparallel β-sheet.[180–185] Large structural rearrangements are induced in the viral spike upon CD4 binding, accompanied by large, unfavorable changes in entropy.[192,193] These changes

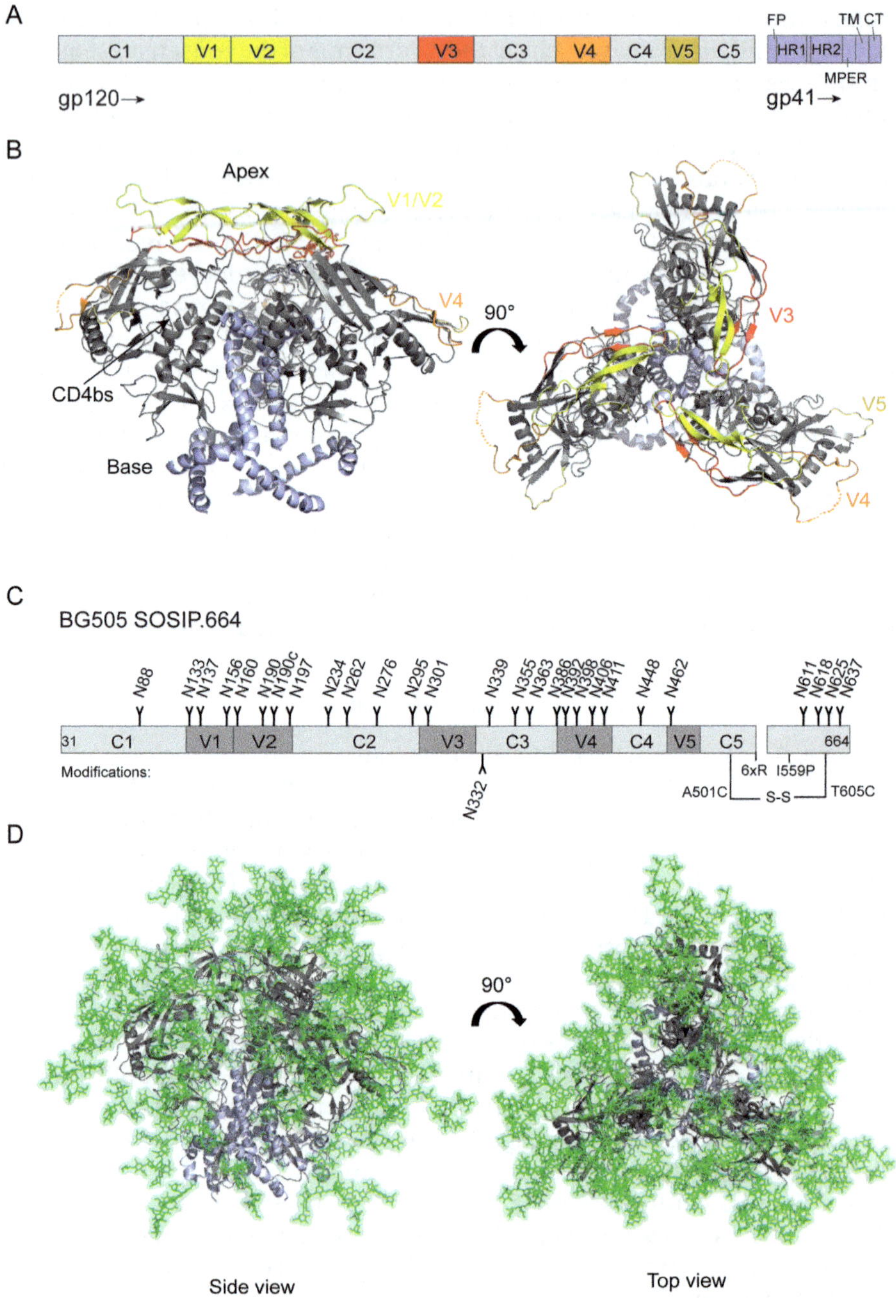

Figure 11.3 Structure of Env. (A) Primary structure of gp120 and gp41. Variable regions (V1–V5) of gp120 are coloured, constant regions (C1–C5) are depicted in grey. Gp41 is shown in light purple. FP, fusion peptide; H1,H2, heptad repeat helices 1 and 2; MPER, membrane proximal external region; TM, transmembrane domain; CT, cytoplasmic tail.

include the formation and stabilization of the co-receptor binding site, consisting of the newly formed bridging sheet and the V3 loop. In this context, the V3 loop has been described as the primary determinant of HIV-1 cell tropism.[194] Co-receptor binding then triggers the insertion of the N-terminal fusion peptide of gp41 into the target cell's lipid membrane, followed by the rearrangement of the HR1 and HR2 α-helices forming a hairpin-like intermediate within each gp41 subunit and taken together a six-helix bundle that drives membrane fusion.[195]

In order to improve trimer stability and to gain more detailed information, numerous different Env constructs *i.e.* trimers with a removed cleavage site, have been created and analyzed. However, it is now known that constructs corresponding to uncleaved trimers do not represent the native conformation.[186] A rather recent breakthrough has been the development of the 'SOSIP' format containing several subtle modifications to generate a stable and soluble trimer mimic. These were generated by introducing an artificial disulfide bond between gp120 and the ectodomain of gp41 (SOS), which has been truncated at position 664 and thereby lacks the MPER and C-terminal domain of Env.[196] Stability was also increased by mutating an isoleucine residue to proline (IP) within the N-terminal heptad repeat region (H1) of gp41 (combined: SOSIP.664 or simply SOSIP).[197,198] Soluble trimers have clear advantages compared with membrane-bound versions regarding ease of production and protein handling. Furthermore, they led to substantial progression in the high-resolution structural elucidation of Env.

Currently the most promising and most analyzed recombinant trimer mimic in vaccine development, BG505 SOSIP.664, is based on clade A transmitted/founder strain BG505 (Figure 11.3C). Compared with the original wild-type BG505 gp160, the sequence has been codon-optimized, and epitopes of antibodies that depend on the N332 glycan have been introduced by a T332N point mutation.[196] It also contains a more susceptible furin cleavage site (Arg_6). This trimer displays native-like antigenic and structural properties.[58,196,199] The prefusion structure of BG505 SOSIP.664 was first resolved by X-ray crystallography and cryo-EM at 4.7 Å and 5.8 Å, respectively.[189,190] Further studies included higher-resolution X-ray structures, that gave more complete information of gp41, and a cryo-EM structure of a fully glycosylated trimer at 4.36 Å.[191,200,201] All in all, structural studies on BG505 SOSIP.664 have provided the most complete and detailed picture of Env to date. The conserved gp41 fusion machinery resides at the trimer base, whereas the more variable gp120 head sits on top of gp41 (Figure 11.3B). The HR1 helices

(B) Structure of a soluble BG505 SOSIP.664 Env trimer in its closed, prefusion conformation. The variable regions and gp41 are coloured as in A. Left: side view; Right: Top view. Images were made in Pymol using PDB ID 4NCO. (C) Schematic representation of the BG505 SOSIP.664 construct including PNGSs and incorporated modifications, adapted from Behrens *et al.* 2016.[232] (D) Model of a fully glycosylated soluble BG505 SOSIP.664 trimer.[232] Glycans are shown in green. The model was made in Coot and Pymol on the basis of PDB ID 5ACO.

of gp41 reside in a three helix arrangement in the core of the six-helix bundle and the HR2 helices form a membrane-proximal collar.[191] Both helices are shorter in the pre-fusion conformation compared with the post-fusion conformation.[186] The trimer apex is comprised of the highly glycosylated, flexible and variable loops, V1, V2 and V3, which make stabilizing inter-protomer contacts and mask the underlying conserved regions.[186] The unglycosylated, conserved CD4 binding site (CD4bs) is located within a hydrophobic pocket between the inner and outer domain close to the protomer interface. CD4- and CD4bs-targeting antibodies need to approach Env from a particular angle in order to bind and successfully avoid surrounding glycans (in particular Asn276) and variable loops V1–V3.[186] The structure of a membrane-bound, fully glycosylated Clade B isolate JR–FL trimer (lacking its cytoplasmic tail) was resolved at high resolution of 4.2 Å by cryo-EM.[188] A similar topology compared with BG505 SOSIP.664 was revealed, despite some differences at the trimer apex and the N-terminal region of HR1.[188]

Env is covered by a dense shield of glycans (Figure 11.3D) that, on the one hand, are a key element when considering structural and antigenic properties of Env, but on the other hand, make structural studies, in particular X-ray crystallography, very challenging. Accessible glycans are therefore often enzymatically removed before crystallization, or proteins are made in cell lines producing solely oligomannose-type glycans, such as GnTI-knockout cell lines (or both).[186,202] Very recently, three structures (including BG505 SOSIP.664) of fully glycosylated soluble Env trimers from clades A, B and G, at 3.4–3.7 Å resolution, have greatly enlarged our knowledge on the interlocking glycan network of Env.[187] Although these trimers were also produced to carry a uniform oligomannose-type *N*-glycosylation pattern, extensive insight into the site-specific processing states of individual glycans as well as information on their arrangement on the trimer surface is provided.[187]

Site-specific *N*-glycosylation of HIV-1 Env has also been extensively studied by mass spectrometry analysis. This approach, despite the very complicated target, can give fine structural details on the processing state and microheterogeneity, including complex glycosylation patterns, at individual glycan sites. Studies have been undertaken using different approaches on a large variety of Env constructs. We will review and discuss these findings in Section 11.2.4.

11.2.3 The Glycan Shield

Glycan structures on the surface of viruses usually function in disease transmission *via* interaction with host-cell receptors. They also help pathogens to hide their variable protein surface from immune recognition. The HIV-1 viral spike is characterized by an unusually high abundance of glycosylation that leads to steric constraints within the glycan shield and hence a large presence of oligomannose-type glycans, which show lower structural variation compared with host-cell glycoproteins.

The number of *N*-linked glycans in gp120 ranges from 18 to 33, with a median of 25 sites.[203] The extensive glycosylation of Env has previously been reported to shield the underlying protein surface from immune surveillance.[204] The glycan shield constantly evolves to provide steric hindrances to newly produced neutralizing antibodies. Such escape mutants show non-random shifts in the position of their *N*-linked glycans.[115] Although the *N*-linked glycans are produced using the host cell's glycosylation machinery (Figure 11.2), studies on recombinantly produced gp120 have revealed an unusually high and conserved population of oligomannose-type glycans—$Man_{5-9}GlcNAc_2$.[164,166,167,205,206] During glycan processing, ER and early Golgi α-mannosidases are responsible for trimming the *N*-linked glycan precursor $Man_9GlcNAc_2$ to enable the subsequent addition of other glycan moieties leading to hybrid and complex-type glycans. However, the high density of glycans on gp120 limits the ability of these α-mannosidases to access and process their glycan substrates, a phenomenon demonstrated by *in vitro* digestion experiments with recombinant enzymes and gp120 glycoforms as well as by the effect of PNGS point mutations on the processing of neighboring glycan sites.[167,207] The processing fate of individual glycans is determined by their position on the trimer. Generally, oligomannose-type glycans form clusters on the surface of Env, which together constitute the so-called 'intrinsic mannose patch' (IMP). The IMP has been shown to be a conserved feature of all HIV-1 clades.[166,167,208] Glycan analyses of recombinant trimers and virion-associated Env showed an even higher oligomannose content compared with gp120 monomers.[166,167,209] This has led to the concept of an additional, 'trimer-associated mannose patch' (TAMP), thought to arise due to additional inter-protomer glycan–glycan and glycan–protein interactions.[210]

Furthermore, *N*-glycosylation of Env, especially the abundance of oligomannose-type glycans, has been reported to be a structure-driven process, and hence can serve as an indication for trimers with a native-like conformation.[168] Glycan analysis of the trimeric clade A Env mimic, BG505 SOSIP.664, showed an extensive population of oligomannose-type glycans on gp120 compared with a significantly higher abundance of processed, complex-type glycoforms on gp41 (Figure 11.4).[168] Gp41 is thus generally more susceptible to cell-directed glycosylation processing compared with gp120. Analysis of uncleaved BG505 gp140 by EM, surface plasmon resonance (SPR) and hydrogen–deuterium exchange–mass-spectrometry (HDX–MS) indicates a non-native, irregular conformation of these trimers with dangling gp120 subunits attached to a gp41 core.[174,211,212] The non-native conformation of Env is also visible in the total glycan profiles of these uncleaved trimers, which show a much higher degree of glycan processing as the glycans are more accessible to processing enzymes compared with fully-cleaved BG505 SOSIP.664.[168] This highlights the importance of furin cleavage upon shaping the structural integrity of the trimer and also indicates post-furin α-mannosidase activity.[168] Trimerization of Env and cleavage by the Golgi-resident furin must occur early enough in the secretory pathway to allow formation of a native-like structure that protects from further α-mannosidase processing.[168]

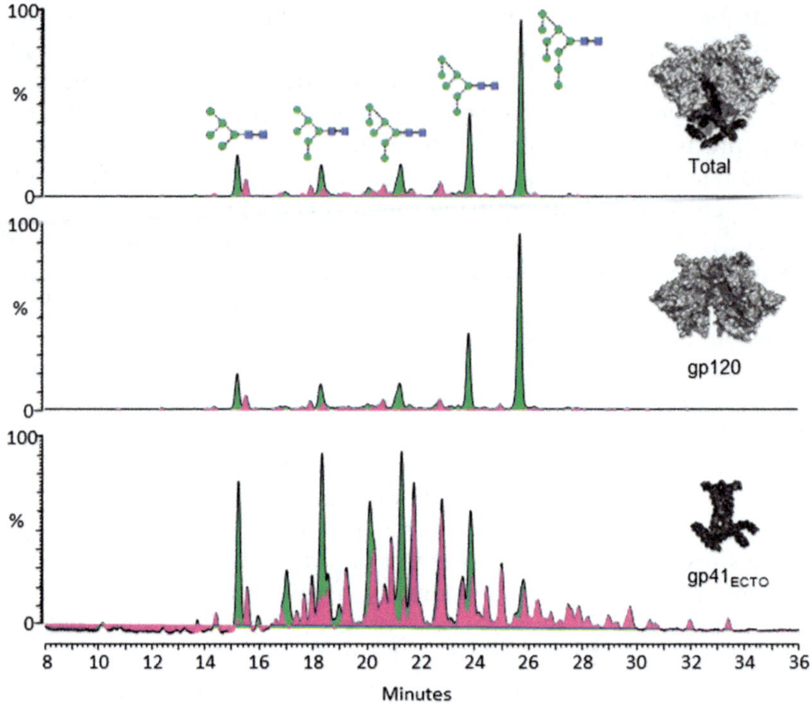

Figure 11.4 Oligomannose-dominated glycosylation profile of BG505 SOSIP.664 trimers. Hydrophilic interaction chromatography–ultra performance liquid chromatography (HILIC-UPLC) profile of released and fluorescently labelled *N*-linked glycans from BG505 SOSIP.664 trimers stably produced in HEK293T cells and purified by 2G12-affinity chromatography–size-exclusion chromatography. Oligomannose-type and hybrid glycans (green) were identified by their sensitivity to Endo H digestion. Complex glycans, pink. Adapted from Behrens *et al.* 2016.[232]

Overall, the glycan profiles of recombinant Env trimers can be used as a good indication of a proper folding state.

Sialic acids are terminal modifications on complex-type glycans (Figure 11.2) that show particular effects on the immune system. These effects seem to be linkage-specific with α2,6-linked sialic acids exerting an anti-inflammatory phenotype.[213] Evidence for this comes from the immunosuppressive activity of modified immunoglobulins, as well as the immune-modulatory role of CD22 present on the surface of B cells.[213,214] In HIV-1, the contribution of the sialic acid modification to the antiviral immune response has yet to be delineated. However, analysis of Env isolated from virus propagated in peripheral blood mononuclear cells (PBMC) showed that α2,6-linked sialic acids were predominant.[215] One established role of sialylation of *N*-linked glycans is the complement-mediated enhancement of HIV-1 infection due to the recruitment of the complement-regulatory factor H to gp41.[104,216] Factor H recognizes sialic acids as "self" and accelerates the decay of the C3 convertase C3b. It is also a cofactor for the serine protease factor I in the cleavage of activated C3 convertases leading to their deactivation.[217] As a consequence, desialylation of Env promotes activation of the alternative complement pathway.[218,219] Besides modulating the activation of the humoral immune system, sialic acids are also proposed to modulate the cellular immune response. Specifically, α2,6-linked sialic acids are known to engage with inhibitory Siglecs on multiple immune cells.[220] In addition to these potential effects in natural HIV-1 biology, the presence of α2,6-linkages has consequences that need to be considered for recombinant models of the viral spike. Commonly used expression systems are usually dominated by α2,3-linked sialic acid moieties, which may have implications for our understanding of bNAb epitopes involving complex-type glycans.[215]

The role of *O*-linked glycosylation in Env biology is uncertain. *O*-glycosylation generally occurs on extended proline-rich regions, where tandem repeats of serine or threonine in proline-rich regions can become heavily *O*-glycosylated and thus form mucin-like domains. HIV-1 has no such regions. However, serine or threonine present on extended peptide loops can be a substrate for the *O*-GalNAc transferase. There have been reports of *O*-linked glycans being identified on recombinant Env constructs at position T499 and T606.[206,221–223] However, data regarding the presence of *O*-glycans on the native virus are conflicting; Stansell and colleagues did not identify *O*-linked glycans at T499, whereas Yang and colleagues identified the core 1 structure Galβ1,3–GalNAc (termed T antigen) at this position.[222,224] The strongest evidence that Env is indeed *O*-glycosylated comes from the observation that anti-Tn antibodies weakly neutralize HIV-1.[225] The Tn epitope consists of a single *N*-acetylgalactosamine (GalNAc) linked to either a serine or a threonine. The abundance of any such *O*-linked epitope may be heavily influenced by the origin of the viral producer cell.

11.2.4 Site-specific *N*-Linked Glycan Analysis

In order to further understand the fine structure of the glycan shield as well as to elucidate the epitopes of glycan-binding broadly neutralizing antibodies, a number of site-specific analyses targeting HIV-1 Env *N*-glycosylation have been performed. These include work on various gp120 and gp140 constructs, membrane-bound material and gp120 derived from virions, using a variety of different methods, most commonly mass spectrometry. Figure 11.5 gives an overview of the results obtained thus far in terms of their classification into mainly oligomannose, mixed or complex sites. The first study was conducted more than 25 years ago by Leonard and colleagues, who analyzed recombinant monomeric HIV-1_{IIIB} gp120 by a combination of enzymatic treatment and reverse-phase HPLC (RP–HPLC). They found all 24 PNGSs to be fully occupied, with 11 sites containing oligomannose/hybrid glycans.[164] Since then, improvements in mass spectrometry have led to successive studies providing a more detailed insight into the individual microheterogeneity and occupancy of each site.[165,205,206,221,223,224,226–232,395] Env shows varying degrees of processing. As can be seen in Figure 11.5, higher glycan processing generally happens towards the C- and N-terminus of Env.

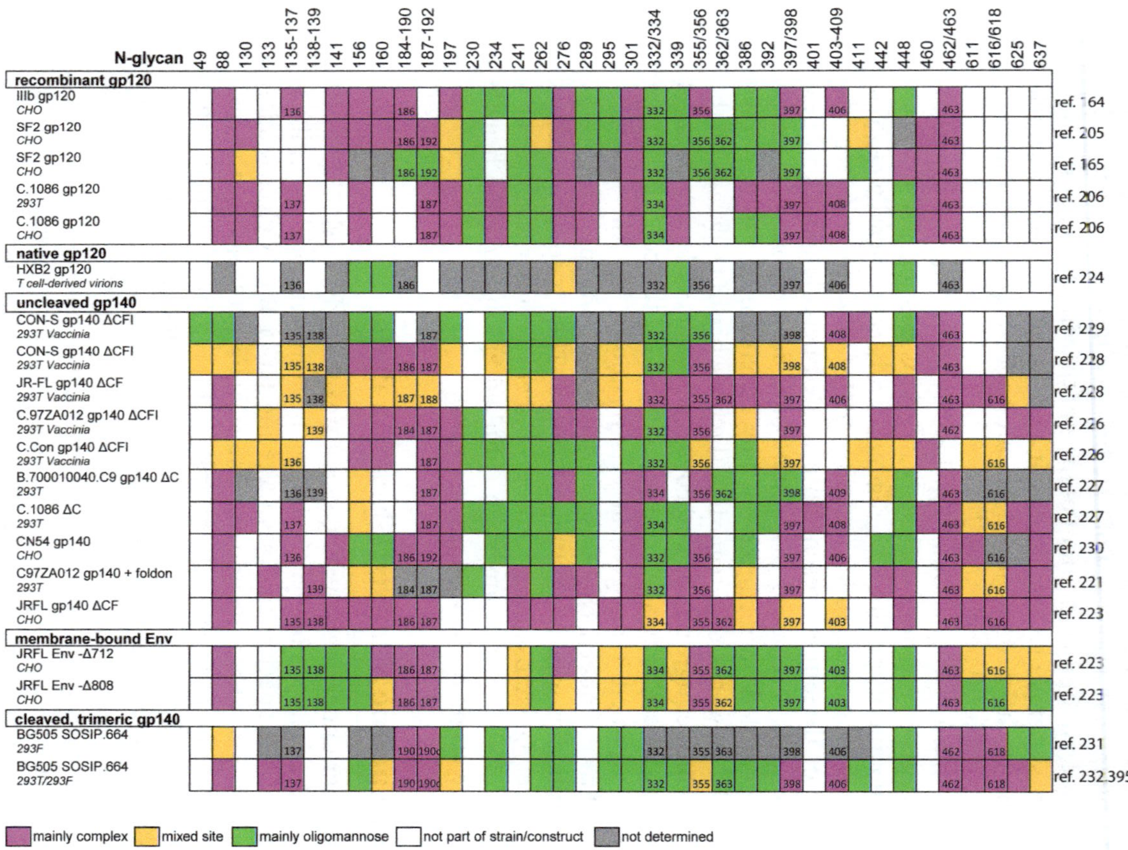

Figure 11.5 Summary of results of site-specific studies on Env *N*-glycosylation. Numbers above indicate the *N*-linked glycan position according to HXB2 numbering. Left: Strain and production system. Green, mainly oligomannose-type glycans were identified; pink, mainly complex glycans were identified; orange, mixed site; grey, not determined; white, not part of construct/strain. Go and colleagues, focused on the number of different glycan structures identified on each site rather than their abundance. Their data was grouped as follows: Green, mostly oligomannose structures; orange, more than 20% oligomannose-type structures; pink, rest. ΔC/ΔCF/ΔCFI, deletion of cleavage site (C), fusion domain (F) and immunodominant region of gp41 (I).

Underprocessed, oligomannose-type glycans mainly reside within the C2 and C3 region, where the intrinsic mannose patch on the outer domain is centered on an asparagine at position 332 (Asn332; N332). As well as monomeric constructs, trimeric Envs are among the analyzed constructs, which, although produced and purified differently, give preliminary insights into *N*-linked glycan sites that change from complex to oligomannose in the context of trimerization. Comparing the results of the first (recombinant gp120) and last listed studies (fully cleaved trimer), for example, the *N*-linked glycan sites N156 and N160 in the V1–V2 region as well as N276 in the C2 region change from complex to less processed glycans, hence supporting the idea of a TAMP.[164,232]

One of the limitations of site-analysis methodologies has been the lack of quantification when determining the relative abundance of different glycans at a particular site.[232] Behrens and colleagues presented the first semi-quantitative study performed on a trimeric construct resembling the native spike.[232] Figure 11.6 shows a model of a fully glycosylated Env based on a cryo-EM structure which reveals a mosaic of sites bearing complex, mixed

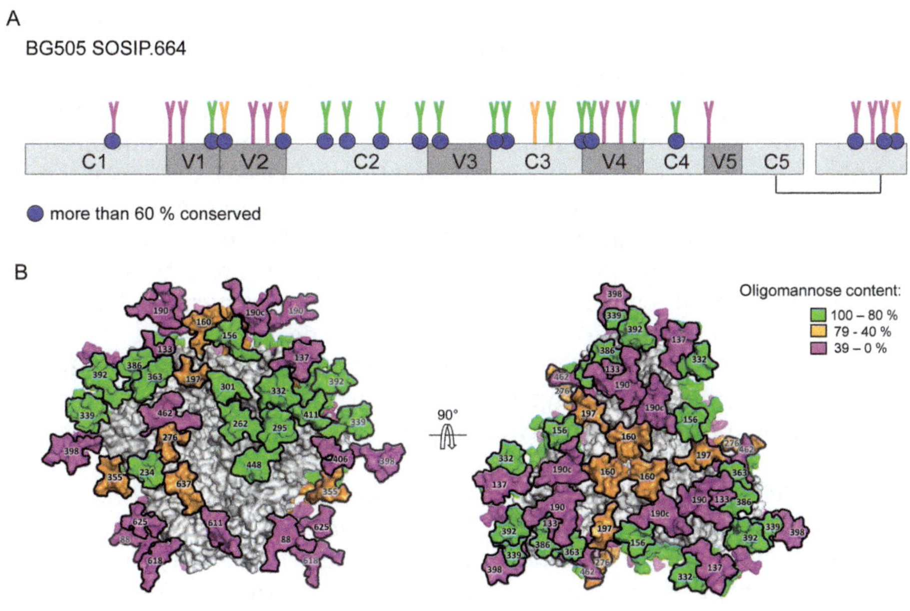

Figure 11.6 Site-specific glycosylation of BG505 SOSIP.664. Green, oligomannose-type glycan; orange, mixed site; pink, complex-type glycan. (A) Overview of *N*-linked glycans on BG505 SOSIP.664 with PNGSs that are conserved to more than 60% across all HIV-1 clades highlighted with a blue dot.[235] (B) Model of a fully glycosylated Env trimer, based on PDB ID 5ACO, glycans modelled as described in/figure based on Behrens *et al.*, 2016 and Behrens *et al.*, 2017.[232,395]

or oligomannose-type glycans. An underprocessed interlocking network of oligomannose-type glycans wraps around the outer domain of Env.[232] The trimer displays an extensive mannose patch, centered on the N332 glycan, a key component for the epitopes of many bNAbs and the so-called 'supersite' of vulnerability.[65] At the trimer apex N156 and N160 from all three protomers form a ring of minimally processed glycans. In contrast, the trimer base shows a much higher degree of glycan processing, particularly the highly processed N88 glycan and the gp41 glycans. Gp41 is much less glycosylated than gp120 with four PNGSs on average. Behrens and colleagues suspected N625 to be only partially occupied, which is in close agreement with previous reports on partial or alternative occupancy of gp41 sites.[227,230,231,233] Overall, results from site-analysis studies of Env confirm the dominance of oligomannose-type glycans, but also highlight regions occupied by highly processed glycans.

11.2.5 Glycans in Immune Escape

The HIV-1 glycan shield constantly evolves to escape the immune system.[115] *N*-glycosylation sites can be shifted or deleted as a result of mutations in the viral genome sequence. Nevertheless, the extent to which PNGSs are able to migrate is limited by structural features of the trimeric spike such as the non-glycosylated CD4 binding site and inter-protomer interfaces. In fact, the total number of PNGSs on Env has remained relatively constant, in spite of years of viral evolution.[234] In BG505 SOSIP.664, for example, 17 out of the 28 *N*-linked glycan sites are highly conserved (Figure 11.6A); particularly conserved are PNGSs near the trimer base such as N88, N611, N625 and N637 which occur in more than 90% of HIV-1 strains.[235] Also, almost all identified oligomannose-type glycans on the trimeric Env mimic are among the most conserved PNGSs (Figure 11.6A), indicating essential structural features of the virus that present possible targets for vaccine design.

Glycan shifts can be caused by just a single nucleotide change. In the Nx(S/T/N)x(S/T) motif, for example, exchanging the middle serine or threonine to asparagine moves the PNGS two positions along the protein sequence. As the codon triplets for S, N and half of the T codons are located next to each other on the codon table, this exchange can be achieved by simply changing the nucleotide triplet ACC (T) to AAC (N).[236] One *N*-linked glycan site that has been reported to be vulnerable to this type of evolutionary migration is N332.[237] A significant number of clade C founder viruses lack an *N*-linked glycan at position N332 and instead contain a glycan site at N334. During the course of infection, broadly neutralizing antibodies can arise that target this particular glycan site. In one study, the virus has been shown to respond by first shifting N334 to N332, then after around two years of infection, the PNGS was shown to migrate back to N334 thereby thwarting the immune system once again.[237]

Together, N332 and N334 are present in 93% of all HIV-1 viruses and hence present a highly conserved feature of the glycan shield.[235] The glycan structure of these two sites has almost exclusively been described as oligomannose (Figure 11.5). On the native Env mimic BG505 SOSIP.664, N332 is almost entirely found to bear $Man_9GlcNAc_2$ glycans, displaying the smallest possible amount of structural variation.[232] Interestingly, broadly neutralizing antibody PGT128 that targets the high-mannose patch, in particular N332, is able to tolerate a shift to N334 in some viral isolates.[237] Similar observations have been reported for PGT121-7.[238] The fact that the immune system is able to overcome certain viral escape mutations within the glycan shield due to promiscuous glycan site recognition by bNAbs raises hopes that targeting the glycan shield could be a productive strategy in vaccine design.

11.3 A Target for Broadly Neutralizing Antibodies

The initial enthusiasm that followed the discovery of broadly neutralizing antibodies in a small subset of HIV-1 infected patients was quickly dampened as suspicions arose that many of these antibodies were actually self-reactive, and therefore might be largely eliminated from the B cell reservoir.[239] For example, the self-nature of the glycan shield underscored the potential difficulty in eliciting antibodies targeting Env. However, there have subsequently been waves of discovery of new potent and particularly broad bNAbs. Importantly, the majority of these bNAbs utilize *N*-glycosylation as a significant feature of their epitopes.[240] In parallel to these findings, it is now known that the formation of bNAbs within HIV-1 infected individuals is not as rare as originally thought, with up to around 30% exhibiting notably active antibodies arising on average two years after infection, though in rare cases bNAb development as early as one year after infection has also been reported.[241–243] Unfortunately, patients that develop these bNAbs do not benefit from them themselves as the viral load and diversity is already too high by the time they evolve.[244] The observation that these antibodies do not arise very quickly, supports the notion that they operate at the edge of immune tolerance mechanisms and rely on rare mutations to achieve binding maturation. These features, which are described in detail below, clearly pose barriers for vaccine strategies designed to rapidly elicit bNAbs.

The time window and location for neutralization for any vaccine-induced antibody response is very narrow as it needs to prevent, from the outset, the establishment of a latent viral infection. However, numerous studies have shown that these bNAbs are able to provide passive protection from viral challenge with chimeric simian/human immunodeficiency virus (SHIV) in non-human primates, which underlines and justifies efforts to elicit these antibodies through rationally designed vaccine strategies.[69,70,73,74,245] In order to achieve this, it is likely that we require a more complete understanding of several fundamental areas such as the nature of the B cell maturation pathway and how germinal centers can be rendered more productive, how improved mimics of the envelope spike can be engineered, how bNAb epitopes can be rendered more immunogenic and how bNAb precursor germline B cells can be activated in prime-and-boost vaccination strategies.[246,247]

11.3.1 Sites of Vulnerability

All broadly neutralizing antibodies isolated before 2016 (about 100 according to a bNAb database[248]) target defined, conserved regions of vulnerability on the HIV-1 Env glycoprotein (Figure 11.7). These regions include the mannose patch, the CD4 binding site, the MPER of gp41, glycan epitopes on the variable loops V1–V2 and V3 and the bridging region between the two subunits.[249,250] Many of these bNAbs recognize epitopes that are completely or partially composed of *N*-linked glycans.[54,55,57–68] Despite its potency in defending the virus against antibody-mediated neutralization, the dense HIV-1 glycan shield does contain vulnerabilities that can be exploited by the adaptive immune system.

Glycan-recognizing bNAbs target three main regions on Env: the N332 glycan (and the V3 loop), the N160 glycan (and the V1–V2 loops) and glycans at the gp120–gp41 interface.[210] The first bNAb identified to bind the HIV-1 glycan shield was 2G12.[60,251] It has been shown to exclusively recognize conserved glycans on the gp120 outer domain, including N295, N332, N339 and N392.[252] Binding of multiple *N*-linked glycans with high avidity is mediated

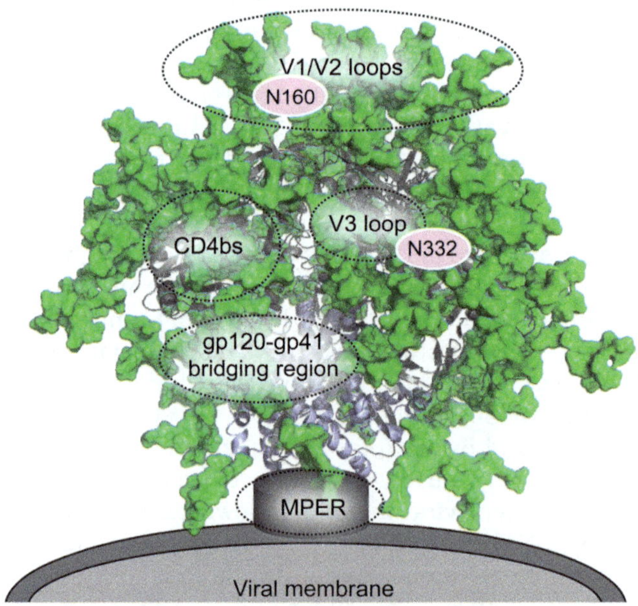

Figure 11.7 Sites of immune vulnerability on the HIV-1 envelope spike. HIV-1 bNAbs target five main regions on Env: The CD4bs (*e.g.* VRC01, ANC131, CH103); the MPER (*e.g.* 4E10, 2F5); the V1/V2 glycans—in particular N160 (*e.g.* PG9, PG16, CH01); the V3 loop and glycan N332 (*e.g.* PGT128, PGT125 and PGT138) and the gp120–gp41 bridging region including proximal glycans (*e.g.* 35 022, PGT151). This figure is based on a model of fully glycosylated, trimeric Env (based on PDB ID 5ACO) as described in Behrens *et al.*, 2016.[232]

through a unique domain-exchanged antibody structure that allows the formation of a multivalent surface.[57] The N332 glycan in the heart of the oligomannose patch on the outer domain of gp120 is also included in the epitopes of many other bNAbs, and hence has been dubbed the "supersite of vulnerability" on the envelope spike.[54,55,61,65] Unlike 2G12, all other identified bNAbs targeting the glycan shield show non-domain-exchanged, Y-type antibody structures and contain both protein and glycan components as part of their epitopes.[253] They have also been shown to have much higher breadths and potencies compared with 2G12. For example, PGT128 is able to neutralize 72% of circulating HIV-1 strains with a median concentration giving 50% inhibition (IC_{50}) of 0.02 μg ml^{-1} (compared with 2G12—32% of viruses with a median IC_{50} of 2.38 μg ml^{-1}).[54,55,65] A common feature for N332-binding bNAbs (apart from 2G12) is the use of long third complementarity determining region of the antibody heavy chain (CDRH3) loops, which they employ to penetrate the glycan shield in order to interact with the V3/V4 loops, N332 and neighboring glycans.[54,65,189] Long CDRH3 loops are also observed among N160-recognizing antibodies. For example PG9 and PG16 form a large "hammerhead" CDRH3 loop that contains tyrosine sulfation sites and is used to reach through the glycan shield and bind both the N160 and N156 glycans and the V1 loop.[66,67,254] The bNAb PGT151 targets an epitope at the gp120–gp41 interface that relies on the presence of complex glycans on gp41 and protein epitopes on gp120.[59]

Overall, broadly neutralizing antibodies targeting HIV-1 glycans have overcome tolerance mechanisms to self glycans, but also the naturally weak nature of glycan–protein interactions and the protected protein surface of Env by the development of several unique characteristics. We will highlight these features in detail in the next section.

11.3.2 Unusual Features of Broadly Neutralizing Antibodies

Broadly neutralizing antibodies targeting HIV-1 commonly display one or more of the following rather unusual characteristics:

- Long CDRH3
- Unusual post-translational modification (*e.g.* Tyrosine sulfation)
- Domain swapping (2G12)
- Autoreactivity and polyreactivity
- Extensive somatic hypermutation

Many HIV-1 bNAbs, especially those targeting V1–V2 and V3, the subunit bridging region and MPER epitopes, contain long and charged CDRH3 sequences.[255] Whereas the typical length of a human B cell CDRH3 is between 8 and 16 residues, much longer sequences have been reported for bNAbs.[256] This generally permits the antibody to reach through the glycan shield in order to target more vulnerable protein regions. For example, PGT145, which binds the N160 glycan at the trimer apex, has a 33-residue-long CDRH3, and the above mentioned "hammerhead" CDRH3 of PG9 and PG16 contains 30 residues that permit penetration of the glycan shield in order to reach the trimer apex.[66,257–259] Antibody 2F5, which targets the MPER, uses its long CDRH3 region (24 residues) to bind a linear epitope and extend a hydrophobic patch into the viral membrane, showing a different structural application for the same antibody maturation phenomenon.[258,260] Long CDRH3s are mainly attributed to original light chain recombination events selecting only a limited subset of germline genes, rather than to somatic hypermutation during antibody affinity maturation.[255,261] It was recently shown that long CDRH3 sequences are in fact present in the naïve human B cell repertoire, supporting the idea of targeting bNAbs with this characteristic feature.[262] Furthermore, rather unusual post-translational modifications have been reported for bNAbs. For example, the CDRH3 of certain bNAbs carries a tyrosine sulfation, which has been reported to contribute to antibody binding and neutralization.[257,263]

A property that is relatively common among HIV-1 bNAbs is polyreactivity and autoreactivity.[264,265] It has been shown that bNAbs are generally more polyreactive and autoreactive than non-neutralizing antibodies.[266] This is striking as both characteristics are typically strongly deselected during B cell development; most naïve B cells are in fact polyreactive or autoreactive, but only 5% of mature cells are.[267,268] It has also been shown that B cells are capable of regaining polyreactivity and autoreactivity during their maturation in germinal centers (see below).[269] High polyreactivity is said to be strongly linked to the presence of long CDRH3s.[270] Polyreactivity allows bNAbs to undergo bivalent heteroligation between one high-affinity and one low-affinity epitope on HIV-1 and therefore increases the apparent binding affinity.[265] In this context, it is also important to note that glycan-binding bNAb 2G12, that recognizes Manα1,2–Manα1,2–Man residues on the outer domain of Env, has been described as selectively binding to yeast mannans, which display extended arrays of the same glycan epitope.[57,60,271]

One explanation for the enhanced presence of autoreactive bNAbs could be the hypothesis that conserved, neutralizing HIV-1 epitopes mimic host proteins in order to avoid strong humoral responses.[264] Examples of autoreactive bNAbs are the four CD4bs-specific antibodies VRC01, VRC02, CH1063 and CH106, that all also recognize the human ubiquitin ligase E3A.[266] Avidity for the ligase correlates with breadth of neutralization of the antibody.[266] MPER-recognizing antibodies 2F5 and 4E10 are self-reactive to human kynureninase and splicing factor 3b subunit 3, respectively.[272] Evidence that bNAbs to these epitopes might be impeded by the immune system comes from the hindered B cell development and activation in the corresponding knock-in mice through tolerance mechanisms.[273–275] Overall, the development of bNAbs carrying autoreactivity and polyreactivity may be dependent on prolonged exposure to the antigen, as multiple B cell tolerance checkpoints need to be bypassed. Hence, their induction in healthy humans through vaccination might turn out to be impossible.[255,270,276]

Another common feature among bNAbs is the presence of extensive somatic hypermutations.[55,277–279] Normally, somatic hypermutations during antibody maturation are limited to the CDR regions, however, mutation in a HIV-1-targeting bNAb can also occur in the typically conserved framework regions (FWR) of the antibody.[280] These mutations have been shown to contribute to antibody binding.[280] There is also evidence of a positive correlation between the number of mutations and the development of breadth and potency of the antibody for the PGT121 family.[281] However, the presence of extensive mutations, including in the FWR, may be more of a hallmark of the length of time of bNAb development, rather than a trait of bNAb activity *per se*, as a partially germline-reverted form of bNAb VRC01 still shows considerable potency.[282,283]

11.3.3 Development of Broadly Neutralizing Antibodies

Naïve B cells circulate in the blood and lymphoid tissues and become activated by binding of antigens to their germline B cell receptors (BCR). These antigens can be either free or bound to antigen-presenting cells. B cell activation can be modulated and depends on a variety of additional factors such as T helper cells or PAMPs. Following a T-cell-dependent activation, B cells migrate to draining lymph nodes and their development can take one of two paths. One option is that the B cells are recruited into extrafollicular responses, where they grow as low-affinity antibody-producing plasmablasts.[283] These responses generally arise early during infection and are associated with hardly any somatic hypermutation.[284] The other option is the formation of a germinal center within the lymph nodes, where B cells proliferate and undergo class switching and affinity maturation through somatic hypermutation. The latter two are caused by the action of a B-cell-specific, intrinsically expressed, activation-induced cytidine deaminase. Follicular T helper cell (T_{FH}) and follicular dendritic cells (which secrete cytokines influencing B cell development) select affinity maturated B cells, which then leave the germinal center as memory cells or long-lived, circulating, antibody-secreting plasma cells. Memory B cells can re-enter the germinal center at a later stage when reactivated.[283] Plasmablasts and plasma cells are both sources of circulating antigen-specific antibodies in the blood serum. In terms of an HIV-1 infection, the primarily bone marrow-resident plasma cells are considered to be the main source of HIV-1-specific antibodies.[285] A strong correlation between HIV-1 Env-specific plasma cells and HIV-1-specific circulating antibodies has recently been described, whereas a correlation between bNAbs and memory B cells remains elusive.[285] As most bNAbs are cloned from isolated HIV-1-specific memory B cells, these findings indicate that, although more difficult to assess, the bone marrow should be included as an important compartment for bNAb screening as well as evaluation of vaccine effects.[283,285]

During an HIV-1 infection, the first humoral immune responses are IgG and IgM antibodies targeting gp41, which are developed within weeks after transmission. In contrast, anti-gp120 antibodies appear after an additional few weeks.[112] These initial IgG and IgM antibodies are capable of binding the virus, however, they only have a low effect on acute-phase viremia.[112] Autologous neutralizing antibody responses have been shown to occur only several months after viral transmission.[115,116,286,287] Although these antibodies may show some cross-reactivity, a broad immune response that targets conserved epitopes on Env is usually only established after at least two years of infection in a subset of patients.[241,288] The development of antibody breadth is thought to be associated with the initial viral load and HIV-1 Env diversity, albeit not with disease progression in general.[289] One hypothesis for the early development of antibody breadth is that it occurs coincidentally with antibody affinity maturation following an early acquisition of the target antigen.[290] The virus and the B cells co-evolve and "race" against each other in terms of antibody maturation and viral escape mutants. Interestingly, in cases where more than one bNAb has been isolated from the same HIV-1 patient, the antibodies often evolved from the same B cell lineage.[291] A recent study analyzed the longitudinal evolution of the VRC01 lineage of bNAbs in one individual over 15 years of infection.[292] Three diverse, multi-clade families of bNAbs could be isolated from this patient that all recognized the same CD4bs epitope. However, each family had developed a different structural solution to target the antigen. Remarkably, all three families probably originated from the same ancestor VRC01 B cell, indicating a high importance of early germline B cell selection for bNAb development. Additionally, it was shown that the VRC01 lineage developed with similar exceptionally high mutation rates to the virus during the course of infection.[292] Taken together, the results of this study give new insights into the evolutionary path of bNAbs while also demonstrating the complexity of their development.

11.3.4 Mechanisms of Neutralization

Broadly neutralizing antibodies are thought to act upon HIV-1 infection in a variety of different ways. As they directly bind to free virions in the plasma, they are able to neutralize the virus before host-cell entry. This may happen by blocking viral attachment or fusion and the associated conformational changes within gp41.[293,294] Direct, cell-free binding prevents further infection and also accelerates viral clearance in the plasma and possibly in tissues as well.[295] Next, some broadly neutralizing antibodies can inhibit cell-to-cell transmission, both between lymphocytes and from lymphocytes to dendritic cells, by targeting viruses present at virological synapses.[296–298] Additionally, anti-gp120 antibodies (not anti-gp41 antibodies) have been shown to efficiently block viral transfer from dendritic cells and macrophages to T cells.[299,300] Fc-mediated effector functions play an important role in inhibiting viral cell-to-cell transmission and in passive immunotherapy in general.[296] These effector functions include engagement of the complement system leading to cell lysis, triggering the destruction of infected cells through antibody-dependent cellular cytotoxicity (ADCC) and also the formation of immune complexes that improve the antigen presentation by dendritic cells.[295] It is not yet fully understood which of the Fc-mediated activities plays the major role in the action of bNAbs, however, they are required for the antibody to exhibit its full potency both for clearing and preventing HIV-1 infections.[295] Moreover, bNAbs were shown to protect from mucosal transmission of HIV-1 in humanized mice and from rectal or vaginal viral challenge with SHIV in non-human primates.[296] It is not yet completely clear whether the reason for this is the inhibition of viral transcytosis or the intracellular neutralization of virions in epithelial cells.[296] Other factors affecting passive protection of bNAbs could be their ability to block viral cell-to-cell transmission or Fc-mediated effector functions.[296]

11.3.5 Application in Therapy and Cure Research

The main downside of current antiretroviral therapies is their inability to target latent viral reservoirs, hence making them not curative. The yet-unsolved goals of cure research are to find either a sterilizing cure, where the viral reservoirs are fully eliminated, or a functional cure that would enable the immune response to control the infection in the absence of therapy.[295] The application of broadly neutralizing antibodies plays an important role in cure research. Passive immunotherapy not only has the capability to protect from viral challenge–as described earlier–but also significantly lowered the viral load in humanized mice, non-human primates and in the first human clinical trials.[295] Decline of viremia and prevention of viral escape in humanized mice has been shown to be achievable through a combination of three potent bNAbs.[301] Strikingly, application of a single bNAb is sufficient to control viremia if the immunotherapy is preceded by antiretroviral therapy.[301] Studies in non-human primates surprisingly lead to even more promising results; application of just the glycan-dependent bNAb PGT121, for instance, was shown to control viremia to a level below the detection limit in chronically infected rhesus monkeys, as long as antibody titers remained detectable.[302] Similar results were shown for the CD4 binding site-directed antibody 3BNC117 and the V3-targeting 10-1074.[303] The bNAb 3BNC117 has since been tested in a phase I clinical trial in humans; it led to a significant reduction of viremia, providing a promising basis for further exploration of bNAb therapy in humans.[304]

Latently infected immune cells, among them resting memory $CD4^+$ T cells, have a rather long half-lifetime of from 6 to 44 months.[295] As current antiretroviral therapies are not capable of clearing the latent virus reservoirs, the consensus proposed strategy is to first activate viral replication in latent cells and then kill them in the presence of HAART—"kick and kill".[305] There have been several attempts to stimulate latent reservoirs, for example using histone deacetylase inhibitors; despite some progress, new ideas and approaches are still needed.[295] A promising candidate for the "kill" part of this strategy is the passive transfer of bNAbs, especially because reactivated cells have been proposed to be resistant to virus-specific and autologous cytotoxic $CD8^+$ T cell responses.[306,307]

The bNAbs might be able to block released viruses and directly kill reactivated cells.[295] Initial studies in mice showed some promising results.[308]

Another potential advantage of bNAbs is their wider capability of penetrating tissues compared with HAART.[74] Certain tissues, such as the GALT, show higher levels of viral transmission.[33,35] It has been suggested that a high multiplicity of cell-to-cell spread renders HIV-1 in these areas less sensitive to antiretroviral therapies making antibodies (that are relatively highly concentrated in tissue surrounding fluids) a possible alternative, especially as some bNAbs have been shown to be able to prevent viral cell-to cell transmission.[295,296,309]

Furthermore, multiple rational attempts to improve the efficacies of broadly neutralizing antibodies have been investigated. These include engineering strategies to improve breadth, potency, *in vivo* half-life or Fc-binding *e.g.* by altering the CDRH3 loop, construction of chimeric glycan-dependent bNAbs or production of Fc nonfucosylated antibodies.[67,310–312] Additionally, bispecific antibody formats containing variable regions from different bNAbs have been explored.[311] For example, a bispecific IgG of VRC07 and PG9-16 (PG9-16 is a chimeric antibody based on PG9) displayed superior breadth and potency compared with its parent antibodies.[313] Antibody engineering and bispecific formats thus display very promising options for improving bNAbs for use in prevention and treatment of HIV-1.

Overall, great progress has been made employing bNAbs in cure research. Nevertheless, one of the main remaining hurdles is to identify an efficient way to activate latent reservoirs. Only then can subsequent viral elimination strategies, which probably include bNAbs, be successful.

11.4 Strategies for Vaccine Design

11.4.1 Challenges

An effective HIV-1 vaccine will ideally stimulate the elicitation of broadly neutralizing antibodies and also activate effective cytotoxic T lymphocyte (CTL) responses. As traditional and empirical vaccination approaches were either not applicable or did not result in effective immunization, current HIV-1 vaccine research focuses on rational immunogen design, though many challenges remain that must first be overcome. In order to prevent infection, it is crucial that a vaccine mediates sterilizing immunity and thus prevents the establishment of a latent viral reservoir. This is particularly challenging as the window of action for inhibiting HIV-1 infection is unusually short, whereas for most other vaccines protection against disease is usually sufficient. Moreover, a very broad immune response is required, that has the potential to tackle the enormous amount of genetically different, highly mutating HIV-1 strains. Sterilizing, broadly acting immunity will hopefully be provided through bNAbs, but immunogens eliciting an effective T cell response have the potential for long-term control of viremia and therefore should not be neglected.[314]

Many bNAbs target glycans on the surface of Env. However, the generation of antibodies to glycans has to overcome multiple hurdles as glycans are rather poor immunogens.[210] Firstly, carbohydrate–protein interactions are generally weaker than protein–protein interactions, requiring enhancement through avidity effects. Furthermore, the microheterogeneity, including varying, dynamic glycan conformations at single *N*-glycosylation sites, also weakens the immune response to individual glycans.[210] Additionally, as *N*-glycosylation is a common post-translational modification of many mammalian proteins, an intrinsic immune tolerance towards *N*-linked glycans is to be expected, and thus there may be autoreactivity and negative selection towards these antibodies *in vivo*.[315]

In the quest to elicit a broadly neutralizing antibody response by vaccination, it remains to be determined which strategies, antigens, vectors and adjuvants will lead to the most promising effects. Several obstacles and questions remain to be answered and overcome,[291] including concerns about the epitope dominance of non-neutralizing antibody epitopes or epitopes on variable, non-conserved regions of Env; concerns about the activation of BCRs if the germline antibody has low or no affinity to the antigen of its mature version; concerns about how to trigger the proper affinity maturation of bNAbs; as well as concerns about immune tolerance towards bNAbs containing unusual or unfavorable characteristics.[291] Immunogen design strategies aiming to induce either bNAbs or CTL responses as part of a complementary vaccination approach have recently been reviewed in great detail.[314,316] Here, we will focus on immunogen design and vaccination strategies to induce bNAbs.

11.4.2 Rational Immunogen Design to Induce Broadly Neutralizing Antibodies

HIV-1 vaccination research is dominated by a so-called "reverse vaccine engineering" or "antibody-guided" strategy, in which known bNAbs are used to rationally design immunogens aimed to induce similar broad and potent B cell responses.[317,318] Structural information about the bNAb epitopes gained by, for example, co-crystallization of a protective antibody with its antigen is usually crucial for this strategy.[318] Thus far many different potential HIV-1 Env immunogens have been designed and evaluated (Figure 11.8).

Peptide immunogens are among the earliest designed HIV-1 immunogens and are based on linear Env epitopes such as the V3 region and the MPER. Their employment in immunogenicity studies resulted at best in the elicitation of Tier 1 neutralizing antibodies (NAbs), and are thus most probably not sufficient to replace the complex conformational epitopes most bNAbs bind to.[319–323] Multiple other immunogen design strategies try to master the presentation of viral epitopes in a more native-like way in order to retain their antigenic character. Among these are multiple studies on a variety of Env epitopes presented on protein scaffolds, which so far have not yet succeeded

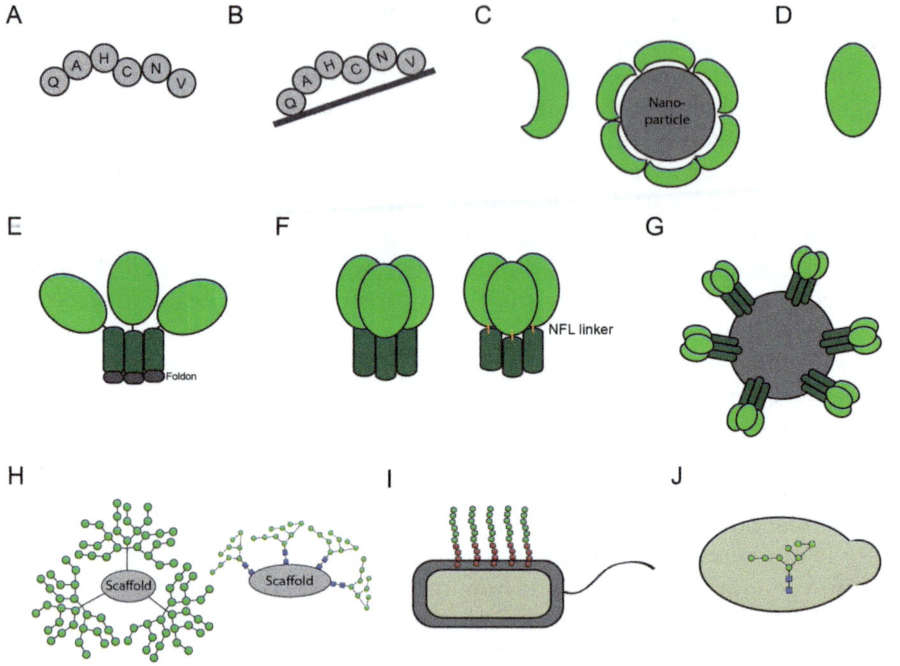

Figure 11.8 Overview of a selection of HIV-1 immunogens. (A) Peptide immunogens that contain linear epitopes.[319–323] (B) Peptide epitope scaffolds have been reported for the V3 region, the MPER and the CD4bs.[324–328] (C) Engineered outer domain (eOD) constructs. Monomeric (left) and modified, nanoparticle-multimerized eODs (right) have been investigated as potential immunogens.[246,247,331,387–389] (D) Gp120-based immunogens include monomeric gp120, hyperglycosylated gp120 and gp120 core constructs (deleted V1–V3 loops).[142,147,211,330,333,337,390–392] (E) Uncleaved gp140 trimer, usually stabilized by a foldon trimerization domain, engages a non-native trimer structure.[174,211,336–340] (F) Native-like, soluble trimers. Shown are a fully, cleaved trimer (left) and native flexibly-linked (NFL) trimer (right).[189–191,320,344] (G) Particulate-presented Env trimers can include presentation on nanoparticles, liposomes and virus-like particles.[346,348,393,394] (H) Synthetic carbohydrate antigens include for example oligosaccharides on small-molecule scaffolds including $Man_9GlcNAc_2$ clusters (left) and Man_9 dendrimers (right).[354] (I) Lipooligosaccharide expressed by the Gram-negative bacterium *Rhizobium radiobacter* is an analogue of the D1 arm of oligomannose as present in the 2G12 epitope.[368,369] (J) Engineered yeast has been employed to produce modified oligomannose *N*-glycosylation structures to mimic the 2G12 epitope.[370,371]

in inducing a sufficiently effective neutralizing antibody response in animal experiments.[324–328] Many bNAbs bind to glycans as part of their epitopes, though mimicking the glycosylation state at a particular *N*-glycosylation site poses an additional challenge. Glycan-based immunogen design is a large research area on its own that we will outline further below, however, larger immunogens that represent partial to almost complete envelope spikes, may impose similar steric hindrance phenomena to glycan-processing enzymes and hence manage to mimic the emergence of the oligomannose patches. Engineered, germline-targeting outer domain (eOD) constructs, gp120 core proteins and monomeric gp120 have been evaluated as vaccine candidates.[141,147,246,247,320,329–331] However, none of these strategies have yet shown a breakthrough in Tier 2 NAb induction. The limited immune response is thought to be attributed to less restricted access to epitopes on eOD and gp120 constructs that lead to antibody binding angles, particularly at the CD4bs, that are not compatible with the more restricted access to epitopes presented on a native trimer.[211,320,332] Furthermore, the interference of non-neutralizing epitopes on gp120, arising from the presence of the inner domain, which is usually buried within the trimer, as well as the lack of trimer-specific bNAb epitopes (*e.g.* PG9/16 and PGT151), are probable explanations for the failure to induce Tier 2 neutralizing antibodies.[56,58,320] However, BG505 gp120 displays a noteworthy exception as this monomer managed to induce an autologous Tier 2 neutralizing antibody response in rabbits and is also known to be unusually immunogenic.[333,334] All in all, while these constructs may find utility in priming the immune response, these results support the development of soluble trimeric mimics of Env trimer in order to properly present bNAb epitopes and avoid the possible immunodominance of non-neutralizing epitopes. Additionally, multiple neutralizing epitopes are presented allowing a polyclonal antibody response against different epitopes (thus providing barriers to emerging viral resistance mutants).[335]

Soluble gp140 trimers are generated by truncating gp41 leaving gp120 and a $gp41_{ECTO}$ domain that lacks the transmembrane region and the cytoplasmic tail, though these soluble trimers are very unstable, thus making it necessary to further stabilize the interaction between gp120 and gp41. One approach is to create uncleaved gp140 constructs by removing the furin cleavage site between gp120 and gp41 and to induce trimerization by including a C-terminal trimerization motif (*e.g.* a foldon domain).[336–339] However, results of multiple studies have indicated that these trimers adopt a non-native conformation with the gp120 subunits dangling from the trimerized gp41 and exposing non-NAb epitopes.[174,211,340] Additionally, the formation of aberrant disulfides as well as an unusually high abundance of complex-type glycans has been reported for uncleaved gp140 constructs, indicating protein misfolding and a non-native representation of the HIV-1 glycan shield.[168,221,341] Uncleaved gp140 immunogens were not able to induce NAbs against primary Tier 2 viruses.[336,337,342]

A next generation of HIV-1 immunogens was initiated with the development of the soluble, native-like Env SOSIP.664 trimers that we described in detail in Section 11.2.2 with the example of BG505 SOSIP.664. This soluble trimer enabled the determination of the first high-resolution cryo-EM and crystallography structures, which was an important step in the area of structure-based vaccine design.[189–191,320] BG505 SOSIP.664, as well as the clade B trimer, B41 SOSIP.664 successfully raised antibodies against the corresponding autologous Tier 2 virus.[333] However, a study in mice revealed the immunodominance of a non-neutralizing epitope near the trimer base, indicating that further immunogen improvement to obscure these unwanted epitopes might be necessary.[343] One disadvantage of the SOSIP constructs is the requirement for furin cleavage for native-like trimer folding. In an alternative strategy, cleavage-independent trimers have been generated that contain a flexible peptide linker between the gp120 and gp41 subunits—termed native flexibly linked (NFL) trimers.[344] These trimers also engage a native-like structure when analyzed using cryo-EM, but their immunogenicity still remains to be determined.[344]

Despite the advances in immunogen design brought by the soluble, native-like trimers, there is a large chance that vaccination with a recombinant protein will not be sufficiently immunogenic to induce a strong antibody response.[345] Virus-like particles (VLPs), liposomes and inorganic or self-assembling protein-based nanoparticles are potential particulate delivery systems to display HIV-1 Env antigens.[320,346] This allows for repetitive presentation of the antigen, which has several immunogenic advantages and will probably increase the longevity and potency of the antibody response.[320] These include, amongst others, an increased avidity of low-affinity bNAb epitopes, the potential to cross-link B cell receptors and the binding of natural, circulating IgM to repetitively displayed antigens through high-avidity interactions, leading to the activation of the complement system.[320,345] Furthermore, particulate presentation of Env also probably occludes the non-neutralizing epitopes at the trimer base.[343] Immunogenicity experiments using particulate trimer presentations generally gave promising results and good neutralizing antibody responses. For example, rabbit studies using membrane-bound trimers on VLPs occasionally showed promising induction of a Tier 2 NAb antibody responses that displayed some cross-neutralizing activity against viruses lacking the N197 glycan.[347] The presentation of eOD constructs on nanoparticles enabled, in contrast to BG505 SOSIP.664, the activation of bNAb class VRC01 precursors in VRC01-heavy chain knock-in mice.[246] Additionally, increased immunogenicity in rabbits was observed using ferritin nanoparticle-presented BG505 SOSIP.664 compared with the trimers on their own.[348] Overall, particle-associated Env immunogens have a significant potential that should be further developed and investigated.

11.4.3 Chemical Biology of Carbohydrate-Based Immunogens

Many efforts have been undertaken to design carbohydrate-based HIV-1 immunogens. Their development has been generally motivated by, and concentrates on, the mimicry of the bNAb 2G12 glycan epitope. 2G12 recognizes a cluster of α1,2-linked mannoses on the outer domain of Env *via* its unique domain-exchanged structure.[60,349] The rationale is that an immunogen containing a multivalent array of α1,2-linked mannoses might be capable of inducing 2G12-like antibodies to this specific epitope if it is presented in a sufficiently immunogenic context.[315,350,351] Chemically synthesized, multivalent oligomannose-glycan antigens have been created by clustering carbohydrates on rationally designed peptide,[352–355] dendrimer,[356,357] carbohydrate,[358,359] DNA,[360–363] steroid,[364] gold nanoparticle[365] and biomacromolecular[366,367] scaffolds. Many of these constructs have been reported to bind to 2G12, however none of the antisera has so far been reported to show HIV-1 neutralization activity.[354,363]

Another approach is the exploitation of naturally-derived carbohydrate antigens as mimics of the 2G12 epitope. For example, bacterial lipooligosaccharides have been investigated as potential immunogens.[368,369] Similar to the chemically designed immunogens, no neutralizing anti-HIV-1 activity was detected in raised antibody sera.[368] Further attempts to place key components of the epitope in a wider immunogenic context also include yeast strains genetically engineered to manipulate *N*-glycosylation. Knock-out mutants of glycan-processing enzymes in *Saccharomyces cerevisiae* were used to generate mannans displaying extended arrays of α1,2-linked mannoses, as well as mannans that homogenously display the self-glycan $Man_8GlcNAc_2$ on its glycoproteins.[370,371] Antigenicity and immunogenicity studies using whole yeast cells of these mutants generally yielded antibody responses that were able to bind to oligomannose-type glycans but which often failed to bind gp120.[351,370,371] Similarly, HIV-1 neutralization was reported to be either extremely weak or only detected for viruses that had been produced in the presence of kifunensine to force retention of $Man_9GlcNAc_2$ at all *N*-linked glycans.[351,370] A likely explanation is that these antibodies possibly recognize oligomannose structures that are not present on native virions and are localized outside the 2G12 epitope.[351,372] Similar results have been reported for glycoprotein immunogens that were produced in genetically engineered yeast.[372,373]

Questions and uncertainties remain regarding whether or not there are any dangers of bypassing natural tolerance mechanisms to oligomannose glycans by immunogen development. However, as the innate immune system is highly sensitive to cell-surface mannose residues, it is unlikely that mimicking this will have any untoward effects in the adaptive immune system. The sensitivity of the innate immune system to oligomannose is illustrated by its chronic activation and lupus-like pathology shown in mice deficient in Golgi α-mannosidase II.[374] Furthermore, passive transfer of bNAbs exhibiting specificity towards oligomannose-type glycans is well-tolerated.

11.4.4 Vaccination Strategies

The development of native-like trimer mimics as subunit vaccines will rely on the selection and usage of suitable adjuvants and delivery vectors that help to render a humoral and cellular immune response more effective. In terms of adjuvants, several new compounds, including Toll-like receptor agonists, are in clinical development to provide additional options alongside the most widely used aluminum hydroxide (Alum) and Freund's adjuvants.[375] The detailed discussion of vectors and adjuvants constitutes a whole separate area of vaccine design and is not within the scope of this chapter. However, BG505 SOSIP.664 trimers have successfully been shown to be stable when formulated in ISCOMATRIX adjuvant.[333,376] Here, we would like to highlight the current main vaccination strategies to induce the production of bNAbs.

The rationally designed, native-like soluble trimers with occluded non-neutralizing epitopes that display multiple, correct bNAb epitopes have the potential to raise a broad, polyclonal antibody response. Ideally this will prevent the formation of viral escape mutants as well as act upon a panel of different HIV-1 subtypes. However, as the so far best characterized immunogen BG505 SOSIP.664 only managed to induce autologous NAbs, it is likely that a single Env trimer won't be sufficient for a broadly acting vaccine.[320,333] Simultaneous or sequential vaccination with recombinant trimers from different clades represents a promising alternative strategy to be explored.[320]

Sequential immunization is also the basis of novel prime–boost strategies being developed to specifically target the naïve and intermediate mutation versions of germline bNAb ancestors.[246,247,377] Recombinant Env immunogens usually show little to no binding to the unmutated, germline-reverted versions of bNAbs and their B cell receptors.[184,277,378,379] As described in Section 11.3.3, bNAbs co-evolve together with the highly mutating virus, indicating that it might be necessary to mimic this progress using specifically designed vaccination steps.[276] These would probably start with the activation of the relevant naïve B cell receptor, followed by immunogens that guide antibody development along the appropriate and desired path.[320] Env immunogen improvements to specifically engage germline precursors include trimers with specifically eliminated PNGSs, as well as germline-targeting outer domain constructs.[247,331,378] These strategies will need to be further evaluated in animal and human trials.

Another approach to positively influence the evolution of bNAbs is the induction of more productive germinal centers together with an enhanced T_{FH} response.[283] As described in Section 11.3.3, T_{FH}s are thought to have a crucial role in bNAb development. However, little is known about the factors that can drive T_{FH} responses during vaccination.[283] Perhaps a simpler option to positively influence bNAb development in germinal centers would be to deliver cytokines that mediate T_{FH}-like help, such as interleukin-21 and/or interleukin-12, as adjuvants.[283,380,381] Moreover, the members of the tumor necrosis factor family B cell activating factor (BAFF) and proliferation-inducing ligand (APRIL) seem to be important for productive germinal centers and have shown promising results in vaccine development studies.[283,382,383] The thus induced larger and more prolonged germinal centers might enable the production of B cells that elicit more mutated and higher-affinity bNAbs.[283]

11.5 Conclusion and Outlook

The development of a vaccine against HIV-1 presents a formidable challenge. The antigenic diversity of the virus has rendered all previously developed vaccine strategies ineffective. However, there have been several areas of advancement which give grounds for optimism. Central among these has been the discovery of bNAbs that are potent at titers achievable by typical vaccination regimens. That these bNAbs are highly mutated has also been shown to be more closely associated with the length of time of bNAb development rather than the requirement for substantial mutations for specificity and function. A successful vaccine is likely to include the breakdown of self-tolerance to the self 'glycan shield' and the long stable presentation of conformationally specific epitopes able to sequentially walk B cell development from distant specificities to a bNAb-mimicking phenotype. In total, it is likely that adjuvants combined with a range of staggered and co-administered HIV-1 trimer mimics will have to orchestrate the development of a protective antibody response.

References

1. UNAIDSinfo, AIDS info, http://aidsinfo.unaids.org/, accessed 15/03/2016.
2. D. R. Burton, R. Ahmed, D. H. Barouch, S. T. Butera, S. Crotty, A. Godzik, D. E. Kaufmann, M. J. McElrath, M. C. Nussenzweig, B. Pulendran, C. N. Scanlan, W. R. Schief, G. Silvestri, H. Streeck, B. D. Walker, L. M. Walker, A. B. Ward, I. A. Wilson and R. Wyatt, *Cell Host Microbe*, 2012, **12**, 396.
3. E. Fee and T. M. Brown, *Am. J. Public Health*, 2006, **96**, 982.
4. M. S. Gottlieb, H. M. Schlanker, P. T. Fan, A. Saxon and J. D. Weismann, *MMWR (CDC)*, 1981, **30**, 1.
5. F. Barré-Sinoussi, A. L. Ross and J.-F. Delfraissy, *Nat. Rev. Microbiol.*, 2013, **11**, 877.
6. UNAIDS, *Fact Sheet 2015*, 2015.
7. K. A. Dilley, N. Ni, O. A. Nikolaitchik, J. Chen, A. Galli and W.-S. Hu, *J. Virol.*, 2011, **85**, 10499.
8. R. Marlink, P. Kanki, I. Thior, K. Travers, G. Eisen, T. Siby, I. Traore, C. C. Hsieh, M. C. Dia and E. H. Gueye, *Science*, 1994, **265**, 1587.
9. L. Buonaguro, M. L. Tornesello and F. M. Buonaguro, *J. Virol.*, 2007, **81**, 10209.
10. J.-C. Plantier, M. Leoz, J. E. Dickerson, F. De Oliveira, F. Cordonnier, V. Lemée, F. Damond, D. L. Robertson and F. Simon, *Nat. Med.*, 2009, **15**, 871.
11. L. A. Lasky, J. E. Groopman, C. W. Fennie, P. M. Benz, D. J. Capon, D. J. Dowbenko, G. R. Nakamura, W. M. Nunes, M. E. Renz and P. W. Berman, *Science*, 1986, **233**, 209.
12. A. G. Dalgleish, P. C. Beverley, P. R. Clapham, D. H. Crawford, M. F. Greaves and R. A. Weiss, *Nature*, 1984, **312**, 763.
13. D. Klatzmann, E. Champagne, S. Chamaret, J. Gruest, D. Guetard, T. Hercend, J. C. Gluckman and L. Montagnier, *Nature*, 1984, **312**, 767.
14. H. Deng, R. Liu, W. Ellmeier, S. Choe, D. Unutmaz, M. Burkhart, P. Di Marzio, S. Marmon, R. E. Sutton, C. M. Hill, C. B. Davis, S. C. Peiper, T. J. Schall, D. R. Littman and N. R. Landau, *Nature*, 1996, **381**, 661.

15. T. Dragic, V. Litwin, G. P. Allaway, S. R. Martin, Y. Huang, K. A. Nagashima, C. Cayanan, P. J. Maddon, R. A. Koup, J. P. Moore and W. A. Paxton, *Nature*, 1996, **381**, 667.
16. T. Zhu, H. Mo, N. Wang, D. S. Nam, Y. Cao, R. A. Koup and D. D. Ho, *Science*, 1993, **261**, 1179.
17. A. B. van't Wout, N. A. Kootstra, G. A. Mulder-Kampinga, N. Albrecht-van Lent, H. J. Scherpbier, J. Veenstra, K. Boer, R. A. Coutinho, F. Miedema and H. Schuitemaker, *J. Clin. Invest.*, 1994, **94**, 2060.
18. R. Blumenthal, S. Durell and M. Viard, *J. Biol. Chem.*, 2012, **287**, 40841.
19. A. M. Lever and K. T. Jeang, *Int. J. Hematol.*, 2006, **84**, 23.
20. E. O. Freed, *Nat. Rev. Microbiol.*, 2015, **13**, 484.
21. D. Finzi, M. Hermankova, T. Pierson, L. M. Carruth, C. Buck, R. E. Chaisson, T. C. Quinn, K. Chadwick, J. Margolick, R. Brookmeyer, J. Gallant, M. Markowitz, D. D. Ho, D. D. Richman and R. F. Siliciano, *Science*, 1997, **278**, 1295.
22. T. W. Chun, L. Stuyver, S. B. Mizell, L. A. Ehler, J. A. Mican, M. Baseler, A. L. Lloyd, M. A. Nowak and A. S. Fauci, *Proc. Natl. Acad. Sci. U. S. A.*, 1997, **94**, 13193.
23. R. F. Siliciano and W. C. Greene, *Cold Spring Harbor Perspect. Med.*, 2011, **1**, a007096.
24. L. B. Cohn, I. T. Silva, T. Y. Oliveira, R. A. Rosales, E. H. Parrish, G. H. Learn, B. H. Hahn, J. L. Czartoski, M. J. McElrath, C. Lehmann, F. Klein, M. Caskey, B. D. Walker, J. D. Siliciano, R. F. Siliciano, M. Jankovic and M. C. Nussenzweig, *Cell*, 2015, **160**, 420.
25. F. Hladik and M. J. McElrath, *Nat. Rev. Immunol.*, 2008, **8**, 447.
26. Y. Ganor and M. Bomsel, *Am. J. Reprod. Immunol.*, 2011, **65**, 284.
27. D. L. Sodora, A. Gettie, C. J. Miller and P. A. Marx, *AIDS Res. Hum. Retroviruses*, 1998, **14**, 119.
28. B. F. Keele, E. E. Giorgi, J. F. Salazar-Gonzalez, J. M. Decker, K. T. Pham, M. G. Salazar, C. Sun, T. Grayson, S. Wang, H. Li, X. Wei, C. Jiang, J. L. Kirchherr, F. Gao, J. A. Anderson, L. H. Ping, R. Swanstrom, G. D. Tomaras, W. A. Blattner, P. A. Goepfert, J. M. Kilby, M. S. Saag, E. L. Delwart, M. P. Busch, M. S. Cohen, D. C. Montefiori, B. F. Haynes, B. Gaschen, G. S. Athreya, H. Y. Lee, N. Wood, C. Seoighe, A. S. Perelson, T. Bhattacharya, B. T. Korber, B. H. Hahn and G. M. Shaw, *Proc. Natl. Acad. Sci. U. S. A.*, 2008, **105**, 7552.
29. S. B. Joseph, R. Swanstrom, A. D. Kashuba and M. S. Cohen, *Nat. Rev. Microbiol.*, 2015, **13**, 414.
30. J. F. Salazar-Gonzalez, M. G. Salazar, B. F. Keele, G. H. Learn, E. E. Giorgi, H. Li, J. M. Decker, S. Wang, J. Baalwa, M. H. Kraus, N. F. Parrish, K. S. Shaw, M. B. Guffey, K. J. Bar, K. L. Davis, C. Ochsenbauer-Jambor, J. C. Kappes, M. S. Saag, M. S. Cohen, J. Mulenga, C. A. Derdeyn, S. Allen, E. Hunter, M. Markowitz, P. Hraber, A. S. Perelson, T. Bhattacharya, B. F. Haynes, B. T. Korber, B. H. Hahn and G. M. Shaw, *J. Exp. Med.*, 2009, **206**, 1273.
31. A. J. McMichael, P. Borrow, G. D. Tomaras, N. Goonetilleke and B. F. Haynes, *Nat. Rev. Immunol.*, 2010, **10**, 11.

32. S. Mehandru, M. A. Poles, K. Tenner-Racz, A. Horowitz, A. Hurley, C. Hogan, D. Boden, P. Racz and M. Markowitz, *J. Exp. Med.*, 2004, **200**, 761.
33. X. Wang, T. Rasmussen, B. Pahar, B. Poonia, X. Alvarez, A. A. Lackner and R. S. Veazey, *Blood*, 2007, **109**, 1174.
34. J. M. Brenchley, D. A. Price and D. C. Douek, *Nat. Immunol.*, 2006, **7**, 235.
35. J. M. Brenchley, T. W. Schacker, L. E. Ruff, D. A. Price, J. H. Taylor, G. J. Beilman, P. L. Nguyen, A. Khoruts, M. Larson, A. T. Haase and D. C. Douek, *J. Exp. Med.*, 2004, **200**, 749.
36. J. Arthos, C. Cicala, E. Martinelli, K. Macleod, D. Van Ryk, D. Wei, Z. Xiao, T. D. Veenstra, T. P. Conrad, R. A. Lempicki, S. McLaughlin, M. Pascuccio, R. Gopaul, J. McNally, C. C. Cruz, N. Censoplano, E. Chung, K. N. Reitano, S. Kottilil, D. J. Goode and A. S. Fauci, *Nat. Immunol.*, 2008, **9**, 301.
37. N. Wagner, J. Lohler, E. J. Kunkel, K. Ley, E. Leung, G. Krissansen, K. Rajewsky and W. Muller, *Nature*, 1996, **382**, 366.
38. U. H. von Andrian and C. R. Mackay, *N. Engl. J. Med.*, 2000, **343**, 1020.
39. M. K. Gorny, R. Pan, C. Williams, X. H. Wang, B. Volsky, T. O'Neal, B. Spurrier, J. M. Sampson, L. Li, M. S. Seaman, X. P. Kong and S. Zolla-Pazner, *Virology*, 2012, **427**, 198.
40. B. Rodriguez, A. K. Sethi, V. K. Cheruvu, W. Mackay, R. J. Bosch, M. Kitahata, S. L. Boswell, W. C. Mathews, D. R. Bangsberg, J. Martin, C. C. Whalen, S. Sieg, S. Yadavalli, S. G. Deeks and M. M. Lederman, *JAMA*, 2006, **296**, 1498.
41. T. W. Schacker, J. P. Hughes, T. Shea, R. W. Coombs and L. Corey, *Ann. Intern. Med.*, 1998, **128**, 613.
42. P. Borrow, H. Lewicki, B. H. Hahn, G. M. Shaw and M. B. Oldstone, *J. Virol.*, 1994, **68**, 6103.
43. R. A. Koup, J. T. Safrit, Y. Cao, C. A. Andrews, G. McLeod, W. Borkowsky, C. Farthing and D. D. Ho, *J. Virol.*, 1994, **68**, 4650.
44. R. S. Veazey, M. DeMaria, L. V. Chalifoux, D. E. Shvetz, D. R. Pauley, H. L. Knight, M. Rosenzweig, R. P. Johnson, R. C. Desrosiers and A. A. Lackner, *Science*, 1998, **280**, 427.
45. J. J. Mattapallil, D. C. Douek, B. Hill, Y. Nishimura, M. Martin and M. Roederer, *Nature*, 2005, **434**, 1093.
46. D. H. Nguyen and J. E. Hildreth, *J. Virol.*, 2000, **74**, 3264.
47. R. Chan, P. D. Uchil, J. Jin, G. Shui, D. E. Ott, W. Mothes and M. R. Wenk, *J. Virol.*, 2008, **82**, 11228.
48. B. Brugger, B. Glass, P. Haberkant, I. Leibrecht, F. T. Wieland and H. G. Krausslich, *Proc. Natl. Acad. Sci. U. S. A.*, 2006, **103**, 2641.
49. M. Pino, I. Erkizia, S. Benet, E. Erikson, M. T. Fernandez-Figueras, D. Guerrero, J. Dalmau, D. Ouchi, A. Rausell, A. Ciuffi, O. T. Keppler, A. Telenti, H. G. Krausslich, J. Martinez-Picado and N. Izquierdo-Useros, *Retrovirology*, 2015, **12**, 37.
50. N. Izquierdo-Useros, M. Lorizate, P. J. McLaren, A. Telenti, H. G. Krausslich and J. Martinez-Picado, *PLoS Pathog.*, 2014, **10**, e1004146.

51. M. S. Motswaledi, I. Kasvosve and O. O. Oguntibeju, *Biomed Res. Int.*, 2013, **2013**, 758682.
52. S. J. Neil, A. McKnight, K. Gustafsson and R. A. Weiss, *Blood*, 2005, **105**, 4693.
53. M. Arendrup, J. E. Hansen, H. Clausen, C. Nielsen, L. R. Mathiesen and J. O. Nielsen, *AIDS*, 1991, **5**, 441.
54. R. Pejchal, K. J. Doores, L. M. Walker, R. Khayat, P.-S. Huang, S.-K. Wang, R. L. Stanfield, J.-P. Julien, A. Ramos, M. Crispin, R. Depetris, U. Katpally, A. Marozsan, A. Cupo, S. Maloveste, Y. Liu, R. McBride, Y. Ito, R. W. Sanders, C. Ogohara, J. C. Paulson, T. Feizi, C. N. Scanlan, C.-H. Wong, J. P. Moore, W. C. Olson, A. B. Ward, P. Poignard, W. R. Schief, D. R. Burton and I. A. Wilson, *Science*, 2011, **334**, 1097.
55. L. M. Walker, M. Huber, K. J. Doores, E. Falkowska, R. Pejchal, J.-P. Julien, S.-K. Wang, A. Ramos, P.-Y. Chan-Hui, M. Moyle, J. L. Mitcham, P. W. Hammond, O. A. Olsen, P. Phung, S. Fling, C.-H. Wong, S. Phogat, T. Wrin, M. D. Simek, W. C. Koff, I. A. Wilson, D. R. Burton and P. Poignard, *Nature*, 2011, **477**, 466.
56. L. M. Walker, S. K. Phogat, P.-Y. Chan-Hui, D. Wagner, P. Phung, J. L. Goss, T. Wrin, M. D. Simek, S. Fling, J. L. Mitcham, J. K. Lehrman, F. H. Priddy, O. A. Olsen, S. M. Frey, P. W. Hammond, S. Kaminsky, T. Zamb, M. Moyle, W. C. Koff, P. Poignard and D. R. Burton, *Science*, 2009, **326**, 285.
57. D. A. Calarese, C. N. Scanlan, M. B. Zwick, S. Deechongkit, Y. Mimura, R. Kunert, P. Zhu, M. R. Wormald, R. L. Stanfield, K. H. Roux, J. W. Kelly, P. M. Rudd, R. A. Dwek, H. Katinger, D. R. Burton and I. A. Wilson, *Science*, 2003, **300**, 2065.
58. C. Blattner, J. H. Lee, K. Sliepen, R. Derking, E. Falkowska, A. T. de la Pena, A. Cupo, J. P. Julien, M. van Gils, P. S. Lee, W. Peng, J. C. Paulson, P. Poignard, D. R. Burton, J. P. Moore, R. W. Sanders, I. A. Wilson and A. B. Ward, *Immunity*, 2014, **40**, 669.
59. E. Falkowska, K. M. Le, A. Ramos, K. J. Doores, J. H. Lee, C. Blattner, A. Ramirez, R. Derking, M. J. van Gils, C.-H. Liang, R. McBride, B. von Bredow, S. S. Shivatare, C.-Y. Wu, P.-Y. Chan-Hui, Y. Liu, T. Feizi, M. B. Zwick, W. C. Koff, M. S. Seaman, K. Swiderek, J. P. Moore, D. Evans, J. C. Paulson, C.-H. Wong, A. B. Ward, I. A. Wilson, R. W. Sanders, P. Poignard and D. R. Burton, *Immunity*, 2014, **40**, 657.
60. C. N. Scanlan, R. Pantophlet, M. R. Wormald, E. Ollmann Saphire, R. Stanfield, I. A. Wilson, H. Katinger, R. A. Dwek, P. M. Rudd and D. R. Burton, *J. Virol.*, 2002, **76**, 7306.
61. H. Mouquet, L. Scharf, Z. Euler, Y. Liu, C. Eden, J. F. Scheid, A. Halper-Stromberg, P. N. P. Gnanapragasam, D. I. R. Spencer, M. S. Seaman, H. Schuitemaker, T. Feizi, M. C. Nussenzweig and P. J. Bjorkman, *Proc. Natl. Acad. Sci. U. S. A.*, 2012, **109**, 3268.
62. L. Scharf, J. F. Scheid, J. H. Lee, A. P. West, C. Chen, H. Gao, P. N. P. Gnanapragasam, R. Mares, M. S. Seaman, A. B. Ward, M. C. Nussenzweig and P. J. Bjorkman, *Cell Rep.*, 2014, **7**, 785.

63. F. Garces, D. Sok, L. Kong, R. McBride, H. J. Kim, K. F. Saye-Francisco, J.-P. Julien, Y. Hua, A. Cupo, J. P. Moore, J. C. Paulson, A. B. Ward, D. R. Burton and I. A. Wilson, *Cell*, 2014, **159**, 69.
64. J. Huang, B. H. Kang, M. Pancera, J. H. Lee, T. Tong, Y. Feng, I. S. Georgiev, G.-Y. Chuang, A. Druz, N. A. Doria-Rose, L. Laub, K. Sliepen, M. J. van Gils, A. T. de la Peña, R. Derking, P.-J. Klasse, S. A. Migueles, R. T. Bailer, M. Alam, P. Pugach, B. F. Haynes, R. T. Wyatt, R. W. Sanders, J. M. Binley, A. B. Ward, J. R. Mascola, P. D. Kwong and M. Connors, *Nature*, 2014, **515**, 138.
65. L. Kong, J. H. Lee, K. J. Doores, C. D. Murin, J. P. Julien, R. McBride, Y. Liu, A. Marozsan, A. Cupo, P. J. Klasse, S. Hoffenberg, M. Caulfield, C. R. King, Y. Hua, K. M. Le, R. Khayat, M. C. Deller, T. Clayton, H. Tien, T. Feizi, R. W. Sanders, J. C. Paulson, J. P. Moore, R. L. Stanfield, D. R. Burton, A. B. Ward and I. A. Wilson, *Nat. Struct. Mol. Biol.*, 2013, **20**, 796.
66. J. S. McLellan, M. Pancera, C. Carrico, J. Gorman, J.-P. Julien, R. Khayat, R. Louder, R. Pejchal, M. Sastry, K. Dai, S. O'Dell, N. Patel, S. Shahzad-ul-Hussan, Y. Yang, B. Zhang, T. Zhou, J. Zhu, J. C. Boyington, G.-Y. Chuang, D. Diwanji, I. Georgiev, Y. D. Kwon, D. Lee, M. K. Louder, S. Moquin, S. D. Schmidt, Z.-Y. Yang, M. Bonsignori, J. A. Crump, S. H. Kapiga, N. E. Sam, B. F. Haynes, D. R. Burton, W. C. Koff, L. M. Walker, S. Phogat, R. Wyatt, J. Orwenyo, L.-X. Wang, J. Arthos, C. A. Bewley, J. R. Mascola, G. J. Nabel, W. R. Schief, A. B. Ward, I. A. Wilson and P. D. Kwong, *Nature*, 2011, **480**, 336.
67. M. Pancera, S. Shahzad-Ul-Hussan, N. A. Doria-Rose, J. S. McLellan, R. T. Bailer, K. Dai, S. Loesgen, M. K. Louder, R. P. Staupe, Y. Yang, B. Zhang, R. Parks, J. Eudailey, K. E. Lloyd, J. Blinn, S. M. Alam, B. F. Haynes, M. N. Amin, L.-X. Wang, D. R. Burton, W. C. Koff, G. J. Nabel, J. R. Mascola, C. A. Bewley and P. D. Kwong, *Nat. Struct. Mol. Biol.*, 2013, **20**, 804.
68. N. T. Freund, J. A. Horwitz, L. Nogueira, S. A. Sievers, L. Scharf, J. F. Scheid, A. Gazumyan, C. Liu, K. Velinzon, A. Goldenthal, R. W. Sanders, J. P. Moore, P. J. Bjorkman, M. S. Seaman, B. D. Walker, F. Klein and M. C. Nussenzweig, *PLoS Pathog.*, 2015, **11**, e1005238.
69. A. J. Hessell, P. Poignard, M. Hunter, L. Hangartner, D. M. Tehrani, W. K. Bleeker, P. W. H. I. Parren, P. A. Marx and D. R. Burton, *Nat. Med.*, 2009, **15**, 951.
70. A. J. Hessell, E. G. Rakasz, P. Poignard, L. Hangartner, G. Landucci, D. N. Forthal, W. C. Koff, D. I. Watkins and D. R. Burton, *PLoS Pathog.*, 2009, **5**, e1000433.
71. A. J. Hessell, E. G. Rakasz, D. M. Tehrani, M. Huber, K. L. Weisgrau, G. Landucci, D. N. Forthal, W. C. Koff, P. Poignard, D. I. Watkins and D. R. Burton, *J. Virol.*, 2010, **84**, 1302.
72. J. R. Mascola, M. G. Lewis, G. Stiegler, D. Harris, T. C. VanCott, D. Hayes, M. K. Louder, C. R. Brown, C. V. Sapan, S. S. Frankel, Y. Lu, M. L. Robb, H. Katinger and D. L. Birx, *J. Virol.*, 1999, **73**, 4009.
73. J. R. Mascola, G. Stiegler, T. C. VanCott, H. Katinger, C. B. Carpenter, C. E. Hanson, H. Beary, D. Hayes, S. S. Frankel, D. L. Birx and M. G. Lewis, *Nat. Med.*, 2000, **6**, 207.

74. B. Moldt, E. G. Rakasz, N. Schultz, P. Y. Chan-Hui, K. Swiderek, K. L. Weisgrau, S. M. Piaskowski, Z. Bergman, D. I. Watkins, P. Poignard and D. R. Burton, *Proc. Natl. Acad. Sci. U. S. A.*, 2012, **109**, 18921.
75. M. Altfeld and M. Gale Jr, *Nat. Immunol.*, 2015, **16**, 554.
76. D. Sauter and F. Kirchhoff, *Curr. Opin. HIV AIDS*, 2016, **11**, 173.
77. M. R. Jakobsen, D. Olagnier and J. Hiscott, *Curr. Opin. HIV AIDS*, 2015, **10**, 96.
78. A. Silvin and N. Manel, *Curr. Opin. Immunol.*, 2015, **32**, 54.
79. K. Drickamer, *Curr. Opin. Struct. Biol.*, 1999, **9**, 585.
80. Y. van Kooyk and G. A. Rabinovich, *Nat. Immunol.*, 2008, **9**, 593.
81. M. J. Robinson, D. Sancho, E. C. Slack, S. LeibundGut-Landmann and C. Reis e Sousa, *Nat. Immunol.*, 2006, **7**, 1258.
82. L. de Witte, A. Nabatov, M. Pion, D. Fluitsma, M. A. de Jong, T. de Gruijl, V. Piguet, Y. van Kooyk and T. B. Geijtenbeek, *Nat. Med.*, 2007, **13**, 367.
83. E. Cassol, L. Cassetta, C. Rizzi, D. Gabuzda, M. Alfano and G. Poli, *AIDS*, 2013, **27**, 707.
84. T. B. Geijtenbeek, D. S. Kwon, R. Torensma, S. J. van Vliet, G. C. van Duijnhoven, J. Middel, I. L. Cornelissen, H. S. Nottet, V. N. KewalRamani, D. R. Littman, C. G. Figdor and Y. van Kooyk, *Cell*, 2000, **100**, 587.
85. A. M. G. van der Aar, S. I. Gringhuis and T. B. H. Geijtenbeek, in *HIV Glycans in Infection and Immunity*, ed. R. Pantophlet, Springer, 2014, vol. 1, ch. 2.
86. C. Dong, A. M. Janas, J. H. Wang, W. J. Olson and L. Wu, *J. Virol.*, 2007, **81**, 11352.
87. S. G. Turville, J. J. Santos, I. Frank, P. U. Cameron, J. Wilkinson, M. Miranda-Saksena, J. Dable, H. Stossel, N. Romani, M. Piatak Jr., J. D. Lifson, M. Pope and A. L. Cunningham, *Blood*, 2004, **103**, 2170.
88. D. S. Kwon, G. Gregorio, N. Bitton, W. A. Hendrickson and D. R. Littman, *Immunity*, 2002, **16**, 135.
89. K. Hijazi, Y. Wang, C. Scala, S. Jeffs, C. Longstaff, D. Stieh, B. Haggarty, G. Vanham, D. Schols, J. Balzarini, I. M. Jones, J. Hoxie, R. Shattock and C. G. Kelly, *PLoS One*, 2011, **6**, e28307.
90. S. G. Turville, P. U. Cameron, A. Handley, G. Lin, S. Pohlmann, R. W. Doms and A. L. Cunningham, *Nat. Immunol.*, 2002, **3**, 975.
91. C. G. Figdor, Y. van Kooyk and G. J. Adema, *Nat. Rev. Immunol.*, 2002, **2**, 77.
92. N. S. Stambach and M. E. Taylor, *Glycobiology*, 2003, **13**, 401.
93. J. Valladeau, O. Ravel, C. Dezutter-Dambuyant, K. Moore, M. Kleijmeer, Y. Liu, V. Duvert-Frances, C. Vincent, D. Schmitt, J. Davoust, C. Caux, S. Lebecque and S. Saeland, *Immunity*, 2000, **12**, 71.
94. F. Hladik, P. Sakchalathorn, L. Ballweber, G. Lentz, M. Fialkow, D. Eschenbach and M. J. McElrath, *Immunity*, 2007, **26**, 257.
95. Q. Yu, R. Yu and X. Qin, *Cell. Mol. Immunol.*, 2010, **7**, 334.
96. M. M. Aasa-Chapman, S. Holuigue, K. Aubin, M. Wong, N. A. Jones, D. Cornforth, P. Pellegrino, P. Newton, I. Williams, P. Borrow and A. McKnight, *J. Virol.*, 2005, **79**, 2823.

97. B. L. Sullivan, E. J. Knopoff, M. Saifuddin, D. M. Takefman, M. N. Saarloos, B. E. Sha and G. T. Spear, *J. Immunol.*, 1996, **157**, 1791.
98. G. Senaldi, M. Peakman, T. McManus, E. T. Davies, D. E. Tee and D. Vergani, *J. Infect. Dis.*, 1990, **162**, 1227.
99. C. F. Ebenbichler, N. M. Thielens, R. Vornhagen, P. Marschang, G. J. Arlaud and M. P. Dierich, *J. Exp. Med.*, 1991, **174**, 1417.
100. J. S. Haurum, S. Thiel, I. M. Jones, P. B. Fischer, S. B. Laursen and J. C. Jensenius, *AIDS*, 1993, **7**, 1307.
101. M. L. Hart, M. Saifuddin and G. T. Spear, *J. Gen. Virol.*, 2003, **84**, 353.
102. G. T. Spear, M. R. Zariffard, J. Xin and M. Saifuddin, *Immunology*, 2003, **110**, 80.
103. J. Schmitz, J. P. Zimmer, B. Kluxen, S. Aries, M. Bogel, I. Gigli and H. Schmitz, *J. Clin. Invest.*, 1995, **96**, 1520.
104. C. Pinter, A. G. Siccardi, L. Lopalco, R. Longhi and A. Clivio, *AIDS Res. Hum. Retroviruses*, 1995, **11**, 971.
105. Z. Bajtay, C. Speth, A. Erdei and M. P. Dierich, *J. Immunol.*, 2004, **173**, 4775.
106. N. Thieblemont, N. Haeffner-Cavaillon, A. Ledur, J. L'Age-Stehr, H. W. Ziegler-Heitbrock and M. D. Kazatchkine, *Clin. Exp. Immunol.*, 1993, **92**, 106.
107. C. M. Coleman, C. S. Gelais and L. Wu, *Adv. Exp. Med. Biol.*, 2013, **762**, 109.
108. L. Wu and V. N. KewalRamani, *Nat. Rev. Immunol.*, 2006, **6**, 859.
109. E. L. Turnbull, M. Wong, S. Wang, X. Wei, N. A. Jones, K. E. Conrod, D. Aldam, J. Turner, P. Pellegrino, B. F. Keele, I. Williams, G. M. Shaw and P. Borrow, *J. Immunol.*, 2009, **182**, 7131.
110. N. Goonetilleke, M. K. Liu, J. F. Salazar-Gonzalez, G. Ferrari, E. Giorgi, V. V. Ganusov, B. F. Keele, G. H. Learn, E. L. Turnbull, M. G. Salazar, K. J. Weinhold, S. Moore, N. Letvin, B. F. Haynes, M. S. Cohen, P. Hraber, T. Bhattacharya, P. Borrow, A. S. Perelson, B. H. Hahn, G. M. Shaw, B. T. Korber and A. J. McMichael, *J. Exp. Med.*, 2009, **206**, 1253.
111. D. E. Kaufmann, P. M. Bailey, J. Sidney, B. Wagner, P. J. Norris, M. N. Johnston, L. A. Cosimi, M. M. Addo, M. Lichterfeld, M. Altfeld, N. Frahm, C. Brander, A. Sette, B. D. Walker and E. S. Rosenberg, *J. Virol.*, 2004, **78**, 4463.
112. G. D. Tomaras, N. L. Yates, P. Liu, L. Qin, G. G. Fouda, L. L. Chavez, A. C. Decamp, R. J. Parks, V. C. Ashley, J. T. Lucas, M. Cohen, J. Eron, C. B. Hicks, H. X. Liao, S. G. Self, G. Landucci, D. N. Forthal, K. J. Weinhold, B. F. Keele, B. H. Hahn, M. L. Greenberg, L. Morris, S. S. Karim, W. A. Blattner, D. C. Montefiori, G. M. Shaw, A. S. Perelson and B. F. Haynes, *J. Virol.*, 2008, **82**, 12449.
113. E. T. Crooks, P. L. Moore, M. Franti, C. S. Cayanan, P. Zhu, P. Jiang, R. P. de Vries, C. Wiley, I. Zharkikh, N. Schulke, K. H. Roux, D. C. Montefiori, D. R. Burton and J. M. Binley, *Virology*, 2007, **366**, 245.
114. P. L. Moore, E. T. Crooks, L. Porter, P. Zhu, C. S. Cayanan, H. Grise, P. Corcoran, M. B. Zwick, M. Franti, L. Morris, K. H. Roux, D. R. Burton and J. M. Binley, *J. Virol.*, 2006, **80**, 2515.

115. X. Wei, J. M. Decker, S. Wang, H. Hui, J. C. Kappes, X. Wu, J. F. Salazar-Gonzalez, M. G. Salazar, J. M. Kilby, M. S. Saag, N. L. Komarova, M. A. Nowak, B. H. Hahn, P. D. Kwong and G. M. Shaw, *Nature*, 2003, **422**, 307.
116. D. D. Richman, T. Wrin, S. J. Little and C. J. Petropoulos, *Proc. Natl. Acad. Sci. U. S. A.*, 2003, **100**, 4144.
117. L. Stamatatos, L. Morris, D. R. Burton and J. R. Mascola, *Nat. Med.*, 2009, **15**, 866.
118. E. Landais, X. Huang, C. Havenar-Daughton, B. Murrell, M. A. Price, L. Wickramasinghe, A. Ramos, C. B. Bian, M. Simek, S. Allen, E. Karita, W. Kilembe, S. Lakhi, M. Inambao, A. Kamali, E. J. Sanders, O. Anzala, V. Edward, L. G. Bekker, J. Tang, J. Gilmour, S. L. Kosakovsky-Pond, P. Phung, T. Wrin, S. Crotty, A. Godzik and P. Poignard, *PLoS Pathog.*, 2016, **12**, e1005369.
119. C. A. Muro-Cacho, G. Pantaleo and A. S. Fauci, *J. Immunol.*, 1995, **154**, 5555.
120. M. A. Fischl, D. D. Richman, M. H. Grieco, M. S. Gottlieb, P. A. Volberding, O. L. Laskin, J. M. Leedom, J. E. Groopman, D. Mildvan and R. T. Schooley, *et al.*, *N. Engl. J. Med.*, 1987, **317**, 185.
121. F. Clavel and A. J. Hance, *N. Engl. J. Med.*, 2004, **350**, 1023.
122. J. M. Cuevas, R. Geller, R. Garijo, J. Lopez-Aldeguer and R. Sanjuan, *PLoS Biol.*, 2015, **13**, e1002251.
123. J. M. Coffin, *Science*, 1995, **267**, 483.
124. B. D. Preston, B. J. Poiesz and L. A. Loeb, *Science*, 1988, **242**, 1168.
125. J. D. Roberts, K. Bebenek and T. A. Kunkel, *Science*, 1988, **242**, 1171.
126. L. M. Mansky and H. M. Temin, *J. Virol.*, 1995, **69**, 5087.
127. J. J. Holland, J. C. De La Torre and D. A. Steinhauer, *Curr. Top. Microbiol. Immunol.*, 1992, **176**, 1.
128. E. J. Arts and D. J. Hazuda, *Cold Spring Harbor Perspect. Med.*, 2012, **2**, a007161.
129. E. K. Ngemu, C. Khayeka-Wandabwa, E. J. Kweka, J. K. Choge, E. Anino and E. Oyoo-Okoth, *BMC Res. Notes*, 2014, **7**, 52.
130. A. C. Collier, R. W. Coombs, D. A. Schoenfeld, R. L. Bassett, J. Timpone, A. Baruch, M. Jones, K. Facey, C. Whitacre, V. J. McAuliffe, H. M. Friedman, T. C. Merigan, R. C. Reichman, C. Hooper and L. Corey, *N. Engl. J. Med.*, 1996, **334**, 1011.
131. A. D. Kelleher, A. Carr, J. Zaunders and D. A. Cooper, *J. Infect. Dis.*, 1996, **173**, 321.
132. B. Autran, G. Carcelain, T. S. Li, C. Blanc, D. Mathez, R. Tubiana, C. Katlama, P. Debre and J. Leibowitch, *Science*, 1997, **277**, 112.
133. M. M. Lederman, E. Connick, A. Landay, D. R. Kuritzkes, J. Spritzler, M. St Clair, B. L. Kotzin, L. Fox, M. H. Chiozzi, J. M. Leonard, F. Rousseau, M. Wade, J. D. Roe, A. Martinez and H. Kessler, *J. Infect. Dis.*, 1998, **178**, 70.
134. J. M. Baeten, D. Donnell, P. Ndase, N. R. Mugo, J. D. Campbell, J. Wangisi, J. W. Tappero, E. A. Bukusi, C. R. Cohen, E. Katabira, A. Ronald, E.

Tumwesigye, E. Were, K. H. Fife, J. Kiarie, C. Farquhar, G. John-Stewart, A. Kakia, J. Odoyo, A. Mucunguzi, E. Nakku-Joloba, R. Twesigye, K. Ngure, C. Apaka, H. Tamooh, F. Gabona, A. Mujugira, D. Panteleeff, K. K. Thomas, L. Kidoguchi, M. Krows, J. Revall, S. Morrison, H. Haugen, M. Emmanuel-Ogier, L. Ondrejcek, R. W. Coombs, L. Frenkel, C. Hendrix, N. N. Bumpus, D. Bangsberg, J. E. Haberer, W. S. Stevens, J. R. Lingappa, C. Celum and E. P. S. T. Partners Pr, *N. Engl. J. Med.*, 2012, **367**, 399.

135. S. Mehta, R. D. Moore and N. M. Graham, *AIDS*, 1997, **11**, 1665.
136. M. J. Atkinson and J. J. Petrozzino, *AIDS Patient Care STDs*, 2009, **23**, 903.
137. B. J. Morris and J. D. Klausner, *Isr. J. Health Policy Res.*, 2015, **4**, 40.
138. S. S. Abdool Karim and C. Baxter, *Curr. Top. Microbiol. Immunol.*, 2014, **383**, 97.
139. T. W. Chun, D. Engel, M. M. Berrey, T. Shea, L. Corey and A. S. Fauci, *Proc. Natl. Acad. Sci. U. S. A.*, 1998, **95**, 8869.
140. Y. Z. Cohen and R. Dolin, *Ther. Adv. Vaccines*, 2013, **1**, 99.
141. N. M. Flynn, D. N. Forthal, C. D. Harro, F. N. Judson, K. H. Mayer and M. F. Para, *J. Infect. Dis.*, 2005, **191**, 654.
142. P. Pitisuttithum, P. Gilbert, M. Gurwith, W. Heyward, M. Martin, F. van Griensven, D. Hu, J. W. Tappero and K. Choopanya, *J. Infect. Dis.*, 2006, **194**, 1661.
143. G. Gray, S. Buchbinder and A. Duerr, *Curr. Opin. HIV AIDS*, 2010, **5**, 357.
144. M. J. McElrath, S. C. De Rosa, Z. Moodie, S. Dubey, L. Kierstead, H. Janes, O. D. Defawe, D. K. Carter, J. Hural, R. Akondy, S. P. Buchbinder, M. N. Robertson, D. V. Mehrotra, S. G. Self, L. Corey, J. W. Shiver and D. R. Casimiro, *Lancet*, 2008, **372**, 1894.
145. S. P. Buchbinder, D. V. Mehrotra, A. Duerr, D. W. Fitzgerald, R. Mogg, D. Li, P. B. Gilbert, J. R. Lama, M. Marmor, C. Del Rio, M. J. McElrath, D. R. Casimiro, K. M. Gottesdiener, J. A. Chodakewitz, L. Corey and M. N. Robertson, *Lancet*, 2008, **372**, 1881.
146. B. D. Walker and D. R. Burton, *Science*, 2008, **320**, 760.
147. S. Rerks-Ngarm, P. Pitisuttithum, S. Nitayaphan, J. Kaewkungwal, J. Chiu, R. Paris, N. Premsri, C. Namwat, M. de Souza, E. Adams, M. Benenson, S. Gurunathan, J. Tartaglia, J. G. McNeil, D. P. Francis, D. Stablein, D. L. Birx, S. Chunsuttiwat, C. Khamboonruang, P. Thongcharoen, M. L. Robb, N. L. Michael, P. Kunasol and J. H. Kim, *N. Engl. J. Med.*, 2009, **361**, 2209.
148. G. E. Gray, F. Laher, E. Lazarus, B. Ensoli and L. Corey, *Curr. Opin. Virol.*, 2016, **17**, 104.
149. S. M. Hammer, M. E. Sobieszczyk, H. Janes, S. T. Karuna, M. J. Mulligan, D. Grove, B. A. Koblin, S. P. Buchbinder, M. C. Keefer, G. D. Tomaras, N. Frahm, J. Hural, C. Anude, B. S. Graham, M. E. Enama, E. Adams, E. DeJesus, R. M. Novak, I. Frank, C. Bentley, S. Ramirez, R. Fu, R. A. Koup, J. R. Mascola, G. J. Nabel, D. C. Montefiori, J. Kublin, M. J. McElrath, L. Corey and P. B. Gilbert, *N. Engl. J. Med.*, 2013, **369**, 2083.

150. A. McMichael, L. J. Picker, J. P. Moore and D. R. Burton, *Nat. Med.*, 2013, **19**, 1576.
151. M. P. D'Souza and O. O. Yang, *AIDS*, 2015, **29**, 395.
152. Y. Li, L. Luo, D. Y. Thomas and C. Y. Kang, *Virology*, 2000, **272**, 417.
153. Y. Li, L. Luo, D. Y. Thomas and C. Y. Kang, *Virology*, 1994, **204**, 266.
154. O. K. Haffar, D. J. Dowbenko and P. W. Berman, *J. Cell Biol.*, 1988, **107**, 1677.
155. M. A. Checkley, B. G. Luttge and E. O. Freed, *J. Mol. Biol.*, 2011, **410**, 582.
156. R. Kornfeld and S. Kornfeld, *Annu. Rev. Biochem.*, 1985, **54**, 631.
157. F. E. Ware, A. Vassilakos, P. A. Peterson, M. R. Jackson, M. A. Lehrman and D. B. Williams, *J. Biol. Chem.*, 1995, **270**, 4697.
158. A. Helenius and M. Aebi, *Annu. Rev. Biochem.*, 2004, **73**, 1019.
159. S. Pollock, R. A. Dwek, D. R. Burton and N. Zitzmann, *AIDS*, 2008, **22**, 1961.
160. P. B. Fischer, M. Collin, G. B. Karlsson, W. James, T. D. Butters, S. J. Davis, S. Gordon, R. A. Dwek and F. M. Platt, *J. Virol.*, 1995, **69**, 5791.
161. P. B. Fischer, G. B. Karlsson, T. D. Butters, R. A. Dwek and F. M. Platt, *J. Virol.*, 1996, **70**, 7143.
162. R. A. Dwek, T. D. Butters, F. M. Platt and N. Zitzmann, *Nat. Rev. Drug Discovery*, 2002, **1**, 65.
163. A. Land, D. Zonneveld and I. Braakman, *FASEB J.*, 2003, **17**, 1058.
164. C. K. Leonard, M. W. Spellman, L. Riddle, R. J. Harris, J. N. Thomas and T. J. Gregory, *J. Biol. Chem.*, 1990, **265**, 10373.
165. J. M. Cutalo, L. J. Deterding and K. B. Tomer, *J. Am. Soc. Mass Spectrom.*, 2004, **15**, 1545.
166. C. Bonomelli, K. J. Doores, D. C. Dunlop, V. Thaney, R. a. Dwek, D. R. Burton, M. Crispin and C. N. Scanlan, *PLoS One*, 2011, **6**, e23521.
167. K. J. Doores, C. Bonomelli, D. J. Harvey, S. Vasiljevic, R. A. Dwek, D. R. Burton, M. Crispin and C. N. Scanlan, *Proc. Natl. Acad. Sci. U. S. A.*, 2010, **107**, 13800.
168. L. K. Pritchard, S. Vasiljevic, G. Ozorowski, G. E. Seabright, A. Cupo, R. Ringe, H. J. Kim, R. W. Sanders, K. J. Doores, D. R. Burton, I. A. Wilson, A. B. Ward, J. P. Moore and M. Crispin, *Cell Rep.*, 2015, **11**, 1604.
169. P. M. Rudd and R. A. Dwek, *Crit. Rev. Biochem. Mol. Biol.*, 1997, **32**, 1.
170. A. Pinter, W. J. Honnen, S. A. Tilley, C. Bona, H. Zaghouani, M. K. Gorny and S. Zolla-Pazner, *J. Virol.*, 1989, **63**, 2674.
171. P. L. Earl, R. W. Doms and B. Moss, *Proc. Natl. Acad. Sci. U. S. A.*, 1990, **87**, 648.
172. G. Thomas, *Nat. Rev. Mol. Cell Biol.*, 2002, **3**, 753.
173. J. M. McCune, L. B. Rabin, M. B. Feinberg, M. Lieberman, J. C. Kosek, G. R. Reyes and I. L. Weissman, *Cell*, 1988, **53**, 55.
174. R. P. Ringe, R. W. Sanders, A. Yasmeen, H. J. Kim, J. H. Lee, A. Cupo, J. Korzun, R. Derking, T. van Montfort, J.-P. Julien, I. A. Wilson, P. J. Klasse, A. B. Ward and J. P. Moore, *Proc. Natl. Acad. Sci. U. S. A.*, 2013, **110**, 18256.
175. M. Boge, S. Wyss, J. S. Bonifacino and M. Thali, *J. Biol. Chem.*, 1998, **273**, 15773.

176. P. Zhu, J. Liu, J. Bess Jr., E. Chertova, J. D. Lifson, H. Grise, G. A. Ofek, K. A. Taylor and K. H. Roux, *Nature*, 2006, **441**, 847.
177. E. Yuste, J. D. Reeves, R. W. Doms and R. C. Desrosiers, *J. Virol.*, 2004, **78**, 6775.
178. J. Schiller and B. Chackerian, *PLoS Pathog.*, 2014, **10**, e1004254.
179. J. S. Klein and P. J. Bjorkman, *PLoS Pathog.*, 2010, **6**, e1000908.
180. P. D. Kwong, R. Wyatt, J. Robinson, R. W. Sweet, J. Sodroski and W. A. Hendrickson, *Nature*, 1998, **393**, 648.
181. C. C. Huang, M. Tang, M. Y. Zhang, S. Majeed, E. Montabana, R. L. Stanfield, D. S. Dimitrov, B. Korber, J. Sodroski, I. A. Wilson, R. Wyatt and P. D. Kwong, *Science*, 2005, **310**, 1025.
182. T. Zhou, L. Xu, B. Dey, A. J. Hessell, D. Van Ryk, S. H. Xiang, X. Yang, M. Y. Zhang, M. B. Zwick, J. Arthos, D. R. Burton, D. S. Dimitrov, J. Sodroski, R. Wyatt, G. J. Nabel and P. D. Kwong, *Nature*, 2007, **445**, 732.
183. M. Pancera, S. Majeed, Y. E. Ban, L. Chen, C. C. Huang, L. Kong, Y. D. Kwon, J. Stuckey, T. Zhou, J. E. Robinson, W. R. Schief, J. Sodroski, R. Wyatt and P. D. Kwong, *Proc. Natl. Acad. Sci. U. S. A.*, 2010, **107**, 1166.
184. T. Zhou, I. Georgiev, X. Wu, Z. Y. Yang, K. Dai, A. Finzi, Y. D. Kwon, J. F. Scheid, W. Shi, L. Xu, Y. Yang, J. Zhu, M. C. Nussenzweig, J. Sodroski, L. Shapiro, G. J. Nabel, J. R. Mascola and P. D. Kwong, *Science*, 2010, **329**, 811.
185. Y. D. Kwon, A. Finzi, X. Wu, C. Dogo-Isonagie, L. K. Lee, L. R. Moore, S. D. Schmidt, J. Stuckey, Y. Yang, T. Zhou, J. Zhu, D. A. Vicic, A. K. Debnath, L. Shapiro, C. A. Bewley, J. R. Mascola, J. G. Sodroski and P. D. Kwong, *Proc. Natl. Acad. Sci. U. S. A.*, 2012, **109**, 5663.
186. A. B. Ward and I. A. Wilson, *Trends Biochem. Sci.*, 2015, **40**, 101.
187. G. B. E. Stewart-Jones, C. Soto, T. Lemmin, G.-Y. Chuang, A. Druz, R. Kong, P. V. Thomas, K. Wagh, T. Zhou, A.-J. Behrens, T. Bylund, C. W. Choi, J. R. Davison, I. S. Georgiev, M. G. Joyce, Y. D. Kwon, M. Pancera, J. Taft, Y. Yang, B. Zhang, S. S. Shivatare, V. S. Shivatare, C.-C. D. Lee, C.-Y. Wu, C. A. Bewley, D. R. Burton, W. C. Koff, M. Connors, M. Crispin, U. Baxa, B. T. Korber, C.-H. Wong, J. R. Mascola and P. D. Kwong, *Cell*, 2016, **165**, 1.
188. J. H. Lee, G. Ozorowski and A. B. Ward, *Science*, 2016, **351**, 1043.
189. J. P. Julien, A. Cupo, D. Sok, R. L. Stanfield, D. Lyumkis, M. C. Deller, P. J. Klasse, D. R. Burton, R. W. Sanders, J. P. Moore, A. B. Ward and I. A. Wilson, *Science*, 2013, **342**, 1477.
190. D. Lyumkis, J. P. Julien, N. de Val, A. Cupo, C. S. Potter, P. J. Klasse, D. R. Burton, R. W. Sanders, J. P. Moore, B. Carragher, I. A. Wilson and A. B. Ward, *Science*, 2013, **342**, 1484.
191. M. Pancera, T. Zhou, A. Druz, I. S. Georgiev, C. Soto, J. Gorman, J. Huang, P. Acharya, G. Y. Chuang, G. Ofek, G. B. Stewart-Jones, J. Stuckey, R. T. Bailer, M. G. Joyce, M. K. Louder, N. Tumba, Y. Yang, B. Zhang, M. S. Cohen, B. F. Haynes, J. R. Mascola, L. Morris, J. B. Munro, S. C. Blanchard, W. Mothes, M. Connors and P. D. Kwong, *Nature*, 2014, **514**, 455.

192. D. G. Myszka, R. W. Sweet, P. Hensley, M. Brigham-Burke, P. D. Kwong, W. A. Hendrickson, R. Wyatt, J. Sodroski and M. L. Doyle, *Proc. Natl. Acad. Sci. U. S. A.*, 2000, **97**, 9026.
193. P. D. Kwong, M. L. Doyle, D. J. Casper, C. Cicala, S. A. Leavitt, S. Majeed, T. D. Steenbeke, M. Venturi, I. Chaiken, M. Fung, H. Katinger, P. W. I. H. Parren, J. Robinson, D. Van Ryk, L. Wang, D. R. Burton, E. Freire, R. Wyatt, J. Sodroski, W. A. Hendrickson and J. Arthos, *Nature*, 2002, **420**, 678.
194. S. S. Hwang, T. J. Boyle, H. K. Lyerly and B. R. Cullen, *Science*, 1991, **253**, 71.
195. R. W. Doms and J. P. Moore, *J. Cell Biol.*, 2000, **151**, F9.
196. R. W. Sanders, R. Derking, A. Cupo, J. P. Julien, A. Yasmeen, N. de Val, H. J. Kim, C. Blattner, A. T. de la Pena, J. Korzun, M. Golabek, K. de Los Reyes, T. J. Ketas, M. J. van Gils, C. R. King, I. A. Wilson, A. B. Ward, P. J. Klasse and J. P. Moore, *PLoS Pathog.*, 2013, **9**, e1003618.
197. J. M. Binley, R. W. Sanders, B. Clas, N. Schuelke, A. Master, Y. Guo, F. Kajumo, D. J. Anselma, P. J. Maddon, W. C. Olson and J. P. Moore, *J. Virol.*, 2000, **74**, 627.
198. R. W. Sanders, M. Vesanen, N. Schuelke, A. Master, L. Schiffner, R. Kalyanaraman, M. Paluch, B. Berkhout, P. J. Maddon, W. C. Olson, M. Lu and J. P. Moore, *J. Virol.*, 2002, **76**, 8875.
199. R. Khayat, J. H. Lee, J.-P. Julien, A. Cupo, P. J. Klasse, R. W. Sanders, J. P. Moore, I. A. Wilson and A. B. Ward, *J. Virol.*, 2013, **87**, 9865.
200. Y. D. Kwon, M. Pancera, P. Acharya, I. S. Georgiev, E. T. Crooks, J. Gorman, M. G. Joyce, M. Guttman, X. Ma, S. Narpala, C. Soto, D. S. Terry, Y. Yang, T. Zhou, G. Ahlsen, R. T. Bailer, M. Chambers, G. Y. Chuang, N. A. Doria-Rose, A. Druz, M. A. Hallen, A. Harned, T. Kirys, M. K. Louder, S. O'Dell, G. Ofek, K. Osawa, M. Prabhakaran, M. Sastry, G. B. Stewart-Jones, J. Stuckey, P. V. Thomas, T. Tittley, C. Williams, B. Zhang, H. Zhao, Z. Zhou, B. R. Donald, L. K. Lee, S. Zolla-Pazner, U. Baxa, A. Schon, E. Freire, L. Shapiro, K. K. Lee, J. Arthos, J. B. Munro, S. C. Blanchard, W. Mothes, J. M. Binley, A. B. McDermott, J. R. Mascola and P. D. Kwong, *Nat. Struct. Mol. Biol.*, 2015, **22**, 522.
201. J. H. Lee, N. de Val, D. Lyumkis and A. B. Ward, *Structure*, 2015, **23**, 1943.
202. V. T. Chang, M. Crispin, A. R. Aricescu, D. J. Harvey, J. E. Nettleship, J. A. Fennelly, C. Yu, K. S. Boles, E. J. Evans, D. I. Stuart, R. A. Dwek, E. Y. Jones, R. J. Owens and S. J. Davis, *Structure*, 2007, **15**, 267.
203. B. Korber, B. Gaschen, K. Yusim, R. Thakallapally, C. Kesmir and V. Detours, *Br. Med. Bull.*, 2001, **58**, 19.
204. R. Wyatt, P. D. Kwong, E. Desjardins, R. W. Sweet, J. Robinson, W. A. Hendrickson and J. G. Sodroski, *Nature*, 1998, **393**, 705.
205. X. Zhu, C. Borchers, R. J. Bienstock and K. B. Tomer, *Biochemistry*, 2000, **39**, 11194.
206. E. P. Go, H.-X. Liao, S. M. Alam, D. Hua, B. F. Haynes and H. Desaire, *J. Proteome Res.*, 2013, **12**, 1223.

207. L. K. Pritchard, D. I. Spencer, L. Royle, C. Bonomelli, G. E. Seabright, A. J. Behrens, D. W. Kulp, S. Menis, S. A. Krumm, D. C. Dunlop, D. J. Crispin, T. A. Bowden, C. N. Scanlan, A. B. Ward, W. R. Schief, K. J. Doores and M. Crispin, *Nat. Commun.*, 2015, **6**, 7479.
208. K. B. Alexandre, E. S. Gray, B. E. Lambson, P. L. Moore, I. A. Choge, K. Mlisana, S. S. Karim, J. McMahon, B. O'Keefe, R. Chikwamba and L. Morris, *Virology*, 2010, **402**, 187.
209. D. Eggink, M. Melchers, M. Wuhrer, T. van Montfort, A. K. Dey, B. A. Naaijkens, K. B. David, V. Le Douce, A. M. Deelder, K. Kang, W. C. Olson, B. Berkhout, C. H. Hokke, J. P. Moore and R. W. Sanders, *Virology*, 2010, **401**, 236.
210. M. Crispin and K. J. Doores, *Curr. Opin. Virol.*, 2015, **11**, 63.
211. A. Yasmeen, R. Ringe, R. Derking, A. Cupo, J. P. Julien, D. R. Burton, A. B. Ward, I. A. Wilson, R. W. Sanders, J. P. Moore and P. J. Klasse, *Retrovirology*, 2014, **11**, 41.
212. I. S. Georgiev, M. G. Joyce, Y. Yang, M. Sastry, B. Zhang, U. Baxa, R. E. Chen, A. Druz, C. R. Lees, S. Narpala, A. Schon, J. Van Galen, G. Y. Chuang, J. Gorman, A. Harned, M. Pancera, G. B. Stewart-Jones, C. Cheng, E. Freire, A. B. McDermott, J. R. Mascola and P. D. Kwong, *J. Virol.*, 2015, **89**, 5318.
213. P. R. Crocker, J. C. Paulson and A. Varki, *Nat. Rev. Immunol.*, 2007, **7**, 255.
214. R. M. Anthony, F. Nimmerjahn, D. J. Ashline, V. N. Reinhold, J. C. Paulson and J. V. Ravetch, *Science*, 2008, **320**, 373.
215. L. K. Pritchard, D. J. Harvey, C. Bonomelli, M. Crispin and K. J. Doores, *J. Virol.*, 2015, **89**, 8932.
216. H. Stoiber, R. Schneider, J. Janatova and M. P. Dierich, *Immunobiology*, 1995, **193**, 98.
217. V. P. Ferreira, M. K. Pangburn and C. Cortes, *Mol. Immunol.*, 2010, **47**, 2187.
218. R. L. Hirsch, J. S. Wolinsky and J. A. Winkelstein, *Arch. Virol.*, 1986, **87**, 181.
219. D. T. Fearon, *Proc. Natl. Acad. Sci. U. S. A.*, 1978, **75**, 1971.
220. Z. Zou, A. Chastain, S. Moir, J. Ford, K. Trandem, E. Martinelli, C. Cicala, P. Crocker, J. Arthos and P. D. Sun, *PLoS One*, 2011, **6**, e24559.
221. E. P. Go, D. Hua and H. Desaire, *J. Proteome Res.*, 2014, **13**, 4012.
222. E. Stansell, M. Panico, K. Canis, P. C. Pang, L. Bouche, D. Binet, M. J. O'Connor, E. Chertova, J. Bess, J. D. Lifson, S. M. Haslam, H. R. Morris, R. C. Desrosiers and A. Dell, *PLoS One*, 2015, **10**, e0124784.
223. E. P. Go, A. Herschhorn, C. Gu, L. Castillo-Menendez, S. Zhang, Y. Mao, H. Chen, H. Ding, J. K. Wakefield, D. Hua, H. X. Liao, J. C. Kappes, J. Sodroski and H. Desaire, *J. Virol.*, 2015, **89**, 8245.
224. W. Yang, P. Shah, S. Toghi Eshghi, S. Yang, S. Sun, M. Ao, A. Rubin, J. B. Jackson and H. Zhang, *Anal. Chem.*, 2014, **86**, 6959.
225. J. E. Hansen, B. Jansson, G. J. Gram, H. Clausen, J. O. Nielsen and S. Olofsson, *Arch. Virol.*, 1996, **141**, 291.

226. E. P. Go, Q. Chang, H.-X. Liao, L. L. Sutherland, S. M. Alam, B. F. Haynes and H. Desaire, *J. Proteome Res.*, 2009, **8**, 4231.
227. E. P. Go, G. Hewawasam, H.-X. Liao, H. Chen, L.-H. Ping, J. A. Anderson, D. C. Hua, B. F. Haynes and H. Desaire, *J. Virol.*, 2011, **85**, 8270.
228. E. P. Go, J. Irungu, Y. Zhang, D. S. Dalpathado, H.-X. Liao, L. L. Sutherland, S. M. Alam, B. F. Haynes and H. Desaire, *J. Proteome Res.*, 2008, **7**, 1660.
229. J. Irungu, E. P. Go, Y. Zhang, D. S. Dalpathado, H.-X. Liao, B. F. Haynes and H. Desaire, *J. Am. Soc. Mass Spectrom.*, 2008, **19**, 1209.
230. M. Pabst, M. Chang, J. Stadlmann and F. Altmann, *Biol. Chem.*, 2012, **393**, 719.
231. M. Guttman, N. K. Garcia, A. Cupo, T. Matsui, J. P. Julien, R. W. Sanders, I. A. Wilson, J. P. Moore and K. K. Lee, *Structure*, 2014, **22**, 974.
232. A.-J. Behrens, S. Vasiljevic, L. K. Pritchard, D. J. Harvey, R. S. Andev, S. A. Krumm, W. B. Struwe, A. Cupo, A. Kumar, N. Zitzmann, G. E. Seabright, H. B. Kramer, D. I. Spencer, L. Royle, J. H. Lee, P. J. Klasse, D. R. Burton, I. A. Wilson, A. B. Ward, R. W. Sanders, J. P. Moore, K. J. Doores and M. Crispin, *Cell Rep.*, 2016, **14**, 2695.
233. R. S. Depetris, J. P. Julien, R. Khayat, J. H. Lee, R. Pejchal, U. Katpally, N. Cocco, M. Kachare, E. Massi, K. B. David, A. Cupo, A. J. Marozsan, W. C. Olson, A. B. Ward, I. A. Wilson, R. W. Sanders and J. P. Moore, *J. Biol. Chem.*, 2012, **287**, 24239.
234. M. Zhang, B. Gaschen, W. Blay, B. Foley, N. Haigwood, C. Kuiken and B. Korber, *Glycobiology*, 2004, **14**, 1229.
235. L. Alamos, *HIV Database*, http://www.hiv.lanl.gov/, accessed 15/03/2016.
236. L. Kong, R. L. Stanfield and I. A. Wilson, in *HIV Glycans in Infection and Immunity*, ed. R. Pantophlet, Springer, 2014, vol. 1, ch. 5.
237. P. L. Moore, E. S. Gray, C. K. Wibmer, J. N. Bhiman, M. Nonyane, D. J. Sheward, T. Hermanus, S. Bajimaya, N. L. Tumba, M.-R. Abrahams, B. E. Lambson, N. Ranchobe, L. Ping, N. Ngandu, Q. Abdool Karim, S. S. Abdool Karim, R. I. Swanstrom, M. S. Seaman, C. Williamson and L. Morris, *Nat. Med.*, 2012, **18**, 1688.
238. D. Sok, K. J. Doores, B. Briney, K. M. Le, K. L. Saye-Francisco, A. Ramos, D. W. Kulp, J. P. Julien, S. Menis, L. Wickramasinghe, M. S. Seaman, W. R. Schief, I. A. Wilson, P. Poignard and D. R. Burton, *Sci. Transl. Med.*, 2014, **6**, 236ra63.
239. C. Doyle-Cooper, K. E. Hudson, A. B. Cooper, T. Ota, P. Skog, P. E. Dawson, M. B. Zwick, W. R. Schief, D. R. Burton and D. Nemazee, *J. Immunol.*, 2013, **191**, 3186.
240. M. Crispin and T. A. Bowden, *Nat. Struct. Mol. Biol.*, 2013, **20**, 771.
241. I. Mikell, D. N. Sather, S. A. Kalams, M. Altfeld, G. Alter and L. Stamatatos, *PLoS Pathog.*, 2011, **7**, e1001251.
242. N. A. Doria-Rose, R. M. Klein, M. G. Daniels, S. O'Dell, M. Nason, A. Lapedes, T. Bhattacharya, S. A. Migueles, R. T. Wyatt, B. T. Korber, J. R. Mascola and M. Connors, *J. Virol.*, 2010, **84**, 1631.

243. D. N. Sather, J. Armann, L. K. Ching, A. Mavrantoni, G. Sellhorn, Z. Caldwell, X. Yu, B. Wood, S. Self, S. Kalams and L. Stamatatos, *J. Virol.*, 2009, **83**, 757.
244. Z. Euler, M. J. van Gils, E. M. Bunnik, P. Phung, B. Schweighardt, T. Wrin and H. Schuitemaker, *J. Infect. Dis.*, 2010, **201**, 1045.
245. R. Shibata, T. Igarashi, N. Haigwood, A. Buckler-White, R. Ogert, W. Ross, R. Willey, M. W. Cho and M. A. Martin, *Nat. Med.*, 1999, **5**, 204.
246. P. Dosenovic, L. von Boehmer, A. Escolano, J. Jardine, N. T. Freund, A. D. Gitlin, A. T. McGuire, D. W. Kulp, T. Oliveira, L. Scharf, J. Pietzsch, M. D. Gray, A. Cupo, M. J. van Gils, K. H. Yao, C. Liu, A. Gazumyan, M. S. Seaman, P. J. Bjorkman, R. W. Sanders, J. P. Moore, L. Stamatatos, W. R. Schief and M. C. Nussenzweig, *Cell*, 2015, **161**, 1505.
247. J. G. Jardine, T. Ota, D. Sok, M. Pauthner, D. W. Kulp, O. Kalyuzhniy, P. D. Skog, T. C. Thinnes, D. Bhullar, B. Briney, S. Menis, M. Jones, M. Kubitz, S. Spencer, Y. Adachi, D. R. Burton, W. R. Schief and D. Nemazee, *Science*, 2015, **349**, 156.
248. CHAVI-ID, *bNAber*, www.bnaber.org, accessed 15/03/2016, 2016.
249. B. F. Haynes, *Curr. Opin. Immunol.*, 2015, **35**, 39.
250. D. R. Burton and J. R. Mascola, *Nat. Immunol.*, 2015, **16**, 571.
251. A. Trkola, M. Purtscher, T. Muster, C. Ballaun, A. Buchacher, N. Sullivan, K. Srinivasan, J. Sodroski, J. P. Moore and H. Katinger, *J. Virol.*, 1996, **70**, 1100.
252. C. D. Murin, J.-P. Julien, D. Sok, R. L. Stanfield, R. Khayat, A. Cupo, J. P. Moore, D. R. Burton, I. A. Wilson and A. B. Ward, *J. Virol.*, 2014, **88**, 10177.
253. K. J. Doores, *FEBS J.*, 2015, **282**, 4679.
254. J.-P. Julien, J. H. Lee, A. Cupo, C. D. Murin, R. Derking, S. Hoffenberg, M. J. Caulfield, C. R. King, A. J. Marozsan, P. J. Klasse, R. W. Sanders, J. P. Moore, I. A. Wilson and A. B. Ward, *Proc. Natl. Acad. Sci. U. S. A.*, 2013, **110**, 4351.
255. L. Yu and Y. Guan, *Front. Immunol.*, 2014, **5**, 250.
256. F. Wang, D. C. Ekiert, I. Ahmad, W. Yu, Y. Zhang, O. Bazirgan, A. Torkamani, T. Raudsepp, W. Mwangi, M. F. Criscitiello, I. A. Wilson, P. G. Schultz and V. V. Smider, *Cell*, 2013, **153**, 1379.
257. R. Pejchal, L. M. Walker, R. L. Stanfield, S. K. Phogat, W. C. Koff, P. Poignard, D. R. Burton and I. A. Wilson, *Proc. Natl. Acad. Sci. U. S. A.*, 2010, **107**, 11483.
258. A. P. West Jr., L. Scharf, J. F. Scheid, F. Klein, P. J. Bjorkman and M. C. Nussenzweig, *Cell*, 2014, **156**, 633.
259. D. Sok, M. J. van Gils, M. Pauthner, J. P. Julien, K. L. Saye-Francisco, J. Hsueh, B. Briney, J. H. Lee, K. M. Le, P. S. Lee, Y. Hua, M. S. Seaman, J. P. Moore, A. B. Ward, I. A. Wilson, R. W. Sanders and D. R. Burton, *Proc. Natl. Acad. Sci. U. S. A.*, 2014, **111**, 17624.
260. G. Ofek, K. McKee, Y. Yang, Z. Y. Yang, J. Skinner, F. J. Guenaga, R. Wyatt, M. B. Zwick, G. J. Nabel, J. R. Mascola and P. D. Kwong, *J. Virol.*, 2010, **84**, 2955.

261. B. S. Briney, J. R. Willis and J. E. Crowe Jr., *PLoS One*, 2012, **7**, e36750.
262. J. R. Willis, J. A. Finn, B. Briney, G. Sapparapu, V. Singh, H. King, C. C. LaBranche, D. C. Montefiori, J. Meiler and J. E. Crowe Jr., *Proc. Natl. Acad. Sci. U. S. A.*, 2016, **113**, 4446.
263. M. Pancera, J. S. McLellan, X. Wu, J. Zhu, A. Changela, S. D. Schmidt, Y. Yang, T. Zhou, S. Phogat, J. R. Mascola and P. D. Kwong, *J. Virol.*, 2010, **84**, 8098.
264. B. F. Haynes, M. A. Moody, L. Verkoczy, G. Kelsoe and S. M. Alam, *Hum. Antibodies*, 2005, **14**, 59.
265. H. Mouquet, J. F. Scheid, M. J. Zoller, M. Krogsgaard, R. G. Ott, S. Shukair, M. N. Artyomov, J. Pietzsch, M. Connors, F. Pereyra, B. D. Walker, D. D. Ho, P. C. Wilson, M. S. Seaman, H. N. Eisen, A. K. Chakraborty, T. J. Hope, J. V. Ravetch, H. Wardemann and M. C. Nussenzweig, *Nature*, 2010, **467**, 591.
266. M. Liu, G. Yang, K. Wiehe, N. I. Nicely, N. A. Vandergrift, W. Rountree, M. Bonsignori, S. M. Alam, J. Gao, B. F. Haynes and G. Kelsoe, *J. Virol.*, 2015, **89**, 784.
267. H. Wardemann, S. Yurasov, A. Schaefer, J. W. Young, E. Meffre and M. C. Nussenzweig, *Science*, 2003, **301**, 1374.
268. M. Tsuiji, S. Yurasov, K. Velinzon, S. Thomas, M. C. Nussenzweig and H. Wardemann, *J. Exp. Med.*, 2006, **203**, 393.
269. T. Tiller, M. Tsuiji, S. Yurasov, K. Velinzon, M. C. Nussenzweig and H. Wardemann, *Immunity*, 2007, **26**, 205.
270. H. Mouquet and M. C. Nussenzweig, *Cell. Mol. Life Sci.*, 2012, **69**, 1435.
271. D. C. Dunlop, A. Ulrich, B. J. Appelmelk, D. R. Burton, R. A. Dwek, N. Zitzmann and C. N. Scanlan, *AIDS*, 2008, **22**, 2214.
272. G. Yang, T. M. Holl, Y. Liu, Y. Li, X. Lu, N. I. Nicely, T. B. Kepler, S. M. Alam, H. X. Liao, D. W. Cain, L. Spicer, J. L. VandeBerg, B. F. Haynes and G. Kelsoe, *J. Exp. Med.*, 2013, **210**, 241.
273. Y. Chen, J. Zhang, K. K. Hwang, H. Bouton-Verville, S. M. Xia, A. Newman, Y. B. Ouyang, B. F. Haynes and L. Verkoczy, *J. Immunol.*, 2013, **191**, 1260.
274. L. Verkoczy, Y. Chen, H. Bouton-Verville, J. Zhang, M. Diaz, J. Hutchinson, Y. B. Ouyang, S. M. Alam, T. M. Holl, K. K. Hwang, G. Kelsoe and B. F. Haynes, *J. Immunol.*, 2011, **187**, 3785.
275. L. Verkoczy, M. Diaz, T. M. Holl, Y. B. Ouyang, H. Bouton-Verville, S. M. Alam, H. X. Liao, G. Kelsoe and B. F. Haynes, *Proc. Natl. Acad. Sci. U. S. A.*, 2010, **107**, 181.
276. B. F. Haynes, G. Kelsoe, S. C. Harrison and T. B. Kepler, *Nat. Biotechnol.*, 2012, **30**, 423.
277. J. F. Scheid, H. Mouquet, B. Ueberheide, R. Diskin, F. Klein, T. Y. K. Oliveira, J. Pietzsch, D. Fenyo, A. Abadir, K. Velinzon, A. Hurley, S. Myung, F. Boulad, P. Poignard, D. R. Burton, F. Pereyra, D. D. Ho, B. D. Walker, M. S. Seaman, P. J. Bjorkman, B. T. Chait and M. C. Nussenzweig, *Science*, 2011, **333**, 1633.

278. X. Wu, T. Zhou, J. Zhu, B. Zhang, I. Georgiev, C. Wang, X. Chen, N. S. Longo, M. Louder, K. McKee, S. O'Dell, S. Perfetto, S. D. Schmidt, W. Shi, L. Wu, Y. Yang, Z. Y. Yang, Z. Yang, Z. Zhang, M. Bonsignori, J. A. Crump, S. H. Kapiga, N. E. Sam, B. F. Haynes, M. Simek, D. R. Burton, W. C. Koff, N. A. Doria-Rose, M. Connors, J. C. Mullikin, G. J. Nabel, M. Roederer, L. Shapiro, P. D. Kwong and J. R. Mascola, *Science*, 2011, **333**, 1593.
279. X. Xiao, W. Chen, Y. Feng and D. S. Dimitrov, *Viruses*, 2009, **1**, 802.
280. F. Klein, R. Diskin, J. F. Scheid, C. Gaebler, H. Mouquet, I. S. Georgiev, M. Pancera, T. Zhou, R. B. Incesu, B. Z. Fu, P. N. Gnanapragasam, T. Y. Oliveira, M. S. Seaman, P. D. Kwong, P. J. Bjorkman and M. C. Nussenzweig, *Cell*, 2013, **153**, 126.
281. D. Sok, U. Laserson, J. Laserson, Y. Liu, F. Vigneault, J. P. Julien, B. Briney, A. Ramos, K. F. Saye, K. Le, A. Mahan, S. Wang, M. Kardar, G. Yaari, L. M. Walker, B. B. Simen, E. P. St John, P. Y. Chan-Hui, K. Swiderek, S. H. Kleinstein, G. Alter, M. S. Seaman, A. K. Chakraborty, D. Koller, I. A. Wilson, G. M. Church, D. R. Burton and P. Poignard, *PLoS Pathog.*, 2013, **9**, e1003754.
282. I. S. Georgiev, R. S. Rudicell, K. O. Saunders, W. Shi, T. Kirys, K. McKee, S. O'Dell, G. Y. Chuang, Z. Y. Yang, G. Ofek, M. Connors, J. R. Mascola, G. J. Nabel and P. D. Kwong, *J. Immunol.*, 2014, **192**, 1100.
283. S. Sadanand, T. J. Suscovich and G. Alter, *Annu. Rev. Med.*, 2015, **67**, 185.
284. I. C. MacLennan, K. M. Toellner, A. F. Cunningham, K. Serre, D. M. Sze, E. Zuniga, M. C. Cook and C. G. Vinuesa, *Immunol. Rev.*, 2003, **194**, 8.
285. J. M. Montezuma-Rusca, S. Moir, L. Kardava, C. M. Buckner, A. Louie, L. J. Kim, B. H. Santich, W. Wang, O. R. Fankuchen, G. Diaz, J. R. Daub, S. D. Rosenzweig, T. W. Chun, Y. Li, R. C. Braylan, K. R. Calvo and A. S. Fauci, *J. Immunol.*, 2015, **194**, 2561.
286. M. M. Aasa-Chapman, A. Hayman, P. Newton, D. Cornforth, I. Williams, P. Borrow, P. Balfe and A. McKnight, *AIDS*, 2004, **18**, 371.
287. E. S. Gray, P. L. Moore, I. A. Choge, J. M. Decker, F. Bibollet-Ruche, H. Li, N. Leseka, F. Treurnicht, K. Mlisana, G. M. Shaw, S. S. Karim, C. Williamson and L. Morris, *J. Virol.*, 2007, **81**, 6187.
288. E. S. Gray, M. C. Madiga, T. Hermanus, P. L. Moore, C. K. Wibmer, N. L. Tumba, L. Werner, K. Mlisana, S. Sibeko, C. Williamson, S. S. Abdool Karim and L. Morris, *J. Virol.*, 2011, **85**, 4828.
289. A. Piantadosi, D. Panteleeff, C. A. Blish, J. M. Baeten, W. Jaoko, R. S. McClelland and J. Overbaugh, *J. Virol.*, 2009, **83**, 10269.
290. D. N. Sather, S. Carbonetti, D. C. Malherbe, F. Pissani, A. B. Stuart, A. J. Hessell, M. D. Gray, I. Mikell, S. A. Kalams, N. L. Haigwood and L. Stamatatos, *J. Virol.*, 2014, **88**, 12968.
291. J. R. Mascola and B. F. Haynes, *Immunol. Rev.*, 2013, **254**, 225.
292. X. Wu, Z. Zhang, C. A. Schramm, M. G. Joyce, Y. D. Kwon, T. Zhou, Z. Sheng, B. Zhang, S. O'Dell, K. McKee, I. S. Georgiev, G. Y. Chuang, N. S. Longo, R. M. Lynch, K. O. Saunders, C. Soto, S. Srivatsan, Y. Yang, R. T. Bailer, M. K. Louder, J. C. Mullikin, M. Connors, P. D. Kwong, J. R. Mascola and L. Shapiro, *Cell*, 2015, **161**, 470.

293. Y. Li, S. O'Dell, L. M. Walker, X. Wu, J. Guenaga, Y. Feng, S. D. Schmidt, K. McKee, M. K. Louder, J. E. Ledgerwood, B. S. Graham, B. F. Haynes, D. R. Burton, R. T. Wyatt and J. R. Mascola, *J. Virol.*, 2011, **85**, 8954.
294. J. Chen, G. Frey, H. Peng, S. Rits-Volloch, J. Garrity, M. S. Seaman and B. Chen, *J. Virol.*, 2014, **88**, 1249.
295. A. Halper-Stromberg and M. C. Nussenzweig, *J. Clin. Invest.*, 2016, **162**, 415.
296. H. Mouquet, *Trends Immunol.*, 2014, **35**, 549.
297. N. Martin, S. Welsch, C. Jolly, J. A. Briggs, D. Vaux and Q. J. Sattentau, *J. Virol.*, 2010, **84**, 3516.
298. M. Malbec, F. Porrot, R. Rua, J. Horwitz, F. Klein, A. Halper-Stromberg, J. F. Scheid, C. Eden, H. Mouquet, M. C. Nussenzweig and O. Schwartz, *J. Exp. Med.*, 2013, **210**, 2813.
299. B. Su, K. Xu, A. Lederle, M. Peressin, M. E. Biedma, G. Laumond, S. Schmidt, T. Decoville, A. Proust, M. Lambotin, V. Holl and C. Moog, *Blood*, 2012, **120**, 3708.
300. C. J. Duncan, J. P. Williams, T. Schiffner, K. Gartner, C. Ochsenbauer, J. Kappes, R. A. Russell, J. Frater and Q. J. Sattentau, *J. Virol.*, 2014, **88**, 2025.
301. J. A. Horwitz, A. Halper-Stromberg, H. Mouquet, A. D. Gitlin, A. Tretiakova, T. R. Eisenreich, M. Malbec, S. Gravemann, E. Billerbeck, M. Dorner, H. Buning, O. Schwartz, E. Knops, R. Kaiser, M. S. Seaman, J. M. Wilson, C. M. Rice, A. Ploss, P. J. Bjorkman, F. Klein and M. C. Nussenzweig, *Proc. Natl. Acad. Sci. U. S. A.*, 2013, **110**, 16538.
302. D. H. Barouch, J. B. Whitney, B. Moldt, F. Klein, T. Y. Oliveira, J. Liu, K. E. Stephenson, H. W. Chang, K. Shekhar, S. Gupta, J. P. Nkolola, M. S. Seaman, K. M. Smith, E. N. Borducchi, C. Cabral, J. Y. Smith, S. Blackmore, S. Sanisetty, J. R. Perry, M. Beck, M. G. Lewis, W. Rinaldi, A. K. Chakraborty, P. Poignard, M. C. Nussenzweig and D. R. Burton, *Nature*, 2013, **503**, 224.
303. M. Shingai, Y. Nishimura, F. Klein, H. Mouquet, O. K. Donau, R. Plishka, A. Buckler-White, M. Seaman, M. Piatak Jr., J. D. Lifson, D. S. Dimitrov, M. C. Nussenzweig and M. A. Martin, *Nature*, 2013, **503**, 277.
304. M. Caskey, F. Klein, J. C. Lorenzi, M. S. Seaman, A. P. West Jr., N. Buckley, G. Kremer, L. Nogueira, M. Braunschweig, J. F. Scheid, J. A. Horwitz, I. Shimeliovich, S. Ben-Avraham, M. Witmer-Pack, M. Platten, C. Lehmann, L. A. Burke, T. Hawthorne, R. J. Gorelick, B. D. Walker, T. Keler, R. M. Gulick, G. Fatkenheuer, S. J. Schlesinger and M. C. Nussenzweig, *Nature*, 2015, **522**, 487.
305. A. R. Cillo and J. W. Mellors, *Curr. Opin. Virol.*, 2016, **18**, 14.
306. R. B. Jones, R. O'Connor, S. Mueller, M. Foley, G. L. Szeto, D. Karel, M. Lichterfeld, C. Kovacs, M. A. Ostrowski, A. Trocha, D. J. Irvine and B. D. Walker, *PLoS Pathog.*, 2014, **10**, e1004287.
307. K. Deng, M. Pertea, A. Rongvaux, L. Wang, C. M. Durand, G. Ghiaur, J. Lai, H. L. McHugh, H. Hao, H. Zhang, J. B. Margolick, C. Gurer, A. J. Murphy, D. M. Valenzuela, G. D. Yancopoulos, S. G. Deeks, T. Strowig,

P. Kumar, J. D. Siliciano, S. L. Salzberg, R. A. Flavell, L. Shan and R. F. Siliciano, *Nature*, 2015, **517**, 381.
308. A. Halper-Stromberg, C. L. Lu, F. Klein, J. A. Horwitz, S. Bournazos, L. Nogueira, T. R. Eisenreich, C. Liu, A. Gazumyan, U. Schaefer, R. C. Furze, M. S. Seaman, R. Prinjha, A. Tarakhovsky, J. V. Ravetch and M. C. Nussenzweig, *Cell*, 2014, **158**, 989.
309. A. Sigal, J. T. Kim, A. B. Balazs, E. Dekel, A. Mayo, R. Milo and D. Baltimore, *Nature*, 2011, **477**, 95.
310. B. Moldt, M. Shibata-Koyama, E. G. Rakasz, N. Schultz, Y. Kanda, D. C. Dunlop, S. L. Finstad, C. Jin, G. Landucci, M. D. Alpert, A. S. Dugast, P. W. Parren, F. Nimmerjahn, D. T. Evans, G. Alter, D. N. Forthal, J. E. Schmitz, S. Iida, P. Poignard, D. I. Watkins, A. J. Hessell and D. R. Burton, *J. Virol.*, 2012, **86**, 6189.
311. S. A. Sievers, L. Scharf, A. P. West Jr. and P. J. Bjorkman, *Curr. Opin. HIV AIDS*, 2015, **10**, 151.
312. R. Diskin, J. F. Scheid, P. M. Marcovecchio, A. P. West Jr., F. Klein, H. Gao, P. N. Gnanapragasam, A. Abadir, M. S. Seaman, M. C. Nussenzweig and P. J. Bjorkman, *Science*, 2011, **334**, 1289.
313. M. Asokan, R. S. Rudicell, M. Louder, K. McKee, S. O'Dell, G. Stewart-Jones, K. Wang, L. Xu, X. Chen, M. Choe, G. Chuang, I. S. Georgiev, M. G. Joyce, T. Kirys, S. Ko, A. Pegu, W. Shi, J. P. Todd, Z. Yang, R. T. Bailer, S. Rao, P. D. Kwong, G. J. Nabel and J. R. Mascola, *J. Virol.*, 2015, **89**, 12501.
314. A. J. McMichael and W. C. Koff, *Nat. Immunol.*, 2014, **15**, 319.
315. C. N. Scanlan, J. Offer, N. Zitzmann and R. A. Dwek, *Nature*, 2007, **446**, 1038.
316. J. K. Mann and T. Ndung'u, *Virol. J.*, 2015, **12**, 1.
317. D. R. Burton, *Nat. Rev. Immunol.*, 2002, **2**, 706.
318. D. W. Kulp and W. R. Schief, *Curr. Opin. Virol.*, 2013, **3**, 322.
319. K. Javaherian, A. J. Langlois, G. J. LaRosa, A. T. Profy, D. P. Bolognesi, W. C. Herlihy, S. D. Putney and T. J. Matthews, *Science*, 1990, **250**, 1590.
320. K. Sliepen and R. W. Sanders, *Expert Rev. Vaccines*, 2015, **15**, 349.
321. S. M. Dennison, L. L. Sutherland, F. H. Jaeger, K. M. Anasti, R. Parks, S. Stewart, C. Bowman, S. M. Xia, R. Zhang, X. Shen, R. M. Scearce, G. Ofek, Y. Yang, P. D. Kwong, S. Santra, H. X. Liao, G. Tomaras, N. L. Letvin, B. Chen, S. M. Alam and B. F. Haynes, *PLoS One*, 2011, **6**, e27824.
322. M. B. Zwick, *AIDS*, 2005, **19**, 1725.
323. M. K. Hart, T. J. Palker, T. J. Matthews, A. J. Langlois, N. W. Lerche, M. E. Martin, R. M. Scearce, C. McDanal, D. P. Bolognesi and B. F. Haynes, *J. Immunol.*, 1990, **145**, 2677.
324. T. Zhou, J. Zhu, Y. Yang, J. Gorman, G. Ofek, S. Srivatsan, A. Druz, C. R. Lees, G. Lu, C. Soto, J. Stuckey, D. R. Burton, W. C. Koff, M. Connors and P. D. Kwong, *PLoS One*, 2014, **9**, e99881.
325. G. Ofek, F. J. Guenaga, W. R. Schief, J. Skinner, D. Baker, R. Wyatt and P. D. Kwong, *Proc. Natl. Acad. Sci. U. S. A.*, 2010, **107**, 17880.

326. M. Totrov, X. Jiang, X. P. Kong, S. Cohen, C. Krachmarov, A. Salomon, C. Williams, M. S. Seaman, R. Abagyan, T. Cardozo, M. K. Gorny, S. Wang, S. Lu, A. Pinter and S. Zolla-Pazner, *Virology*, 2010, **405**, 513.
327. R. L. Stanfield, J. P. Julien, R. Pejchal, J. S. Gach, M. B. Zwick and I. A. Wilson, *J. Mol. Biol.*, 2011, **414**, 460.
328. K. Chakraborty, V. Durani, E. R. Miranda, M. Citron, X. Liang, W. Schleif, J. G. Joyce and R. Varadarajan, *Biochem. J.*, 2006, **399**, 483.
329. I. S. Georgiev, M. Gordon Joyce, T. Zhou and P. D. Kwong, *Curr. Opin. HIV AIDS*, 2013, **8**, 382.
330. B. Dey, M. Pancera, K. Svehla, Y. Shu, S. H. Xiang, J. Vainshtein, Y. Li, J. Sodroski, P. D. Kwong, J. R. Mascola and R. Wyatt, *J. Virol.*, 2007, **81**, 5579.
331. J. Jardine, J. P. Julien, S. Menis, T. Ota, O. Kalyuzhniy, A. McGuire, D. Sok, P. S. Huang, S. MacPherson, M. Jones, T. Nieusma, J. Mathison, D. Baker, A. B. Ward, D. R. Burton, L. Stamatatos, D. Nemazee, I. A. Wilson and W. R. Schief, *Science*, 2013, **340**, 711.
332. K. Tran, C. Poulsen, J. Guenaga, N. de Val, R. Wilson, C. Sundling, Y. Li, R. L. Stanfield, I. A. Wilson, A. B. Ward, G. B. Karlsson Hedestam and R. T. Wyatt, *Proc. Natl. Acad. Sci. U. S. A.*, 2014, **111**, E738.
333. R. W. Sanders, M. J. van Gils, R. Derking, D. Sok, T. J. Ketas, J. A. Burger, G. Ozorowski, A. Cupo, C. Simonich, L. Goo, H. Arendt, H. J. Kim, J. H. Lee, P. Pugach, M. Williams, G. Debnath, B. Moldt, M. J. van Breemen, G. Isik, M. Medina-Ramirez, J. W. Back, W. C. Koff, J. P. Julien, E. G. Rakasz, M. S. Seaman, M. Guttman, K. K. Lee, P. J. Klasse, C. LaBranche, W. R. Schief, I. A. Wilson, J. Overbaugh, D. R. Burton, A. B. Ward, D. C. Montefiori, H. Dean and J. P. Moore, *Science*, 2015, **349**, 154.
334. S. Hoffenberg, R. Powell, A. Carpov, D. Wagner, A. Wilson, S. Kosakovsky Pond, R. Lindsay, H. Arendt, J. Destefano, S. Phogat, P. Poignard, S. P. Fling, M. Simek, C. Labranche, D. Montefiori, T. Wrin, P. Phung, D. Burton, W. Koff, C. R. King, C. L. Parks and M. J. Caulfield, *J. Virol.*, 2013, **87**, 5372.
335. B. F. Haynes, G. M. Shaw, B. Korber, G. Kelsoe, J. Sodroski, B. H. Hahn, P. Borrow and A. J. McMichael, *Cell Host Microbe*, 2016, **19**, 292.
336. M. N. Forsell, W. R. Schief and R. T. Wyatt, *Curr. Opin. HIV AIDS*, 2009, **4**, 380.
337. J. M. Kovacs, J. P. Nkolola, H. Peng, A. Cheung, J. Perry, C. A. Miller, M. S. Seaman, D. H. Barouch and B. Chen, *Proc. Natl. Acad. Sci. U. S. A.*, 2012, **109**, 12111.
338. J. M. Kovacs, E. Noeldeke, H. J. Ha, H. Peng, S. Rits-Volloch, S. C. Harrison and B. Chen, *Proc. Natl. Acad. Sci. U. S. A.*, 2014, **111**, 18542.
339. X. Yang, J. Lee, E. M. Mahony, P. D. Kwong, R. Wyatt and J. Sodroski, *J. Virol.*, 2002, **76**, 4634.
340. R. P. Ringe, A. Yasmeen, G. Ozorowski, E. P. Go, L. K. Pritchard, M. Guttman, T. A. Ketas, C. A. Cottrell, I. A. Wilson, R. W. Sanders, A. Cupo, M. Crispin, K. K. Lee, H. Desaire, A. B. Ward, P. J. Klasse and J. P. Moore, *J. Virol.*, 2015, **89**, 12189.

341. E. P. Go, Y. Zhang, S. Menon and H. Desaire, *J. Proteome Res.*, 2011, **10**, 578.
342. M. Kim, Z. S. Qiao, D. C. Montefiori, B. F. Haynes, E. L. Reinherz and H. X. Liao, *AIDS Res. Hum. Retroviruses*, 2005, **21**, 58.
343. J. K. Hu, J. C. Crampton, A. Cupo, T. Ketas, M. J. van Gils, K. Sliepen, S. W. de Taeye, D. Sok, G. Ozorowski, I. Deresa, R. Stanfield, A. B. Ward, D. R. Burton, P. J. Klasse, R. W. Sanders, J. P. Moore and S. Crotty, *J. Virol.*, 2015, **89**, 10383.
344. S. K. Sharma, N. de Val, S. Bale, J. Guenaga, K. Tran, Y. Feng, V. Dubrovskaya, A. B. Ward and R. T. Wyatt, *Cell Rep.*, 2015, **11**, 539.
345. M. F. Bachmann and G. T. Jennings, *Nat. Rev. Immunol.*, 2010, **10**, 787.
346. L. Zhao, A. Seth, N. Wibowo, C. X. Zhao, N. Mitter, C. Yu and A. P. Middelberg, *Vaccine*, 2014, **32**, 327.
347. E. T. Crooks, T. Tong, B. Chakrabarti, K. Narayan, I. S. Georgiev, S. Menis, X. Huang, D. Kulp, K. Osawa, J. Muranaka, G. Stewart-Jones, J. Destefano, S. O'Dell, C. LaBranche, J. E. Robinson, D. C. Montefiori, K. McKee, S. X. Du, N. Doria-Rose, P. D. Kwong, J. R. Mascola, P. Zhu, W. R. Schief, R. T. Wyatt, R. G. Whalen and J. M. Binley, *PLoS Pathog.*, 2015, **11**, e1004932.
348. K. Sliepen, G. Ozorowski, J. A. Burger, T. van Montfort, M. Stunnenberg, C. LaBranche, D. C. Montefiori, J. P. Moore, A. B. Ward and R. W. Sanders, *Retrovirology*, 2015, **12**, 82.
349. R. W. Sanders, M. Venturi, L. Schiffner, R. Kalyanaraman, H. Katinger, K. O. Lloyd, P. D. Kwong and J. P. Moore, *J. Virol.*, 2002, **76**, 7293.
350. C. N. Scanlan, G. E. Ritchie, K. Baruah, M. Crispin, D. J. Harvey, B. B. Singer, L. Lucka, M. R. Wormald, P. Wentworth Jr., N. Zitzmann, P. M. Rudd, D. R. Burton and R. A. Dwek, *J. Mol. Biol.*, 2007, **372**, 16.
351. C. Agrawal-Gamse, R. J. Luallen, B. Liu, H. Fu, F. H. Lee, Y. Geng and R. W. Doms, *J. Virol.*, 2011, **85**, 470.
352. J. G. Joyce, I. J. Krauss, H. C. Song, D. W. Opalka, K. M. Grimm, D. D. Nahas, M. T. Esser, R. Hrin, M. Feng, V. Y. Dudkin, M. Chastain, J. W. Shiver and S. J. Danishefsky, *Proc. Natl. Acad. Sci. U. S. A.*, 2008, **105**, 15684.
353. I. J. Krauss, J. G. Joyce, A. C. Finnefrock, H. C. Song, V. Y. Dudkin, X. Geng, J. D. Warren, M. Chastain, J. W. Shiver and S. J. Danishefsky, *J. Am. Chem. Soc.*, 2007, **129**, 11042.
354. L. X. Wang, *Curr. Opin. Chem. Biol.*, 2013, **17**, 997.
355. S. Horiya, J. K. Bailey, J. S. Temme, Y. V. Guillen Schlippe and I. J. Krauss, *J. Am. Chem. Soc.*, 2014, **136**, 5407.
356. S. K. Wang, P. H. Liang, R. D. Astronomo, T. L. Hsu, S. L. Hsieh, D. R. Burton and C. H. Wong, *Proc. Natl. Acad. Sci. U. S. A.*, 2008, **105**, 3690.
357. A. Kabanova, R. Adamo, D. Proietti, F. Berti, M. Tontini, R. Rappuoli and P. Costantino, *Glycoconjugate J.*, 2010, **27**, 501.
358. L. X. Wang, J. Ni, S. Singh and H. Li, *Chem. Biol.*, 2004, **11**, 127.
359. J. Ni, H. Song, Y. Wang, N. M. Stamatos and L. X. Wang, *Bioconjugate Chem.*, 2006, **17**, 493.

360. I. S. MacPherson, J. S. Temme, S. Habeshian, K. Felczak, K. Pankiewicz, L. Hedstrom and I. J. Krauss, *Angew. Chem., Int. Ed. Engl.*, 2011, **50**, 11238.
361. J. S. Temme, M. G. Drzyzga, I. S. MacPherson and I. J. Krauss, *Chemistry*, 2013, **19**, 17291.
362. J. S. Temme, I. S. MacPherson, J. F. DeCourcey and I. J. Krauss, *J. Am. Chem. Soc.*, 2014, **136**, 1726.
363. S. Horiya, I. S. MacPherson and I. J. Krauss, *Nat. Chem. Biol.*, 2014, **10**, 990.
364. H. Li and L. X. Wang, *Org. Biomol. Chem.*, 2004, **2**, 483.
365. M. Marradi, P. Di Gianvincenzo, P. M. Enriquez-Navas, O. M. Martinez-Avila, F. Chiodo, E. Yuste, J. Angulo and S. Penades, *J. Mol. Biol.*, 2011, **410**, 798.
366. R. D. Astronomo, H. K. Lee, C. N. Scanlan, R. Pantophlet, C. Y. Huang, I. A. Wilson, O. Blixt, R. A. Dwek, C. H. Wong and D. R. Burton, *J. Virol.*, 2008, **82**, 6359.
367. R. D. Astronomo, E. Kaltgrad, A. K. Udit, S. K. Wang, K. J. Doores, C. Y. Huang, R. Pantophlet, J. C. Paulson, C. H. Wong, M. G. Finn and D. R. Burton, *Chem. Biol.*, 2010, **17**, 357.
368. B. E. Clark, K. Auyeung, E. Fregolino, M. Parrilli, R. Lanzetta, C. De Castro and R. Pantophlet, *Chem. Biol.*, 2012, **19**, 254.
369. R. L. Stanfield, C. De Castro, A. M. Marzaioli, I. A. Wilson and R. Pantophlet, *Glycobiology*, 2015, **25**, 412.
370. D. C. Dunlop, C. Bonomelli, F. Mansab, S. Vasiljevic, K. J. Doores, M. R. Wormald, A. S. Palma, T. Feizi, D. J. Harvey, R. A. Dwek, M. Crispin and C. N. Scanlan, *Glycobiology*, 2010, **20**, 812.
371. R. J. Luallen, J. Lin, H. Fu, K. K. Cai, C. Agrawal, I. Mboudjeka, F. H. Lee, D. Montefiori, D. F. Smith, R. W. Doms and Y. Geng, *J. Virol.*, 2008, **82**, 6447.
372. H. Zhang, H. Fu, R. J. Luallen, B. Liu, F. H. Lee, R. W. Doms and Y. Geng, *Vaccine*, 2015, **33**, 5140.
373. R. J. Luallen, H. Fu, C. Agrawal-Gamse, I. Mboudjeka, W. Huang, F. H. Lee, L. X. Wang, R. W. Doms and Y. Geng, *J. Virol.*, 2009, **83**, 4861.
374. M. Crispin, V. T. Chang, D. J. Harvey, R. A. Dwek, E. J. Evans, D. I. Stuart, E. Y. Jones, J. M. Lord, R. A. Spooner and S. J. Davis, *J. Biol. Chem.*, 2009, **284**, 21684.
375. S. Apostolico Jde, S. B. Boscardin, M. M. Yamamoto, J. N. de Oliveira-Filho, J. Kalil, E. Cunha-Neto and D. S. Rosa, *PLoS One*, 2016, **11**, e0145637.
376. A. B. Morelli, D. Becher, S. Koernig, A. Silva, D. Drane and E. Maraskovsky, *J. Med. Microbiol.*, 2012, **61**, 935.
377. F. Garces, J. H. Lee, N. de Val, A. Torrents de la Pena, L. Kong, C. Puchades, Y. Hua, R. L. Stanfield, D. R. Burton, J. P. Moore, R. W. Sanders, A. B. Ward and I. A. Wilson, *Immunity*, 2015, **43**, 1053.
378. A. T. McGuire, S. Hoot, A. M. Dreyer, A. Lippy, A. Stuart, K. W. Cohen, J. Jardine, S. Menis, J. F. Scheid, A. P. West, W. R. Schief and L. Stamatatos, *J. Exp. Med.*, 2013, **210**, 655.

379. A. T. McGuire, J. A. Glenn, A. Lippy and L. Stamatatos, *J. Virol.*, 2014, **88**, 2645.
380. D. Zotos, J. M. Coquet, Y. Zhang, A. Light, K. D'Costa, A. Kallies, L. M. Corcoran, D. I. Godfrey, K. M. Toellner, M. J. Smyth, S. L. Nutt and D. M. Tarlinton, *J. Exp. Med.*, 2010, **207**, 365.
381. C. S. Ma, S. Suryani, D. T. Avery, A. Chan, R. Nanan, B. Santner-Nanan, E. K. Deenick and S. G. Tangye, *Immunol. Cell Biol.*, 2009, **87**, 590.
382. S. Gupta, E. S. Clark, J. M. Termini, J. Boucher, S. Kanagavelu, C. C. LeBranche, S. Abraham, D. C. Montefiori, W. N. Khan and G. W. Stone, *J. Virol.*, 2015, **89**, 4158.
383. P. Schneider, *Curr. Opin. Immunol.*, 2005, **17**, 282.
384. S. Campbell and A. Rein, *J. Virol.*, 1999, **73**, 2270.
385. T. A. Bowden, K. Baruah, C. H. Coles, D. J. Harvey, X. Yu, B.-D. Song, D. I. Stuart, A. R. Aricescu, C. N. Scanlan, E. Y. Jones and M. Crispin, *J. Am. Chem. Soc.*, 2012, **134**, 17554.
386. D. J. Harvey, A. H. Merry, L. Royle, M. P. Campbell, R. A. Dwek and P. M. Rudd, *Proteomics*, 2009, **9**, 3796.
387. M. G. Joyce, M. Kanekiyo, L. Xu, C. Biertumpfel, J. C. Boyington, S. Moquin, W. Shi, X. Wu, Y. Yang, Z. Y. Yang, B. Zhang, A. Zheng, T. Zhou, J. Zhu, J. R. Mascola, P. D. Kwong and G. J. Nabel, *J. Virol.*, 2013, **87**, 2294.
388. S. Bhattacharyya, R. E. Rajan, Y. Swarupa, U. Rathore, A. Verma, R. Udaykumar and R. Varadarajan, *J. Biol. Chem.*, 2010, **285**, 27100.
389. Y. Qin, M. Banasik, S. Kim, A. Penn-Nicholson, H. H. Habte, C. LaBranche, D. C. Montefiori, C. Wang and M. W. Cho, *Virology*, 2014, **462–463**, 363.
390. X. Wu, Z.-Y. Yang, Y. Li, C.-M. Hogerkorp, W. R. Schief, M. S. Seaman, T. Zhou, S. D. Schmidt, L. Wu, L. Xu, N. S. Longo, K. McKee, S. O'Dell, M. K. Louder, D. L. Wycuff, Y. Feng, M. Nason, N. Doria-Rose, M. Connors, P. D. Kwong, M. Roederer, R. T. Wyatt, G. J. Nabel and J. R. Mascola, *Science*, 2010, **329**, 856.
391. J. Ingale, K. Tran, L. Kong, B. Dey, K. McKee, W. Schief, P. D. Kwong, J. R. Mascola and R. T. Wyatt, *J. Virol.*, 2014, **88**, 14002.
392. R. Pantophlet, I. A. Wilson and D. R. Burton, *J. Virol.*, 2003, **77**, 5889.
393. S. Pejawar-Gaddy, J. M. Kovacs, D. H. Barouch, B. Chen and D. J. Irvine, *Bioconjugate Chem.*, 2014, **25**, 1470.
394. D. P. Leaman, J. H. Lee, A. B. Ward and M. B. Zwick, *J. Virol.*, 2015, **89**, 6725.
395. A. J. Behrens, D. J. Harvey, E. Milne, A. Cupo, A. Kumar, N. Zitzmann, W. B. Struwe, J. P. Moore and M. Crispin, Molecular architecture of the cleavage-dependent mannose patch on a soluble HIV-1 envelope glycoprotein trimer, *J. Virol.*, 2017, **91**, e01894–e01816.

CHAPTER 12

Design, Synthesis and Evaluation of Mucin Glycopeptide Based Cancer Vaccine

ZHANYI SUN[a], PUGUANG CHEN[a], YANFANG LIU[a] AND YANMEI LI*[a]

[a]Key Laboratory of Bioorganic Phosphorus Chemistry & Chemical Biology (Ministry of Education), Department of Chemistry, Tsinghua University, 100084, P.R. China
*E-mail: liym@mail.tsinghua.edu.cn

12.1 Introduction

Cancer has become one of the most serious threats to human health with its high morbidity and mortality worldwide. Currently, cancer treatments mainly include surgery, chemotherapy, radiation therapy, immunotherapy, targeted therapy, hormone therapy, stem cell transplantation, precision medicine, *etc.*[1] In the process of understanding the interaction between cancer microenvironments and the immune system of the host, some positive results indicate that active immunotherapy is an efficient treatment to achieve strong and long-term immune responses.[2] Monoclonal antibodies,[3]

Chemical Biology No. 2
Chemical Biology of Glycoproteins
Edited by Zhongping Tan and Lai-Xi Wang

Published by the Royal Society of Chemistry, www.rsc.org

adoptive cell transfer,[4] immune checkpoint inhibitors,[5] and cancer vaccines[6] are several primary active immunotherapies for cancer treatment.

Cancer vaccines function by activating or restoring the abilities of the immune system, including preventive vaccines and therapeutic vaccines. The process of action of therapeutic vaccines covers the following steps. First of all, vaccines activate the immune system through tumor antigens, tumor tissue extract, or tumor cells. Then, the immune system produces specific immune response against tumor cells in order to control tumor growth and restrict tumor metastasis. Finally, the immune system removes tumor cells to treat the tumor thoroughly.[7]

According to the different types of tumor antigens, anti-tumor vaccine can be divided into tumor cell vaccines, DNA or RNA vaccines, dendritic cell (DC) vaccines and tumor (glyco) peptide vaccines. At present, Gardasil, Cervarix, DC-Vax, *etc.*, anti-tumor vaccines have come into the market. At the same time a large number of anti-tumor vaccines are undergoing clinical trials.[8] Herein we mainly focus on the mucin-based glycopeptide antitumor vaccine, including its design, synthesis and evaluation.

12.2 Mucins as the Target of Anti-Tumor Vaccine

12.2.1 Tumor Antigens

Tumor antigens can be divided into tumor-specific antigens and tumor-associated antigens. Tumor-specific antigens refer to antigens only expressed on the surface of tumor cells which are not expressed on normal cells. They include virus-induced tumor antigens, spontaneous tumor antigens and tumor antigens induced under harsh physical or chemical conditions.[9] Tumor-associated antigens refers to a series of antigens not only expressed on tumor cells but also on some normal cells. The important difference is that tumor-associated antigens are present in trace amounts on normal cells but are highly overexpressed during the proliferation of tumor cells. Tumor-associated antigens are principally glycoproteins or glycolipids on the cell surface. Unlike normal cells, tumor cells can overexpress tumor-associated antigens which could be recognized by the body's immune system. The process is of significance for tumor diagnosis and therapy. Tumor-associated antigens include embryonic antigens and tissue-specific differentiation antigens as well as overexpressed antigens. Embryonic antigens are a normal component of embryonic tissue, which gradually reduce after embryonic development, and are re-expressed in the malignant cells. For example, two embryonic antigens, alpha-fetoprotein and carcinoembryonic antigen, have been applied to monitor routine physical examination. Tissue-specific differentiation antigens are expressed variably at different stages of cell differentiation and maturation, such as mucin glycopeptide antigens and melanoma cells antigens. The overexpression of antigen is mainly the product of the expression of tumor suppressor genes and proto-oncogenes, such as p53 and p185. Here we only discuss one of the tumor-associated antigens, mucin glycoprotein, overexpressed in epithelial tumor cells.

12.2.2 Mucins

Mucins (MUC) are a kind of high-molecular-weight glycoprotein whose molecular weight is up to 40 million Daltons. Mucins are widely distributed on the surface of mucous membranes of the body. Usually mucins are expressed in various types of epithelial cells, especially those in adverse circumstance in the human body, such as the interface between air and water in the respiratory system, gastric acid environment in stomach and the intestinal epithelial cells as well as some other secretory organs (liver, pancreas, gallbladder, breast, *etc.*).[10] Mucins can build a highly ordered structure on the cell surface which can produce a molecular micro-environment.[11] The micro-environment can protect cells under various harsh circumstances, for example, different kinds of digestive enzymes, low pH, different ion concentration and the presence of bacteria and virus.[12] Also, mucins can play the role of receptors and sensors on the cell surface.

According to their structural characteristics and biosynthetic pathway, mucins can be divided into two types: membrane-bound mucins and secreted mucins. Membrane-bound mucins include MUC1, MUC3A/B, MUC4, MUC12, MUC13, MUC15, MUC16, MUC17 and MUC20, while secreted mucins include MUC2, MUC5AC, MUC5B, MUC6, MUC7 and MUC19.[13]

The common structural features of mucins are that they include large numbers of variable number of tandem repeats (VNTR) motifs on the extracellular portion. The VNTR motifs of different mucin proteins contain very similar amino acid residues, which are rich in serine, threonine and proline. Both serine and threonine are possible *O*-linked glycosylation sites.

Mucins have consanguineous relation with the production and development of tumors. Results of studies have indicated that the expression of mucins is abnormally increased in the early stages of pancreatic cancer.[14,15] Mucins on the cell surface are very likely to help cancer cells to establish a local environment favorable to survival, which can be a great benefit to the processes of invasion, tumor metastasis and cell division in an unfavorable external environment.

12.2.3 MUC1 in Normal Cells

MUC1 is the most widely distributed and studied mucin, and was also the first to be cloned.[16] Mucins are encoded by different gene sequences. MUC1 is a small mucin, compared with other membrane-bound mucins which are usually large and have complex structures.[17] It is composed of peptide scaffolds and *O*-linked glycosylic side chains. MUC1 is bulky in volume and the molecular weight is over 500 kDa. In normal tissue, MUC1 is heavily glycosylated. The oligosaccharide part accounts for approximately 50–90% of the whole MUC1 molecule. There are three domains in MUC1: a large extracellular domain, a hydrophobic transmembrane domain and a short cytoplasmic domain.[18] MUC1 is extended extracellularly from the cell membrane and the whole molecule is strip-shaped. The huge extracellular domain contains 1000–2200 amino acid residues, mainly including

His-Gly-Val-Thr*-Ser*-Ala-Pro-Asp-Thr*-Arg-Pro-Ala-Pro-Gly-Ser*-Thr*-Ala-Pro-Pro-Ala

Figure 12.1 VNTR of MUC1 (*: potential *O*-linked glycosylation sites).

VNTRs.[19] Each VNTR in the extracellular domain of MUC1 is composed of 20 amino acids residues (see Figure 12.1). Because of the polymorphism of the allelic genes, the number of VNTR can be freely varied from 20 to 120. Therefore, the length of the extracellular portion can be extended to 200–500 nm, much longer than the polysaccharide–protein complex (10 nm) localized on the cell exterior. VNTR of MUC1 with 20 amino acid residues includes five proline residues. Also in the 20-amino-acids VNTR, there are five possible *O*-linked glycosylation sites, including two serine (Ser) residues and three threonine (Thr) residues.

It was initially considered that the functions of MUC1 were solely protecting cells from the invasion of extracellular substances, maintaining cell viscosity and in lubrication and recognition. After the gene for MUC1 was successfully cloned, more functions of MUC1 glycoprotein were discovered.[20–22] MUC1 played an important role in the renewal and differentiation of epithelial cells, maintaining the integrity of the epithelial cells and the occurrence and metastasis of tumors.

12.2.4 MUC1 in Tumor Cells

In common with its function in normal cells, MUC1 plays a critical role in tumor cells. Studies have shown that MUC1 could influence the biological function of tumor cells at least in the following areas:

(a) Control of the local microenvironment of tumor cell proliferation. MUC1 could control the local molecular environment around the cells in the process of tumor cell growth and differentiation. Through this MUC1 could protect cells, promoting cell growth and differentiation in adverse environments. Also MUC1 is able to help establish a better environment for the invasion and metastasis of tumor cells.[23]

(b) Regulation of differentiation and proliferation of tumor cells. The cytoplasmic domain of MUC1 can regulate the differentiation and proliferation of tumor cells through ligand–receptor interactions and cell signaling.[24]

(c) Benefiting tumor invasion and metastasis. MUC1 can form into filaments at a high density on the surface of tumor cell membrane, block the interaction between ligand immobilized on the cell surface and its receptor and reduce the cell–cell interactions mediated by the integration into the extracellular matrix, thereby making it easy for tumor cells to invade. Some epitopes of MUC1 can be a ligand of cadherin and interact with cadherin in the vascular endothelial cells which are damaged or infected with inflammation, thereby enhancing the adhesion between the tumor cells and endothelial cells. The adhesion makes it easy for tumor cells to pass through the vessel wall and is favorable to the metastasis of tumors.[25,26] Also, the sialic acids highly expressed in the side chain of MUC1 on tumor cells are negative charged, which makes the tumor cells repel each other. As a result of the repulsion, tumor cells are easier to transfer. The cytoplasmic domain of MUC1 can be phosphorylated, which may affect cell adhesion through affecting signal transduction.[27]

(d) Helping tumor cells to evade damage from the immune system. MUC1 in normal tissues is isolated from immune cells and glycosylated, making it hard for it to be recognized by the immune system. However, MUC1 in tumor cells is widely distributed and abnormally abundantly expressed on the cell surface, which can conceal interactions between receptors and antigens. In this way, MUC1 can protect tumor cells, allowing them to escape immune surveillance of the host. In addition, MUC1 can also weaken immune recognition through affecting the adhesion and metastasis of tumor cells. Some soluble MUC1 secreted by tumor cells into the blood circulation, can also induce apoptosis of activated T cells, thereby weakening the immune suppression of tumor cells.[28,29]

12.2.5 Differences in the Glycosylation Types of the MUC1 in Tumor Cells

Compared with those in normal cells, MUC1 in tumor cells are distinctly different. In one respect, MUC1 in tumor cells is not only expressed at the outermost end of the cell but also distributed in various positions in the tumor cells. Also, the levels of MUC1 expressed in tumor cells are 100 times higher than those in normal cells.[30] Moreover, the types of MUC1 *O*-linked glycosylation in tumor cells change a lot. Since expression level and activity of some glycosyltransferase have changed, the process of translational modification of complex *O*-linked oligosaccharides changes consequently. The outer ends of complex oligosaccharide chains are truncated, only leaving a short core structure modified with sialic acid.

The glycosylation types of MUC1 in tumor cells have changed greatly. 30 years ago, disaccharide T antigen (Thomsen–Friedenreich T antigen) and the precursor monosaccharide Tn antigen were found in some tumor epithelial cells' membrane-associated glycoproteins.[31] Further studies have shown that the changes in oligosaccharides on the membrane-associated glycoprotein are due to some changes in the type or activity of the enzyme in tumor cells. For example, in patients with breast cancer, due to lack or low activity of β-1,6-*N*-acetylglucosamine transferase, the *O*-linked oligosaccharide remained on the core structure without further modification of sialic acids.[32] In addition, as the sialyltransferase activities of galactose-α-3 and *N*-acyl galactosamine β-6 increased, the core structure would be sialylated and produce some tumor-associated oligosaccharide structures, including the Tn antigen, T antigen and STn antigen (See Section 12.3.3.1).[33]

Another feature of *O*-linked oligosaccharides of MUC1 on tumor cell is that the number of glycosylation sites increases. Results of studies have indicated that in epithelial cells of normal breast tissue, the average number of glycosylation sites of each MUC1 VNTR was 2.5, while analysis of T47D tumor cells showed that the five potential glycosylation sites are all glycosylated.[34]

12.2.6 Immunological Foundation for MUC1 as a Tumor Antigen

In normal cells, MUC1 are coated by a great deal of *O*-linked oligosaccharides, making it difficult for them to be recognized by the immune system. However, in tumor cells, the expression level of MUC1 is greatly upregulated. MUC1 may also be distributed in various positions on the cell membrane. Also, *O*-linked oligosaccharides on MUC1 have abnormally changed, the complex oligosaccharides are truncated, leaving only about two or three monosaccharide units. These abnormal changes will ultimately affect the immunological effects of MUC1 as follows: Oligosaccharide side chains of MUC1 are shortened, and some originally hidden epitopes on MUC1 peptide scaffolds are exposed, which can be identified by the immune system. Meanwhile, the expression levels of MUC1 are increased. In such a situation, it is easier to trigger the immune response.[35] In addition, these abnormal oligosaccharides (such as Tn and STn) are a kind of tumor-associated carbohydrate antigens. In the epithelial cells of tumor tissues, these antigens are discontinuously distributed on MUC1 peptide chains.[36,37] Therefore, both the peptide epitopes and the oligosaccharide epitopes can be recognized by the immune system.

In the mid-1980s, MUC1 could be detected in the majority of serum of patients with breast cancer. Therefore, at first MUC1 were studied as a marker for clinical diagnosis of tumors.[38] Subsequently, several specific lymphocytes and antibodies against MUC1 were found. By monitoring circulating immune complexes, the humoral immune response against MUC1 on the cell surface was observed from patients with benign and malignant breast cancer.[39] Also, antibodies specifically recognizing MUC1 peptide chains were found in ulcerative colitis patients.[40] In addition, B-cell immune responses and IgM immunoglobulin against MUC1 were observed in the sera of patients with breast and pancreatic cancer.[41] Some types of IgG antibodies against MUC1 were also detected in patients with colon cancer.[40] Patients at an early stage of breast cancer can prolong the survival time by obtaining MUC1-associated humoral immune therapy.[42] In lymph nodes of patients with ovarian cancer and breast cancer, some MUC1-specific cytotoxic T lymphocytes were found.[43] In patients with breast cancer, ovarian cancer or pancreatic cancer, some MUC1-VNTR-specific cytotoxic T lymphocytes were also found.

The observed humoral and cellular immune responses against MUC1 indicate that MUC1 on the surface of tumor cells can become antigens available in tumor immunotherapy. Therefore, abnormal MUC1 can be utilized to activate an immune response.

12.3 Overview of Fully Synthetic Vaccines

12.3.1 Structures of Fully Synthetic Vaccines

Compared with the traditional vaccines such as whole-pathogen-based vaccines and recombinant-protein-based vaccines, fully synthetic vaccines have homogeneous components and confirmable structures. As shown in Figure 12.2, fully synthetic vaccines are generally composed of antigen structures and immune-stimulating structures. The fully synthetic vaccines are produced by the chemical conjugation of different components. The efficiency of vaccines can be increased by the design and optimization of vaccine structures. Therefore, fully synthetic vaccines can simulate natural antigen structures and break the immune tolerance of these structures.

12.3.2 Immune Response to Fully Synthetic Vaccines

When fully synthetic vaccines are injected into the body, antigen structures can be recognized by B-lymphocytes through the B-cell receptors (BCRs). B-cells can be partly activated and produce low-affinity IgM antibodies.[44] The activation of helper T-cells (Th-cells) is necessary to generate the isotype switch to high-affinity IgG antibodies.[45,46] For the activation of helper T-cells, antigen-presenting cells (APCs) will capture the vaccines. After internalization and proteolytic cleavage, short helper-T epitope peptides are presented on the surface of APCs as a complex with major histocompatibility complex II (MHC II) molecules. Then, these APCs will migrate to the lymph nodes and the MHC II complex will be recognized by $CD4^+$ T-cells. The activated $CD4^+$ T-cells can stimulate B-cells to differentiate into plasma cells and to proliferate and secrete high-affinity IgG antibodies through producing cytokines and co-stimulatory molecules (see Figure 12.3). Therefore, to improve the efficiency of fully synthetic vaccines, helper T-cells' epitope peptides need to be included in the immune stimulating structures.

In addition to the activation of B-cells and T-cells, the innate immune responses can provide the danger signal to enhance the adaptive immune responses. The immunostimulants such as Toll-like receptors (TLRs) ligands

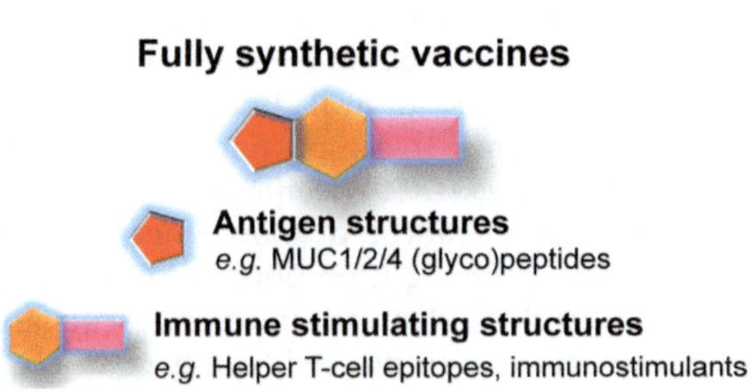

Figure 12.2 Structures of fully synthetic vaccines.

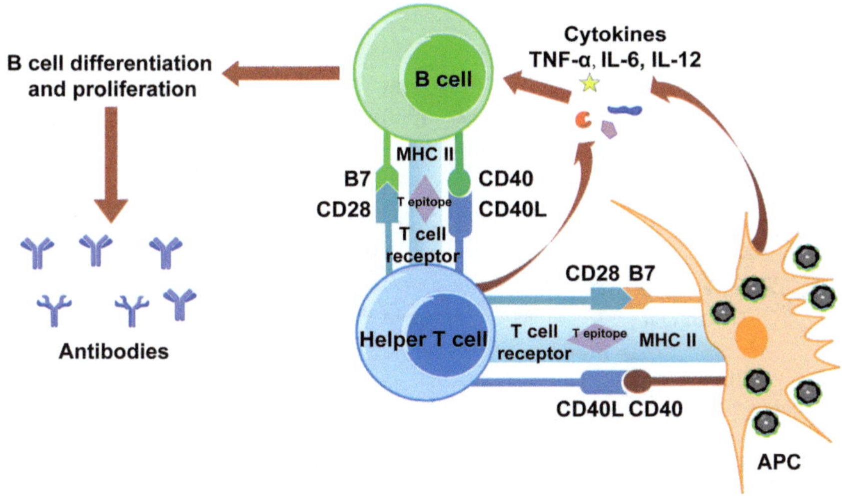

Figure 12.3 Immune response to fully synthetic vaccines.

can be recognized by the receptors of APCs. The activated APCs can produce cytokines such as TNF-α, IL-6 and IL-12, which can trigger the adaptive immune responses (see Figure 12.3).[47–49] Therefore, immunostimulants are very crucial for the design of fully synthetic vaccines.

12.3.3 Antigen Structures of Fully Synthetic Vaccines

Antigen structures are the most important parts of fully synthetic vaccines. Tumor-associated antigens include carbohydrates, peptides and glycopeptides *etc.*

12.3.3.1 Tumor-Associated Carbohydrate Antigens (TACAs)

In tumor cells, the activities of glycosyltransferases are declining and the activities of sialyltranseferases are elevated.[50,51] So the tumor cells overexpress truncated and sialyl carbohydrates. Tumor-associated carbohydrates (see Figure 12.4) are usually linked to lipids or proteins and can be used as the tumor vaccine antigens. For example, Globo-H was synthesized and used for the development of anti-tumor vaccines. Globo-H was conjugated with keyhole limpet hemocyanin (KLH) and injected into mice. The results indicated that this vaccine could induce a significant IgG antibody response.[52]

12.3.3.2 Tumor-Associated MUC1 Peptide Antigens

Mucin 1 glycoproteins with heavy glycosylation play an important role in normal cells. N-terminal ectodomain contains many 20-amino acids-long VNTR with five potential *O*-glycosylation sites. In tumor cells, MUC1 is

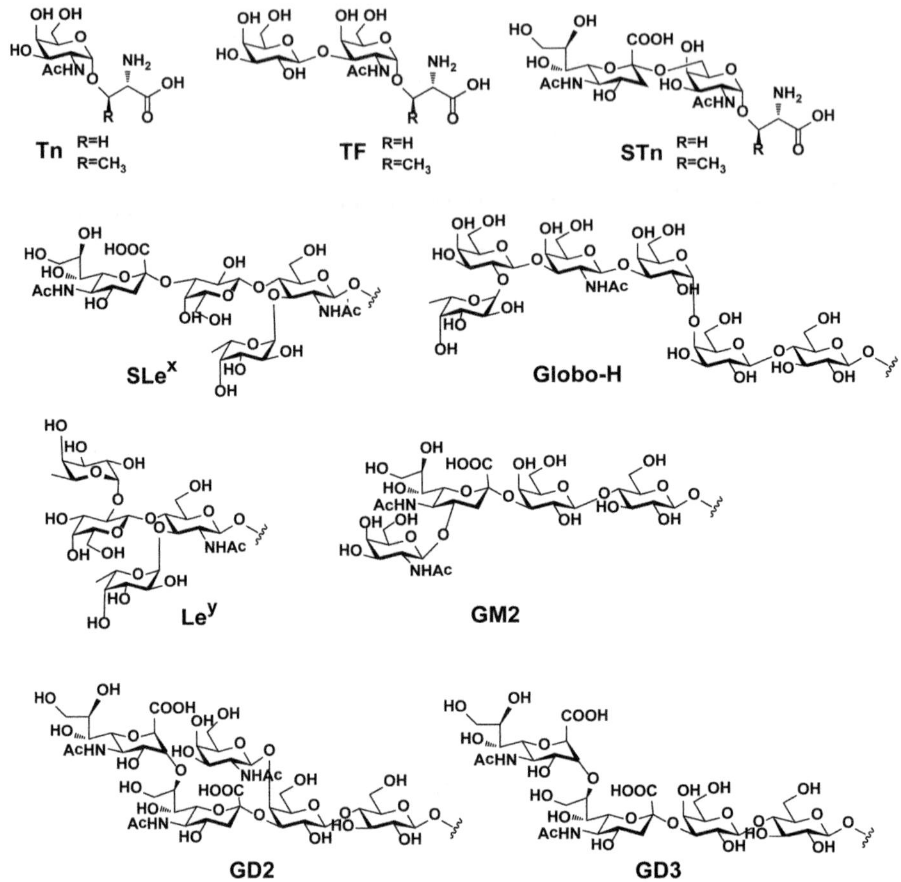

Figure 12.4 Structures of TACAs.

overexpressed and loses the polarity of expression. In addition, due to aberrant glycosylation, the carbohydrate chains of MUC1 are shortened and the peptide epitopes are exposed. Therefore, the MUC1 peptide backbone can be used as a tumor-associated peptide antigen. For example, viral fusion expression vectors such as modified Ankara-expressed MUC1 and IL-2 are utilized in the treatment of prostate cancer and other MUC1-positive tumors.[53,54] Polypeptide vaccines such as five VNTRs have been used to treat prostate cancer,[55] conjugates combined MUC1 peptide with KLH for the treatment of breast cancer,[56] MUC1–lipopeptide BLP25 for treatment of non-small cell lung cancer[57] and multivalent carbohydrate vaccine for the treatment of epithelial ovarian cancer, fallopian tube cancer and peritoneal cancer.[58] At the same time, a number of pre-clinical oncology immunological research projects targeting MUC1 are under way and some of them have achieved some progress.[59,60]

12.3.3.3 Tumor-Associated MUC1 Glycopeptide Antigens

The above-mentioned two kinds of antigens have some limitations. The effects of vaccines based on TACAs or tumor-associated peptide antigens are not satisfactory. In tumor cells, MUC1 glycoproteins contain both carbohydrate antigens and peptide antigens (see Figure 12.5). In addition, these two kinds of antigens can interact with each other. Therefore, MUC1 glycopeptides are considered to be the best antigens for fully synthetic anti-tumor vaccines.[61] In this chapter, we mainly focus on glycopeptide vaccines. The following sections will go into details about this.

12.3.4 Immune-Stimulating Structures of Fully Synthetic Vaccines

Immune-stimulating structures are indispensable for fully synthetic vaccines. Tumor-associated carbohydrate, peptide and glycopeptide antigens are poorly immunogenic. Immune-stimulating structures such as carrier proteins, helper T-cell epitope peptides and immunostimulants can be used to improve the immunogenicity of antigens.

12.3.4.1 Carrier Proteins

Carrier proteins are a class of exogenous proteins with high molecular weights, such as bovine serum albumin (BSA), ovalbumin (OVA), tetanus

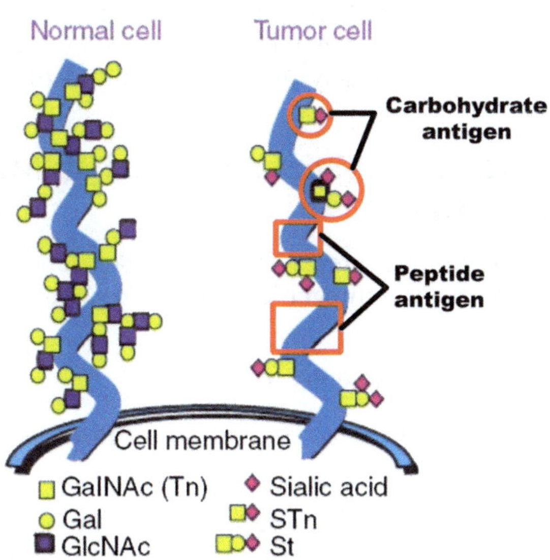

Figure 12.5 Tumor-associated MUC1 antigens. Due to the aberrant glycosylation of tumor cells, the carbohydrate chains of MUC1 are shortened and the peptide epitopes are exposed.

toxin (TT), CRM197, *etc.* They are very immunogenic, which could greatly improve the immunogenicity of antigens after antigens have been conjugated with immunogens. Many different antigens have been conjugated with carrier proteins to produce effective immunogenic vaccines, such as *Haemophilus influenzae* type b (Hib) vaccine, meningococcal vaccine and pneumococcal vaccine,[62–64] as well as mucin glycopeptide antigen.

12.3.4.2 Helper T-Cell Epitope Peptides

Helper T-cells play an important role in inducing an efficient immune response. Helper T-cells activated by helper T-cell epitope peptides can stimulate B-cells to differentiate and proliferate to produce high-affinity IgG antibodies.

Helper T-cell epitope peptides are generally linear determinants and contain 10–20 amino acids. Pan DR epitope (PADRE), recognized by the human HLA-DR receptors, is a universal helper T-cell epitope and can be widely applied in fully synthetic vaccines.[65] In addition, carrier proteins such as tetanus toxoid and ovalbumin are also sources of helper T-cell epitope peptides.

12.3.4.3 Immunostimulants

Immunostimulants can effectively induce the innate immune response and enhance the adaptive immune response. Toll-like receptors (TLRs) are very crucial for the regulation of the innate immune response. TLRs can recognize pathogen-associated molecular patterns and activate APCs to produce cytokines. TLR ligands such as Pam_3CysSK_4 and unmethylated CpG oligodeoxynucleotide (CpG ODN) have been extensively used in fully synthetic vaccines. Figure 12.6 shows the widely applicable immunostimulants.

12.4 Fully Synthetic Vaccines Based on Mucin Glycopeptides

In recent years, numerous strategies have been developed using chemical synthesis to improve the immunogenicity of mucin glycopeptides.

12.4.1 Carrier Protein Vaccines

12.4.1.1 Carrier Protein Vaccines Based on Bovine Serum Albumin

Bovine serum albumin (BSA) is a serum albumin protein from cows and consisting of 583 amino acids. There are nearly 60 free amino groups in BSA, which is of benefit for conjugation. BSA is a model carrier protein for immunological studies because of its good solubility and conjugation reactivity. Kunz and coworkers firstly conjugated MUC1 glycopeptide with BSA by

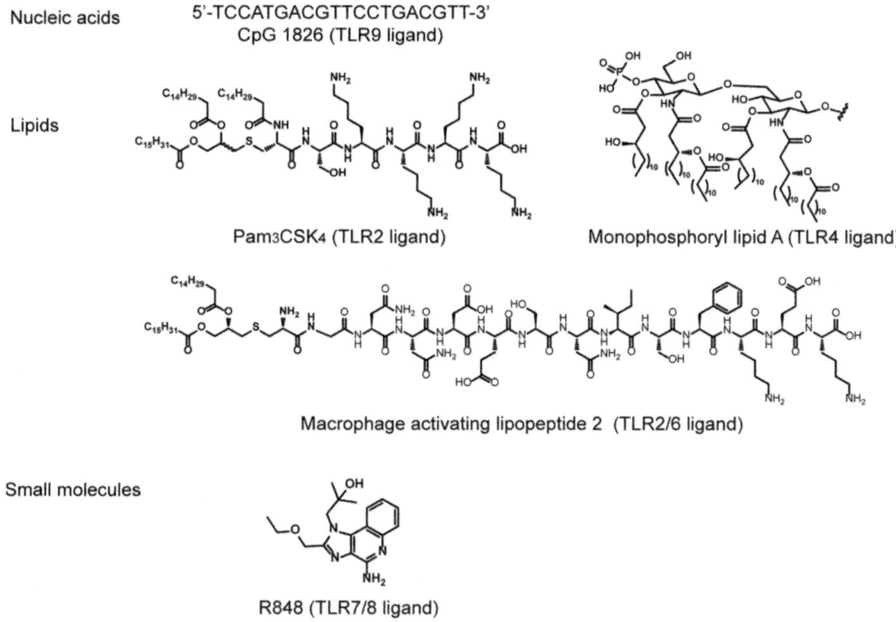

Figure 12.6 Several kinds of immunostimulants.

thio–ene chemistry.[66] They synthesized MUC1 glycopeptide antigen with a sulfydryl substituted polyethylene glycol linkage by fluorenylmethyloxycarbonyl (Fmoc) solid-phase peptide synthesis (SPPS). At the same time, the BSA was modified with an alkenyl group by *N*-(dimethylaminopropyl)-*N*′-ethylcarbodiimide hydrochloride-mediated amide coupling. Then, the thiol-modified MUC1 glycopeptide reacted with alkene-modified BSA under ultraviolet light. After characterization by matrix-assisted laser desorption ionization–time of flight (MALDI–TOF) mass spectroscopy, the results indicated that each BSA protein was conjugated with, on average, nine MUC1 glycopeptide molecules (see Figure 12.7).

However, the conjugation efficiency of thiol–ene chemistry is insufficient and more effective bioconjugation reactions should be explored. Kunz and co-workers conjugated BSA with MUC1 glycopeptide by squaric acid chemistry.[67] The MUC1 glycopeptides were extended with an amino-group-modified polyethylene glycol linker. The two-ester bond of the squaric acid diethyl ester could be nucleophilicly attacked by amino group at different basic pH values. So the squaric acid diethyl ester was first reacted with MUC1 glycopeptide at pH 8.0 and then coujugated with the amino group of lysine residues in the BSA at pH 9.0. The MALDI–TOF MS results indicated that each protein was covalently linked with, on average' nearly seven MUC1 glycopeptide molecules (see Figure 12.8).

After different glycosylated MUC1–BSA conjugates hade been obtained, mice were immunized with synthetic vaccines with complete Freund's adjuvants (CFA) and incomplete Freund's adjuvants (IFA) three times. One week

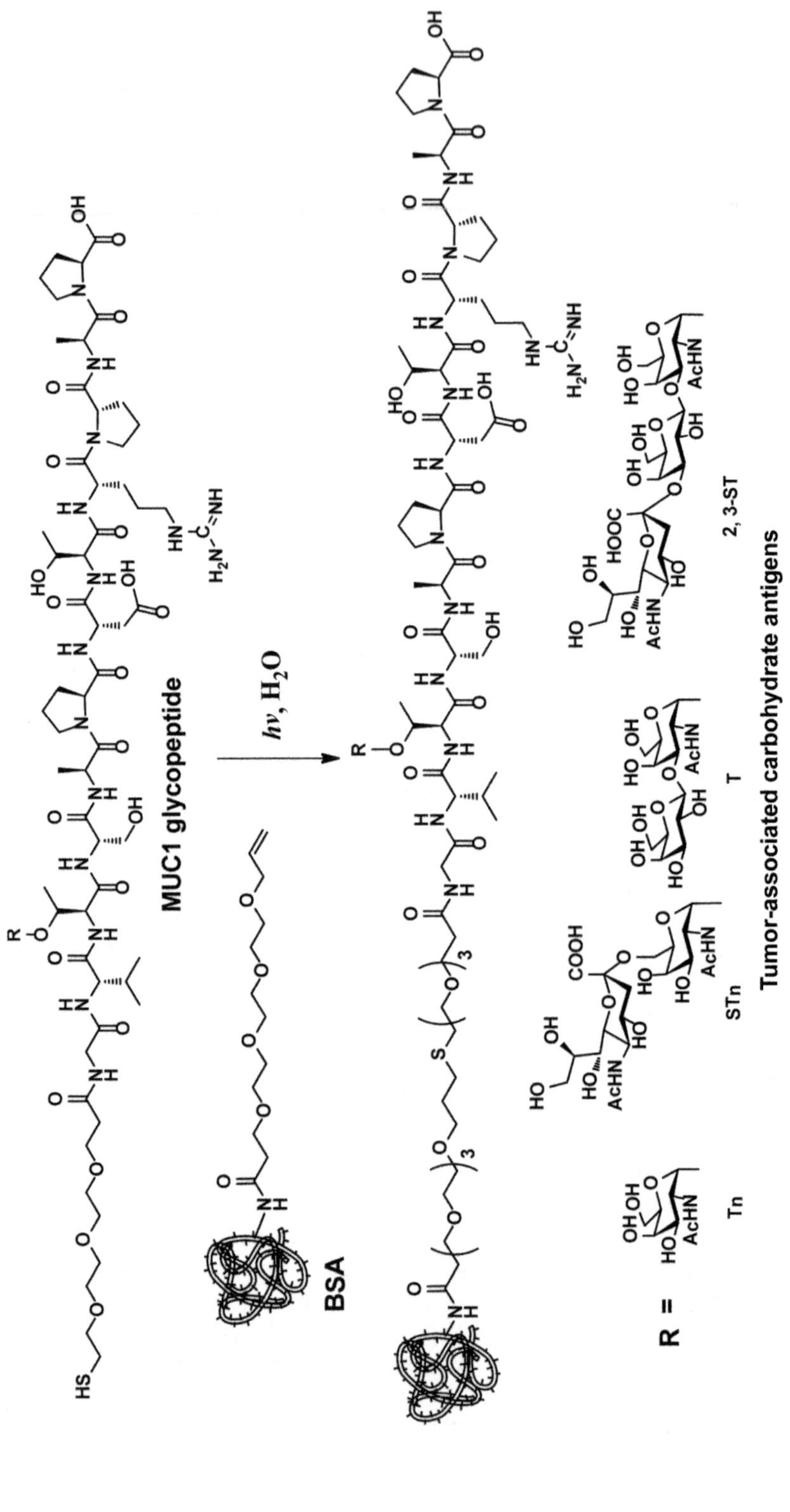

Figure 12.7 Synthesis of BSA–MUC1 glycopeptide conjugated vaccine by thiol–ene chemistry.[66]

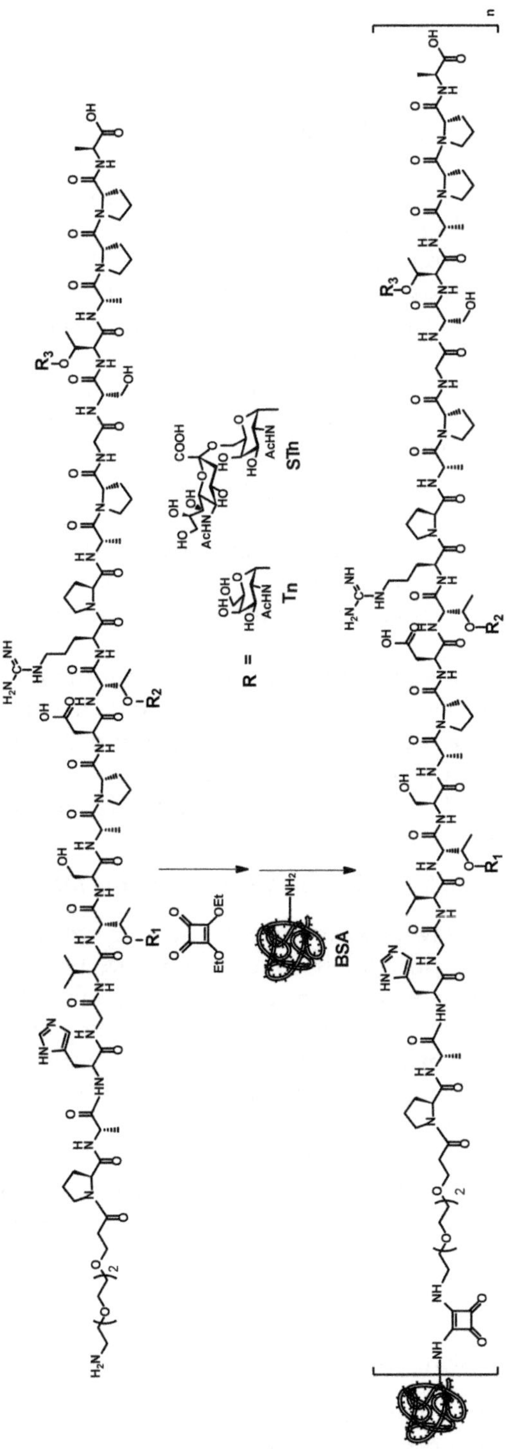

Figure 12.8 Synthesis of BSA–MUC1 glycopeptide antitumor vaccine by squaric acid ligation.[67]

after the third injection, the mice were killed and the sera were collected. The antibody titers were measured by enzyme-linked immunosorbent assay (ELISA) and the results indicated that the immunogenicity of MUC1 glycopeptides was dramatically improved after these glycopeptide antigens were conjugated with BSA. The vaccines effectively stimulated the mouse immune system and elicited strong antibodies against MUC1 glycopeptides.[68] To further characterize the immunological response, the antibody isotypes were measured by ELISA. The high titer of IgG1 antibody indicated that the immune response was antigen-specific. Moreover, the flow cytometry (fluorescence-activated cell sorting (FACS)) analysis demonstrated that the antibody could effectively recognize and bind with the MUC1-expressing tumor cells. Very recently, a tumor-associated MUC1 monoclonal antibody GGSK-1/30 was produced using MUC1–tetanus toxoid (TT) conjugate vaccine. This monoclonal antibody was able to specifically distinguish between human cancer epithelial cells and normal cells, which enables the diagnosis of epithelial cancers.[69]

Li and co-workers used similar strategy to conjugate different glycosylated MUC1 glycopeptides with BSA by squaric acid chemistry.[70] The results of the immunological studies revealed that the glycosylation pattern influences the immunogenicity of glycopeptide antigens. Meanwhile, glycosylation at Thr9 and sialylation could improve the immune response against chemical synthetic vaccines.[71]

MUC4 glycopeptide was identified as another ideal tumor-associated antigen for constructing cancer vaccines. Westerlind and co-workers synthesized the MUC4 glycopeptide antigen bearing with TACAs and then conjugated these antigens with BSA. Immunological studies revealed that these vaccines could elicit strong MUC4 specific immune responses and high titers of antibody.[72]

12.4.1.2 *Carrier Protein Vaccines Based on Other Proteins*

However, BSA has the drawbacks of toxicity and side effects as a carrier protein, so that there is still no synthetic vaccine approved by the FDA based on BSA protein. Kunz and co-workers used tetanus toxin (TT) for the construction of antitumor vaccines.[73,74] The results of immunological studies demonstrated that TT-based synthetic vaccine could also effectively stimulate the mouse immune system and elicited high titers of antibodies and significant amounts of IgG1 antibody.

Besides BSA and TT, there are some other kinds of carrier proteins, such as KLH, ovalbumin (OVA) and CRM197. These three carrier proteins have been approved by FDA for constructing synthetic vaccines, but have not been used in mucin-based antitumor vaccines. So they might be the optimal carrier proteins for clinical trials of mucin based antitumor vaccines.

12.4.2 Two-Component Vaccines

Nevertheless, carrier proteins are very immunogenic and much bigger than epitopes, so that carrier protein might suppress the immune response against the antigens and reduce the efficiency of vaccines. To tackle this problem, other strategies of constructing vaccines should be examined.

12.4.2.1 Two-Component Vaccines Based on T-Helper Cell Epitope

Kunz and co-workers conjugated a T-helper cell epitope from ovalbumin with the MUC1 glycopeptide to prepare two-component antitumor vaccines (see Figure 12.9).[75] The results of the immunological studies revealed that the two-component vaccines could also effectively stimulate the mouse immune system and elicit strong antigen-specific immune responses. Li and co-workers conjugated three different known universal T-helper cell epitope peptides (**P2**, **P4** and **P30**) derived from TT with the MUC1 glycopeptide antigens to form antitumor vaccines (see Figure 12.10).[76] By immunological evaluation, they identified that the **P30** is an ideal T-helper cell epitope, which could dramatically improve the glycopeptide immunogenicity and evoke a strong immune response in mice.

During the immunization process, the multivalent antigen could improve the efficiency of antigen uptake, processing and presentation. To further improve the immune response against MUC1 glycopeptide, Kunz and co-workers coupled the MUC1 glycopeptide and T-helper cell epitope peptide with a water-soluble polymer by copper-catalyzed alkyne–azide cycloaddition (CuAAC) as potential antitumor vaccines (see Figure 12.11).[77] Immunological evaluations demonstrated the improved antigen specific immune response in mice. More recently, they conjugated the MUC1 glycopeptide with three different T-helper cell epitopes by squaric acid ligation (see Figure 12.12).[78] The branched-structure vaccine elicited strong immune responses in mice and demonstrated the synergistic effect of three T-helper cell epitopes in activating immune cells.

12.4.2.2 Two-Component Vaccines Based on Immunostimulants

Besides T-helper cell epitopes, some known stimulators could activate the immune system *via* pattern recognition receptors (PRRs). These immune stimulators could be build-in adjuvants in chemical synthetic vaccines. Kunz and co-workers coupled the MUC1 glycopeptide with the agonist of Toll-like receptor 1 and 2 heterodimers (TLR1/2), Pam_3CysSK_4 lipopeptide, as fully synthetic antitumor vaccines (see Figure 12.13).[79] The immunological studies indicated that the Pam_3CysSK_4 lipopeptide could activate the innate immune system and evoked a strong immune response against MUC1 glycopeptide. To further prove this concept, Payne and co-workers synthesized perglycosylated MUC1 glycopeptide with five TACAs and then conjugated these glycopeptides with Pam_3CysSK_4 lipopeptide to form an antitumor vaccine.[80]

As previously mentioned, there is multivalent effect in immune recognition and processing. Li and co-workers modified the Pam_3CysSK_4 lipopeptide with four alkyne groups and extended MUC1 glycopeptides

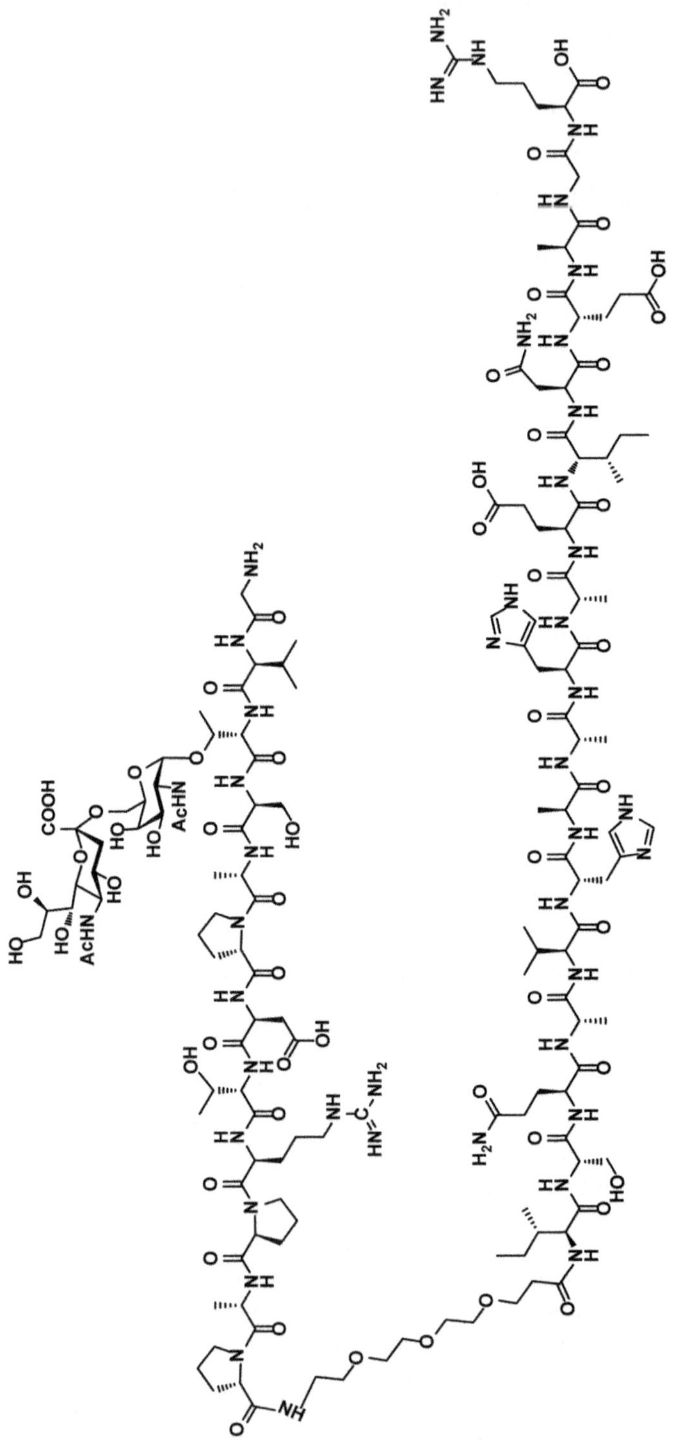

Figure 12.9 Chemical structure of two-component antitumor vaccine with T-helper cell epitope.[75]

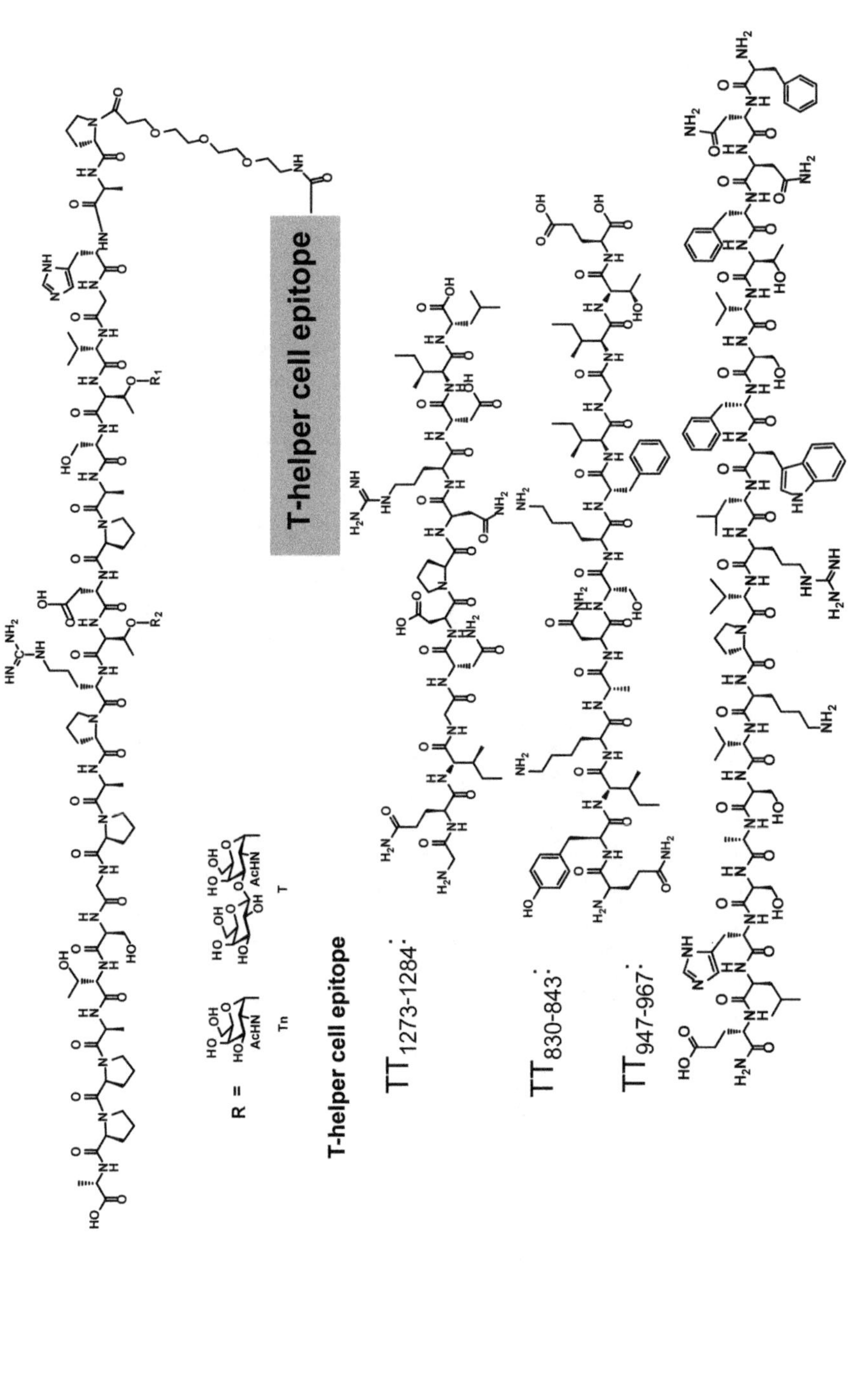

Figure 12.10 Chemical structure of two-component antitumor vaccines and T-helper cell epitopes.[76]

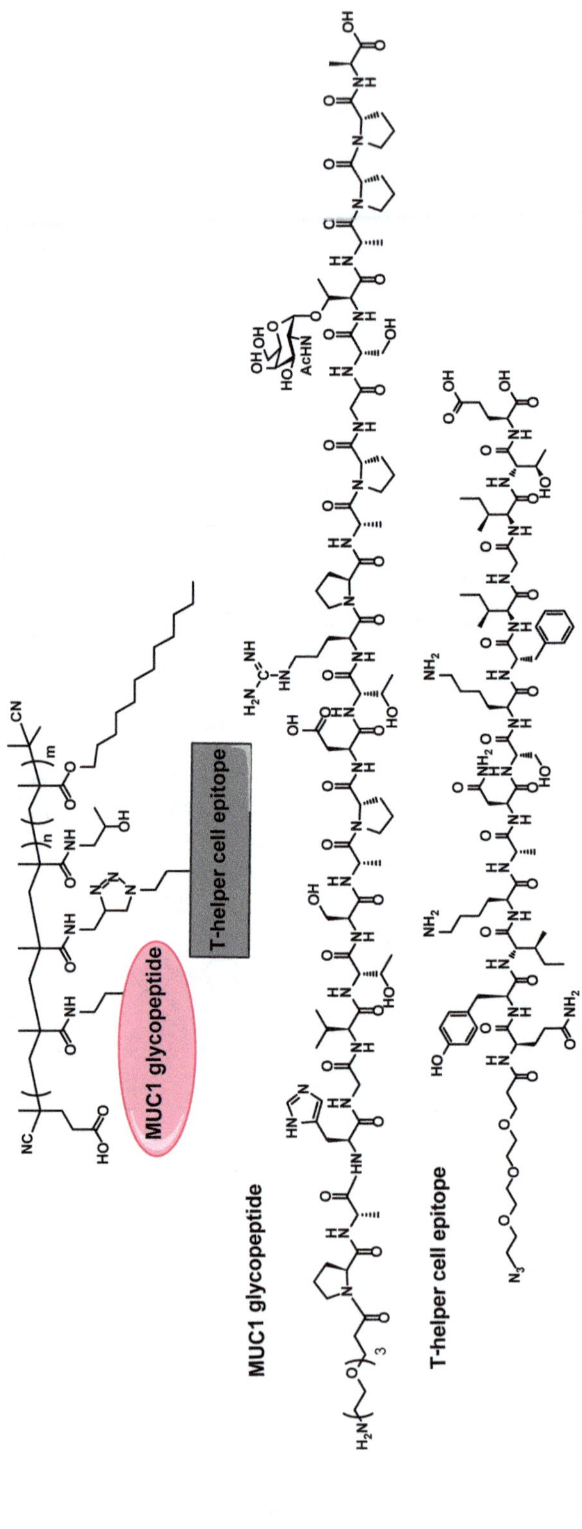

Figure 12.11 Chemical structure of multivalent two-component antitumor vaccines with polymers.[77]

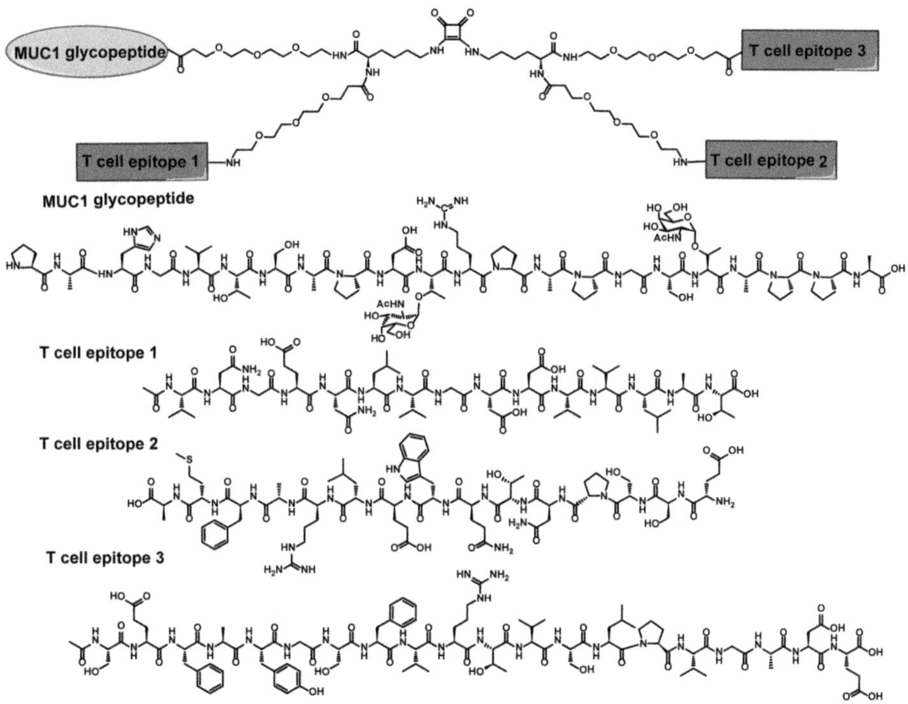

Figure 12.12 Chemical structure of two-component antitumor vaccines with three kinds of T-helper cell epitopes.[78]

with azide group-modified linkers. The multivalent antitumor vaccines were constructed by cycloaddition of the alkyne part and azido part with click chemistry (see Figure 12.14).[81] Immunological evaluations demonstrated the multivalent effect of immune system activation and the strong immune response evoked by multivalent synthetic antitumor vaccine. Moreover, IgG1 was the dominant antibody elicited by multivalent vaccine, which is associated with Th1-cell-mediated cellular immunity and beneficial for killing tumor cells.[82]

In addition to the TLR2 ligand Pam_3CysSK_4, Payne and co-workers applied macrophage activating lipopeptide 2 (MALP2, a TLR2/6 ligand) as an immunostimulant to increase the immunogenicity of MUC1 glycopeptides (see Figure 12.15).[83] They synthesized Tn- and T-perglycosylated three-component MUC1 glycopeptides using pentafluorophenyl esters as the *N*-acyl donor. After the immunization in mice, antisera were collected for immune evaluation. ELISA results indicated that the cross-reactivity of antibodies induced by Tn-perglycosylated three-component MUC1 glycopeptides was similar to those of unglycosylated and T-containing MUC1 glycopeptides. However, specific cytotoxic $CD4^+$ and $CD8^+$ T cell responses were not detected by cytokine staining *in vitro* and the cytotoxic T lymphocyte (CTL) assay *in vivo*.

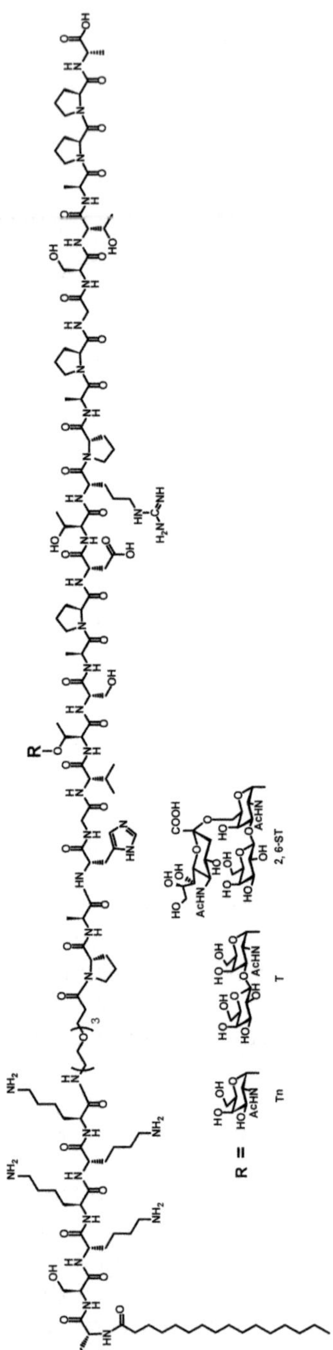

Figure 12.13 Chemical structure of two-component antitumor vaccines with Pam_3CysSK_4.[79]

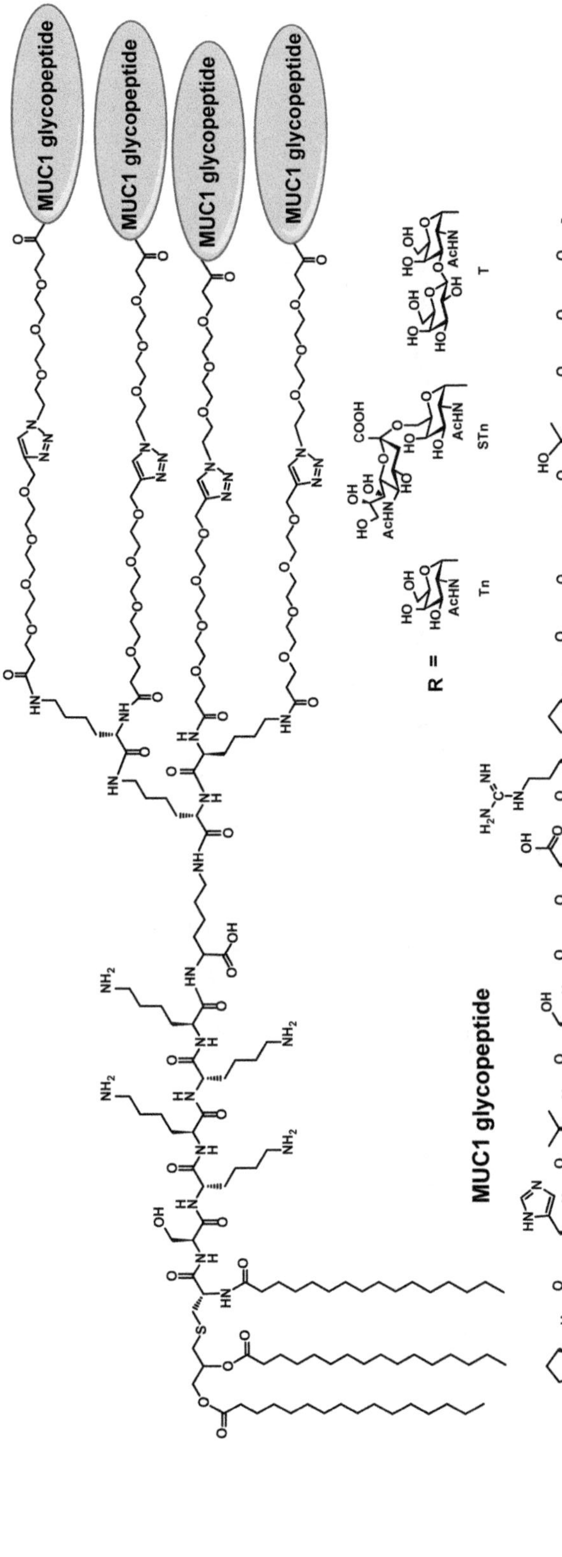

Figure 12.14 Chemical structure of multivalent two-component antitumor vaccines with Pam_3CysSK_4.[81]

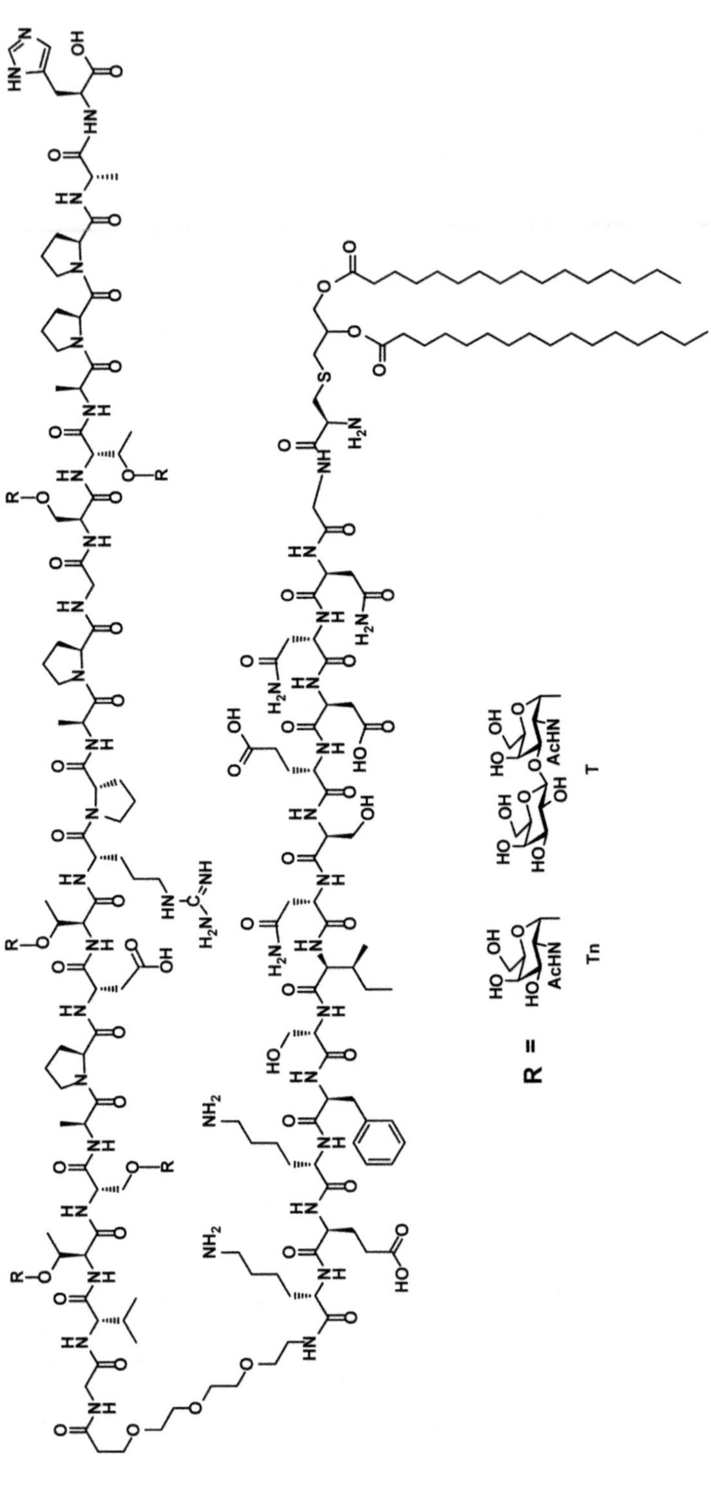

Figure 12.15 Chemical structure of multivalent vaccines with MALP2.[83]

12.4.2.3 Other Strategies for Two-Component Vaccines

Recently, Westerlind and co-workers extended the T-helper cell-epitope-based two-component MUC1 vaccines with mercapto modified polyethylene glycol (PEG) linker and then conjugated these antigens with gold nanoparticle (AuNP) to construct vaccine candidates. Immunological evaluation demonstrated that the AuNP-based antitumor vaccine could greatly improve the immunogenicity of MUC1 glycopeptide and induce strong antigen-specific immune responses.[84]

Similarly, Barchi and co-workers designed and synthesized AuNPs coated with tumor-associated MUC4 glycopeptide antigens and an adjuvant peptide derived from C3d protein as potential antitumor vaccines. The results of immunological studies indicated that this novel cancer vaccine could activate the immune system and produce high titers of both IgM and IgG antibodies in mice.[85]

12.4.3 Three-Component Vaccines

In order to further improve the effect of vaccines, the strategy of using three-component vaccines was developed. These three-component vaccines contain a B-cell epitope, a helper T-cell epitope and an immunostimulant. Helper T-cell epitopes and immunostimulants can simultaneously activate helper T-cells and APCs. B cells can differentiate and proliferate to secrete high-affinity IgG antibodies against B-cell epitopes.

12.4.3.1 Three-Component Vaccines Based on TLR2 Ligands

The lipopeptide Pam_3CysSK_4, a ligand of TLR2, plays a crucial role in the activation of the innate immune response.[86] Boons and co-workers developed a three-component vaccine composed by the lipopeptide Pam_3CysSK_4, the helper T-cell epitope derived from polio virus and MUC1 tumor-associated glycopeptide (see Figure 12.16a).[87] Native chemical ligation (NCL) was used

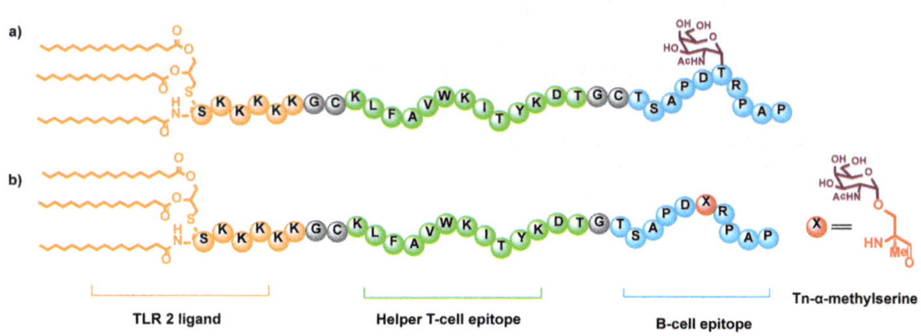

Figure 12.16 Three-component vaccines based on native chemical ligation. (a) A three-component vaccine containing two cysteine residues.[87] (b) A three-component vaccine containing one cysteine residue and a Tn-α-methylserine.[91]

for the conjugation of three components. Three components were respectively synthesized by SPPS. To improve the efficiency of NCL, they developed the liposome-mediated NCL. MUC1 glycopeptide containing a cysteine at the N-terminal, helper T-cell epitope peptide extended with the N-terminally protected cysteine and the C-terminal thioester and the lipopeptide extended with the C-terminal thioester were conjugated by sequential liposome-mediated NCL.

Next, the three-component glycopeptide vaccines containing Pam_3Cys and/or Pam_2Cys were incorporated into the phospholipid-based unilamellar vesicles.[88,89] The liposome vaccines were used to immunize mice intraperitoneally with or without the potent saponin immunoadjuvant QS-21 four times every two weeks. Anti-MUC1 antibody titers were measured by ELISA. The results indicated that covalent three-component vaccines induced exceptionally high titers of MUC1 glycopeptide antibodies. The isotype of IgG antibody indicated that the immune response was a Th2 response.

To further study the immune response to three-component MUC1 glycopeptide vaccines, a new three-component vaccine containing one cysteine was synthesized.[90] These vaccines were used to immunize MUC1.Tg mice that express human MUC1, three times at biweekly intervals. After the last immunization, the mice were challenged with MUC1 tumor cells. Tumor weight was measured and results indicated that the three-component vaccines reduced the tumor burden in MUC1.Tg mice. Results of ELISA indicated that the three-component vaccine could also induce high titers of antibodies which could induce robust antibody-dependent cell-mediated cytotoxicity (ADCC). In addition to the humoral immunity, the authors evaluated the cellular immunity. $CD8^+$ T-cells were isolated from lymph nodes. Then, DCs and MUC1 glycopeptides were added. After incubation, the cytotoxic T cell responses induced by three-component vaccines were much stronger than those induced by other vaccines. To study the immune activity of Pam_3CysSK_4, three-component vaccines were incubated with primary DCs. The results demonstrated that compounds containing Pam_3CysSK_4 could induce the secretion of cytokines such as TNF-α, IL-6 and IL-12.

Corzana and co-workers synthesized a three-component vaccine containing Tn-α-methylserine following the above methods (see Figure 12.16b).[91] They found the immune response elicited by the vaccine with unnatural Tn amino acid was not superior to that elicited by the natural threonine vaccine. The results demonstrated that MUC1 glycopeptide vaccines with unnatural amino acids must simulate the conformation of glycosidic bond between Tn and threonine.

Li and co-workers developed a new thioether-ligation strategy to synthesize self-adjuvanting three-component vaccines that consist of the lipopeptide Pam_3CysSK_4, the helper T-cell epitope derived from the tetanus toxoid and 20-amino acid MUC1 glycopeptides (see Figure 12.17).[92] MUC1 glycopeptides and helper T-cell epitope peptides were synthesized by SPPS. These peptides were extended with a triethylene glycol spacer containing a thiol group at the N-terminal. The lipopeptide was also synthesized by SPPS and

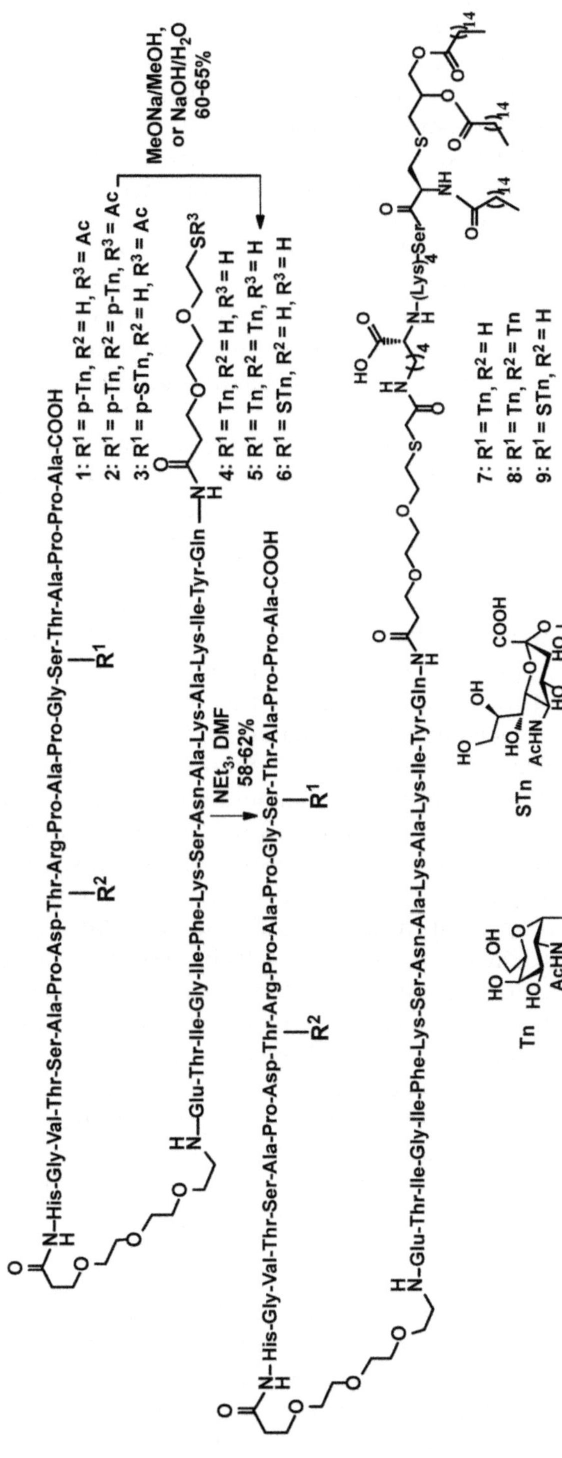

Figure 12.17 The synthesis of a three-component vaccine based on the thioether-ligation strategy.[92]

a bromoacetyl group was linked to a lysine at the C-terminal. Then, the lipopeptide was conjugated to MUC1 glycopeptides with triethylamine (Et_3N). After purification by reverse phase-high performance liquid chromatography (RP-HPLC), the thioether-conjugated three-component vaccines were used to immunize mice, with or without an external adjuvant, at intervals of 14 days. Seven days after the fifth immunization, the sera were collected and used for the immune evaluation. ELISA results indicated that antibody titers induced by three-component vaccines without an external adjuvant were extremely high. Antibody isotypes were analyzed by ELISA. Three-component vaccines induced predominantly high levels of IgG 1 and IgG 3. Flow cytometry (FACS) showed the induced sera could effectively bind to human breast tumor cells (MCF-7). In addition, the antibodies induced by three-component vaccines could initiate the obvious killing of tumor cells through complement-dependent cytotoxicity (CDC).

Though three-component vaccines could elicit a robust immune response against MUC1 glycopeptides, the synthesis and purification of these vaccines were very complex. Li and co-workers developed a novel noncovalent strategy to construct chemically synthesized three-component vaccine (see Figure 12.18).[93] MUC1 glycopeptides and helper T-cell epitope peptide from tetanus toxoid were synthesized following SPPS and naphthalene was decorated at the N-terminal. Lipopeptide Pam_3CysSK_4 was functionalized with methyl viologen (MV^{2+}). After purification, these two components were added to cucurbit[8]uril (CB[8]) in phosphate-buffered saline buffer. The non-covalent three-component vaccines were prepared through host–guest interactions. Transmission electron microscopy (TEM) and dynamic light scattering (DLS) assays indicated that these vaccines could self-assemble into 240 nm nanoparticles. After the immunization in mice, ELISA results indicated that non-covalent three-component vaccines could elicit a robust immune response without an external adjuvant. Antibodies produced in response to non-covalent three-component vaccines could bind to the MCF-7 tumor cells and efficiently mediate the killing of tumor cells by activation of CDC.

To improve the efficiency of synthesis of three-component vaccines, Payne and co-workers developed a strategy for the construction of three-component vaccines by using pentafluorophenyl esters as the *N*-acyl donor.[94,95] They separately synthesized perglycosylated MUC1 glycopeptides, helper T-cell epitope peptide from tetanus toxoid and lipopeptide Pam_3CysSK_4. Helper T-cell epitope peptide and Pam_3CysSK_4 were employed with pentafluorophenyl ester and these three components could be efficiently conjugated with hydroxybenzotriazole (HOBt) and *N*,*N*-diisopropylethylamine (DIPEA). After purification, these three-component vaccines could self-assemble into 12–20 nm micelles or particles. Then, the immunogenicity of three-component vaccines in mice was evaluated. The results indicated that three-component vaccines could trigger robust antibody responses without an external adjuvant. The analysis of antibody isotype indicated that the immune response was a type-2 helper T-cell (Th2) phenotype. The antisera induced by three-component vaccines could bind to MUC1 glycopeptide epitopes on MCF-7 cells.

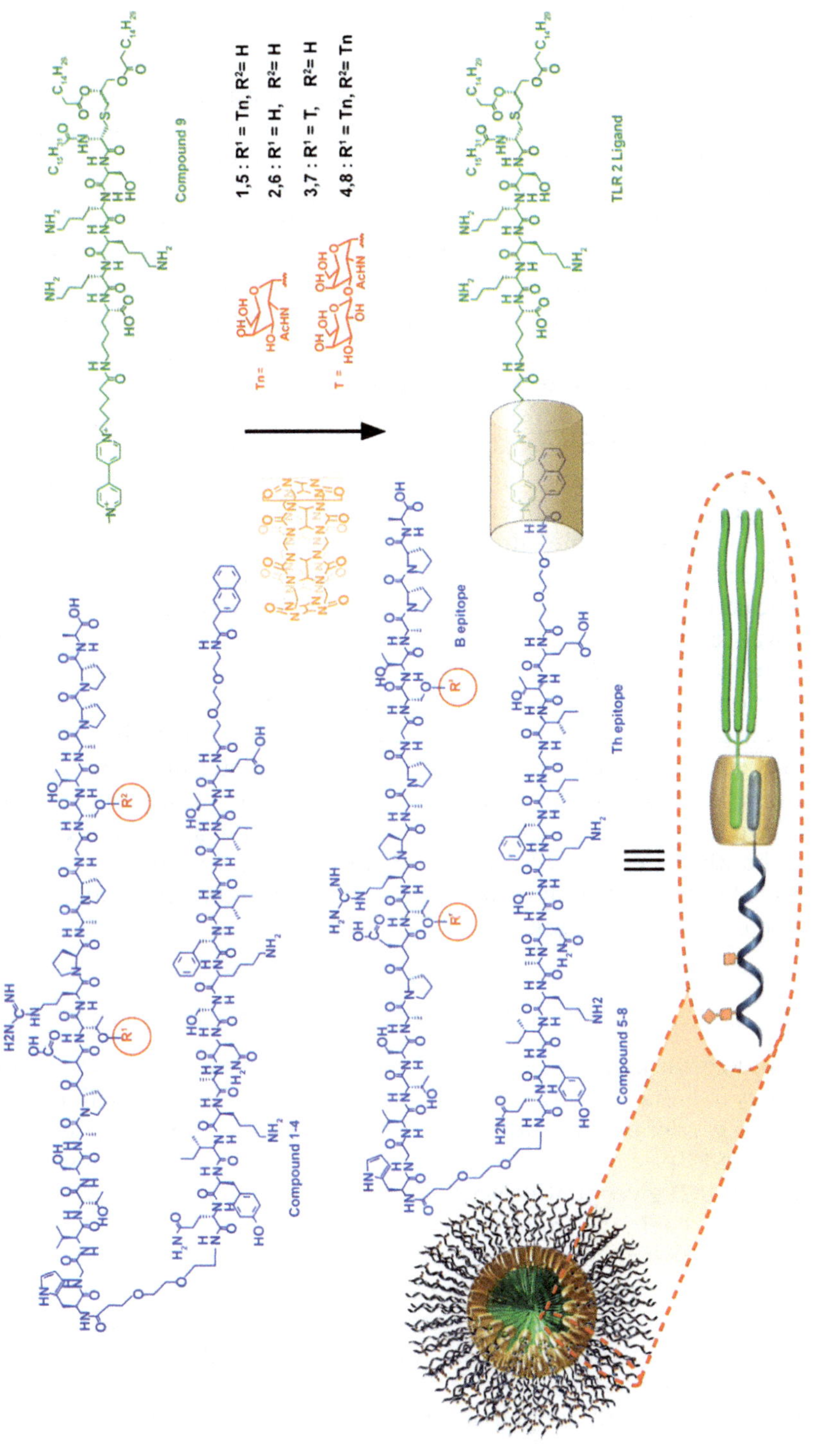

Figure 12.18 A three-component vaccine based on the non-covalent strategy.[93]

12.4.3.2 Three-Component Vaccines Based on TLR9 Ligands

CpG, which can trigger a robust immunostimulatory response, can be recognized by TLR9 expressed on APCs.[96] CpG can stimulate APCs to secrete various cytokines such as IL-6, IL-12 and TNF-α. CpG can be applied in fully synthetic vaccines to increase the immunogenicity of antigens.

Boons and co-workers developed a three-component vaccine composed of CpG, a helper T-cell epitope peptide and the MUC1 glycopeptide (see Figure 12.19).[97] CpG was conjugated using the thioether-ligation strategy. However, the results of the immune evaluation indicated that the CpG three-component vaccine did not improve the anti-tumor immunity. The results of the studies indicated that the inbuilt immunostimulant could significantly affect immune responses elicited by MUC1 glycopeptides.

Zentel and co-workers developed a three-component vaccine based on cationic nanohydrogel particles containing CpG, a helper T-cell epitope derived from tetanus toxin and the MUC1 glycopeptide.[98] The cationic nanohydrogel particles decorated with the alkyne groups and the spermine were prepared by reversible addition–fragmentation chain transfer (RAFT) polymerization. The MUC1 glycopeptide and the helper T-cell epitope peptide modified by an azide were synthesized by SPPS. After purification, the peptide was conjugated with nanohydrogel particles by click chemistry using ascorbic acid and copper acetate. Prior to immune evaluation, the three-component nanohydrogel vaccine was incubated with bone-marrow-derived dendritic cells (BMDCs). The results indicated that this vaccine could effectively activate BMDCs and trigger T-cell proliferation. For the immune evaluation, the three-component nanohydrogel vaccine was injected into mice every 21 days without Freund's adjuvants. ELISA results revealed that the three-component nanohydrogel vaccine could induce high titers of antibody. FACS analysis results demonstrated that antibodies could bind sufficiently to tumor cells.

12.4.3.3 Other Strategies for Three-Component Vaccines

Sucheck and co-workers developed a new approach based on an L-rhamnose (Rha) moiety to enhance the immunogenicity of vaccines. They used the liposomes as the carrier of Rha and the MUC1 glycopeptide containing lipopeptide Pam_3CysSK_4 (see Figure 12.20).[99,100] Rha was chemically conjugated to cholesterol. MUC1 glycopeptide was linked to the lipopeptide Pam_3CysSK_4 by click chemistry. Then, the three-component vaccine based on liposomes was prepared by the extrusion method with 1,2-dipalmitoyl-*sn*-glycero-3-phosphocholine (DPPC), MUC1 glycolipopeptide and cholesterol–Rha. Following the above immune methods, the liposome-based vaccine induced higher antibody titers compared with those induced in mice that did not receive immunization with RhaFicoll.

Li and co-workers developed a novel strategy for the synthesis of a three-component glycopeptide vaccine based on the interaction of negative

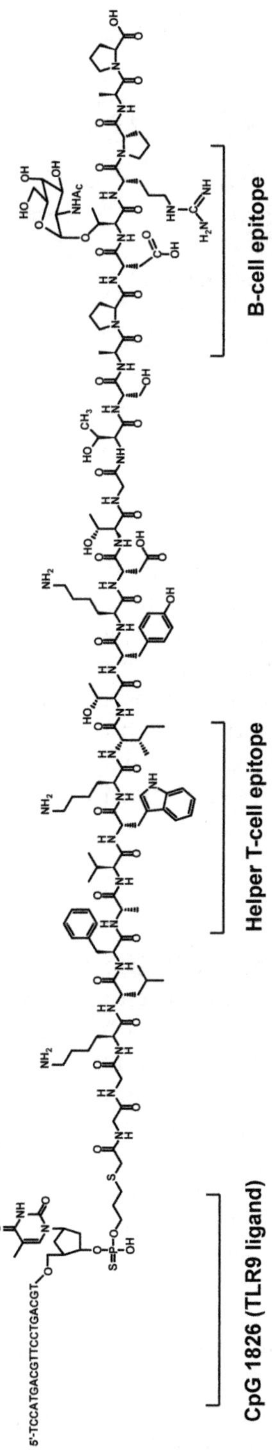

Figure 12.19 A three-component vaccine composed of CpG, a helper T-cell epitope peptide and the MUC1 glycopeptide.[97]

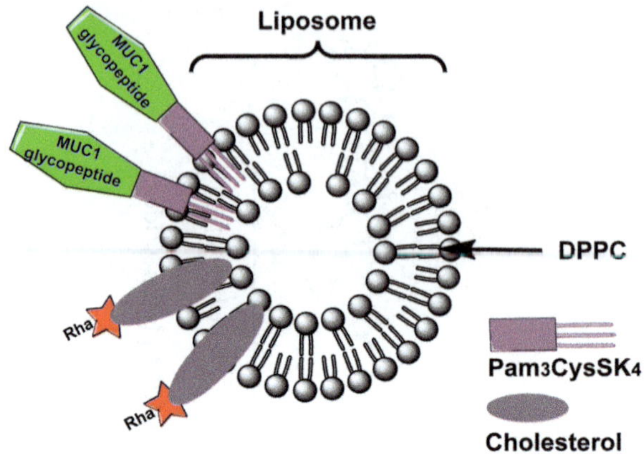

Figure 12.20 A three-component vaccine based on the L-rhamnose liposome strategy.[99,100]

and positive charges.[101] Seven lysine residues were separately extended to the N-terminal of MUC1 glycopeptide and helper T-cell epitope peptide from tetanus toxoid. γ-polyglutamic acid (γ-PGA) was chosen as the immunostimulant to provide the negative charge. These three components were mixed and the three-component vaccine was prepared. This vaccine could effectively activate APCs to produce cytokines such as IL-6 and IL-12 *in vitro*. For the immune evaluation, the vaccine was injected into mice. Results indicated that this vaccine could induce a robust immune response against MUC1 glycopeptide.

12.5 Conclusions

Cancer immunotherapy has gradually become a research hotspot in recent years. Fully synthetic glycopeptide vaccines are very promising and a large number of MUC1 glycopeptide vaccines have been developed. Carrier proteins have been used to improve the immunogenicity of MUC1 glycopeptides. Two-component and three-component MUC1 glycopeptide vaccines containing different helper T-cell epitopes and/or immunostimulants have also been developed. Though these glycopeptide vaccines could induce strong immune responses in mice, no vaccines have appeared on the market. The efficiency and safety of glycopeptide vaccines still need to be studied in clinical trials. With the development of medical immunology, efficient immunostimulants and strategies for vaccine design will be developed. Efficient immunostimulants can stimulate APCs to produce cytokines and improve the immune response. Strategies for vaccine design such as cluster effect and size effect can improve the stability of vaccines *in vivo* and permit them to be effectively recognized by APCs. Through further research, we can expect that fully synthetic glycopeptide vaccines could be a new option for cancer treatment.

Acknowledgements

The authors thank the National Natural Science Foundation of China (21332006) and the Sino-German Center for Research (GZ561) for financial support.

References

1. http://www.cancer.gov/about-cancer/treatment/types, last accessed April 2007.
2. I. Mellman, G. Coukos and G. Dranoff, *Nature*, 2011, **480**, 480–489.
3. G. P. Adams and L. M. Weiner, *Nat. Biotechnol.*, 2005, **23**, 1147–1157.
4. S. A. Rosenberg, N. P. Restifo, J. C. Yang, R. A. Morgan and M. E. Dudley, *Nat. Rev. Cancer*, 2008, **8**, 299–308.
5. D. M. Pardoll, *Nat. Rev. Cancer*, 2012, **12**, 252–264.
6. O. J. Finn, *Nat. Rev. Immunol.*, 2003, **3**, 630–641.
7. S. P. Gaddy and O. J. Finn, *Crit. Rev. Oncol. Hematol.*, 2008, **67**, 93–102.
8. S. A. Rosenberg, J. C. Yang and N. P. Restifo, *Nat. Med.*, 2004, **10**, 909–915.
9. T. Seremet, F. Brasseur and P. G. Coulie, *Cancer J.*, 2011, **17**, 325–330.
10. J. F. Forstner, *Digestion*, 1978, **17**, 234–263.
11. I. Carlstedt, J. K. Sheehan, A. P. Corfield and J. T. Gallagher, *Essays Biochem.*, 1985, **20**, 40–76.
12. B. E. Gustafsson, *Scand. J. Gastroenterol., Suppl.*, 1982, **77**, 117–131.
13. S. K. Linden, P. Sutton, N. G. Karlsson, V. Korolik and M. A. McGuckin, *Mucosal Immunol.*, 2008, **1**, 183–197.
14. R. Lev, *Lab. Invest.*, 1965, **14**, 2080.
15. P. B. Hukill and R. A. Vidone, *Lab. Invest.*, 1965, **14**, 1624.
16. A. P. Spicer, G. Parry, S. Patton and S. J. Gendler, *J. Biol. Chem.*, 1991, **266**, 15099–15109.
17. J. R. Gum, *Am. J. Respir. Cell Mol. Biol.*, 1992, **7**, 557–564.
18. T. Becker, S. Dziadek, S. Wittrock and H. Kunz, *Curr. Cancer Drug Targets*, 2006, **6**, 491–517.
19. Y. Li and P. J. Cozzi, *Curr. Cancer Drug Targets*, 2007, **7**, 259–271.
20. S. J. Gendler, J. M. Burchell, T. Duhig, D. Lamport, R. White, M. Parker and J. Taylorpapadimitriou, *Proc. Natl. Acad. Sci. U. S. A.*, 1987, **84**, 6060–6064.
21. S. J. Gendler, C. A. Lancaster, J. Taylorpapadimitriou, T. Duhig, N. Peat, J. Burchell, L. Pemberton, E. Lalani and D. Wilson, *J. Biol. Chem.*, 1990, **265**, 15286–15293.
22. M. S. Lan, S. K. Batra, W. N. Qi, R. S. Metzgar and M. A. Hollingsworth, *J. Biol. Chem.*, 1990, **265**, 15294–15299.
23. K. M. McDermott, P. R. Crocker, A. Harris, M. D. Burdick, Y. Hinoda, T. Hayashi, K. Imai and M. A. Hollingsworth, *Int. J. Cancer*, 2001, **94**, 783–791.
24. R. J. Quin and M. A. McGuckin, *Int. J. Cancer*, 2000, **87**, 499–506.
25. K. Zhang, D. Baeckstrom, H. Brevinge and G. C. Hansson, *J. Cell. Biochem.*, 1996, **60**, 538–549.

26. D. Nath, A. Hartnell, L. Happerfield, D. W. Miles, J. Burchell, J. Taylor-Papadimitriou and P. R. Crocker, *Immunology*, 1999, **98**, 213–219.
27. J. Taylor-Papadimitriou, J. Burchell, D. W. Miles and M. Dalziel, *Biochim. Biophys. Acta, Mol. Basis Dis.*, 1999, **1455**, 301–313.
28. Y. J. Kim, L. Borsig, H. L. Han, N. M. Varki and A. Varki, *Am. J. Pathol.*, 1999, **155**, 461–472.
29. P. A. Cohen, L. M. Peng, J. Kjaergaard, G. E. Plautz, J. H. Finke, G. K. Koski, B. J. Czerniecki and S. Y. Shu, *Crit. Rev. Immunol.*, 2001, **21**, 215–248.
30. V. Apostolopoulos and I. F. C. McKenzie, *Crit. Rev. Immunol.*, 1994, **14**, 293–309.
31. G. F. Springer, *Science*, 1984, **224**, 1198–1206.
32. I. Brockhausen, J. M. Yang, J. Burchell, C. Whitehouse and J. Taylorpapadimitriou, *Eur. J. Biochem.*, 1995, **233**, 607–617.
33. C. Whitehouse, J. Burchell, S. Gschmeissner, I. Brockhausen, K. O. Lloyd and J. TaylorPapadimitriou, *J. Cell Biol.*, 1997, **137**, 1229–1241.
34. S. Mueller, S. Goletz, N. Packer, A. Gooley, A. M. Lawson and F. G. Hanisch, *J. Biol. Chem.*, 1997, **272**, 24780–24793.
35. A. Girling, J. Bartkova, J. Burchell, S. Gendler, C. Gillett and J. Taylorpapadimitriou, *Int. J. Cancer*, 1989, **43**, 1072–1076.
36. S. H. Itzkowitz, M. Yuan, C. K. Montgomery, T. Kjeldsen, H. K. Takahashi, W. L. Bigbee and Y. S. Kim, *Cancer Res.*, 1989, **49**, 197–204.
37. S. Ogata, R. Koganty, M. Reddish, B. M. Longenecker, A. L. Chen, C. Perez and S. H. Itzkowitz, *Glycoconjugate J.*, 1998, **15**, 29–35.
38. J. Hilkens, V. Kroezen, J. M. G. Bonfrer, M. Dejongbakker and P. F. Bruning, *Cancer Res.*, 1986, **46**, 2582–2587.
39. S. vonMensdorffPouilly, M. M. Gourevitch, P. Kenemans, A. A. Verstraeten, S. V. Litvinov, G. J. vanKamp, S. Meijer, J. Vermorken and J. Hilgers, *Eur. J. Cancer*, 1996, **32A**, 1325–1331.
40. Y. Hinoda, N. Nakagawa, H. Nakamura, Y. Makiguchi, F. Itoh, M. Adachi, T. Yabana, K. Imai and A. Yachi, *Immunol. Lett.*, 1993, **35**, 163–168.
41. A. Rughetti, V. Turchi, C. A. Ghetti, G. Scambia, P. B. Panici, G. Roncucci, S. Mancuso, L. Frati and M. Nuti, *Cancer Res.*, 1993, **53**, 2457–2459.
42. S. von Mensdorff-Pouilly, A. A. Verstraeten, P. Kenemans, F. G. M. Snijdewint, A. Kok, G. J. Van Kamp, M. A. Paul, P. J. Van Diest, S. Meijer and J. Hilgers, *J. Clin. Oncol.*, 2000, **18**, 574–583.
43. K. R. Jerome, D. L. Barnd, K. M. Bendt, C. M. Boyer, J. Taylorpapadimitriou, I. F. C. McKenzie, R. C. Bast and O. J. Finn, *Cancer Res.*, 1991, **51**, 2908–2916.
44. A. L. DeFranco, *Immunol. Rev.*, 2000, **176**, 5–9.
45. J. Stavnezer, *Adv. Immunol.*, 1996, **61**, 79–146.
46. T. Honjo, K. Kinoshita and M. Muramatsu, *Annu. Rev. Immunol.*, 2002, **20**, 165–196.
47. B. Beutler, K. Hoebe, X. Du and R. J. Ulevitch, *J. Leukocyte Biol.*, 2003, **74**, 479–485.
48. E. Lien and R. R. Ingalls, *Crit. Care Med.*, 2002, **30**, S1–S11.
49. W. Check, *Am. Soc. Microbiol. News*, 2004, **70**, 317–322.

50. I. Brockhausen, J. M. Yang, J. Burchell, C. Whitehouse and J. Taylor-Papadimitriou, *Eur. J. Biochem.*, 1995, **233**, 607–617.
51. C. Whitehouse, J. Burchell, S. Gschmeissner, I. Brockhausen, K. O. Lloyd and J. Taylor-Papadimitriou, *J. Cell Biol.*, 1997, **137**, 1229–1241.
52. S. J. Danishefsky, Y. K. Shue, M. N. Chang and C. H. Wong, *Acc. Chem. Res.*, 2015, **48**, 643–652.
53. R. Ramlau, E. Quoix, J. Rolski, M. Pless, H. Lena, E. Levy, M. Krzakowski, D. Hess, E. Tartour, M. P. Chenard, J. M. Limacher, N. Bizouarne, B. Acres, C. Halluard and T. Velu, *J. Thorac. Oncol.*, 2008, **3**, 735–744.
54. C. Rochlitz, R. Figlin, P. Squiban, M. Salzberg, M. Pless, R. Herrmann, E. Tartour, Y. X. Zhao, N. Bizouarne, M. Baudin and B. Acres, *J. Gene Med.*, 2003, **5**, 690–699.
55. R. K. Ramanathan, K. M. Lee, J. McKolanis, E. Hitbold, W. Schraut, A. J. Moser, E. Warnick, T. Whiteside, J. Osborne, H. Kim, R. Day, M. Troetschel and O. J. Finn, *Cancer Immunol. Immunother.*, 2005, **54**, 254–264.
56. T. Gilewski, S. Adluri, G. Ragupathi, S. L. Zhang, T. J. Yao, K. Panageas, M. Moynahan, A. Houghton, L. Norton and P. O. Livingston, *Clin. Cancer Res.*, 2000, **6**, 1693–1701.
57. C. Butts, N. Murray, A. Maksymiuk, G. Goss, E. Marshall, D. Soulieres, Y. Cormier, P. Ellis, A. Price, R. Sawhney, M. Davis, J. Mansi, C. Smith, D. Vergidis, P. Ellis, M. MacNeil and M. Palmer, *J. Clin. Oncol.*, 2005, **23**, 6674–6681.
58. P. J. Sabbatini, G. Ragupathi, C. Hood, C. A. Aghajanian, M. Juertzka, A. Lasonos, M. L. Hensley, M. K. Spassova, O. Ouerfelli, D. R. Spriggs, W. P. Tew, J. Konner, H. Clausen, N. Abu Rustum, S. J. Dansihefsky and P. O. Livingston, *Clin. Cancer Res.*, 2007, **13**, 4170–4177.
59. C. K. Tang and V. Apostolopoulos, *Expert Rev. Vaccines*, 2008, **7**, 951–962.
60. B. Hu, J. Wang, Y. Guo, T. Chen, W. Ni, H. Yuan, N. Zhang, F. Xie and G. Tai, *Int. Immunopharmacol.*, 2016, **33**, 108–118.
61. D. W. Kufe, *Nat. Rev. Cancer*, 2009, **9**, 874–885.
62. D. M. Granoff, M. H. Rathore, S. J. Holmes, P. D. Granoff and A. H. Lucas, *Vaccine*, 1993, **11**, S46–S51.
63. M. Burrage, A. Robinson, R. Borrow, N. Andrews, J. Southern, J. Findlow, S. Martin, C. Thornton, D. Goldblatt, M. Corbel, D. Sesardic, K. Cartwight, P. Richmond and E. Miller, *Infect. Immun.*, 2002, **70**, 4946–4954.
64. J. C. Butler, R. F. Breiman, J. F. Campbell, H. B. Lipman, C. V. Broome and R. R. Facklam, *JAMA, J. Am. Med. Assoc.*, 1993, **270**, 1826–1831.
65. J. Alexander, J. Sidney, S. Southwood, J. Ruppert, C. Oseroff, A. Maewal, K. Snoke, H. M. Serra, R. T. Kubo, A. Sette and H. M. Grey, *Immunity*, 1994, **1**, 751–761.
66. S. Wittrock, T. Becker and H. Kunz, *Angew. Chem., Int. Ed.*, 2007, **46**, 5226–5230.
67. S. Dziadek, D. Kowalczyk and H. Kunz, *Angew. Chem., Int. Ed.*, 2005, **44**, 7624–7630.

68. U. Westerlind, A. Hobel, N. Gaidzik, E. Schmitt and H. Kunz, *Angew. Chem., Int. Ed.*, 2008, **47**, 7551–7556.
69. B. Palitzsch, N. Gaidzik, N. Stergiou, S. Stahn, S. Hartmann, B. Gerlitzki, N. Teusch, P. Flemming, E. Schmitt and H. Kunz, *Angew. Chem., Int. Ed.*, 2016, **55**, 2894–2898.
70. H. Cai, Z. H. Huang, L. Shi, P. Zou, Y. F. Zhao, H. Kunz and Y. M. Li, *Eur. J. Org. Chem.*, 2011, 3685–3689.
71. H. Cai, Z. H. Huang, L. Shi, Z. Y. Sun, Y. F. Zhao, H. Kunz and Y. M. Li, *Angew. Chem., Int. Ed.*, 2012, **51**, 1719–1723.
72. H. Cai, B. Palitzsch, S. Hartmann, N. Stergiou, H. Kunz, E. Schmitt and U. Westerlind, *ChemBioChem*, 2015, **16**, 959–967.
73. A. Hoffmann-Roder, A. Kaiser, S. Wagner, N. Gaidzik, D. Kowalczyk, U. Westerlind, B. Gerlitzki, E. Schmitt and H. Kunz, *Angew. Chem., Int. Ed.*, 2010, **49**, 8498–8503.
74. N. Gaidzik, A. Kaiser, D. Kowalczyk, U. Westerlind, B. Gerlitzki, H. P. Sinn, E. Schmitt and H. Kunz, *Angew. Chem., Int. Ed.*, 2011, **50**, 9977–9981.
75. S. Dziadek, A. Hobel, E. Schmitt and H. Kunz, *Angew. Chem., Int. Ed.*, 2005, **44**, 7630–7635.
76. H. Cai, M. S. Chen, Z. Y. Sun, Y. F. Zhao, H. Kunz and Y. M. Li, *Angew. Chem., Int. Ed.*, 2013, **52**, 6106–6110.
77. L. Nuhn, S. Hartmann, B. Palitzsch, B. Gerlitzki, E. Schmitt, R. Zentel and H. Kunz, *Angew. Chem., Int. Ed.*, 2013, **52**, 10652–10656.
78. B. Palitzsch, S. Hartmann, N. Stergiou, M. Glaffig, E. Schmitt and H. Kunz, *Angew. Chem., Int. Ed.*, 2014, **53**, 14245–14249.
79. A. Kaiser, N. Gaidzik, T. Becker, C. Menge, K. Groh, H. Cai, Y. M. Li, B. Gerlitzki, E. Schmitt and H. Kunz, *Angew. Chem., Int. Ed.*, 2010, **49**, 3688–3692.
80. B. L. Wilkinson, L. R. Malins, C. K. Y. Chun and R. J. Payne, *Chem. Commun.*, 2010, **46**, 6249–6251.
81. H. Cai, Z. H. Huang, L. Shi, Y. F. Zhao, H. Kunz and Y. M. Li, *Chem.–Eur. J.*, 2011, **17**, 6396–6406.
82. H. Cai, Z. Y. Sun, M. S. Chen, Y. F. Zhao, H. Kunz and Y. M. Li, *Angew. Chem., Int. Ed.*, 2014, **53**, 1699–1703.
83. D. M. McDonald, B. L. Wilkinson, L. Corcilius, M. Thaysen-Andersen, S. N. Byrneb and R. J. Payne, *Chem. Commun.*, 2014, **50**, 10273–10276.
84. H. Cai, F. Degliangeli, B. Palitzsch, B. Gerlitzki, H. Kunz, E. Schmitt, R. Fiammengo and U. Westerlind, *Bioorg. Med. Chem.*, 2016, **24**, 1132–1135.
85. R. P. Brinas, A. Sundgren, P. Sahoo, S. Morey, K. Rittenhouse-Olson, G. E. Wilding, W. Deng and J. J. Barchi, *Bioconjugate Chem.*, 2012, **23**, 1513–1523.
86. R. Spohn, U. Buwitt-Beckmann, R. Brock, G. Jung, A. J. Ulmer and K. H. Wiesmüller, *Vaccine*, 2004, **22**, 2494–2499.
87. S. Ingale, T. Buskas and G. J. Boons, *Org. Lett.*, 2006, **8**, 5785–5788.
88. S. Ingale, M. A. Wolfert, J. Gaekwad, T. Buskas and G. J. Boons, *Nat. Chem. Biol.*, 2007, **3**, 663–667.

89. S. Ingale, M. A. Wolfert, J. Gaekwad, T. Buskas and G. J. Boons, *ChemBioChem*, 2009, **10**, 455–463.
90. V. Lakshminarayanan, P. Thompson, M. A. Wolfert, T. Buskas, J. M. Bradley, L. B. Pathangey, C. S. Madsen, P. A. Cohen, S. J. Gendler and G. J. Boons, *Proc. Natl. Acad. Sci. U. S. A.*, 2012, **109**, 261–266.
91. N. Martínez-Sáez, N. T. Supekar, M. A. Wolfert, I. A. Bermejo, R. Hurtado-Guerrero, J. L. Asensio, J. Jiménez-Barbero, J. H. Busto, A. Avenoza, G. J. Boons, J. M. Peregrina and F. Corzana, *Chem. Sci.*, 2016, **7**, 2294–2301.
92. H. Cai, Z. Y. Sun, Z. H. Huang, L. Shi, Y. F. Zhao, H. Kunz and Y. M. Li, *Chem.–Eur. J.*, 2013, **19**, 1962–1970.
93. Y. Gao, Z. Y. Sun, Z. H. Huang, P. G. Chen, Y. X. Chen, Y. F. Zhao and Y. M. Li, *Chem.–Eur. J.*, 2014, **20**, 13541–13546.
94. B. L. Wilkinson, S. Day, L. R. Malins, V. Apostolopoulos and R. J. Payne, *Angew. Chem., Int. Ed.*, 2011, **50**, 1635–1639.
95. B. L. Wilkinson, S. Day, R. Chapman, S. Perrier, V. Apostolopoulos and R. J. Payne, *Chem.–Eur. J.*, 2012, **18**, 16540–16548.
96. A. Krieg, *Annu. Rev. Immunol.*, 2002, **20**, 709–760.
97. A. M. Abdel-Aal, V. Lakshminarayanan, P. Thompson, N. Supekar, J. M. Bradley, M. A. Wolfert, P. A. Cohen, S. J. Gendler and G. J. Boons, *ChemBioChem*, 2014, **15**, 1508–1513.
98. S. Hartmann, L. Nuhn, B. Palitzsch, M. Glaffig, N. Stergiou, B. Gerlitzki, E. Schmitt, H. Kunz and R. Zentel, *Adv. Healthcare Mater.*, 2015, **11**, 522–527.
99. S. Sarkar, A. C. D. Salyer, K. A. Wall and S. J. Sucheck, *Bioconjugate Chem.*, 2013, **24**, 363–375.
100. P. Karmakar, K. Lee, S. Sarkar, K. A. Wall and S. J. Sucheck, *Bioconjugate Chem.*, 2016, **27**, 110–120.
101. Y. F. Liu, Z. Y. Sun, P. G. Chen, Z. H. Huang, Y. Gao, L. Shi, Y. F. Zhao, Y. X. Chen and Y. M. Li, *Bioconjugate Chem.*, 2015, **26**, 1439–1442.

CHAPTER 13

Selective Chemical Glycosylation of Therapeutic Proteins

QUN ZHOU[a]

[a]Protein Engineering, Biologics Research, Sanofi, 49 New York Avenue, Framingham, Massachusetts 01701, USA
*E-mail: qun.zhou@Sanofi.com, qun.zhou@genzyme.com

13.1 Introduction

Glycosylation plays an important role in the structure and function of glycoproteins.[1] It maintains overall protein stability and is an important attribute for most protein therapeutics.[2] For example, antibodies lacking *N*-glycans at the N297 site often have reduced thermal stability, while those containing non-fucosylated glycans at the same site showed strong antibody-dependent cellular cytotoxicity.[3] With therapeutic proteins, glycosylation has an even more significant effect on their efficacy in treating patients in clinics. When it is not optimal, the protein often showed fast clearance or antigenicity. In contrast, optimal glycosylation could lead to increased serum half-life, with improved protein biodistribution to disease-affected tissue, thereby improving its pharmacokinetics and pharmacodynamics.[4–6] Due to the strong relevance of glycosylation to therapeutic efficacy, many efforts at rational manipulation of the glycosylation through glycoengineering have been made

Chemical Biology No. 2
Chemical Biology of Glycoproteins
Edited by Zhongping Tan and Lai-Xi Wang

Published by the Royal Society of Chemistry, www.rsc.org

aiming to prepare therapeutic glycoproteins with optimal glycosylation using either: (1) an *in vivo* recombinant approach through expression systems or (2) *in vitro* remodeling through enzymatic and chemical modification.[7] This chapter focuses on the work using *in vitro* chemical and enzymatic glycosylation related to generating therapeutic proteins with improved pharmacokinetics or pharmacodynamics.

13.2 Glycosylation for Increasing Pharmacokinetics and Stability

Small therapeutic proteins, especially hormones or other regulatory proteins with molecular weights below 60 kDa, often have short serum half-lives due to renal clearance, fast passage through the endothelial cell barrier of blood vessels and binding to specific receptors. To improve their pharmacokinetics, many approaches have been extensively investigated for remodeling of these proteins with hydrophilic polymers, including conjugation using polyethylene glycol (PEG), a well-established and clinically proven technology.[8,9] Among them, the PEGylation on glycans of therapeutic proteins provides site-specific conjugation instead of random conjugation to lysine or cysteine residues in the protein backbone.[9,10]

GlycoPEGylation is one of the PEGylation methods targeting protein glycans by introducing a PEGylated sialic acid *in vitro*. It was initially developed by Neose Technologies Inc. using a chemoenzymatic approach.[11] As reported in the literature, the modified CMP–sialic acid PEGylated at the 5′-amino position (CMP–SA–5-$NHCOCH_2NHPEG$) can be efficiently transferred with sialyltransferases to glycan acceptors, such as *N*- or *O*-glycans, in glycoproteins. DeFrees and colleagues reported the glycoPEGylation of recombinant therapeutic proteins generated in *Escherichia coli*.[11] The serine or threonine residues of nonglycosylated recombinant proteins were first *O*-glycosylated with a *N*-acetylgalactosamine (GalNAc) by a polypeptide GalNAc transferase (GalNAc-T). The GalNAc was then sialylated by enzymatic transfer of PEGylated sialic acid from CMP–SA–5-$NHCOCH_2NHPEG$ using a sialyltransferase (ST6GalNAc1) (Figure 13.1(A)). The *O*-linked GalNAc in recombinant proteins can also be galactosylated first using core 1 galactosyltransferase 1 (core1 GalT) before they are sialylated by transfer of PEGylated sialic acid from CMP–SA–5-$NHCOCH_2NHPEG$ using a sialyltransferase, ST3Gal1 (Figure 13.1(B)). The glycoPEGylation of *O*-glycans was demonstrated with therapeutic proteins used in clinical settings, including granulocyte colony stimulating factor (G-CSF), interferon-alpha2b (IFN-α2b), and granulocyte/macrophage colony stimulating factor (GM-CSF) with increased serum half-life.

GlycoPEGylation through *O*- and *N*-glycans has also been applied for improving the therapeutic index of several recombinant coagulation factors, including factors VIIa, VIII and IX for more effective treatment of hemophilia (Figure 13.2(A) and (B)). Haemophilia A and haemophilia B are

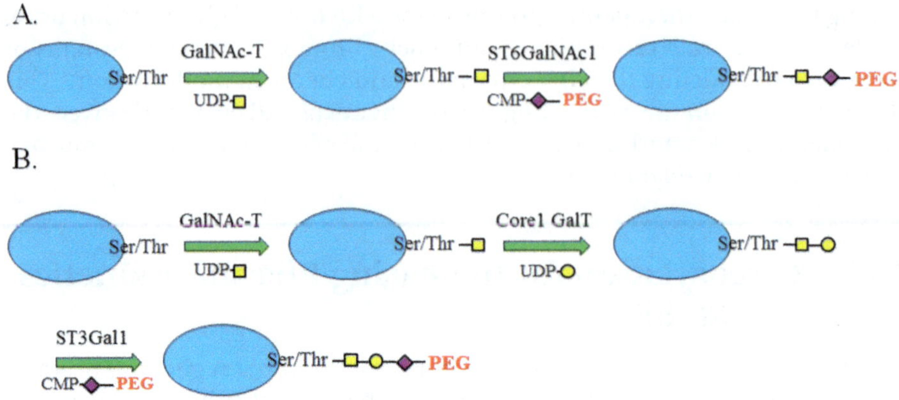

Figure 13.1 *GlycoPEGylation of recombinant proteins expressed in E. coli.* (A). Two-step approach by enzymatic *N*-acetylgalactosamine (GalNAc) addition to an engineered *O*-glycosylation site containing Serine (Ser) or threonine (Thr) with polypeptide GalNAc transferase (GalNAc-T) and then PEGylated sialic acid transfer to GalNAc using sialyltransferase, ST6GalNAc1 and CMP–sialic acid PEG; (B) Three-step approach by enzymatic GalNAc addition to an engineered *O*-glycosylation site containing Ser or Thr with GalNAc-T, galactose attached to GalNAc with a galactosyltransferase, core 1 GalT, and PEGylated sialic acid transfer to galactose using a sialyltansferase, ST3Gal1, and CMP-sialic acid PEG.

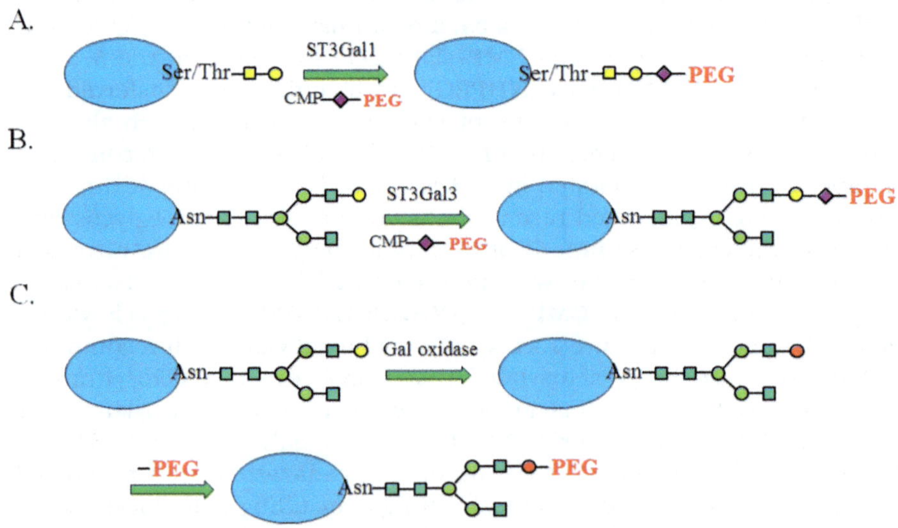

Figure 13.2 *Site-specific PEGylation of recombinant proteins expressed in mammalian cells.* (A) PEGylated sialic acid was transferred to galactose at the non-reducing terminus of *O*-linked glycan using a sialyltransferase, ST3Gal1, and CMP–sialic acid PEG; (B) PEGylated sialic acid was transferred to galactose at the non-reducing terminus of an *N*-linked glycan using a sialyltransferase, ST3Gal3, and CMP-sialic acid PEG; (C) Galactose at the non-reducing terminus was oxidized by galactose oxidase, generating galactose aldehyde which was then conjugated to aminooxy PEG.

genetic disorders characterized by deficiencies in FVIII and FIX, respectively, often resulting in recurrent bleeding with severe morbidity or mortality if untreated. Factor VIIa (FVIIa) is a trypsin-like serine protease which binds to exposed tissue factor (TF) upon vascular injury, activating FIX and FX with a burst of thrombin, fibrin deposition and formation of a haemostatic plug on the surface of activated platelets. Recombinant FVIIa has been used as a treatment for bleeding episodes in haemophilia patients with inhibitors. However, it needs to be administered daily for long-term prevention due to its rapid clearance from serum with a half-life of 2–4 hours. The numerous interactions of FVIIa with the cell surface, TF, FIX and FX made it impossible to perform nonspecific PEGylation of lysine residues in the protein without loss of activity. Thus, FVIIa was glycoPEGylated for selective PEGylation of the *N*-glycans from two *N*-glycosylation sites, N145 and N322, in its light chain and protease domain.[12] The recombinant FVIIa and its glycosylation mutants, in which either of two *N*-glycosylation sites was deleted, was desialylated to expose terminal galactose to which a PEGylated sialic acid was subsequently introduced by a sialyltransferase, ST3Gal3. The enzymatic reaction was kinetically controlled and the number of PEGs added depended on the reaction time. Finally, the unreacted galactose was blocked with sialic acid to avoid clearance through asialoglycoprotein receptor located in the liver. The glycoPEGylated variants were generated using PEG with different sizes or shapes, including, 2, 5, 10 and 20 kDa linear PEG and 40 kDa branched PEG. These glycoPEGylated variants were evaluated *in vitro* for their ability to interact with small synthetic substrates as well as key molecules relevant to function in the coagulation pathway. It was found that modification of FVIIa using glycoPEGylation had only limited effect on its substrate hydrolysis and FX activation. However, the glycoPEGylation altered the ability of FVIIa to interact with TF to some extent, depending on the PEG size, and more importantly, it reduced the rate of antithrombin III (ATIII) inhibition by up to 50%, which could allow for an extended serum half-life. The 10 and 40 kDa PEGylated variants were further investigated.[13–15] They retained the catalytic activity and interacted efficiently with FX and the plasma inhibitors. The molar hydrodynamic volume of the glycoPEGylated protein was larger than the volume of its composites. The measured intrinsic viscosity of rFVIIa was 6.0 ml g^{-1}, while the intrinsic viscosities of the 10 kDa PEG-rFVIIa and 40 kDa PEG-rFVIIa were 29.5 ml g^{-1} and 79.0 ml g^{-1}, respectively. Furthermore, the secondary and tertiary structures of rFVIIa measured by circular dichroism (CD) were maintained upon glycoPEGylation. However, the thermal and kinetic stabilities were affected, as measured by differential scanning calorimetry (DSC) and light scattering (LS).[14] Both the apparent unfolding temperature (T_m) and the temperature of aggregation (T_{agg}) were increased with glycoPEGylation, independent of the size or the shape of the PEG chain. Sen and colleagues found that glycoPEGylated rFVIIa was able to activate FX on TF-expressing cell surfaces at a rate similar to that of non-PEGylated rFVIIa, but its uptake in endothelial cells and fibroblasts was significantly reduced.[16] FVIIa glycoPEGylated with

5, 10, 20 and 40 kDa PEG showed significantly prolonged half-life in mice, dogs, and pigs as measured by rFVIIa clot activity.[17] The half-life ($T_{1/2}$) for rFVIIa containing 5, 10, 20, and 40 kDa PEG was increased by 2.4, 3.9, 5.7, and 10-fold, respectively, compared with rFVIIa in minipigs. The trend of prolonged half-life with increased size of PEG was also observed in mice and dogs. The 10 kDa and 40 kDa PEG-rFVIIa are efficacious in stopping bleeding in the haemophilia A mouse tail-bleeding model. N7–GP, a glycoPEGylated rFVIIa manufactured by enzymatic transfer of 40 kDa PEG conjugated sialic acid (over 85% mono-PEGylated), was tested for safety and pharmacokinetics in a randomized, placebo-controlled, dose-escalation clinical trial. The overall plasma half-life was 15 hours, which was approximately a 4.5-fold increase as compared with that of rFVIIa based on previous reports.[18] There were no serious adverse events, nor were neutralizing antibodies against N7–GP detected.

In addition, rFVIII was glycoPEGylated for increasing serum half-life to prevent bleeding associated with hemophilia A. The PEG conjugates were prepared using an enzymatic transfer of PEG-conjugated sialic acid to a unique *O*-glycan introduced at the FVIII B-domain.[19,20] It was found that all the conjugates with PEG ranging from 10 to 80 kDa maintained their binding to von Willebrand factor (VWF). After cleavage with thrombin, the B-domain and the associated PEG were released, resulting in activated FVIII with the same primary structure as native FVIIIa. *In vivo* in FVIII- and VWF-deficient mice, the serum half-life of the conjugates increased with the size of PEG. rFVIII glycoPEGylated with a branched 40 kDa PEG, N8–GP, had a twofold prolonged half-life in mice, rabbits, and cynomolgus monkeys. It also showed prolonged effects in tail bleeding, $FeCl_3$-induced injury, and joint bleeding models in FVIII-deficient mice. In a human trial with single dose, N8–GP (rFVIII glycoPEGylated with a branched 40 kDa PEG) was found to be well-tolerated and the mean terminal half-life was 1.6-fold longer than that of non-PEGylated rFVIII with no safety concerns.[21] rFIX was also glycoPEGylated through enzymatic transfer of a branched 40 kDa PEG-conjugated sialic acid to the core fucosylated tri- and tetra-antennary complex-type *N*-glycan from either N157 or N167 on the activation peptide.[22] By optimizing reaction conditions, the conjugation stoichiometry could be controlled to yield a product (N9–GP) containing mainly the mono-PEGylated form and a minor proportion of the di-PEGylated form (5%). Release of the activation peptide by physiological activators converted glycoPEGylated rFIX (N9–GP) to native rFIXa with normal kinetics. N9–GP exhibited up to twofold increased *in vivo* recovery and a markedly prolonged half-life in mini-pig (76 hours) and hemophilia B dogs (113 hours) compared with rFIX (16 hours). In the first human trial in patients with hemophilia B, the glycoPEGylated rFIX or N9–GP was found to be well-tolerated with fivefold increased half-life compared with non-PEGylated rFIX in single dose. The results indicate that the glycoPEGylated rFVIII and rFIX have the potential to reduce dosing frequency while providing effective treatment of bleeding episodes.[23]

The glycans in therapeutic proteins were also PEGylated directly to sialic acid or galactose residues pre-oxidized with chemicals or enzymes.[24] The terminal sialic acid residues of recombinant human thyroid-stimulating hormone were oxidized with periodate to generate aldehydes which reacted subsequently with aminooxy PEG. For enzymatic modification, the glycoprotein was first desialylated to expose the galactose before it was oxidized with galactose oxidase to generate aldehydes which were subsequently reacted with aminooxy PEG (Figure 13.2(C)). There was significant increase in serum half-life of mono- or multi-PEGylated recombinant hormone *in vivo*.

In addition to PEG, other hydrophilic polymers, including biodegradable polysaccharides, were also applied in protein conjugation for prolonging serum half-life. Polysialic acid is a naturally occurring biopolymer which is known as colominic acid in bacteria or PSA in neural cell adhesion molecule (NCAM) in human. It is unique with α-2,8-linked sialic acid (5-*N*-acetylneuraminic acid, or NeuNAc) repeat. Polysialic acid was initially demonstrated by Gregoriadis and colleagues as an alternative to PEG in protein conjugation for modulating serum half-life.[25] It has since been used in many proteins, including insulin, IFN-α_{2b}, EPO, asparaginase, β-galactosidase, and aprotinin with improved pharmacokinetics and stability as well as reduced immunogenicity.[26–28] The proteins were polysialylated mainly through reductive amination, in which the periodate oxidized non-reducing end of the polymer was randomly conjugated to lysine residues (Figure 13.3(A)). Constantinou and colleagues reported the polysialylation of antitumor antibody Fab through reductive amination as mentioned above. They found increased serum

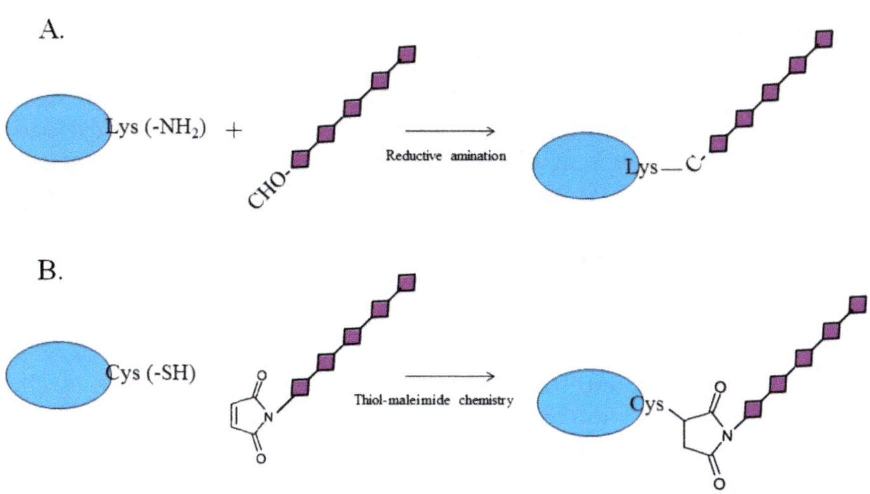

Figure 13.3 *Polysialylation of recombinant proteins.* (A) The lysine residue, Lys, in protein was conjugated to oxidized polysialic-acid-containing aldehyde through reductive amination; (B) The cysteine residue, Cys, in protein was conjugated to polysialic-acid-containing maleimide using thiol-maleimide chemistry.

half-life of polysialylated Fab with higher tumor uptake than the non-conjugated Fab.[29] The polysialylation of an antitumor single-chain Fv fragment (scFv) through random reductive amination and site-specific thiol maleimido coupling was compared.[30] It was shown that polysialylation using both methods increased serum half-life of scFv by 3.4–4.9-fold compared with non-conjugated scFv. However, random reductive amination showed reduced antigen interaction, probably due to conjugation of the polysaccharide to lysine residues critical for binding. In contrast, there was no loss of antigen binding of site-specific thiol maleimido conjugate polysialylated through an engineered C-terminal thiol residue (Figure 13.3(B)). Furthermore, the site-specific conjugate showed a sharper band in electrophoresis and improved tumor uptake *in vivo*. The protein was also polysialylated using a two-step *in vitro* enzymatic modification.[31] The two-step method uses the *Campylobacter jejuni* Cst-II α2,8-sialyltransferase to provide a primer on *N*-glycans of proteins, followed by polysialylation using the *Neisseria meningitidis* α2,8-polysialyltransferase. The human therapeutic protein alpha-1-*antitrypsin* (A1AT), factor IX, and bovine fetuin, were polysialylated using the *in vitro* enzymatic modification. It was found that the polysialylated A1AT maintained its *in vitro* inhibition activity against human neutrophil elastase. The polysialylation of A1AT resulted in an improved pharmacokinetic profile. Clearance rates from the blood were significantly reduced for disialylated (addition of one α2,8-linked sialic acid to a native terminal sialic acid)-A1AT (2.11 ml h^{-1}) and polysialylated-A1AT (0.75 ml h^{-1}), compared with A1AT (14.33 ml h^{-1}). Polysialylation also affected protein biodistribution in mice.

In addition to polysialic acid, other glycans have been used as alternative hydrophilic polymers.[32,33] Styslinger and colleagues reported a method for the rapid site-selective glycosylation of proteins with various-sized carbohydrates, up to a molecular weight of 10 kDa.[33] They used an oxime-ligation to the aldehyde available in the open-chain conformation on the reducing end carbohydrate for the installation of a linker and the azide–alkyne cycloaddition reaction for subsequent addition of maleimide to allow for site-selective glycosylation. Three different maltose-type glycans (with molecular weights ranging from 504 to 1153 Da) and a dextran glycan (molecular weight at approximately 10 kDa) containing maleimide linker were generated without using tedious protecting-group strategies. The glycan linkers were selectively conjugated to free cysteine present on the β chain of bovine hemoglobin (bHb). Glycosylation does not appear to alter the secondary, tertiary, and quaternary structures and haem environment of bHb, as indicated by circular dichroism examination and X-ray crystallography, while the glycosylated bHbs possess a higher affinity for oxygen compared with non-conjugated bHb. Mero and colleagues reported the conjugation of hyaluronic acid (HA) with two model enzymes, trypsin, and RNase A, and with insulin.[32] A hyaluronan–aldehyde derivative was synthesized and conjugated for N-terminal site-selective conjugation. The HA conjugates maintained their activities and thermal stabilities. The HA-conjugated insulin (17% loading) showed a greater effect on blood glucose level than non-conjugated insulin.

Furthermore, 10 kDa dextran was also used in conjugation to improve serum half-life and reduce immunogenicity of proteins.[34–38] The oxidized dextran was coupled to multiple lysine residues in IgG or Fab through reductive amination. Half-life prolongation was also demonstrated with conjugation of ATIII-binding pentasaccharides to insulin.[39] Either N-terminal PheB1 or near-C-terminal LysB29 of insulin derivatized with a maleimdo group was conjugated with an ATIII-binding pentasaccharide-containing thiol group with ethylene glycol spacer using thiol-maleimide chemistry. Although the overall process recovery was only 30 to 40%, there was significantly improved pharmacokinetic profile of the pentasaccharide-conjugated insulin as compared with non-conjugated insulin in rat ($t_{1/2}$ approximately 5 h *vs* 10 min for the conjugate *vs* insulin). The ATIII-binding pentasaccharide-conjugated insulin was also tested in clinical trials and found to be safe and well-tolerated. Its mean terminal elimination half-life was increased dramatically, ranging between three and five hours, compared with non-conjugated insulin (minutes) in humans.[40]

The conjugation of therapeutic proteins with hydrophilic polymers including polysaccharides has improved serum half-life and reduced immunogenicity in mice. Polysaccharides seem to be a valuable alternative to PEG with potential advantages in biodegradability. However, there is more work needed for establishing consistency in conjugation and validation of the conjugates both *in vivo* and in clinical settings. It is also well established that glycosylation is essential for protein stability. Glycoproteins often precipitate or aggregate when their glycans are removed through an *in vitro* enzymatic approach or *in vivo* recombinant approach. Without glycosylation, the protein is also more susceptible to protease digestion.

More work was reported on investigation of whether increased glycosylation benefits protein stability. Rores-Fernandez and colleagues showed that chemical glycosylation reduces moisture-induced protein instabilities.[41] The solid α-chymotrypsin formulation usually aggregates and inactivates under condition of increased relative humidity. It loses its native secondary and tertiary structure rapidly, resulting in the formation of buffer-insoluble aggregates, inactivation, and loss of native conformation upon reconstitution. When the α-chymotrypsin was conjugated with *N*-hydroxysuccinimide (NHS)-activated glycans of differing lengths (lactose and 10 kDa dextran) using lysine–succinimidyl chemistry, it showed a marked decrease in aggregation and an increase in residual activity after incubation under high-relative-humidity conditions. The stabilization effects through chemical glycosylation were independent of glycan size. The effects of chemical glycosylation on physical and chemical stability of cytochrome c (Cyt c) have also been reported.[42] Cyt c is an apoptosis-initiating protein when released into the cytoplasm of eukaryotic cells through caspase activation. The increased long-lasting bioavailability of the protein through chemical glycosylation was investigated. Three glycans of different molecular weights (lactose as well as two dextrans of 1 and 10 kDa) were chemically coupled to surface-exposed lysine residues in Cyt c using succinimidyl chemistry *via* amide bonds.

Five neo-glycoconjugates with various number or size of glycan conjugated per protein [Lac4–Cyt-c, Lac9–Cyt-c, Dex5(10 kDa)–Cyt-c, Dex8(10 kDa)–Cyt-c, and Dex3 (1 kDa)–Cyt-c] were prepared and their structures, activities, and stabilities were determined. The analysis using circular dichroism showed little change in secondary structure, while only minor perturbations of the tertiary structure in glycosylated Cyt c were observed. There were no major changes in Cyt c activity in caspase activation upon chemical glycosylation. However, glycosylation protected Cyt c from detrimental effects of some stresses, such as elevated temperature and humidity, and from proteolytic degradation. In contrast, non-conjugated Cyt c was found to be more susceptible to denaturation by a water–organic solvent interface than its glycoconjugates. These results indicate that chemical glycosylation may be a valuable approach for increasing protein stability during formulation, storage, and, potentially, *in vivo* administration.

In addition, the interleukin-1β and tumor necrosis factor (TNF)-α monoclonal antibodies were conjugated with high-molecular-weight hyaluronic acid (HA; approximately 1.6 MDa) and carboxymethylcellulose (approximately 700 kDa) through NHS chemistry.[43–45] It was found that the conjugates maintained their binding to pro-inflammatory cytokines. The authors suggested that high-molecular-weight polysaccharides can make their antibody conjugates unable to circulate in the bloodstream due to their big sizes when delivered to tissues. The effect of topically applied anti-TNF-α conjugated with high-molecular-weight HA on burn progression and influence on the inflammatory microenvironment was further analyzed. Although HA-conjugated anti-TNF-α decreased macrophage infiltration to a lesser extent than the non-conjugated antibody, the former was more effective than the latter in decreasing nonviable tissue and the related inflammatory markers.[45] The method provided localized delivery of anti-cytokine therapeutics, which may avoid many of the complications associated with systemic delivery. Kordopati and colleagues reported the conjugation of a luteinizing-hormone-releasing hormone (LHRH) analogue containing ε-aminohexanoic acid with carboxylic acid to the amino group of 3-monoamino-β-cyclodextrin. The conformational properties of the conjugate were studied using NMR and molecular modeling. The results indicated that both the phenyl group of tyrosine and the indole group of tryptophan can be encapsulated inside the cyclodextrin cavity through intramolecular but not intermolecular interactions. The intramolecular interaction in the peptide β-cyclodextrin conjugate could improve drug delivery and increase drug stability against enzymatic proteolysis, leading to an improved pharmacological profile.

Based on the evidence described above, glycosylation increased protein stability and showed its potential to be applied to many therapeutic proteins, especially those with low stability. Conjugation with high-molecular-weight polysaccharides may provide a better method for local delivery of therapeutic proteins. However, these approaches need to be further investigated both *in vitro* and *in vivo*. The recovery and scalability of the method should also be further demonstrated.

13.3 Glycosylation for Increased Protein Targeting to Disease-Affected Tissues

There are many glycan-binding proteins (or carbohydrate-binding proteins, lectins) on the surface of mammalian cells, including hepatocytes, endothelial cells, lymphocytes, macrophages, and dendritic cells. Except for a few of them, the physiological functions of these proteins are largely unknown. When binding to the glycan ligands, these protein complexes are taken up into endosomal or lysosomal compartments inside the cells through receptor-mediated endocytosis. Thus, the glycan-binding proteins are potential targets for delivering enzymes or proteins into defined intracellular compartment of specific disease-affected tissues.

Lysosomal storage diseases (or LSDs) are a group of approximately 50 rare inherited metabolic disorders that result from defects in lysosomal function due to genetic mutations of many different lysosomal enzymes. The lack of single lysosomal enzyme often results in accumulation of its substrate, which could be lipid or glycolipid, glycoprotein, or mucopolysaccharide, leading to lysosomal dysfunction.

Gaucher disease is a lysosomal storage disease caused by deficiency of β-glucocerebrosidase (also known as acid β-glucosidase, GCase, D-glucosyl-*N*-acylsphingosine glucohydrolase, EC 3.2.1.45) encoded by the human *GBA1* gene.[46–48] Since GCase cleaves glucosylceramide (or glucocerebroside) into glucose and ceramide, the lack of the enzyme results in the accumulation of its substrate, glucosylceramide, in the lysosomes of cells, particularly macrophages, reticuloendothelial cells, and other phagocytic cells in liver and spleen, leading to enlargement of these organs in patients. There are glycans derived from four *N*-glycosylation sites which contain oligomannose-type (approximately 20%) and complex-type (approximately 80%) glycans, with trisialylated triantennary glycan as the major species in the human placental enzyme.[49,50] To target delivery of the missing enzyme into lysosomes of macrophages, the modification of glycans in purified human placental GCase was investigated for binding to mannose receptor, which is present in macrophage and reticuloendothelial cells.[51,52] GCase was conjugated to a synthetic triantennary mannosyl glycopeptide (Man_3Lys_2) using 1-ethyl-3-(3-dimethylaminopropyl)carbodiimide (EDC) chemistry to its lysine residues with eight to nine mol glycopeptide coupled per mol enzyme.[53] The *in vitro* mannan-sensitive uptake of Man_3Lys_2-conjugated GCase into alveolar macrophages was more than three times that of the non-conjugated enzyme. *In vivo*, Man_3Lys_2-conjugated GCase showed faster clearance from circulation than non-conjugated enzyme in rat. The glycoconjugate and non-conjugate had serum half-lives of 8.5 and 18.5 minutes, respectively. Although non-conjugated placental GCase was internalized by both hepatocytes and non-parenchymal cells in rats due to the presence of both oligomannose (approximately 20%) and non-sialylated complex-type glycans (approximately 30%),[50] there was a fourfold increased uptake of Man_3Lys_2-conjugated enzyme in non-parenchymal cells as compared with the non-conjugated enzyme.[54] GCase was also sequentially deglycosylated and its uptake in hepatocytes and non-parenchymal cells was investigated. When it was desialylated with neuraminidase, its serum half-life was reduced by approximately tenfold. The desialylated enzyme was cleared predominantly by asialoglycoprotein receptor present on hepatocytes, resulting in reduced uptake by Kupffer cells.[55] When GCase was desialylated and degalactosylated to expose *N*-acetylglucosamine (GlcNAc), the clearance of the modified enzyme was 14-times faster in rats than that of the non-modified one.[56] There was approximately 5.7-fold increased uptake of aglactosylated enzyme in non-parenchymal cells compared with that from non-modified enzyme when both were injected at the same dose into rats. In another experiment, sequential treatment of GCase with exoglycosidases (neuraminidase, β-galactosidase, and β-*N*-acetylglucosaminidase) to expose terminal galactose, GlcNAc, or mannose, resulted in greater than tenfold increased clearance rates in rats. Although there was more desialylated enzyme cleared through hepatocytes than unmodified enzyme, the biodistribution of enzyme with terminal GlcNAc and mannose was increased in non-parenchymal cells, such as Kupffer cells and reticuloendothelial cells. The change in the biodistribution of the sequentially deglycosylated enzyme with a particular terminal monosaccharide is consistent with the expression of asialoglycoprotein receptors on hepatocytes and mannose receptors on non-paranchymal cells. Although both terminal GlcNAc and mannose are carbohydrate ligands for mannose receptor, there was more uptake of mannose-terminated enzyme in non-parenchymal cells than the same enzyme with terminal GlcNAc.[57] The liver uptake of Man_3Lys_2-conjugated GCase through chemical conjugation and the same enzyme with terminal mannose generated through modification with exoglycosidases was compared. Interestingly, the biodistribution of exoglycosidase-modified enzyme with terminal mannose residue in non-parenchymal cells was increased by approximately 16-fold, while the enzyme conjugated with Man_3Lys_2 showed only a fourfold increase as compared with unmodified enzyme. The data indicate that the artificial Man_3Lys_2 glycan may not be the optimal ligand for mannose receptor expression on non-parenchymal cells.

Pompe disease (glycogen storage disease type II) is another autosomal recessive lysosomal disorder in cardiac, skeletal and respiratory muscle tissues. It results from lysosomal glycogen accumulation due to a deficiency in the glycogen-degrading lysosomal enzyme, acid α-glucosidase or GAA (EC 3.2.1.3).[58] GAA is a 110 kDa glycoprotein with seven *N*-linked glycosylation sites occupied by mainly two high-mannose-type glycans containing 1 mol per mol mannose 6-phosphate (M6P) and five mainly complex-type glycans.[59–61] The recombinant human α-glucosidase, rhGAA (alglucosidase alfa) is used successfully in clinics in treating Pompe disease.[62–64] However, even at doses which were high relative to other enzyme-replacement therapeutics, the extent of glycogen clearance by the recombinant enzyme in skeletal muscle cells was found to vary among infants.[65] A high dose was also needed in clearing skeletal muscle glycogen in a murine Pompe model, probably due to relatively low amounts of M6P-containing glycans in the recombinant enzyme for cation-independent mannose 6-phosphate receptor (CI-MPR), which is present at the surface of muscle cells mediating uptake of the bisM6P-containing glycoproteins into lysosomes.[60,66] To improve uptake of the enzyme into lysosomes in muscle cells, various enzymatic and chemical conjugation methods were investigated to increase bisM6P glycans in rhGAA.[60,67–69] The recombinant protein was produced as a glycoprotein containing all the high-mannose glycans, which was then enzymatically modified into highly phosphorylated bisM6P glycans with UDP–GlcNAc-dependent GlcNAc-1-phosphotransferase (GlcNAc phosphotransferase) and α-GlcNAc-1-phosphodieser glycosidase (GlcNAc phosphoglycosidase), generating highly phosphorylated rhGAA or HP-GAA.[60,67] Although a significant amount of bisM6P glycans was generated on HP-GAA, the enzymatic modification was not complete, with some unphosphorylated oligomannose glycans remaining. Consistently, there was significantly higher uptake of HP-GAA than of unmodified rhGAA produced from CHO cells into human Pompe fibroblasts.[60] However, HP-GAA showed fast clearance through mannose receptor in macrophages and reticular endothelial cells, probably due to those unphosphorylated oligomannose glycans, leading to less glycogen clearance in heart and quadriceps muscles than unmodified rhGAA. In contrast, rhGAA conjugated with phosphorylated glycans derived from another recombinant lysosomal enzyme, human α-galactosidase, showed increased binding to CI-MPR and L6 myoblast uptake *in vitro*, as well as significantly increased glycogen clearance in the heart and skeletal muscles as compared with rhGAA.[68] rhGAA was also conjugated with synthetic M6P glycan. There were significantly increased CI-MPR binding and L6 myoblast uptake of conjugated GAA *in vitro*. Furthermore, fourfold to fivefold increased potency in glycogen clearance in heart and skeletal muscles was observed with bisM6P glycan-conjugated GAA in GAA knockout mice, or Pompe mice, *in vivo* compared with rhGAA.[68,69]

In addition, chemical glycosylation has been used as lectin-directed enzyme-activated prodrug therapy (LEAPT)[70] (Figure 13.4). The glycans of a α-L-rhamnosidase were deglycosylated with endoglycosidase H and chemically reglycosylated with 2-imino-2-methoxyethyl-1-thioglycoside containing monosaccharide galactose or mannose as well as branched di-galactose, at lysine residues of the enzyme. The conjugation resulted in 10–15 carbohydrates per enzyme molecule as determined by matrix-assisted laser desorption ionization–time of flight (MALDI–TOF) mass spectrometry. The ability of these glycosylated enzymes to be selectively targeted to hepatocytes and to activate the rhamnose-capped prodrug was demonstrated *in vivo*. The enzyme conjugated with galactose showed selective re-distribution from non-hepatocytes (90% of non-conjugate) to 86% localization in hepatocytes, resulting in over 30-fold enhancement in the absolute level of enzyme delivered to hepatocytes. Coadministration of galactose-conjugated enzyme with rhamnose-capped doxorubicin reduced tumor burden in a hepatocellular carcinoma (HepG2) mouse model.

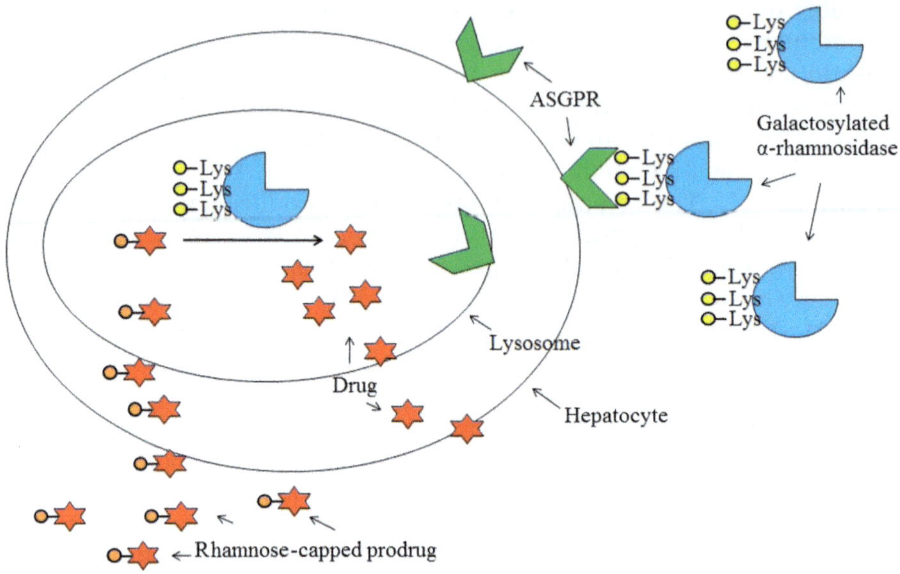

Figure 13.4 *Lectin-directed enzyme-activated prodrug therapy (LEAPT)*. The galactose conjugated non-mammalian glycosidase (α-rhamnosidase) was selectively bound to a lectin, asialoglycoprotein receptor (ASGPR), on the surface of a hepatocyte and delivered into lysosome inside the cell through receptor-mediated endocytosis. The galactosylated α-rhamnosidase taken up releases rhamnose in the prodrug, generating active drug in the lysosome.

Protein targeting through glycosylation provided a unique approach to generating therapeutics for many diseases. It showed efficacy both *in vivo* and in clinical settings. However, there are currently only limited carbohydrate receptors being demonstrated for targeting and the approach was limited to lysosomal targeting. With the advances in understanding of other carbohydrate receptors, the method has potential to be applied to areas beyond the treatment of lysosomal storage diseases.

13.4 Glycosylation for Site-Specific Drug Conjugation

Chemical glycosylation has also been used for site-specific drug conjugation of glycoproteins, specifically antibodies. Antibody–drug conjugation combines the specific recognition of tumor antigen through antibodies with highly potent anticancer drugs. It has been used in clinical settings as an effective therapy for oncology. Two antibody–drug conjugates (ADCs) [brentuximab vedotin (anti-CD30 ADC) and trastuzumab emtansine (anti-HER2 ADC)] showed strong efficacy in treating patients with anaplastic large cell lymphoma–Hodgkin's lymphoma and advanced human epidermal growth factor receptor 2 (HER2)-positive breast cancer, respectively.[71–74] These ADCs were prepared using conventional conjugation approaches which conjugate the optimal four drugs to cysteine residues from partially reduced interchain disulfides or lysine residues. Since there are eight such cysteines and 30–40 lysines in antibody IgG, these approaches can potentially generate conjugates with significant heterogeneity with one hundred to one million different species.[75] Thus, second-generation approaches, including those using chemical glycosylation are being reported for preparing homogeneous and site-specific ADC.

Since the antibody glycans are derived from the well-conserved N297 site and are relatively simple compared with other glycoproteins, they were utilized as unique conjugation sites for anticancer drugs using different methods (Figure 13.5). Except for fucose, which has an effect on antibody binding to Fcγ receptors, the other monosaccharides have minimal effects on antigen or Fcγ receptor interactions.[76] Conjugation of a drug-linker to the N297 glycans has previously been investigated using hydrazone chemistry.[77,78] The glycans, (fucose, galactose, mannose, and GlcNAc) were oxidized with high concentrations of sodium periodate (10 or 12.5 mM) to provide the aldehydes necessary for coupling hydrazide-containing drug-linkers. Since the drug conjugation sites include various monosaccharides oxidized at high periodate concentrations, it was likely that the ADC generated contained linkages of undefined structure and a high degree of heterogeneity. The exposure of the antibody to high concentrations of periodate could also result in undesired oxidation of amino acid residues, including methionine residues in the C_H3 region critical for FcRn binding, as well as potential protein aggregation. Furthermore, the hydrazone bond is not stable in serum, potentially leading to drug-linker loss before the ADC reaches tumor cells and generating off-target toxicity.

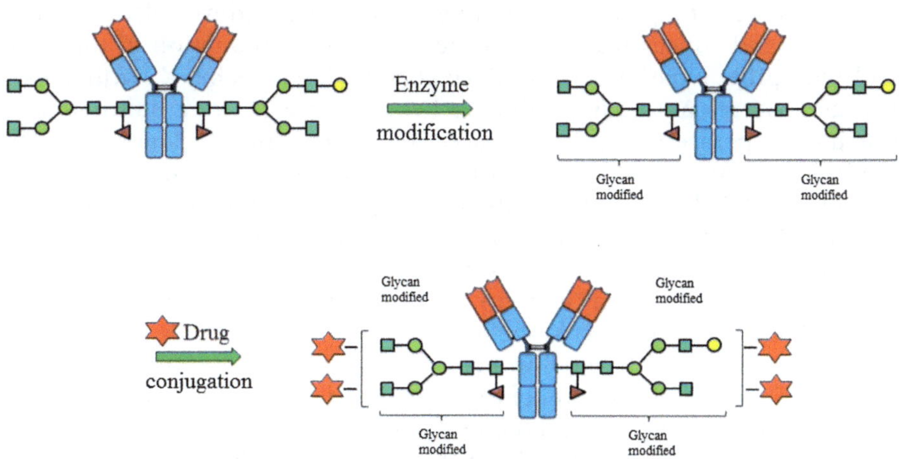

Figure 13.5 *Site-specific antibody–drug conjugation through glycoengineering.* The N297 glycans in antibodies are modified using different glycoengineering methods and then selectively conjugated to anticancer drugs.

Using a similar conjugation chemistry, site-specific conjugation of fucose was investigated by oxidizing antibody N297 glycans with a high concentration (10 mM) of sodium periodate before the oxidized antibody was conjugated with 100-fold molar excess of the hydrazide version of dolastatin.[79] As expected, the conjugates with the hydrazone bond were not stable and they were spontaneously hydrolyzed in phosphate buffered saline (PBS) at pH 7.4 at 37 °C with a half-life of 18 hours. In addition, since fucosylated biantennary glycan with no galactose (G0F) and fucosylated biantennary glycan with one galactose (G1F) are the major glycan species present in the antibody, the specific oxidation of fucose, but not other monosaccharides, such as galactose or GlcNAc, in the antibody, at such a high periodate concentration was not confirmed. On the other hand, site-specific fucose conjugation through metabolically incorporated fucose analogues was reported.[80] The authors screened 200 synthetic fucose analogues for their metabolic incorporation into the antibody. Among those analogues, 6-thiolfucose was found to be efficiently incorporated into the antibody *N*-glycans for subsequent conjugation using thiol-maleimide chemistry. CHO cells expressing either anti-CD70 or anti-CD30 antibodies were treated with 1 mM 6-thiofucose, which was incorporated into antibodies with 60–70% efficiency or 1.2–1.4 thiolfucose per antibody. The purified antibodies containing the fucose analog were reduced and re-oxidized prior to conjugation with 20% molar excess of maleimide containing monomethyl auristatin E (MMAE) drug-linker, leading to an ADC with 1.3 drugs per antibody. It was found that the ADC containing thiolfucose was stable, with at least 85% of the drug-linker remaining on the antibody during a four-day incubation with rat plasma, while 40% of the drug was lost from the conjugate prepared using conventional cysteine conjugation under the same incubation conditions. The thiolfucose ADC also showed high cytotoxicity *in vitro*.[80]

Site-specific conjugation through antibody sialic acid was also explored. Rochefort and colleagues reported the metabolic exploitation of the sialic acid biosynthetic pathway to produce site-specifically labeled antibodies.[81] The anti-CA19-9 antibody producing hybridoma cells were cultured with 100 μM of peracetylated *N*-azidoacetylmannosamine (Az_4ManMAz). The resulting azido-sialic acid containing antibody was conjugated to either phosphine-PEG_3-biotin or DyLight-650-Phosphine *via* a modified Staudinger reaction. The fluorescent-dye-conjugated antibody was able to stain tumor cells after injection into CA19-9-positive tumor xenografts in mice. However, the coupling efficiency was low with approximately 0.1 fluorophore coupled per antibody, probably due to the low amount (approximately 1%) of sialic acid present in antibody and low metabolic labeling of Az_4ManMAz.

The conjugation of drug-linker to antibody sialic acid oxidized with periodate was also investigated based on the fact that sialic acid is the most sensitive monosaccharide to periodate.[82] Since there is only a very low amount of sialic acid present in N297 glycans from recombinant IgG1, the antibody was first sialylated *in vitro* using sialyltransferase together with galactosyltransferase to produce mainly the monosialylated biantennary glycan.

The sialic acids were then oxidized with periodate at a very low concentration (1 mM) to introduce sialic acid aldehyde groups with minimal oxidation of other monosaccharide or methionine residues, resulting in minimal loss of FcRn binding. The oxidized antibody was then conjugated with aminooxy drug-linkers to produce a glyco ADC with approximately 1.6 drug-linkers per antibody molecule, approximating the number of sialic acid residues. Sialic-acid-mediated antibody conjugation was evaluated with three different antibodies, including trastuzumab and two anti-fibroblast activation protein (FAP) antibodies, using two drug-linkers, the aminooxy version of MMAE and PEG8–Dol10. Highly target-selective *in vitro* anti-proliferative and *in vivo* anticancer activities were found with glyco ADCs in a xenograft model using SCID mice that had received implants of HER2-positive SK-OV-3 tumor cells.

Another method for site-specific conjugation of drug-linker to sialic acid has also been reported that used a sialic acid derivative and sialyltransferase *in vitro*.[83] The azide-containing sialic acid derivative was introduced into N297 glycans of an anti-CD22 antibody using sialyltransferase and galactosyltransferase, followed by conjugation with dibenzylcyclooctyonol (DIBO or DBCO)-linked doxorubicin. Although its drug-to-antibody ratio (DAR) was not fully determined, the ADC produced showed significant cytotoxicity *in vitro*.

In addition, the site-specific glycoconjugation of drug-linkers to other monosaccharides, such as galactose and GalNAc was also developed. The sugar derivative containing a conjugatable moiety, including 2-acetonyl-2-deoxy-galactose or 2-*N*-acetyl azide GalNAc, was transferred into the non-reducing termini of the N297 glycans using a mutant galactosyltransferase.[84,85] The galactose or GalNAc derivative was then conjugated to bioactive agents carrying aminooxy or alkyne or cyclooctyne groups, such as aminooxy- or cyclooctyne-containing biotin and fluorescent probes. It was shown that this approach can be applied for site-specific antibody radiolabeling for delivering diagnostic and therapeutic radioisotopes to tumors.[86] Recently, it has been reported that a chemically reactive keto-galactose was added to an anti-HER2 antibody using a mutant galactosyltransferase.[87] The keto-galactose-containing antibody was subsequently conjugated with aminooxy auristatin F, resulting in four drug-linkers per antibody, as determined by mass spectrometry. Interestingly, the conjugation of four drug-linkers to the N297 glycans did not affect the antibody affinity for Fcγ receptors. The ADC showed similar binding to HER2 antigen and exhibited potent and specific cell-killing activity against HER2-positive cancer. The antibody glycans were also trimmed with an endoglycosidase before the exposed core GlcNAc was tagged with azido-modified GalNAc by a galactosyltransferase mutant (Y289L).[88] The bicyclononyne (BCN)-modified drug-linker was then conjugated to azido GalNAc. The method was applied to the conjugation of maytansine-containing drug-linkers to trastuzumab with an approximate DAR of two. The ADC generated showed strong potency *in vitro* and efficacy *in vivo* with desirable plasma stability and scalability up to 5 g.

Site-specific antibody conjugations through coupling of the drug-linker or diagnostic agent to glycans provided a valuable method for generating homogeneous conjugate without any change in the amino acid sequence of the antibody. Thus, the method has less effect on the antibody structure, due to the fact that the drug-linker is not attached to the protein backbone. However, the methods need to be validated more *in vivo* and in clinical settings in the future.

13.5 Conclusion

Various approaches have been applied for chemically glycosylating therapeutic proteins to improve their pharmacokinetics and pharmacodynamics. Proteins were PEGylated selectively through the glycans and their circulation half-lives were prolonged. Other oligosaccharides or polysaccharides were conjugated to proteins as alternative hydrophilic polymers. The effects of their conjugations were demonstrated by increases in the half-life and stability, as well as reductions in immunogenicity. Glycan modification using either enzymatic or chemical conjugation methods also facilitated protein targeting to disease-affected tissues. Furthermore, either anticancer drugs or diagnostic agents can be site-specifically conjugated to monoclonal antibody glycans to generate homogeneous conjugates. Taken together, selective chemical glycosylation has been shown to be an effective approach for generation of better protein therapeutics for multiple challenging diseases.

Acknowledgements

The author would like to thank Dr Huawei Qiu for helpful suggestions and Mr Michael Drzyzga for proof-reading of this chapter.

References

1. *Essentials of Glycobiology*, ed. A. Varki, R. D. Cummings, J. D. Esko, H. H. Freeze, P. Stanley, C. R. Bertozzi, G. W. Hart and M. E. Etzler, Cold Spring Harbor Laboratory Press, New York, 2nd edn, 2009.
2. R. J. Sola and K. Griebenow, *J. Pharm. Sci.*, 2009, **98**, 1223.
3. Y. Mimura, P. Sondermann, R. Ghirlando, J. Lund, S. P. Young, M. Goodall and R. Jefferis, *J. Biol. Chem.*, 2001, **276**, 45539.
4. D. Reusch and M. L. Tejada, *Glycobiology*, 2015, **25**, 1325.
5. L. Liu, *J. Pharm. Sci.*, 2015, **104**, 1866.
6. C. J. Bosques, B. E. Collins, J. W. Meador 3rd, H. Sarvaiya, J. L. Murphy, G. Dellorusso, D. A. Bulik, I. H. Hsu, N. Washburn, S. F. Sipsey, J. R. Myette, R. Raman, Z. Shriver, R. Sasisekharan and G. Venkataraman, *Nat. Biotechnol.*, 2010, **28**, 1153.
7. J. E. Hudak and C. R. Bertozzi, *Chem. Biol.*, 2014, **21**, 16.
8. G. Pasut and F. M. Veronese, *J. Controlled Release*, 2012, **161**, 461.
9. M. E. Giorgi, R. Agusti and R. M. de Lederkremer, *Beilstein J. Org. Chem.*, 2014, **10**, 1433.

10. C. Chen, A. Constantinou and M. Deonarain, *Expert Opin. Drug Delivery*, 2011, **8**, 1221.
11. S. DeFrees, Z. G. Wang, R. Xing, A. E. Scott, J. Wang, D. Zopf, D. L. Gouty, E. R. Sjoberg, K. Panneerselvam, E. C. Brinkman-Van der Linden, R. J. Bayer, M. A. Tarp and H. Clausen, *Glycobiology*, 2006, **16**, 833.
12. H. R. Stennicke, H. Ostergaard, R. J. Bayer, M. S. Kalo, K. Kinealy, P. K. Holm, B. B. Sorensen, D. Zopf and S. E. Bjorn, *Thromb. Haemostasis*, 2008, **100**, 920.
13. B. Plesner, P. Westh, S. Hvidt and A. D. Nielsen, *J. Pharm. Biomed. Anal.*, 2011, **55**, 597.
14. B. Plesner, P. Westh and A. D. Nielsen, *Int. J. Pharm.*, 2011, **406**, 62.
15. S. Ghosh, P. Sen, U. R. Pendurthi and L. V. Rao, *J. Thromb. Haemostasis*, 2008, **6**, 1525.
16. P. Sen, S. Ghosh, M. Ezban, U. R. Pendurthi and L. Vijaya Mohan Rao, *Haemophilia*, 2010, **16**, 339.
17. D. M. Karpf, B. B. Sorensen, M. B. Hermit, H. L. Holmberg, M. Tranholm, B. V. Bysted, A. V. Groth, S. E. Bjorn and H. R. Stennicke, *Thromb. Res.*, 2011, **128**, 191.
18. J. Moss, A. Rosholm and A. Lauren, *J. Thromb. Haemostasis*, 2011, **9**, 1368.
19. H. Agerso, H. R. Stennicke, H. Pelzer, E. N. Olsen, E. P. Merricks, N. A. Defriess, T. C. Nichols and M. Ezban, *Haemophilia*, 2012, **18**, 941.
20. H. R. Stennicke, M. Kjalke, D. M. Karpf, K. W. Balling, P. B. Johansen, T. Elm, K. Ovlisen, F. Moller, H. L. Holmberg, C. N. Gudme, E. Persson, I. Hilden, H. Pelzer, H. Rahbek-Nielsen, C. Jespersgaard, A. Bogsnes, A. A. Pedersen, A. K. Kristensen, B. Peschke, W. Kappers, F. Rode, L. Thim, M. Tranholm, M. Ezban, E. H. Olsen and S. E. Bjorn, *Blood*, 2013, **121**, 2108.
21. A. Tiede, B. Brand, R. Fischer, K. Kavakli, S. R. Lentz, T. Matsushita, C. Rea, K. Knobe and D. Viuff, *J. Thromb. Haemostasis*, 2013, **11**, 670.
22. H. Ostergaard, J. R. Bjelke, L. Hansen, L. C. Petersen, A. A. Pedersen, T. Elm, F. Moller, M. B. Hermit, P. K. Holm, T. N. Krogh, J. M. Petersen, M. Ezban, B. B. Sorensen, M. D. Andersen, H. Agerso, H. Ahmadian, K. W. Balling, M. L. Christiansen, K. Knobe, T. C. Nichols, S. E. Bjorn and M. Tranholm, *Blood*, 2011, **118**, 2333.
23. C. Negrier, K. Knobe, A. Tiede, P. Giangrande and J. Moss, *Blood*, 2011, **118**, 2695.
24. A. Park, D. M. Honey, L. Hou, J. J. Bird, C. Zarazinski, M. Searles, C. Braithwaite, J. S. Kingsbury, J. Kyazike, K. Culm-Merdek, B. Greene, J. E. Stefano, H. Qiu, J. M. McPherson and C. Q. Pan, *Endocrinology*, 2013, **154**, 1373.
25. G. Gregoriadis, B. McCormack, Z. Wang and R. Lifely, *FEBS Lett.*, 1993, **315**, 271.
26. S. Jain, D. H. Hreczuk-Hirst, B. McCormack, M. Mital, A. Epenetos, P. Laing and G. Gregoriadis, *Biochim. Biophys. Acta*, 2003, **1622**, 42.
27. A. I. Fernandes and G. Gregoriadis, *Int. J. Pharm.*, 2001, **217**, 215.
28. I. V. Smirnov, I. I. Vorobiev, A. A. Belogurov, D. D. Genkin, S. M. Deyev and A. G. Gabibov, *Methods Mol. Biol.*, 2015, **1321**, 389.

29. A. Constantinou, A. A. Epenetos, D. Hreczuk-Hirst, S. Jain and M. P. Deonarain, *Bioconjugate Chem.*, 2008, **19**, 643.
30. A. Constantinou, A. A. Epenetos, D. Hreczuk-Hirst, S. Jain, M. Wright, K. A. Chester and M. P. Deonarain, *Bioconjugate Chem.*, 2009, **20**, 924.
31. T. Lindhout, U. Iqbal, L. M. Willis, A. N. Reid, J. Li, X. Liu, M. Moreno and W. W. Wakarchuk, *Proc. Natl. Acad. Sci. U. S. A.*, 2011, **108**, 7397.
32. A. Mero, M. Pasqualin, M. Campisi, D. Renier and G. Pasut, *Carbohydr. Polym.*, 2013, **92**, 2163.
33. T. J. Styslinger, N. Zhang, V. S. Bhatt, N. Pettit, A. F. Palmer and P. G. Wang, *J. Am. Chem. Soc.*, 2012, **134**, 7507.
34. R. Fagnani, M. S. Hagan and R. Bartholomew, *Cancer Res.*, 1990, **50**, 3638.
35. T. E. Wileman, R. L. Foster and P. N. Elliott, *J. Pharm. Pharmacol.*, 1986, **38**, 264.
36. R. Fagnani, S. Halpern and M. Hagan, *Nucl. Med. Commun.*, 1995, **16**, 362.
37. S. D. Mikolajczyk, D. L. Meyer, R. Fagnani, M. S. Hagan, K. L. Law and J. J. Starling, *Bioconjugate Chem.*, 1996, **7**, 150.
38. W. Richter and L. Kagedal, *Int. Arch. Allergy Appl. Immunol.*, 1972, **42**, 885.
39. M. de Kort, B. Gianotten, J. Wisse, E. Bos, M. Eppink, E. Mattaar, G. Vogel, W. Dokter, M. Honing, S. Vonsovi, M. Smit, J. Wijkmans and C. van Boeckel, *ChemMedChem*, 2008, **3**, 1189.
40. A. M. M. Miltenburg, M. Prohn, J. H. M. Kuijk, R. G. Tiessen, M. Kort and R. J. W. Berg, *Br. J. Clin. Pharmacol.*, 2013, **75**, 1221.
41. G. M. Flores-Fernandez, M. Pagan, M. Almenas, R. J. Sola and K. Griebenow, *BMC Biotechnol.*, 2010, **10**, 57.
42. Y. Delgado, M. Morales-Cruz, J. Hernandez-Roman, Y. Martinez and K. Griebenow, *BMC Biochem.*, 2014, **15**, 16.
43. N. R. Washburn, J. E. Prata, E. E. Friedrich, M. H. Ramadan, A. N. Elder and L. T. Sun, *Biomatter*, 2013, **3**, e25597.
44. L. T. Sun, K. S. Buchholz, M. T. Lotze and N. R. Washburn, *Mol. Pharmaceutics*, 2010, **7**, 1769.
45. E. E. Friedrich, L. T. Sun, S. Natesan, D. O. Zamora, R. J. Christy and N. R. Washburn, *J. Biomed. Mater. Res., Part A*, 2014, **102**, 1527.
46. T. Dinur, K. M. Osiecki, G. Legler, S. Gatt, R. J. Desnick and G. A. Grabowski, *Proc. Natl. Acad. Sci. U. S. A.*, 1986, **83**, 1660.
47. J. Sorge, C. West, B. Westwood and E. Beutler, *Proc. Natl. Acad. Sci. U. S. A.*, 1985, **82**, 7289.
48. J. Sorge, T. Gelbart, C. West, B. Westwood and E. Beutler, *Proc. Natl. Acad. Sci. U. S. A.*, 1985, **82**, 5442.
49. A. Berg-Fussman, M. E. Grace, Y. Ioannou and G. A. Grabowski, *J. Biol. Chem.*, 1993, **268**, 14861.
50. S. Takasaki, G. J. Murray, F. S. Furbish, R. O. Brady, J. A. Barranger and A. Kobata, *J. Biol. Chem.*, 1984, **259**, 10112.
51. R. O. Brady, G. J. Murray and N. W. Barton, *J. Inherited Metab. Dis.*, 1994, **17**, 510.
52. P. B. Deegan and T. M. Cox, *Drug Des., Dev. Ther.*, 2012, **6**, 81.

53. T. W. Doebber, M. S. Wu, R. L. Bugianesi, M. M. Ponpipom, F. S. Furbish, J. A. Barranger, R. O. Brady and T. Y. Shen, *J. Biol. Chem.*, 1982, **257**, 2193.
54. G. J. Murray, T. W. Doebber, T. Y. Shen, M. S. Wu, M. M. Ponpipom, R. L. Bugianesi, R. O. Brady and J. A. Barranger, *Biochem. Med.*, 1985, **34**, 241.
55. F. S. Furbish, C. J. Steer, J. A. Barranger, E. A. Jones and R. O. Brady, *Biochem. Biophys. Res. Commun.*, 1978, **81**, 1047.
56. C. J. Steer, F. S. Furbish, J. A. Barranger, R. O. Brady and E. A. Jones, *FEBS Lett.*, 1978, **91**, 202.
57. F. S. Furbish, C. J. Steer, N. L. Krett and J. A. Barranger, *Biochim. Biophys. Acta*, 1981, **673**, 425.
58. F. Martiniuk, M. Mehler, A. Pellicer, S. Tzall, G. La Badie, C. Hobart, A. Ellenbogen and R. Hirschhorn, *Proc. Natl. Acad. Sci. U. S. A.*, 1986, **83**, 9641.
59. M. M. Hermans, H. A. Wisselaar, M. A. Kroos, B. A. Oostra and A. J. Reuser, *Biochem. J.*, 1993, **289**(3), 681.
60. A. J. McVie-Wylie, K. L. Lee, H. Qiu, X. Jin, H. Do, R. Gotschall, B. L. Thurberg, C. Rogers, N. Raben, M. O'Callaghan, W. Canfield, L. Andrews, J. M. McPherson and R. J. Mattaliano, *Mol. Genet. Metab.*, 2008, **94**, 448.
61. R. J. Moreland, X. Jin, X. K. Zhang, R. W. Decker, K. L. Albee, K. L. Lee, R. D. Cauthron, K. Brewer, T. Edmunds and W. M. Canfield, *J. Biol. Chem.*, 2005, **280**, 6780.
62. A. T. van der Ploeg, P. R. Clemens, D. Corzo, D. M. Escolar, J. Florence, G. J. Groeneveld, S. Herson, P. S. Kishnani, P. Laforet, S. L. Lake, D. J. Lange, R. T. Leshner, J. E. Mayhew, C. Morgan, K. Nozaki, D. J. Park, A. Pestronk, B. Rosenbloom, A. Skrinar, C. I. van Capelle, N. A. van der Beek, M. Wasserstein and S. A. Zivkovic, *N. Engl. J. Med.*, 2010, **362**, 1396.
63. A. T. van der Ploeg, R. Barohn, L. Carlson, J. Charrow, P. R. Clemens, R. J. Hopkin, P. S. Kishnani, P. Laforet, C. Morgan, S. Nations, A. Pestronk, H. Plotkin, B. E. Rosenbloom, K. B. Sims and E. Tsao, *Mol. Genet. Metab.*, 2012, **107**, 456.
64. C. Angelini, C. Semplicini, S. Ravaglia, B. Bembi, S. Servidei, E. Pegoraro, M. Moggio, M. Filosto, E. Sette, G. Crescimanno, P. Tonin, R. Parini, L. Morandi, G. Marrosu, G. Greco, O. Musumeci, G. Di Iorio, G. Siciliano, M. A. Donati, F. Carubbi, M. Ermani, T. Mongini and A. Toscano, *J. Neurol.*, 2012, **259**, 952.
65. B. L. Thurberg, C. Lynch Maloney, C. Vaccaro, K. Afonso, A. C. Tsai, E. Bossen, P. S. Kishnani and M. O'Callaghan, *Lab. Invest.*, 2006, **86**, 1208.
66. N. M. Dahms, L. J. Olson and J. J. Kim, *Glycobiology*, 2008, **18**, 664.
67. L. J. Olson, F. C. Peterson, A. Castonguay, R. N. Bohnsack, M. Kudo, R. R. Gotschall, W. M. Canfield, B. F. Volkman and N. M. Dahms, *Proc. Natl. Acad. Sci. U. S. A.*, 2010, **107**, 12493.
68. Y. Zhu, X. Li, J. Kyazike, Q. Zhou, B. L. Thurberg, N. Raben, R. J. Mattaliano and S. H. Cheng, *J. Biol. Chem.*, 2004, **279**, 50336.
69. Y. Zhu, J. L. Jiang, N. K. Gumlaw, J. Zhang, S. D. Bercury, R. J. Ziegler, K. Lee, M. Kudo, W. M. Canfield, T. Edmunds, C. Jiang, R. J. Mattaliano and S. H. Cheng, *Mol. Ther.*, 2009, **17**, 954.

70. M. A. Robinson, S. T. Charlton, P. Garnier, X. T. Wang, S. S. Davis, A. C. Perkins, M. Frier, R. Duncan, T. J. Savage, D. A. Wyatt, S. A. Watson and B. G. Davis, *Proc. Natl. Acad. Sci. U. S. A.*, 2004, **101**, 14527.
71. A. Gualberto, *Expert Opin. Invest. Drugs*, 2012, **21**, 205.
72. T. Siddiqi, S. H. Thomas and R. Chen, *Pharmacogenomics Pers. Med.*, 2014, 7, 79.
73. S. Dhillon, *Drugs*, 2014, **74**, 675.
74. I. E. Krop, S. B. Kim, A. Gonzalez-Martin, P. M. Lorusso, J. M. Ferrero, M. Smitt, R. Yu, A. C. Leung and H. Wildiers, *Lancet Oncol.*, 2014, **15**, 689.
75. L. Wang, G. Amphlett, W. A. Blattler, J. M. Lambert and W. Zhang, *Protein Sci.*, 2005, **14**, 2436.
76. R. Jefferis, *Trends Pharmacol. Sci.*, 2009, **30**, 356.
77. L. M. Hinman, P. R. Hamann, R. Wallace, A. T. Menendez, F. E. Durr and J. Upeslacis, *Cancer Res.*, 1993, **53**, 3336.
78. P. R. Hamann, L. M. Hinman, C. F. Beyer, L. M. Greenberger, C. Lin, D. Lindh, A. T. Menendez, R. Wallace, F. E. Durr and J. Upeslacis, *Bioconjugate Chem.*, 2005, **16**, 346.
79. K. Zuberbuhler, G. Casi, G. J. L. Bernardes and D. Neri, *Chem. Commun.*, 2012, **48**, 7100.
80. N. M. Okeley, B. E. Toki, X. Zhang, S. C. Jeffrey, P. J. Burke, S. C. Alley and P. D. Senter, *Bioconjugate Chem.*, 2013, **24**, 1650.
81. M. M. Rochefort, M. D. Girgis, J. S. Ankeny and J. S. Tomlinson, *Glycobiology*, 2014, **24**, 62.
82. Q. Zhou, J. E. Stefano, C. Manning, J. Kyazike, B. Chen, D. A. Gianolio, A. Park, M. Busch, J. Bird, X. Zheng, H. Simonds-Mannes, J. Kim, R. C. Gregory, R. J. Miller, W. H. Brondyk, P. K. Dhal and C. Q. Pan, *Bioconjugate Chem.*, 2014, **25**, 510.
83. X. Li, T. Fang and G. J. Boons, *Angew. Chem., Int. Ed. Engl.*, 2014, **53**, 7179.
84. E. Boeggeman, B. Ramakrishnan, M. Pasek, M. Manzoni, A. Puri, K. H. Loomis, T. J. Waybright and P. K. Qasba, *Bioconjugate Chem.*, 2009, **20**, 1228.
85. B. Ramakrishnan, E. Boeggeman, M. Pasek and P. K. Qasba, *Methods Mol. Biol.*, 2011, **751**, 281.
86. B. M. Zeglis, C. B. Davis, R. Aggeler, H. C. Kang, A. Chen, B. J. Agnew and J. S. Lewis, *Bioconjugate Chem.*, 2013, **24**, 1057.
87. Z. Zhu, B. Ramakrishnan, J. Li, Y. Wang, Y. Feng, P. Prabakaran, S. Colantonio, M. A. Dyba, P. K. Qasba and D. S. Dimitrov, *mAbs*, 2014, **6**, 1190.
88. R. van Geel, M. A. Wijdeven, R. Heesbeen, J. M. Verkade, A. A. Wasiel, S. S. van Berkel and F. L. van Delft, *Bioconjugate Chem.*, 2015, **26**, 2233.

Subject Index